Impacts of Soil on Plant Growth and Environment

Impacts of Soil on Plant Growth and Environment

Edited by Jamie Hanks

SYRAWOOD
PUBLISHING HOUSE
New York

Published by Syrawood Publishing House,
750 Third Avenue, 9th Floor,
New York, NY 10017, USA
www.syrawoodpublishinghouse.com

Impacts of Soil on Plant Growth and Environment
Edited by Jamie Hanks

International Standard Book Number: 978-1-68286-395-4 (Hardback)

Cataloging-in-publication Data

Impacts of soil on plant growth and environment / edited by Jamie Hanks.
p. cm.
Includes bibliographical references and index.
ISBN 978-1-68286-395-4
1. Plant-soil relationships. 2. Soil science. 3. Growth (Plants). 4. Soils--Environmental aspects. 5. Soil ecology.
I. Hanks, Jamie.
S596.7 .I47 2017
631.422--dc23

Printed in China.

TABLE OF CONTENTS

Preface IX

Chapter 1 **Methane, Carbon Dioxide and Nitrous Oxide Fluxes in Soil Profile under a Winter Wheat-Summer Maize Rotation in the North China Plain** 1
Yuying Wang, Chunsheng Hu, Hua Ming, Oene Oenema, Douglas A. Schaefer, Wenxu Dong, Yuming Zhang, Xiaoxin Li

Chapter 2 **Influence of Different Plant Species on Methane Emissions from Soil in a Restored Swiss Wetland** 15
Gurbir S. Bhullar, Peter J. Edwards, Harry Olde Venterink

Chapter 3 **Climatic Control on Plant and Soil $\delta^{13}C$ along an Altitudinal Transect of Lushan Mountain in Subtropical China: Characteristics and Interpretation of Soil Carbon Dynamics** 20
Baoming Du, Chunjiang Liu, Hongzhang Kang, Penghua Zhu, Shan Yin, Guangrong Shen, Jingli Hou, Hannu Ilvesniemi

Chapter 4 **The Altitudinal Patterns of Leaf C:N:P Stoichiometry are Regulated by Plant Growth Form, Climate and Soil on Changbai Mountain, China** 28
Ning Zhao, Nianpeng He, Qiufeng Wang, Xinyu Zhang, Ruili Wang, Zhiwei Xu, Guirui Yu

Chapter 5 **Green Manure Addition to Soil Increases Grain Zinc Concentration in Bread Wheat** 37
Forough Aghili, Hannes A. Gamper, Jost Eikenberg, Amir H. Khoshgoftarmanesh, Majid Afyuni, Rainer Schulin, Jan Jansa, Emmanuel Frossard

Chapter 6 **Availability and Temporal Heterogeneity of Water Supply Affect the Vertical Distribution and Mortality of a Belowground Herbivore and Consequently Plant Growth** 50
Tomonori Tsunoda, Naoki Kachi, Jun-Ichirou Suzuki

Chapter 7 **Cultivation-Independent Screening Revealed Hot Spots of IncP-1, IncP-7 and IncP-9 Plasmid Occurrence in Different Environmental Habitats** 57
Simone Dealtry, Guo-Chun Ding, Viola Weichelt, Vincent Dunon, Andreas Schlüter, María Carla Martini, María Florencia Del Papa, Antonio Lagares, Gregory Charles Auton Amos, Elizabeth Margaret Helen Wellington, William Hugo Gaze, Detmer Sipkema, Sara Sjöling, Dirk Springael, Holger Heuer, Jan Dirk van Elsas, Christopher Thomas, Kornelia Smalla

Chapter 8 **Can Biochar and Phytoextractors be Jointly used for Cadmium Remediation?** 65
Huanping Lu, Zhian Li, Shenglei Fu, Ana Méndez, Gabriel Gascó, Jorge Paz-Ferreiro

Chapter 9 **A 2-Year Field Study Shows Little Evidence that the Long-Term Planting of Transgenic Insect-Resistant Cotton Affects the Community Structure of Soil Nematodes** 72
Xiaogang Li, Biao Liu

Chapter 10 **Effect of Optimal Daily Fertigation on Migration of Water and Salt in Soil, Root Growth and Fruit Yield of Cucumber (*Cucumis sativus* L.) in Solar-Greenhouse** 87
Xinshu Liang, Yinan Gao, Xiaoying Zhang, Yongqiang Tian, Zhenxian Zhang, Lihong Gao

Chapter 11 **How Do Earthworms, Soil Texture and Plant Composition Affect Infiltration along an Experimental Plant Diversity Gradient in Grassland?** 97
Christine Fischer, Christiane Roscher, Britta Jensen, Nico Eisenhauer, Jussi Baade, Sabine Attinger, Stefan Scheu, Wolfgang W. Weisser, Jens Schumacher, Anke Hildebrandt

Chapter 12 **Effects of Nitrogen Addition and Fire on Plant Nitrogen use in a Temperate Steppe** 109
Hai-Wei Wei, Xiao-Tao Lü, Fu-Mei Lü, Xing-Guo Han

Chapter 13 **Application of Organic Amendments to a Coastal Saline Soil in North China: Effects on Soil Physical and Chemical Properties and Tree Growth** 117
Linlin Wang, Xiangyang Sun, Suyan Li, Tao Zhang, Wei Zhang, Penghui Zhai

Chapter 14 **Spreading Topsoil Encourages Ecological Restoration on Embankments: Soil Fertility, Microbial Activity and Vegetation Cover** 126
Desirée Rivera, Violeta Mejías, Berta M. Jáuregui, Marga Costa-Tenorio Ana Isabel López-Archilla, Begoña Peco

Chapter 15 **Soil Environmental Conditions and Microbial Build-Up Mediate the Effect of Plant Diversity on Soil Nitrifying and Denitrifying Enzyme Activities in Temperate Grasslands** 135
Xavier Le Roux, Bernhard Schmid, Franck Poly, Romain L. Barnard, Pascal A. Niklaus, Nadine Guillaumaud, Maike Habekost, Yvonne Oelmann, Laurent Philippot, Joana Falcao Salles, Michael Schloter, Sibylle Steinbeiss, Alexandra Weigelt

Chapter 16 **Biochar Decelerates Soil Organic Nitrogen Cycling but Stimulates Soil Nitrification in a Temperate Arable Field Trial** 146
Judith Prommer, Wolfgang Wanek, Florian Hofhansl, Daniela Trojan, Pierre Offre, Tim Urich, Christa Schleper, Stefan Sassmann, Barbara Kitzler, Gerhard Soja, Rebecca Clare Hood-Nowotny

Chapter 17 **Element Pool Changes within a Scrub-Oak Ecosystem after 11 Years of Exposure to Elevated CO_2** 162
Benjamin D. Duval, Paul Dijkstra, Bert G. Drake, Dale W. Johnson, Michael E. Ketterer, J. Patrick Megonigal, Bruce A. Hungate

Chapter 18 **Facilitation Drives the Positive Effects of Plant Richness on Trace Metal Removal in a Biodiversity Experiment** 171
Jiang Wang, Yuan Ge, Tong Chen, Yi Bai, Bao Ying Qian, Chong Bang Zhang

Chapter 19 **Soil Carbon Stocks Decrease following Conversion of Secondary Forests to Rubber (*Hevea brasiliensis*) Plantations**......180
Marleen de Blécourt, Rainer Brumme, Jianchu Xu, Marife D. Corre, Edzo Veldkamp

Chapter 20 **Non-Native Plant Litter Enhances Soil Carbon Dioxide Emissions in an Invaded Annual Grassland**......189
Ling Zhang, Hong Wang, Jianwen Zou, William E. Rogers, Evan Siemann

Chapter 21 **Simulated Nitrogen Deposition Reduces CH_4 Uptake and Increases N_2O Emission from a Subtropical Plantation Forest Soil in Southern China**......197
Yongsheng Wang, Shulan Cheng, Huajun Fang, Guirui Yu1, Minjie Xu, Xusheng Dang, Linsen Li, Lei Wang

Chapter 22 **Competition Increases Sensitivity of Wheat (*Triticumaestivum*) to Biotic Plant-Soil Feedback**......207
W. H. Gera Hol, Wietse de Boer, Freddy ten Hooven, Wim H. van der Putten

Chapter 23 **Contributions of Understory and/or Overstory Vegetations to Soil Microbial PLFA and Nematode Diversities in Eucalyptus Monocultures**......213
Jie Zhao, Songze Wan, Chenlu Zhang, Zhanfeng Liu, Lixia Zhou, Shenglei Fu

Chapter 24 **Effects of Straw Incorporation on Soil Organic Matter and Soil Water-Stable Aggregates Content in Semiarid Regions of Northwest China**......221
Peng Zhang, Ting Wei, Zhikuan Jia, Qingfang Han, Xiaolong Ren, Yongping Li

Chapter 25 **Impacts of Agricultural Management and Climate Change on Future Soil Organic Carbon Dynamics in North China Plain**......232
Guocheng Wang, Tingting Li, Wen Zhang, Yongqiang Yu

Chapter 26 **Soil Particle Heterogeneity Affects the Growth of a Rhizomatous Wetland Plant**......242
Lin Huang, Bi-Cheng Dong, Wei Xue, Yi-Ke Peng, Ming-Xiang Zhang, Fei-Hai Yu

Permissions

List of Contributors

Index

PREFACE

Plant growth and well-being depend on a number of factors but the most important among these is soil quality. This book aims to compile the most advanced research on soil profiles and their availability across various environments. Plant growth and subsequent regeneration require regular supply of nutrients as well as safety from pests and diseases. Conservationists, agronomists and botanists will find this book especially helpful. As the field of plant growth is emerging at a rapid pace, the contents of this book will help the readers understand the modern concepts and applications of the subject.

After months of intensive research and writing, this book is the end result of all who devoted their time and efforts in the initiation and progress of this book. It will surely be a source of reference in enhancing the required knowledge of the new developments in the area. During the course of developing this book, certain measures such as accuracy, authenticity and research focused analytical studies were given preference in order to produce a comprehensive book in the area of study.

This book would not have been possible without the efforts of the authors and the publisher. I extend my sincere thanks to them. Secondly, I express my gratitude to my family and well-wishers. And most importantly, I thank my students for constantly expressing their willingness and curiosity in enhancing their knowledge in the field, which encourages me to take up further research projects for the advancement of the area.

Editor

1

Methane, Carbon Dioxide and Nitrous Oxide Fluxes in Soil Profile under a Winter Wheat-Summer Maize Rotation in the North China Plain

Yuying Wang[1], Chunsheng Hu[1]*, Hua Ming[1], Oene Oenema[2], Douglas A. Schaefer[3], Wenxu Dong[1], Yuming Zhang[1], Xiaoxin Li[1]

1 Key Laboratory of Agricultural Water Resources, Center for Agricultural Resources Research, Institute of Genetics and Developmental Biology, Chinese Academy of Sciences, Shijiazhuang, Hebei, China, **2** Department of Soil Quality, Wageningen University, Alterra, Wageningen, The Netherlands, **3** Key Lab of Tropical Forest Ecology, Xishuangbanna Tropical Botanical Garden, Chinese Academy of Sciences, Menglun, Yunnan, China

Abstract

The production and consumption of the greenhouse gases (GHGs) methane (CH_4), carbon dioxide (CO_2) and nitrous oxide (N_2O) in soil profile are poorly understood. This work sought to quantify the GHG production and consumption at seven depths (0–30, 30–60, 60–90, 90–150, 150–200, 200–250 and 250–300 cm) in a long-term field experiment with a winter wheat-summer maize rotation system, and four N application rates (0; 200; 400 and 600 kg N ha^{-1} $year^{-1}$) in the North China Plain. The gas samples were taken twice a week and analyzed by gas chromatography. GHG production and consumption in soil layers were inferred using Fick's law. Results showed nitrogen application significantly increased N_2O fluxes in soil down to 90 cm but did not affect CH_4 and CO_2 fluxes. Soil moisture played an important role in soil profile GHG fluxes; both CH_4 consumption and CO_2 fluxes in and from soil tended to decrease with increasing soil water filled pore space (WFPS). The top 0–60 cm of soil was a sink of atmospheric CH_4, and a source of both CO_2 and N_2O, more than 90% of the annual cumulative GHG fluxes originated at depths shallower than 90 cm; the subsoil (>90 cm) was not a major source or sink of GHG, rather it acted as a 'reservoir'. This study provides quantitative evidence for the production and consumption of CH_4, CO_2 and N_2O in the soil profile.

Editor: Dafeng Hui, Tennessee State University, United States of America

Funding: This research is supported by the "Strategic Priority Research Program of Chinese Academy of Sciences" (Grant No. XDA0505050202 and XDA05050601) and the "National Science & Technology Pillar Program" (Grant No. 2012BAD14B07-5). It is also supported by the "National Basic Research Program of China" (Grant No. 2010CB833501); the "National Natural Science Foundation of China" (Grant No. 30970534) and the "Main Direction Program of Knowledge Innovation of Chinese Academy of Sciences" (Grant No. KSCX2-EW-J-5). The funders had no role in study design, data collection and analysis, decision to publish, or preparation of the manuscript.

Competing Interests: The authors have declared that no competing interests exist.

* E-mail: cshu@sjziam.ac.cn

Introduction

Atmospheric concentrations of carbon dioxide (CO_2), methane (CH_4) and nitrous oxide (N_2O) have increased considerably since the industrial revolution, and are still increasing annually by about 0.5%, 1.1% and 0.3%, respectively [1]. Worldwide concerns about the increased greenhouse gases (GHGs) concentrations in the atmosphere and its effects on our future environment require a better understanding of the cause of these emissions [2]. Agricultural lands occupy 37% of the earth's land surface; about 13.5% of global anthropogenic GHG was emitted from agricultural production [1]. It was estimated that 84% of N_2O and 52% of CH_4 emitted from agriculture activities [3]. In China, agriculture tends to produce more emissions than the global average over the last 30 years due to increased chemical and manure N inputs. Gaining a better understanding of GHG production and emission processes, and developing methods for mitigating emissions from agroecosystems are essential steps in order to mitigate climate change [4].

Agricultural soils are main sources and sinks of GHG emissions, depending on their characteristics and management. Many studies have been conducted to quantify the net fluxes of CO_2, CH_4, and N_2O across the soil/atmosphere interface [3,5–9]. These studies provide an integrative estimate of the net production and consumption of CH_4, CO_2 and N_2O in the soil, but do not provide information on the depth-distribution of CH_4, CO_2 and N_2O production-consumption patterns within soil profiles. It has been suggested that subsurface processes exert a significant control on carbon (C) and nitrogen (N) dynamics and hence on CO_2, CH_4, and N_2O emissions from soil [10], but few studies have elucidated the role of the subsoil so far. Understanding these processes might also provide a better insight into the possibilities and effectiveness of measures to reduce GHG emissions. For example, a temporary accumulation of GHG in the soil profile influences GHG flux patterns at the soil surface over time, and thereby may confuse empirical relationships between agricultural activities and measured GHG emissions [11]. Thus, measurements of CH_4, CO_2 and N_2O concentration profiles may be helpful for increasing the understanding of the net exchanges of these gases between soil and atmosphere.

Though few studies have examined the production and consumption of CO_2 [12,13], CH_4 [14] and N_2O [15] within individual soil horizons and their transports between soil horizons so far, very few studies have made combined measurements of the dynamics of CO_2, CH_4 and N_2O production and emission processes in soil profiles in agro-ecosystems, especially in China [16]. It has been well-established that N fertilizer applications increase crop growth and N_2O emissions, and tend to decrease CH_4 emissions into the atmosphere, but there is little information about the combined effects of N fertilizer application and irrigation on subsoil N_2O, CO_2 and CH_4 production, consumption and transport.

Recently, Wang et al [16] presented bi-weekly measured CH_4, CO_2 and N_2O concentration profiles down to a depth of 300 cm in a winter wheat-summer maize rotation in the North China Plain, with four N application rates (0; 200; 400 and 600 kg N ha^{-1}). Here, we build on the results of that study, and present calculated subsurface fluxes of CH_4, CO_2 and N_2O over a whole-year period. The purpose of this study is to evaluate the effects of seasonal cropping, N applications, irrigation, soil temperature, and soil moisture on net subsurface transport of CO_2, N_2O and CH_4.

Materials and Methods

Site description

The study was conducted at Luancheng Agroecosystem Experimental Station (37°53′N, 114°41′E, elevation 50 m), Chinese Academy of Sciences. This area is at the piedmont of the Taihang Mountains, in the North China Plain. Mean annual precipitation is about 480 mm, 70% of which is in the period from July to September. Annual average air temperature is 12.5 °C. The dominant cropping system in the region is a winter wheat-summer maize double-cropping system (two crops harvested in a single year) without fallow between the crops.

Field experimental design

The field experiment with a randomized complete block design was laid down in a winter wheat (*Triticum aestivum L.* Wheat variety Kenong 199)-summer maize (*Zea mays L.* Maize variety Xianyu 335) double-cropping system in 1998. It had four N fertilizer (urea) application treatments in triplicate: 0 (N0), 200 (N200), 400 (N400) and 600 (N600) kg N ha^{-1} $year^{-1}$. Plot size was 7 m×10 m. Results of the present study refer to the period March 2007 to January 2008. Details on fertilizer application and crop management activities are presented in Table 1. Crops are flood-irrigated with pumped groundwater about five times per year, depending on rainfall distribution.

Soil sampling, analysis and climate data collection

The soil has a silt-loam texture in the upper 90 cm and clay-loam to clay texture at depth of 90–300 cm (Table 2). All soil samples were collected from different depths of the soil profile before the GHG measurements (on 5 December 2006); soil samples were mixed to make a specific representative soil sample for each depth; and all analyses of soil chemical properties in Table 2 were based on the standard methods for soil analyses described by Sparks [17]. Soil bulk density was determined using the cutting ring method. Soil particle size analysis was done by the Bouyoucos Hydrometer Method [18]. Soil pH was measured in a suspension of 5 g soil with 25 ml distilled water after 1 h after shaking.

Soil core samples were collected from different depths (0–30, 30–60, 60–90, 90–150, 150–200, 200–250, 250–300 cm) of the soil profile in the farmland described above on 5 December 2006 (before the GHG measurements), 16 June 2007 (after the winter wheat harvest) and 11 October 2007 (after the summer maize harvest), respectively. Three different sub-samples, taken from a cross-section around the soil auger (3 meter in length), were mixed to make a specific representative soil sample for each depth from each point. The soil profile samples were sealed in dark plastic bags immediately after sampling and stored at 4°C until NO_3 extraction. Samples of soil NO_3-N were extracted with 1 M KCl solution (1:5 w/v) by shaking for 1 h. The extracts were then filtered and the concentrations of NO_3-N in the soil extracts were measured colorimetrically using a UV spectrophotometer (UV-2450, Shimadzu, Japan). Each measurement was replicated three times.

Soil temperature was measured using seven CS107b soil temperature probes (Cambell Scientific Inc., Logan, UT) installed at depths of 30, 60, 90, 150, 200, 250 and 300 cm. Three-meter neutron access tubes were installed at each plot. Soil moisture at seven depths (30, 60, 90, 150, 200, 250 and 300 cm.) was measured using a neutron moisture meter when gas samples were collected. Soil temperature and water content were used to explore the relationships between calculated CO_2, N_2O and CH_4 fluxes and soil water-filled pore space (WFPS) and soil temperature at various depths. Daily rainfall was recorded at a weather station on the experimental site.

Soil gas sampling and measurements

Measurements of CO_2, N_2O and CH_4 concentrations in soil started in March 2007, i.e., 9 years after the start of the field experiment, assuming that by then the CO_2, N_2O and CH_4 production-consumption dynamics in the subsoil had been adjusted to the experimental treatments. Seven subsurface soil air equilibration tubes were installed at each site with sampling ports at 30, 60, 90, 150, 200, 250 and 300 cm in December 2006 (for more details, see reference 16). Soil-air samples were taken twice a week between 9:00 AM and 11:00 AM, using 100 ml plastic syringes connected to the tubes via the three-way stopcocks at the surface. The surface air was concurrently sampled at a height of 5 cm above the soil surface. The gas samples were analyzed by gas chromatography (Agilent GC-6820, Agilent Technologies Inc. Santa Clara, California, US) with separate electron capture detector (ECD at 330°C) for N_2O determination and flame ionization detector (FID at 200°C) for CH4 and CO2 determinations.

Calculations of gas fluxes

The basic method of our study followed that of Campbell [19]. It was assumed that the soil conditions are uniform in horizontal direction, and that the gas diffusion in soil is in one-dimensional vertical flow, that fundamentally follows Fick's law [20,21]:

$$q = -D_p \frac{\delta c}{\delta x} \quad (1)$$

Where q is the gas flux density (g gas m^{-2} soil s^{-1}), D_p is the soil-gas diffusivity (m^3 soil air m^{-1} soil s^{-1}), $\frac{\delta c}{\delta x}$ is the concentration gradient between two soil layers (g gas m^{-3} soil air m^{-1} soil).

D_p was derived from the following equation [22,23]:

$$D_p = (\varepsilon^{10/3}/E^2) \cdot D_0 \quad (2)$$

Where, D_0 is the gas diffusivity (m^2 air s^{-1}). We estimated the diffusion coefficient D_0 of CH_4, CO_2 and N_2O at 298 K and 1 kPa

Table 1. Fertilization treatments (A); and timing of crop management activities (B).

(A) Treatments	**Basal fertilization, applied at wheat sowing (kg·ha^{-1})**			**Supplementary N fertilization (kg·ha^{-1})**	
	N	**P_2O_5**	**K_2O**	**Wheat (in April)**	**Maize (in July)**
N0	0	65	0	0	0
N200	50	65	0	50	100
N400	100	65	0	100	200
N600	150	65	0	150	300
(B) Timing	**Crop management activities**				
	Winter-wheat season				**Summer-maize season**
October 3, 2006	Basal N fertilization and irrigation (60 mm)				
October 10, 2006	Seeding				
April 7, 2007				Supplementary N fertilization and irrigation (94 mm)	
May 19, 2007					Irrigation (60 mm)
June 1, 2007					Seeding
June 14, 2007				Harvest	
June 19, 2007					Irrigation (60 mm)
July 27, 2007					Supplementary N fertilization
July 29, 2007					Irrigation (60 mm)
October 1, 2007					Harvest
October 2, 2007	Basal N fertilization and irrigation				
October 7, 2007	Seeding				

at 1.79×10^{-5}, 1.32×10^{-5}, 1.29×10^{-5} m^2 s^{-1}, respectively, by using a semiempirical equation by Gilliland et al [24]. Parameter ε is the soil air filled porosity (m^3 air m^{-3} soil), and E is the soil porosity (m^3 voids m^{-3} soil).

The Millington-Quirk model was used to compute ε and E [25]:

$$E=1-\frac{\rho_b}{\rho_s} \quad (3)$$

$$\varepsilon=E-\theta \quad (4)$$

Where ρ_b is the dry bulk density (g m^{-3}) at each soil depth (Table 2), ρ_s is the average bulk density of surface soil (2.65 g m^{-3}); θ is the volumetric soil water content which was measured using a neutron moisture meter at each depth.

Calculations of annual cumulative gas fluxes

The annual cumulative emissions were obtained by multiplying the average daily flux from two consecutive measurements within a week by the number of days between the measurements, and then summing the fluxes of these periods to an accumulative flux for the whole year [26]:

$$T=\sum_{i=1}^{n}(X_i\times24) \quad (n=1,2,3...) \quad (5)$$

Where: T (kg ha^{-1}), X_i (kg ha^{-1} h^{-1}) and i are the accumulative GHG emission, the average daily GHG emission rate, and the number of days, respectively.

Data analyses

All data were subjected to statistical analysis (SPSS 13.0). Differences between treatments were analyzed using ANOVA, followed by LSD at the 0.05 probability level. Regression analysis was used to identify relationships between CH_4, CO_2 and N_2O fluxes and the climatic variables.

Results

Concentrations of CH_4, CO_2 and N_2O

Mean concentrations and its standard deviations of CH_4, CO_2 and N_2O at each depth are shown in Figure 1 A, as function of N application rates. Mean CH_4 concentration decreased with soil depth. Ambient air CH_4 concentration in the area was about 2.2 ppmv. At a depth of 30 cm, CH_4 concentration ranged between 1.4 and 1.6 ppmv and at depth of 60 to 300 cm between 0.3 and 0.6 ppmv. There were no clear effects of N fertilizer application on the mean CH_4 concentration (Figure 1 A). Mean CH_4 concentrations decreased significantly at soil depths of 0, 30, 60 and 90 cm ($P<0.05$); changes in mean concentration below a depth of 90 cm were not significant (Figure 1 C).

Mean CO_2 concentration increased with soil depth. At a depth of 30 cm, CO_2 concentration ranged between 8400 and 9900 ppmv and at depth of 60 to 300 cm between 16000 and 21000 ppmv. Mean CO_2 concentrations increased significantly at soil depths of 0, 30 and 60 cm ($P<0.05$); changes in mean concentration below a depth of 60 cm were not significant

Table 2. Soil characteristics at the experimental site in 2007.

Depth (cm)	pH (H_2O)	Sand (%)	Silt (%)	Clay (%)	Textural class[a]	Dry bulk Density (g cm^{-3})	Total organic matter (g kg^{-1})	Total nitrogen (g N kg^{-1})	Available nitrogen (mg N kg^{-1})	Available phosphorus (mg P kg^{-1})	Available potassium (mg K kg^{-1})
0–30	8.5	25	58	17	SSL	1.47	16.7	1.40	148	4.1	79.3
30–60	7.74	22	60	18	L	1.40	10.9	0.85	72.0	2.1	54.4
60–90	7.78	31	55	14	L	1.45	7.8	0.64	70.1	0.44	36.9
90–150	7.76	15	59	26	SCL	1.57	6.5	0.38	49.8	0.17	27.8
150–200	7.74	18	47	35	GCL	1.43	5.4	0.28	38.9	0.13	21.9
200–250	7.77	15	35	50	C	1.51	4.2	0.15	30.1	0.09	13.9
250–300	7.75	12	35	53	C	1.50	3.0	0.86	25.9	0.04	7.1

[a]SSL: Silty sandy loam; L: Loam; SCL: Silty clay loam; GCL: Gravely clay loam; C = Clay.

(Figure 1 C). There were no clear effects of N fertilizer application on the mean CO_2 concentrations.

Concentrations of N_2O were strongly influenced by agricultural management activities such as N application and irrigation. Fertilizer N applications increased the mean N_2O concentrations. Mean N_2O concentrations at depth of 30 to 300 cm ranged from 600 to 1500, 1100 to 1700, 1600 to 2100 and 2500 to 3000 ppbv for the N0, N200, N400 and N600 treatments, respectively (Figure 1 A). Mean N_2O concentrations increased significantly from soil surface to a depth of 30 cm ($P<0.05$), but changes in mean concentration below a depth of 30 cm were not significant (Figure 1 C). Fertilizer N application increased soil NO_3-N content; differences in mean N_2O concentrations were correlated with differences in mean NO_3-N contents in the four fertilizer N treatments (Figure 1 B).

Fluxes of CH_4 in soil

Diffusive fluxes between soil layers and between soil and atmosphere were calculated from the concentration gradients, using equation 1. There was a net influx of atmospheric CH_4 into the top 0–60 cm (Figure 2), suggesting consumption of CH_4 by methanotropic bacteria. Interestingly, the calculated fluxes into the soil were rather similar for the 0–30 and 30–60 cm soil layers, suggesting similar CH_4 uptake rates. Uptake of CH_4 apparently also occurred in the layers 60–90, 90–150 and 150–200 cm during the first one or two months of the measurement period (Figure 2). However, we cannot exclude the possibility that this apparent uptake of CH_4 in the subsoil during the first two months is an artifact related to the installation of the samplers when atmospheric CH_4 may have diffused into the subsoil. Fluxes between soil layers were negligible small during most of the maize growing season (Figure 2).

Annual cumulative fluxes of CH_4 for all soil layers and N fertilizer treatments are shown in Table 3 A. Evidently, the influx of atmospheric CH_4 decreased with soil depth. During the study period, mean calculated uptake was about 176 g CH_4 per ha by the top 30 cm, 252 g CH_4 per ha by the soil layer 30–60 cm, 98 g CH_4 per ha by the layer 60–90 cm and 22 g CH_4 per ha below a depth of 90 cm; mean calculated uptakes in the layers 0–30 and 30–60 cm were both significantly higher than that in the layer 90–300 cm ($P<0.05$) (Figure 1 C). Annual cumulative CH_4 uptake in the layer 0–90 cm (526 g CH_4 per ha per year) contributed about 96% to that in the layer 0–300 cm (547 g CH_4 per ha per year). Annual cumulative uptake in 0–30 cm layer is relatively low compared to literature data [4–6,27].

Fluxes of CO_2 in soil

There was a large efflux of CO_2 from the top 30 cm of soil to the atmosphere from March till June, i.e., during the second half of the wheat growing season, and from August till October, i.e., during the second half of the maize growing season (Figure 3). The same holds for the upward flux from the layer 30–60 to the layer 0–30 cm. These patterns were related to the crop growing seasons of wheat and maize, and to the changes in water filled pore space (WFPS) and soil temperature. There were no clear relationships between N treatments and CO_2 fluxes. Treatment N200 had the smallest flux from the layer 0–30 cm to the atmosphere, but the largest from 30–60 cm to 0–30 cm from April to May. Upward fluxes from the layers 60–90 cm and especially below this layer were much smaller. There were small but significant changes in fluxes in the subsoil at the transition of the winter-wheat growing season to the summer-maize growing season (Figure 3).

Surprisingly, annual cumulative fluxes of CO_2 tended to decrease with increasing N fertilizer application rates (Table 3

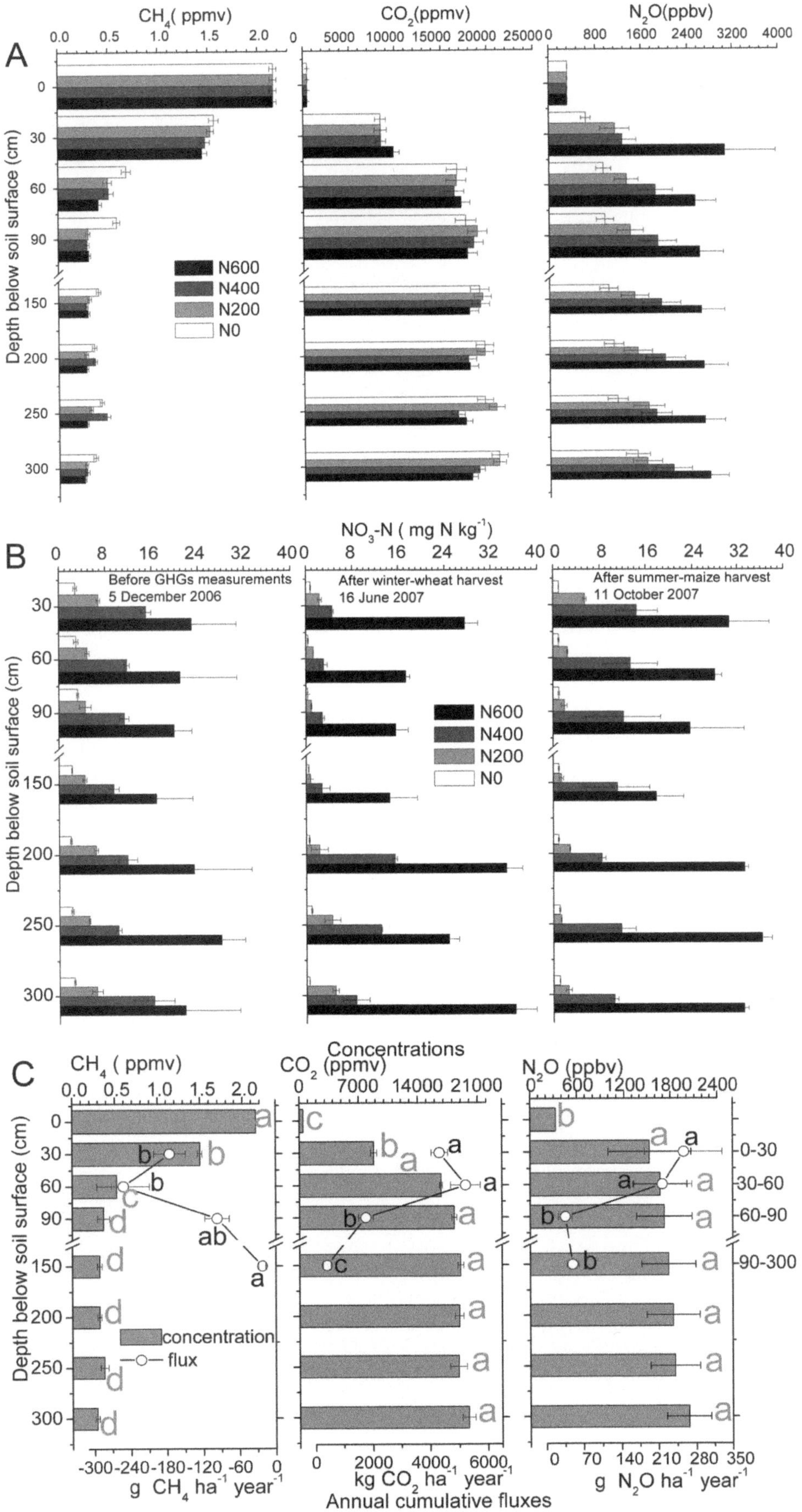
A
CH4(ppmv)
CO2(ppmv)
N2O(ppbv)
Depth below soil surface (cm)
N600
N400
N200
N0
B
NO3-N (mg N kg-1)
Before GHGs measurements
5 December 2006
After winter-wheat harvest
16 June 2007
After summer-maize harvest
11 October 2007
C
Concentrations
concentration
flux
0-30
30-60
60-90
90-300
g CH4 ha-1 year-1
kg CO2 ha-1 year-1
g N2O ha-1 year-1
Annual cumulative fluxes

Figure 1. CH_4, CO_2 and N_2O concentrations (mean ± standard deviations, n = 3) in soil air at various soil depths in a winter wheat-summer maize double cropping rotation receiving 0, 200, 400 and 600 kg of N ha^{-1} $year^{-1}$, in 2007–2008 (A); NO_3-N contents (mean ± standard deviations, n = 3) at various soil depths as function of N fertilizer application rate, in 2007–2008 (B); Profiles of concentration and annual cumulative flux of CH_4, CO_2 and N_2O, in 2007–2008 (mean ± standard deviations, n = 4). Same letters next to the bars indicated no significant differences between slope positions ($P<0.05$). (C). Note the differences in X-axes.

B). Moreover, cumulative upward fluxes were somewhat larger from the layer 30–60 cm (mean 5,227; range 3,800–6,000 kg CO_2 per ha) than from the layer 0–30 cm to the atmosphere (mean 4,331; range 3,800–5,000 kg CO_2 per ha) (Figure 1 C; Table 3 B). This suggests that a relatively large portion of total respiration in soil took place in the layer 30–60 cm. However, we can not exclude the possibility that the calculated CO_2 efflux from the top layer is underestimated, because the concentration gradient in the

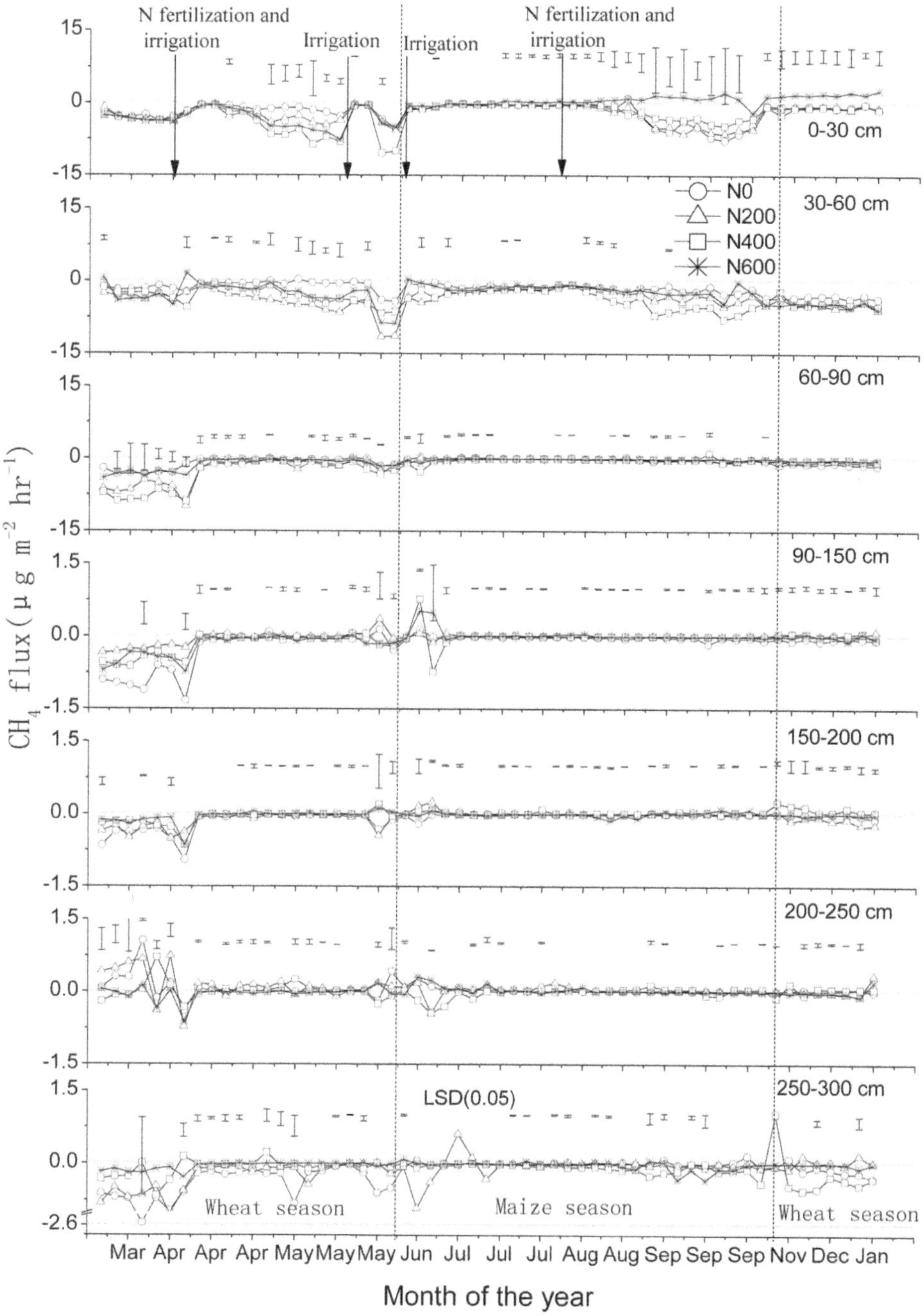

Figure 2. CH_4 flux rates (means ± standard deviations, n = 3) at various soil depths in a winter wheat-summer maize double cropping rotation receiving 0, 200, 400 and 600 kg of N ha^{-1} $year^{-1}$, in 2007–2008. Vertical dashed lines indicate a change in crop. Bars in figures indicate 1 standard deviation (n = 3). Note the differences in Y-axes.

Table 3. Annual cumulative emissions of CH_4 and N_2O (in g ha^{-1} yr^{-1}) and of CO_2 (in kg ha^{-1} yr^{-1}) between soil layers.

(A) CH_4

Treatments	0–30 cm	30–60 cm	60–90 cm	90–150 cm	150–200 cm	200–250cm	250–300 cm	0–300cm
N0	−167 (12) b	−138 (13) a	−63 (2) a	−15 (0.5) c	−9 (0.1) d	5 (0.04) a	−21 (0.3) d	−408
N200	−201 (9) bc	−322 (10) c	−123 (3) b	−2 (0.07) a	−7 (0.5) c	5 (0.08) a	−12 (0.6) c	−662
N400	−228 (3) c	−318 (8) c	−140 (10) b	−6 (0.5) b	−1 (0.1) a	−1 (0.06) b	−10 (0.4) b	−704
N600	−106 (15) a	−231 (5) b	−64 (7) a	−6 (0.5) b	−2 (0.1) b	−1 (0.05) b	−4 (0.1) a	−414

(B) CO_2

Treatments	0–30 cm	30–60 cm	60–90 cm	90–150 cm	150–200 cm	200–250cm	250–300 cm	0–300cm
N0	4,986 (401) a	6,060 (394) a	1,730 (121) b	163 (9) a	92 (1) a	12 (0.1) b	294 (15) a	13,337
N200	3,819 (652) a	6,019 (424) b	2,194 (177) a	194 (19) a	19 (0.5) c	263 (8) a	198 (4) c	12,706
N400	4,640 (225) a	4,955 (174) c	2,100 (112) ab	152 (15) a	−48 (1) d	−1 (0.01) c	251 (12) b	12,049
N600	3,880 (383) a	3,874 (147) c	1,049 (120) c	11 (2) b	33 (0.6) b	−79 (0.5) d	116 (4) d	8,884

(C) N_2O

Treatments	0–30 cm	30–60 cm	60–90 cm	90–150 cm	150–200 cm	200–250cm	250–300 cm	0–300cm
N0	93 (7) c	90 (2) d	23 (3) b	6 (0.3) a	9 (0.3) b	4 (0.4) d	62 (1) a	287
N200	226 (23) b	199 (7) c	24 (1) b	−1 (0.02) b	12 (0.6) a	51 (2) a	−20 (2) d	491
N400	263 (27) b	358 (5) a	23 (2) b	−6 (0.2) d	−2 (0.2) c	20 (1) b	35 (0.5) b	691
N600	447 (1) a	222 (2) b	74 (4) a	−3 (0.1) c	−2 (0.1) c	11 (0.7) c	22 (2) c	771

Values (means with SE in the brackets) followed by the same letter are not significantly different within columns (one-way ANOVA with LSD; $P<0.05$)

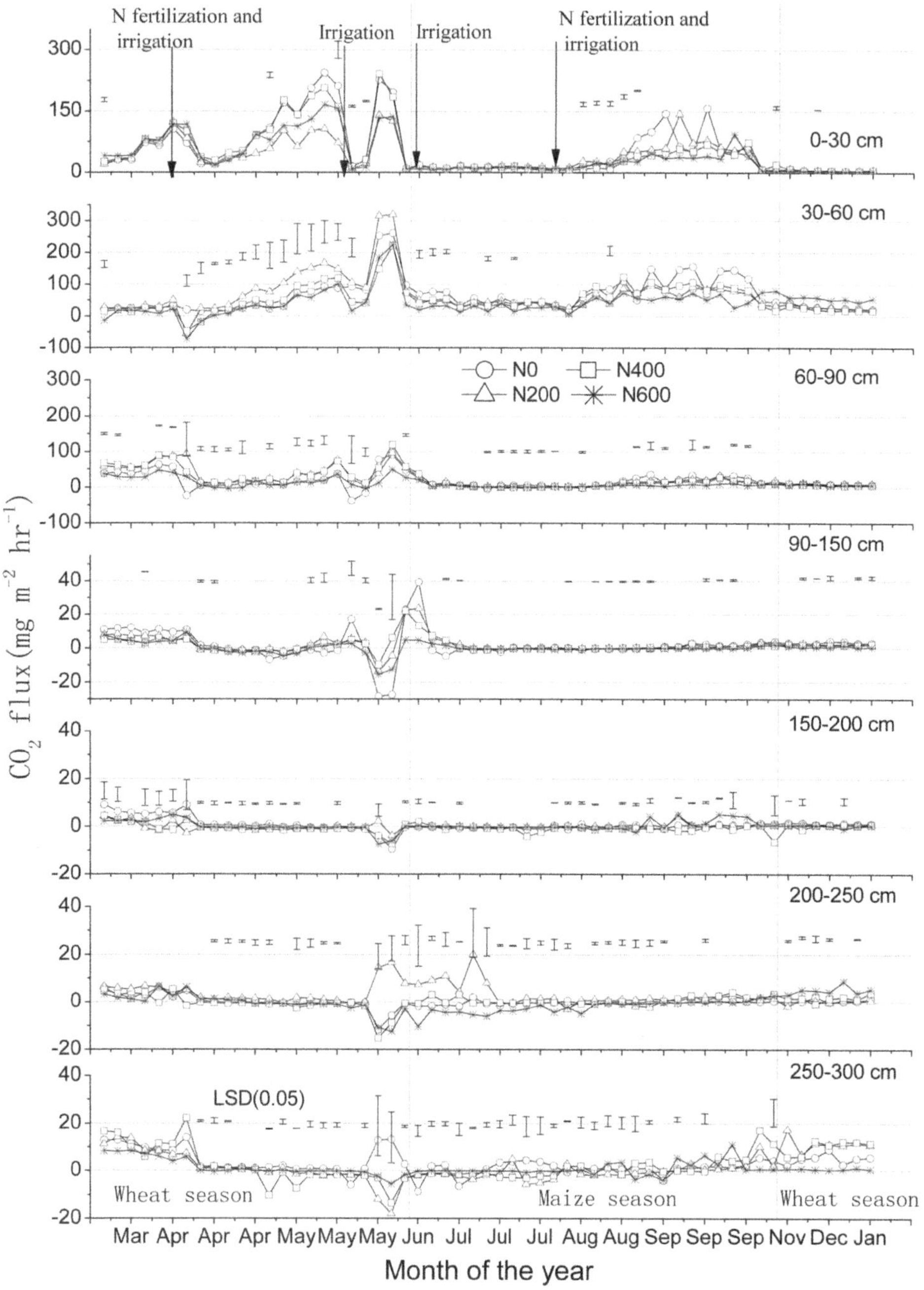

Figure 3. CO_2 flux rates (means ± standard deviations, n = 3) at various soil depths in a winter wheat-summer maize double cropping rotation receiving 0, 200, 400 and 600 kg of N ha^{-1} $year^{-1}$, in 2007–2008. Vertical dashed lines indicate a change in crop. Bars in figures indicate 1 standard deviation (n = 3). Note the differences in Y-axes.

upper 0–30 cm soil layer was averaged, and soil diffusivity may be higher in the top few cm than the bulk of the top 30 cm of soil [28]. Annual cumulative CO_2 flux in the layer 0–90 cm (11,327 kg CO_2 per ha per year) contributed about 97% to that in the layer 0–300 cm (11,744 kg CO_2 per ha per year). Mean calculated fluxes in the layers 0–30, 30–60 and 60–90 cm were all significantly higher than that in the layer 90–300 cm ($P<0.05$); mean annual cumulative fluxes from the soil below 90 cm were very small and in upwards direction (Figure1 C; Table 3 B).

Fluxes of N_2O in soil

Fertilizer application, irrigation and precipitation events triggered an efflux of N_2O from the topsoil to the atmosphere (Figure 4). The peak efflux, associated with the supplemental N fertilizer application and flooding in early April (wheat growing season), was accompanied with significant downward directed fluxes below the topsoil layer (0–30 cm). There was another relatively large efflux of N_2O into the atmosphere during the relatively moist and warm August summer month (maize growing season) (Figures 4 and 5), but this peak was not accompanied with significant downward directed fluxes below the topsoil layer. In the

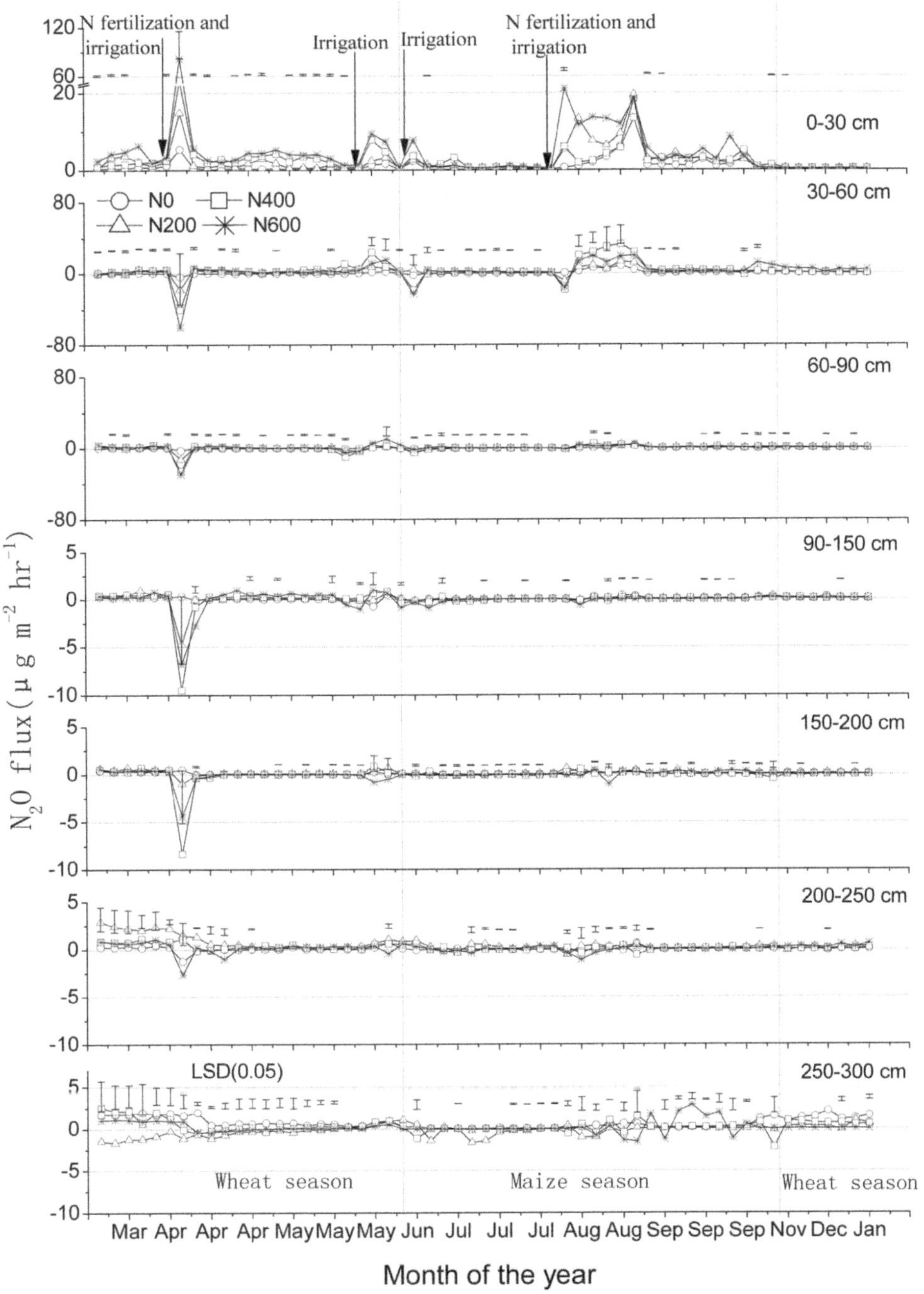

Figure 4. N_2O flux rates (means ± standard deviations, n = 3) at various soil depths in a winter wheat-summer maize double cropping rotation receiving 0, 200, 400 and 600 kg of N ha^{-1} $year^{-1}$, in 2007–2008. Vertical dashed lines indicate a change in crop. Bars in figures indicate 1 standard deviation (n = 3). Note the differences in Y-axes.

subsoil, fluxes were relatively small and directions variable (Figure 4). Essentially all seasonal fluctuations of N_2O flux rates in the subsoil (60–200 cm) seem to be related to fertilizer application, irrigation and rainfall events and changes in WFPS; therefore, there was no clear evidence of N_2O production in the subsoil after excluding these influence of interfering factors [16].

Annual cumulative fluxes of N_2O increased with increasing N fertilizer application rates; calculated total emissions at the soil surface were 93, 226, 263 and 447 g N_2O per ha for the N0, N200, N400 and N600 treatments, respectively; net upward fluxes from the 30–60 cm layer were almost as large (90, 199, 358 and 222 g N_2O per ha for the N0, N200, N400 and N600 treatments, respectively) as the fluxes from the 0–30 cm layer to the atmosphere (Table 3 C). Mean calculated fluxes in the layers 0–30 and 30–60 cm were both significantly higher than those in the layers 60–90 and 90–300 cm ($P<0.05$) (Figure 1 C); mean annual cumulative fluxes from the soil below 90 cm were small but mostly in upwards direction, suggesting that the subsoil was a small source of N_2O, and/or that accumulated N_2O from the previous season contributed to the net upward directed fluxes.

Relations between WFPS and temperature and CH_4, CO_2 and N_2O fluxes

Linear regression relationships between WFPS and CH_4 fluxes (positive) and between WFPS and CO_2 fluxes (negative) were statistically significant ($p<0.05$) for almost all soil layers (Table 4). Uptake of CH_4 by the soil was relatively high when WFPS was relatively low, probably because the diffusion rate of CH_4 into the soil was high when soil was dry, and vice versa [6]. Similarly, the upward transport of CO_2 was low when WFPS was high, and vice versa. This indicates that soil moisture exerted a dominant control on CH_4 and CO_2 fluxes. The linear relationship between WFPS and N_2O flux was also significant for the soil layers 200–250 and 250–300 cm ($p<0.05$), but not for the other layers. Significant downward directed fluxes below the topsoil were only observed down to a deep of 200 cm (Figure 4); note that N_2O fluxes were very low in 200–300 cm soil layer. Apparently, in 200–300 cm deep soil profile, soil moisture exerted a dominant control on nitrification and denitrification processes. But in 0–200 cm soil layer, WFPS was not the dominant controlling factor for the diffusive N_2O flux, likely the combination of WFPS, ammonia, nitrate and metabolizable carbon, because these factors commonly control nitrification and denitrification processes.

Relationships between soil temperature and CH_4, CO_2 and N_2O fluxes showed relatively large scatter (Table 4). Evidently, high temperatures are associated with the summer season, which is relatively moist (Figure 5). The significant relationships ($p<0.05$) between soil temperature and CH_4 fluxes at depth of 60–300 cm may be the result in part of the covariance between WFPS and soil temperature. Fluxes of N_2O in soil were not significantly related to soil temperature (Table 4).

Discussion

Fluxes of CH_4, CO_2 and N_2O at the interface of soil and atmosphere are the net result of production, consumption and transport in the soil [11]. In this study, we inferred fluxes in the soil profile from changes in concentrations with depth and over time, so as to identify soil horizons of CH_4, CO_2 and N_2O production and consumption, and thereby to increase the understanding of the dynamics of the net fluxes at the interface of soil and atmosphere. The study is unique in the sense that the inference of subsurface fluxes of CH_4, CO_2 and N_2O in 300 cm deep soil profiles at high temporal resolution over a full year has not been reported before in such comprehensive manner.

Though the C and N cycles are intimately linked in the biosphere, there were significant differences in the dynamics of CH_4, CO_2 and N_2O production, consumption and transport in the studied soil. The concentration profiles have distinct characteristics (Figure 1 A); the seasonal dynamics were much larger in the topsoil than subsoil. Moreover, the seasonal dynamics in inferred fluxes occurred during distinct periods (Figures 2, 3 and 4), and these were related to changes in WFPS (Figure 5 A), following rainfall and irrigation events. Fertilizer N application affected N_2O fluxes greatly, but not those of CH_4 and CO_2. The soil under the winter wheat-summer maize double cropping system was a net sink of atmospheric CH_4 and a net source of N_2O. It was also a large source of CO_2 but it is unknown whether the efflux compensated the influx of C into the soil via plant growth, as the latter influx was not measured.

The inferred fluxes at the soil-atmosphere interface (Figure 1 C; Table 3) were relatively small compared to those observed in other studies [5,6,29]. Our estimated soil surface fluxes are very likely underestimates because the depth resolution of the gas samplers in the top soil was too low to capture the curvature of the concentration profile properly. Hence, our study may have underestimated the dynamics of the CH_4, CO_2 and N_2O fluxes in the top soil. The depth resolution of the sampling below 90 cm appeared to be adequate. Below, we discuss the dynamics of the CH_4, CO_2 and N_2O fluxes in the soil profile in more detail.

CH_4 flux

Application of fertilizer N has been shown to inhibit CH_4 oxidation in soil [30,31], and several studies noted that non amended soils act as sink of CH_4 [32–34]. In our study, seasonal

Table 4. Linear regressions for the relationship between climatic variables and GHG fluxes.

Climatic variable	Soil depth (cm)	CH_4 ($\mu g\ m^{-2}\ hr^{-1}$)	CO_2($mg\ m^{-2}\ hr^{-1}$)	N_2O ($\mu g\ m^{-2}\ hr^{-1}$)
Soil water filled pore space	0–30	0.853**	−0.775**	−0.149
	30–60	0.645**	−0.372**	0.084
	60–90	0.787**	−0.852**	0.067
	90–150	0.637**	−0.447**	0.093
	150–200	0.771**	−0.268*	0.084
	200–250	−0.146	−0.289*	−0.692**
	250–300	0.763**	−0.532**	−0.579**
Soil temperature	0–30	−0.001	0.042	0.104
	30–60	0.270*	0.356**	0.211
	60–90	0.635**	−0.348**	0.093
	90–150	0.620**	−0.225	0.062
	150–200	0.575**	−0.266*	−0.075
	200–250	−0.122	−0.299*	−0.493**
	250–300	0.473**	−0.323*	−0.263*

Pearson's correlation coefficient, 2-tailed tests of significance.
**Significant correlation at a <0.01.
*Significant correlation at a <0.05.

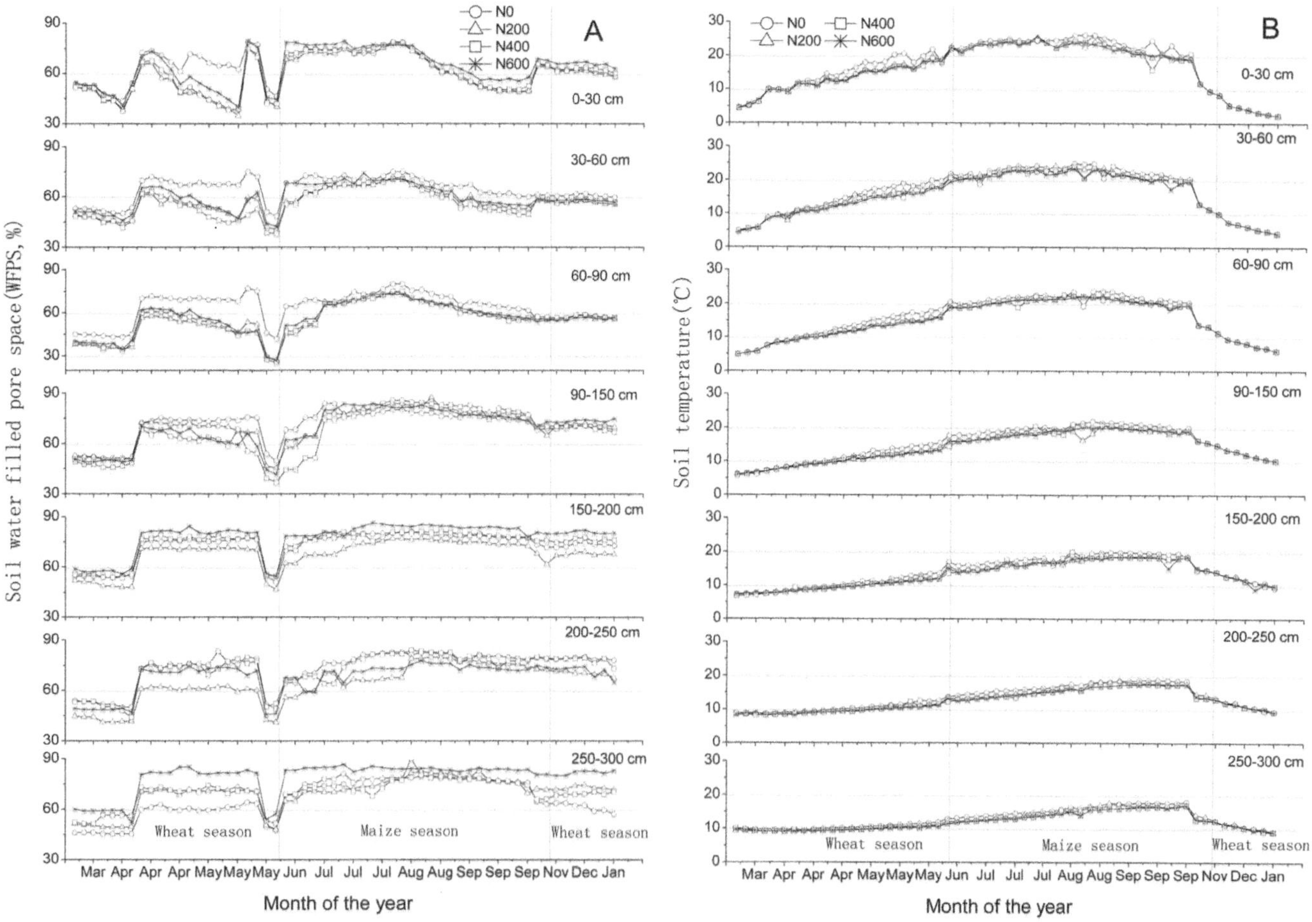

Figure 5. Water-filled pore space (WFPS) at various soil depths in a winter wheat-summer maize double cropping rotation receiving 0, 200, 400 and 600 kg of N ha^{-1} year^{-1}, in 2007–2008. Bars in figures indicate 1 standard deviation (n = 3). (A); Soil temperatures at various soil depths in winter wheat-summer maize double cropping rotation receiving 0, 200, 400 and 600 kg of N ha^{-1} year^{-1}, in 2007–2008.(B)

mean emission rates and annual cumulative fluxes of CH_4 for all soil layers and N fertilizer treatments were consistently directed downward (Figures 1 C and 2; Table 3 A). Though statistical significant differences in cumulative CH_4 fluxes between fertilizer N treatments were observed (Table 3 A), there was no clear trend that an increase in total N application decreased CH_4 uptake by soil. Inferred uptake was higher in the N200 and N400 treatments than in the N0 and N600 treatments at depth of 30 to 90 cm.

The magnitude of methane uptake by soils is largely controlled by diffusion of atmospheric methane into the soil [35], which in turn is strongly influenced by soil moisture [28]. The rate of diffusion of CH_4 in soil was high when WFPS was low. Our results showed a significant negative linear correlation between CH_4 uptake rate and WFPS for almost all layers; and the highest CH_4 uptake rates took place when WFPS was under 70% (Figures 2 and 5 A; Table 4). This is in agreement with the studies by Guo et al, Wu et al and Wang et al [4,6,16]. Inferred downward CH_4 fluxes decreased with depth (Figures 1 C and 2; Table 3 A). It has been reported that methanotrophic activity is most pronounced in the top soil [4,27], but our study suggests that significant uptake took place up to depths of 60 to 90 cm; below 90 cm, inferred fluxes of CH_4 were negligibly small (Figure 1 C).

CO_2 flux

Application of 200 kg fertilizer N per ha per year and more roughly doubled grain yields relative to the control treatment [16], but did not have statistical significant effects on the CO_2 efflux from the soil and the diffusive flux in the soil profile (Figure 3; Table 3 B). Apparently, fertilizer N application affected predominantly aboveground biomass production, and not so much underground biomass production and respiration. Yet, we may have missed some of the topsoil dynamics, also because the incorporation of the stubbles by ploughing was in the top 15 cm of soil only. A relatively large portion of total respiration in soil took place in the layer 30–60 cm; and the total respiration in the layer was significantly higher than those in the layers 60–90 and 90–300 cm ($P<0.05$) (Figure 1 C).

When soil WFPS ranged between 40 and 70% and soil temperature was >10 °C (Figure 5), highest CO_2 fluxes took place at depth of 0–60 cm during the second half of the growing seasons of wheat and maize, i.e., from mid-April to mid-May and from mid-August to mid-September (Figures 3 and 5 A). These elevated CO_2 emissions are attributed to root respiration and to enhanced mineralization of soil organic matter by increased microbial activity [36], but also to changes in the stability and formation of soil aggregates and in the microbial community structure [37]. Sufficient soil moisture is needed to allow and support substrate diffusion to the sites of microbial activity. However, if soil moisture values exceed certain thresholds (which do depend on soil properties such as porosity, bulk density and SOC content) microbial soil respiration can get O_2 limited due to diffusion constrains [6]. In a saturated soil, air is pushed out of soil pore spaces and root respiration further depletes O_2 in the soil air [38,39]. In our study, a significant negative linear correlation was found between CO_2 flux rate and WFPS (40–70%) in all layers (Table 4); and low CO_2 fluxes took place when WFPS exceeded 70%, especially after irrigation or heavy rainfall events, i.e., from June to August (Figures 3 and 5 A). This assertion is also supported by results presented by Davidson et al, Jassal et al and Fang et al [38–40].

Generally, the CO_2 evolution from soil is directly correlated with soil temperature, though within a certain temperature range [41,42], and depending on the presence of active roots [43]. In our study, the relationship of soil temperature and CO_2 fluxes was variable, likely because of the dominant effect of WFPS (Table 4).

N_2O flux

Nitrogen application and irrigation/rainfall are main triggers for increased N_2O concentrations in a soil profile and for increased emissions [15,16]. The top soil was the source of N_2O production. The combined urea applications and irrigations in early April and by the end of July 2007 strongly increased NO_3^- (Figure 1 B), NH_4^+ (data not shown) contents and WFPS (Figure 5 A) in soil, and induced large upward directed fluxes in the upper 0–30 cm soil layer; interestingly, relatively large downward directed fluxes in the subsoil only took place in April (up to the depths of 60 and 90 cm) (Figure 4), we cannot exclude the possibility that the apparent downward directed peaks during the first two months probably related to the soil structure disturbance that resulted from the installation of the samplers in December 2006.

In soil, N_2O is mainly produced by nitrification and denitrification processes. The most important factors controlling these processes are NH_4^+ and NO_3^- contents, O_2 partial pressure, and available carbon to fuel heterotrophic denitrification [44,45]. The rapid increases of WFPS (Figure 5 A), NO_3-N content (Figure 1 B) and N_2O productions in the subsoil (Figures 1 A and 4) would suggest that convective transport contributed to the downward transport of water and solutes (especially in the maize growing season), which is in line with other observations [2,46]. For instance, significant upward directed fluxes of N_2O took place during the relatively moist and warm August (maize growing season) than in the preceding wheat season in 30–60 cm soil layer (Figure 4). Several studies have demonstrated that higher values of soil moisture and temperature result in higher N_2O fluxes [44–46] Also, Li et al found while carrying out a three-year field experiment at the same study sites that significant NO_3- leaching events occur predominantly during August to October (maize growing season) [47]. Zhu et al found while carrying out a four-year field experiment in a hillslope cropland that soil NO_3- concentrations in the subsurface soil (15–30 cm) were higher than in the topsoil (0–15 cm) during most of the maize season, indicating a rapid and effective transport of NO_3- to the subsurface soil following over irrigation or rainfall events [48]. Our results indicate that NO_3-N contents in soil layers after the maize harvest were higher than after the wheat harvest (Figure 1 B); but due to missing measurements, we can not fully rule out that high NH_4^+ content [47] may have contributed to relatively high N_2O emissions in the warm and wet maize season. Although N_2O concentration increased with soil depth, changes in inferred N_2O flux below a depth of 60 cm were relatively small (Figures 1 A and 4). It may be related to the variation of the vertical N_2O concentration gradient; changes in mean N_2O concentration below a depth of 30 cm were not significant (Figure 1 C).

It has been frequently observed that high rates of N_2O emissions take place when WFPS ranges between 30 and 70% [49]. According to Zou et al the N_2O production in dry land soil of Northern China is mostly driven by nitrification [50]. Wang et al suggests that nitrification is likely a main source of the N_2O production in the soil profile when WFPS varied between 45 and 70% at the study site [16]. N_2 starts being emitted through denitrification at a WFPS of 70%, and is the main N gas emitted when WFPS exceeds 75% [51]. This may explain the relatively high inferred N_2O flux from late April to mid-May (WFPS, 40–70%) and the very low flux from late June to late July (before applying nitrogen) (WFPS, >70%) in the layer 0–30 cm (Figures 4 and 5 A).

The accumulated N_2O fluxes were significantly related to N application rate. This was most apparent in the top 30 to 90 cm of soil (Table 3 C). Annual cumulative N_2O flux in the layer 0–90 cm (511 g N_2O per ha per year) contributed about 90% to that in the layer 0–300 cm (560 g N_2O per ha per year). The 90 cm thick cinnamon top soil overlays the so-called Shajiang layer (90–140 cm) with silty clay loam texture [52]. The Shajiang layer has no crop roots, contains many iron-manganese nodules and has high bulk density (Table 2). This compacted subsoil may explain that fertilizer application and irrigation mainly affected N_2O fluxes down to 90 cm (Figure 1 C; Table 3 C). In this study, calculated total emissions in the layer 0–90 cm were 206, 449, 644 and 743 g N_2O per ha for the N0, N200, N400 and N600 treatments, respectively; these fluxes translate into fertilizer-derived emissions of 0.14, 0.10 and 0.07% for the N200, N400 and N600 treatments, respectively. The fertilizer induced emission factors (0.07–0.14%) were lower than the 0.30–0.39% measured by Ding et al [53] over the maize-wheat rotation year in a long-term mineral nitrogen addition field experiments (150–300 kg N ha^{-1} $year^{-1}$, over 20-years) in the North China Plain. A reason for the lower fertilizer induced emission factor is probably related to the likely underestimates of soil surface N_2O fluxes, because the concentration gradient in the upper 0–30 cm soil layer was averaged, and soil diffusivity may be higher in the top few cm than the bulk of the top 30 cm of soil [28]. Furthermore, due to missing measurements we can not fully rule out the indirect N_2O emissions from leaching and atmospheric deposition [47,53].

Conclusions

Our study is one of few that inferred CH_4, CO_2 and N_2O transport between soil layers from changes in CH_4, CO_2 and N_2O concentrations in the upper 300 cm of soil, measured at (bi)-weekly time intervals for one year in a winter wheat-summer maize double crop rotation. The top 30 to 60 cm of soil was a sink of atmospheric CH_4, and a source of both CO_2 and N_2O. There was little or no evidence that the subsoil (>90 cm) acted as a sink or source of GHG; rather it acted as "reservoir".

Nitrogen fertilizer application increased N_2O fluxes but did not affect CH_4 and CO_2 fluxes. The fertilizer-derived N_2O flux was small, likely because our sampling design may have missed N_2O production in the top 15 cm of soil. This holds as well for the CH_4 consumption by soil and the CO_2 emissions from soil; both are likely underestimated. Soil moisture (WFPS) was found to play an important regulating role for CH_4, CO_2 and N_2O fluxes in soil and between soil and atmosphere. Both CH_4 consumption and CO_2 fluxes in and from soil all tended to decrease with increasing WFPS.

More than 90% of the annual cumulative GHG fluxes originated at depths shallower than 90 cm. Mostly because the productive soil of our study site in the North China Plain had two distinct layers (0–90 and >90 cm), with different texture and bulk density. These differences showed up in characteristic differences in GHG concentration profiles and fluxes.

Acknowledgments

The authors would like to thank Luancheng Agroecosystem Experimental Station, Chinese Academy of Sciences for tireless efforts with maintaining the long-term fertilizer experiments. The authors would also like to thank two anonymous reviewers, whose suggestions and comments greatly improved the manuscript.

Author Contributions

Conceived and designed the experiments: CSH YYW HM YMZ XXL WXD OO DAS. Performed the experiments: YYW HM CSH. Analyzed the data: YYW. Wrote the paper: YYW.

References

1. IPCC (2007) Agriculture. In: Metz, B., D.O.R., Bosch P.R. (Eds.), Climate Change 2007: Mitigation, Contribution of Working Group III to the Fourth Assessment Report of the Intergovernmental Panel on Climate Change. Cambridge University Press, Cambridge. United Kingdom and New York, NY, USA.
2. Jassal RS, Black TA, Trofymow AJ, Roy R, Nesic Z (2010) Forest-floor CO_2 and N_2O flux dynamics in a nitrogen-fertilized Pacific Northwest Douglas-fir stand. Geoderma 157(3–4): 118–125.
3. Smith P, Martino D, Cai ZC, Gwary D, Janzen H, et al. (2008) Greenhouse gas mitigation in agriculture. Philosophical Transactions of the Royal Society B: Biological Sciences 363: 789–813.
4. Guo JP, Zhou CD (2007) Greenhouse gas emissions and mitigation measures in Chinese agroecosystems. Agricultural and Forest Meteorology 142: 270–277.
5. Kim Y, Ueyama M, Nakagawa F, Tsunogai U, Harazono Y, et al. (2007) Assessment of winter fluxes of CO_2 and CH_4 in boreal forest soils of central Alaska estimated by the profile method and the chamber method: a diagnosis of methane emission and implications for the regional carbon budget. Tellus 59B: 223–233.
6. Wu X, Yao Z, Brüggemann N, Shen ZY, Wolf B, et al. (2010) Effects of soil moisture and temperature on CO_2 and CH_4 soil-atmosphere exchange of various land use/cover types in a semi-arid grassland in Inner Mongolia, China. Soil Biology & Biochemistry 42: 773–787.
7. Banger K, Tian HQ, Lu CQ (2012) Do nitrogen fertilizers stimulate or inhibit methane emissions from rice fields? Global Change Biology 18: 3259–3267.
8. Sanz-Cobena A, Sánchez-Martín L, García-Torres L, Vallejo A (2012) Gaseous emissions of N_2O and NO and NO_3- leaching from urea applied with urease and nitrification inhibitors to a maize (Zea mays) crop. Agriculture, Ecosystems and Environment 149: 64–73.
9. Sanz-Cobena A, García-Marco S, Quemada M, Gabriel JL, Almendros P, et al. (2014) Do cover crops enhance N_2O, CO_2 or CH_4 emissions from soil in Mediterranean arable systems? Science of the Total Environment 466–467: 164–174.
10. Valentini R, Matteucci G, Dolman AJ, Schulze ED, Rebmann C, et al. (2000) Respiration as the main determinant of carbon balance in European forests. Nature 404: 861–865.
11. Bowden WB, Bormann FH (1986) Transport and loss of nitrous oxide in soil water after forest cutting. Science 233: 867–869.
12. Tang JW, Baldocchi DD, Qi Y, Xu LK (2003) Assessing soil CO_2 efflux using continuous measurements of CO_2 profiles in soils with small solid-state sensors. Agricultural and Forest Meteorology 118: 207–220.
13. Fierer N, Chadwick OA, Trumbore SE (2005) Production of CO_2 in Soil Profiles of a California Annual Grassland. Ecosystems 8: 412–429.
14. Gebert J, Röer IU, Scharff H, Roncato CDL, Cabral AR (2011) Can soil gas profiles be used to assess microbial CH_4 oxidation in landfill covers? Waste Management 31: 987–994.
15. Reth S, Graf W, Gefke O, Schilling R, Seidlitz HK, et al. (2008) Whole-year-round Observation of N_2O Profiles in Soil: A Lysimeter Study.Water Air Soil Pollut: Focus 8:129–137.
16. Wang YY, Hu CS, Ming H, Zhang YM, Li XX, et al. (2013) Concentration profiles of CH_4, CO_2 and N_2O in soils of a wheat–maize rotation cosystem in North China Plain, measured weekly over a whole year. Agriculture, Ecosystems & Environment 164: 260–272.
17. Sparks DL (1996) Methods of soil analysis, part 3–chemical methods. In: Sparks, D.L. (Eds.), SSSA book series, No. 5. SSSA, Inc and American Society of Agronomy, Inc. Madison, WI. pp. 475–1185.
18. Gee GW, Bauder JW (1986) Particle-size analysis. In: Klute, A. (Ed.), Methods of Soil Analysis Part 1-Physical and Mineralogical Methods. American Society of gronomy, Madison, WI. pp. 383–409.
19. Campbell GS (1985) Soil Physics with BASIC. Elsevier Science Publishers BV, Amsterdam. Soil Science Society of America Book Series: 5 Methods of Soil Analysis Part I-Physical and Mineralogical Methods.
20. Marshall TJ (1959) The diffusion of gas through porous media. J Soil Sci 10: 79–82.
21. Rolston DE (1986) 47 Gas Flux, 1103–1109. American Society of Agronomy, Inc. Soil Science Society of America, Inc. Publisher. Madison, Wisconsin USA.
22. Sallam A, Jury WA, Letey J (1984) Measurement of gas-diffusion coefficient under relatively low air-filled porosity. Soil Sci Soc Am J 48: 3–6.
23. Jury WA, Gardner WR, Gardner WH (1991) Soil Physics, fifth ed. Wiley, New York.
24. Gilliland E, Baddour R, Perkinson G, Sladek KJ (1974) Diffusion on surfaces. I. Effect of concentration on the diffusivity of physically adsorbed gases. Ind Eng Chem Fundam 13(2): 95–100.
25. Millington R, Quirk JP (1961) Permeability of porous solids. Trans Faraday Soc 57: 1200–1207.
26. Wang YY, Hu CS, Zhu B, Xiang HY, He XH (2010) Effects of wheat straw application on methane and nitrous oxide emissions from purplish paddy fields. Plant, Soil and Environ 56 (1): 16–22.
27. Stiehl-Braun PA, Powlson DS, Poulton PR, Niklaus PA (2011) Effects of N fertilizers and liming on the micro-scale distribution of soil methane assimilation in the long-term Park Grass experiment at Rothamsted. Soil Biology and Biochemistry 43(5): 1034–1041.
28. Shrestha BM, Sitaula BK, Singh BR, Bajracharya RM (2004) Fluxes of CO_2 and CH_4 in soil profiles of a mountainous watershed of Nepal as influenced by land use, temperature, moisture and substrate Addition. Nutrient Cycling in Agroecosystems 68: 155–164.
29. Chu H, Hosen Y, Yagi K (2004) Nitrogen oxide emissions and soil microbial properties as affected by N-fertilizer management in a Japanese Andisol. Soil Science and Plant Nutrition 50: 287–292.
30. Steudler PA, Bowden RD, Melillo JM, Aber JD (1989) Influence of nitrogen fertilisation on methane uptake in temperate forest soils. Nature 341: 314–316.
31. Kravchenko I, Boeckx P, Galchenko V, Van Cleemput O (2002) Shortand medium-term effects of $NH_4{}^+$on CH_4 and N_2O fluxes in arable soils with a different texture. Soil Biology & Biochemistry 34: 669–678.
32. Flessa H, Beese F (2000) Laboratory estimates of trace gas emissions following surface application and injection of cattle slurry. J Environ Qual 29: 262–268.
33. Sherlock RR, Sommer SG, Khan RZ, Wood CW, Guertal EA, et al. (2002) Emissions of ammonia, methane and nitrous oxide from pig slurry applied to a pasture in New Zealand. J Environ Qual 31: 1491–1501.
34. Rodhe L, Pell M, Yamulki S (2006) Nitrous oxide, methane and ammonia emissions following slurry spreading on grassland. Soil Use Manage 22: 229–237.
35. Koschorreck M, Conrad R (1993) Oxidation of atmospheric methane in soil: measurements in the field, in soil cores and in soil samples. Global Biogeochemical Cycles 7: 109–121.
36. Borken W, Matzner E (2009) Reappraisal of drying and wetting effects on C and N mineralization and fluxes in soils. Global Change Biology 15: 808–824.
37. Denef K, Six J, Bossuyt H, Frey SD, Elliott ET, et al. (2001) Influence of dry-wet cycles on the interrelationship between aggregate, particulate organic matter, and microbial community dynamics. Soil Biology and Biochemistry 33: 1599–1611.
38. Davidson EA, Belk E, Boone RD (1998) Soil water content and temperature as independent or confounded factors controlling soil respiration in a temperate mixed hardwood forest. Global Change Biol 4: 217–227.
39. Jassal RS, Black TA, Drewitt GB, Novak MD, Gaumont-Guay D, et al. (2004) A model of the production and transport of CO_2 in soil: predicting soil CO_2 concentrations and CO_2 efflux from a forest floor. Agric For Meteorol 124: 219–236.
40. Fang YT, Gundersen P, Zhang W, Zhou GY, Christiansen JR, et al. (2009) Soil-atmosphere exchange of N_2O, CO_2 and CH_4 along a slope of an evergreen broad-leaved forest in southern China. Plant and Soil 319: 37–48.
41. Bajracharya RM, Lal R, Kimble JM (2000) Erosion effect on carbon dioxide concentration and carbon flux from an Ohio Alfisol. Soil Sci. Soc. Am. J 64: 694–700.
42. Fang C, Moncrieff JB (2001) The dependence of soil CO_2 efflux on temperature. Soil Biology & Biochemistry 33: 155–165.
43. Kelting DL, Burger JA, Edwards GS (1998) Estimating root respiration, microbial respiration in the rhizosphere, and root-free soil respiration in forest soils. Soil Biology and Biochemistry 30: 961–968.
44. Clough TJ, Sherlock RR, Kelliher FM (2003) Can liming mitigate N_2O fluxes from a urine-amended soil? Aust J Soil Res 41: 439–457.
45. Clough TJ, Kelliher FM, Sherlock RR, Ford CD (2004) Lime and soil moisture effects on nitrous oxide emissions from a Urine Patch. Soil Sci Soc A J 68: 1600–1609.

46. Grandy SA, Robertson PG (2006) Initial cultivation of a temperate-region soil immediately accelerates aggregate turnover and CO_2 and N_2O fluxes. Global Change Biology 12: 1507–1520.
47. Li XX, Hu CS, Delgado JA, Zhang YM, Ouyang ZY (2007) Increased nitrogen use efficiencies as a key mitigation alternative to reduce nitrate leaching in north China plain. Agricultural Water Management 89: 137–147.
48. Zhu B, Wang T, Kuang F, Luo Z, Tang J, et al. (2009) Measurements of nitrate leaching from a hillslope cropland in the Central Sichuan Basin, China. Soil Sci Soc Am J 73: 1419–1426.
49. Dobbie KE, Smith KA (2003) Nitrous oxide emission factors for agricultural soils in Great Britain: the impact of soil water-filled pore space and other controlling variables. Global Change Biology 9: 204–218.
50. Zou GY, Zhang FS, Chen XP, Li XH (2001) Nitrification–denitrification and N_2O emission from arable soil. Soil Environ Sci 10 (4): 273–276 (in Chinese).
51. Davidson EA (1992) Sources of nitric oxide and nitrous oxide following wetting of dry soil. Soil Sci Soc Am J 56: 95–102.
52. Zhu HJ, He YG eds (1992) Soil Geography. Higher Education Press, Beijing, China (In Chinese).
53. Ding WX, Luo JF, Li J, Yu HY, Fan JL, et al. (2013) Effect of long-term compost and inorganic fertilizer application on background N_2O and fertilizer-induced N_2O emissions from an intensively cultivated soil. Science of the Total Environment 465: 115–124.

Influence of Different Plant Species on Methane Emissions from Soil in a Restored Swiss Wetland

Gurbir S. Bhullar[1,2]*, Peter J. Edwards[1], Harry Olde Venterink[1,3]

1 Institute of Integrative Biology, Plant Ecology, ETH Zurich, Zurich, Switzerland, 2 Research Institute of Organic Agriculture (FiBL), Frick, Switzerland, 3 Plant Biology and Nature Management, Vrije Universiteit Brussel, Brussels, Belgium

Abstract

Plants are a major factor influencing methane emissions from wetlands, along with environmental parameters such as water table, temperature, pH, nutrients and soil carbon substrate. We conducted a field experiment to study how different plant species influence methane emissions from a wetland in Switzerland. The top 0.5 m of soil at this site had been removed five years earlier, leaving a substrate with very low methanogenic activity. We found a sixfold difference among plant species in their effect on methane emission rates: *Molinia caerulea* and *Lysimachia vulgaris* caused low emission rates, whereas *Senecio paludosus, Carex flava, Juncus effusus* and *Typha latifolia* caused relatively high rates. *Centaurea jacea, Iris sibirica*, and *Carex davalliana* caused intermediate rates. However, we found no effect of either plant biomass or plant functional groups – based on life form or productivity of the habitat – upon methane emission. Emissions were much lower than those usually reported in temperate wetlands, which we attribute to reduced concentrations of labile carbon following topsoil removal. Thus, unlike most wetland sites, methane production in this site was probably fuelled chiefly by root exudation from living plants and from root decay. We conclude that in most wetlands, where concentrations of labile carbon are much higher, these sources account for only a small proportion of the methane emitted. Our study confirms that plant species composition does influence methane emission from wetlands, and should be considered when developing measures to mitigate the greenhouse gas emissions.

Editor: Fei-Hai Yu, Beijing Forestry University, China

Funding: This work was supported by ETH Zurich [TH-07 07-3]. The funders had no role in study design, data collection and analysis, decision to publish, or preparation of the manuscript.

Competing Interests: The authors have declared that no competing interests exist.

* E-mail: gurbir.bhullar@fibl.org

Introduction

Wetlands are the largest natural source of the important greenhouse gas methane (CH_4), contributing about one-third (80–110Tg yr^{-1}) to global emissions [1–3]. This methane is produced under anoxic conditions by methanogenic microbes (Archaea) [4]. However, the amounts actually emitted from a wetland soil can be significantly influenced by vascular plants [5–8], through their effects upon the production, transport and consumption of methane in soils [4,9,10]. These processes vary greatly among plant species, and their net impact upon methane emissions can range from negative to positive [5,10–12]. Much of our understanding about these processes comes from mesocosm experiments or studies conducted in environments lacking substrate uniformity. To better understand why plant species influence methane emissions differently, comparative studies are needed in wetland ecosystems under homogeneous field conditions.

In an earlier study, we showed that graminoids tended to transport more methane from rhizosphere to atmosphere than forbs [13,14], and other workers have also found plant growth form or functional type to be a factor influencing methane emissions from wetlands [15–17]. However, most of this information has come from mesocosm experiments, and field studies and needs to confirm whether the observed differences are ecologically important. Furthermore, it would be useful to have more quantitative information on this topic as a basis for modelling studies and for designing mitigation strategies [18,19].

Methane emissions from wetlands may also be influenced, either directly or indirectly, by a range of environmental factors such as water table, temperature, pH, nutrients and soil carbon [20–23]. For example, by affecting rates of root exudation and rhizosphere oxidation, factors influencing plant productivity could have an indirect influence upon methane emissions. Indeed, some studies have found a positive relationship of emissions with plant productivity [24,25], while others have reported either no relationship or a negative one [26,27]. In our mesocosm experiments, we found that plant species from less productive habitats caused higher rates of methane emission than species from more fertile habitats [10,28]. This was apparently due to higher rates of organic acid exudation by species from less productive habitats, which increased the carbon available to methanogenic bacteria [10,14].

Our aim in this study was to understand how plant species and/or functional plant groups (based on growth form and productivity indication values) influence methane emissions under field conditions. For this purpose, we needed a site that was as uniform as possible, to avoid possible species effects being confounded by effects due to heterogeneity in soil and hydrological conditions [29]. We therefore selected a wetland site in northern Switzerland where the top 0.5 m of soil had been removed five years previously as part of a restoration scheme, and the area sown with mixtures of native wetland plants. The species sown included both forbs and graminoids, and plants characteristic of habitats of high and low productivity. We assumed that the restoration treatment had reduced soil heterogeneity, so that emissions would be influenced more by the plant studied than by local site differences. The hypotheses were:

i. Plant species differ in their effect on methane emission from wetlands.
ii. Graminoids cause higher methane emissions than forbs.
iii. Species from low productive habitats cause higher emissions than those from high productive habitats.
iv. Methane emissions are negatively related to plant biomass.

Materials and Methods

The experiment was conducted in a wetland (nature reserve Hütwilersee) in Eastern Switzerland (47°36′42.31" N and 8°50′15.35"E) during May 2010. The permission for the conduct of research at this site was granted by Stiftung Seebachtal, CH-Frauenfeld, the foundation owning the land and being set in-charge for the entire restoration project "Seebachtal" by the authorities of canton Thurgau. The vegetation of the study site is composed of both annual and perennial forbs and graminoids. For much of the year, the water table at the study site is at or close to the soil surface, though it may drop to 60 cm in dry periods during the summer. The site is mown once a year in autumn, and the cut material is removed. This management is similar to that practised over the centuries when the site was used as a traditional wet meadow. Preliminary tests conducted in the laboratory showed very low methanogenic activity in soil samples from the site, which increased substantially when glucose was added as a carbon source. This suggests that the unaltered substrate had very low concentrations of labile carbon, presumably as an effect of removing the top 0.5 m of soil five years previously.

The plant species studied were five forbs (*Centaurea jacea* L., *Iris sibirica* L., *Lysimachia vulgaris* L., *Senecio paludosus* L., *Typha latifolia* L.) and four graminoids (*Carex davalliana* SM., *Carex flava* L., *Molinia caerulia* (L.) Moench, *Juncus effuses* L.) (Table 1). Because the area was densely vegetated, there was no opportunity to make reference measurements of methane emission from bare soil. Also, no comparable site without topsoil removal was available for sampling.

The methane measurements were made during the last week of May 2010, using a Photo Acoustic Field Gas-Monitor type 1412 (Innova AirTech Instruments) fitted with a moisture filter [13]. The water table was recorded from the measurement tubes already installed for another experiment running in the same field. It had rained almost daily for the preceding three weeks, and the water table when the measurements were made was close to the soil surface level throughout the site. For each of the nine species, we selected six replicate spots where the plant was growing well. All spots were spread randomly within an area of 0.6 ha. At each of these spots, one or a few individuals (depending upon species) were carefully selected and covered with a transparent Plexiglas chamber (diameter 19.28 cm, height 60.17 cm), which was placed over the aboveground part of the plant(s) and was pushed about a centimetre into the moist soil for making it air tight. The change in methane concentration inside the chamber during a period of three hours was recorded, and methane emission was expressed per square meter (1 m^{-2}) of soil surface (calculated from the diameter of the chamber).

Measurements were always made between 10:00 and 14:00 hours. Weather conditions varied from rainy to partly cloudy, and the ambient air temperature ranged between 11 and 20°C. Methane density values corresponding to air temperature were used for calculating absolute methane emissions (measured in $\mu g\ m^{-2}\ hr^{-1}$). We did not extrapolate these values over longer periods because methane emissions from wetland soils are known to exhibit strong diurnal and seasonal variation [7,30–32]. After each measurement, the total aboveground plant biomass from the measured spot was harvested, and a soil sample was collected from a depth of 0–30 cm. The harvested material was dried at 70°c for 48 hours and weighed to determine aboveground biomass. Soil moisture was determined by drying a portion of the sampled soil at 105°C for 24 hours. Soil pH was determined in distilled water. Nitrate and ammonium concentrations extracted with 1M KCL were determined colorimetrically using a (FIASTAR 5000) flow injection analyser [33]. During the measurement period, soil was almost always saturated with water table between 6 and 11 cm below soil surface and soil moisture (0–30 cm depth) between 86 and 89 percent. Soil pH at all sampling points was almost uniform at close to 6. Soil extractable nitrate was below detection limits, and ammonium was between 8 to 17 mg N kg^{-1} dry soil.

Table 1. Plant species studied in the experiment ranked according to their habitat preference based on Ellenberg & Move fertility indication values.

Species	Species code	Plant type	Ellenberg N-value*	Move N-value**
Molinia caerulea	MC	Graminoid (grass)	1	3.15±1.11
Carex davalliana	CD	Graminoid (Sedge)	2	-
Iris sibirica	IS	Forb (Monocot)	2	-
Carex flava	CF	Graminoid (Sedge)	2	-
Centaurea jacea	CJ	Forb (Dicot)	−99	4.76±1.01
Juncus effuses	JE	Graminoid (Rush)	4	4.79±1.13
Lysimachia vulgaris	LV	Forb (Dicot)	−99	4.88±1.17
Typha latifolia	TL	Forb (Monocot)	8	5.95±0.90
Senecio paludosus	SP	Forb (Dicot)	6	6.09±0.78

*[42]**[43].

Differences among plant species in methane emissions and environmental parameters (temperature, pH, nitrogen, soil moisture) were tested by means of ANOVA and Tukey test. To confirm with the assumption of homogeneity of variance and normality, the data were log-transformed prior to data analysis. We also calculated linear regression between methane emission rates and plant biomass for each species. The data were analysed using statistical software R, version 2.8.1 [34].

Results

Methane emissions varied significantly among plant species (Fig. 1). *Molinia caerulea* was associated with the lowest emissions, while *Senecio paludosus* and *Carex flava* had the highest emissions. Emissions from *Juncus effusus* and *Typha latifolia* also tended to be higher but were statistically not different from other species. Methane emissions were sensitive to changes in air temperature, but this did not affect the comparison among different plant species since the replicates within species were spread over different days and temperatures.

We found neither significant difference between graminoids and forbs ($p = 0.95$), nor between species with low and high Ellenberg N-values (below and above 4; $p = 0.49$). Methane emissions were also not significantly related to aboveground plant biomass; and emissions from different species were not significantly related to any of the measured site factors (Table 2).

Discussion

Methane emissions from natural wetlands have been reported to range between 0 and 660 mg m^{-2} day^{-1} [1,2,30,35]. At our site, they varied from 0.25 to 1.55 µg m^{-2} hr^{-1}, which is lower by a magnitude of thousands than those measured at many other temperate wetlands [27,36]. Moore and Knowles [37] have reported similarly low rates of emission from an ombrotrophic bog in Quebec (Canada), where they were perhaps caused by very slow decomposition of *Sphagnum* litter and suboptimal pH for methanogenesis. At our site, the low methanogenic activity was probably a consequence of removing the topsoil five years previously, thereby reducing concentrations of labile carbon in the substrate. Indeed, in a separate mesocosm study, we found that methane emissions increased dramatically when glucose was added to the substrate obtained from the same field site. In addition, the site was mown each year and the hay removed, so that little aboveground biomass entered the soil as litter. Thus, unlike most wetland sites, methane production in this site was probably fuelled chiefly by root exudation from living plants and from root decay [38]. The very low overall methane emission, which is one of the main results from our study, underlines an important aspect of wetland management and methane mitigation. It suggests that topsoil removal coupled with regular biomass removal can reduce methane emissions for an extended period – although we can only base this conclusion on comparison with other sites since control sites without topsoil removal were not available in our site. Furthermore, it must be noted that we measured methane emission on only one occasion, chosen for the hydrological homogeneity caused by several weeks of continuous rain.

Despite the low emissions, the site was well suited for comparing methane emissions associated with different plant species. In support of our first hypothesis, methane emission from soil at various points varied significantly according to the plant species growing there. Previous studies have also shown that plant species differ in their influence on methane emissions, though these were conducted either in mesocosms [10,13,17,22] or under rather heterogeneous field conditions [8,11] where microtopographic or other factors might have affected the results [5,36]. In our field site, we found no correlation between environmental parameters and methane emissions from different plant species, presumably because of the rather uniform site conditions [20,23].

In an earlier greenhouse experiment comparing a large set of wetland plants, we found that graminoids had a greater capacity to transport methane from soil to atmosphere (chimney effect) than forbs [13]. This effect appeared to be largely a result of differences between the root systems in the two functional groups [39,40]. No such relationship was evident in the current study, however,

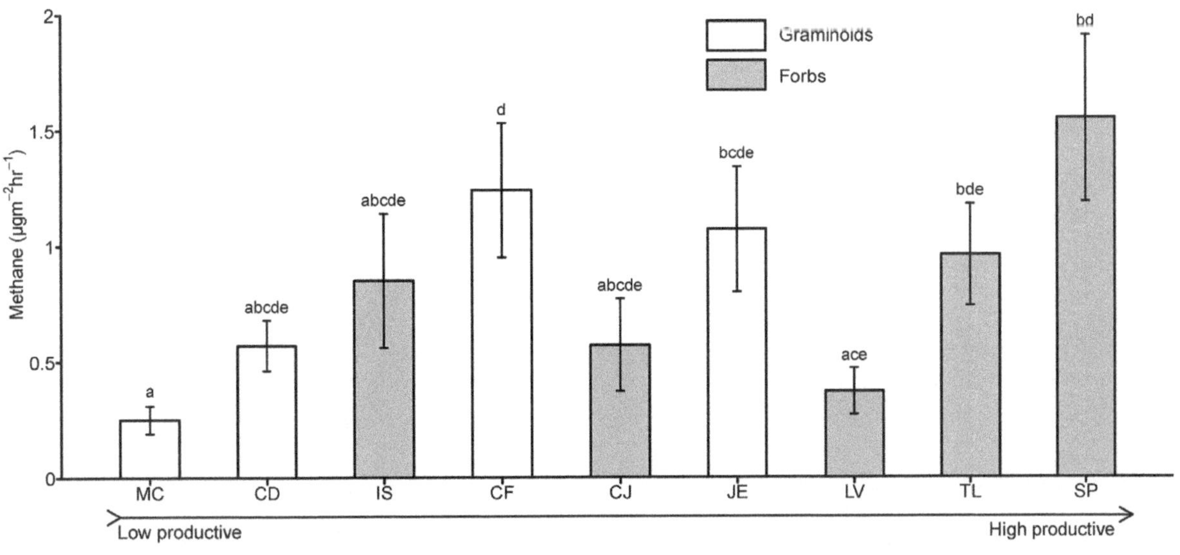

Figure 1. Mean CH_4 emission rates from different plant species, arranged in order of increasing fertility indication [42,43] from left to right on x-axis (see Table 1). MC: *Molinia caerulia* (L.) Moench, CD: *Carex davalliana* SM., IS: *Iris sibirica* L., CF: *Carex flava* L., CJ: *Centaurea jacea* L., JE: *Juncus effuses* L., LV: *Lysimachia vulgaris* L., TL: *Typha latifolia* L., SP: *Senecio paludosus* L. Error bars represent original data whereas statistics are based on log transformed data.

Table 2. Results of linear regression between methane emission and environmental factors.

	Based on all data points (log-transformed)		Based on mean value per species	
Factor	R squared	p-value	R squared	p-value
Plant Biomass	0.005	0.61	0.011	0.78
Soil pH	0.004	0.65	0.049	0.56
Soil extractable NH_4– Nitrogen	0.009	0.49	0.120	0.35
Air Temperature	0.088 *	0.03	0.001	0.92
Soil moisture	0.001	0.86	0.112	0.37
Water table	0.012	0.42	0.075	0.47

Soil variables were measured in the top 30 cm.
*P≤0.05; **P≤0.01; ***P≤0.001.*

presumably because too little methane was produced to detect any differences in transport capacity.

Results of this field study do not support our third and fourth hypothesis, which were based on mesocosm experiments showing higher emissions caused by species of less productive habitats and decreasing emissions with increase in plant productivity [10,26–28]. Higher methane emission from mesocosms with species adapted to less productive habitats was attributed to higher carbon exudation rates in these species [10]. Because our study site was depleted of labile carbon, we had expected to find similar effects in the field, arguing that under these conditions differences in exudation rates among species would be detectable as differences in methane emissions. But we found neither a difference among plant species from low and high productive habitats nor a significant relationship between methane emissions and plant biomass (Table 2). Some field studies have shown a positive correlation between methane emissions and vegetation biomass or productivity [12,24,25], while mesocosm experiments have shown the opposite effect [10,27,28]. This difference may have arisen because the amount of methane actually emitted is the resultant of several processes, but which of these dominates may vary according to the conditions. For example, under conditions at our site, the addition of carbon through root exudation was probably the primary controlling mechanism, which in turn depended upon current photosynthetic activity of the vegetation [24,41]. In contrast, when plants are grown under controlled conditions such as in mesocosm experiments, the rhizosphere oxidation by the plant roots might be of relatively greater significance [26], and therefore plants with larger biomass are able to exert a relatively larger influence through radial oxygen loss by occupying all the rhizosphere space available in the mesocosm. Thus, mesocosm experiments do not necessarily reflect processes operating in the field.

In conclusion, this study demonstrates that plant species vary in their capacity to influence methane emissions even under conditions where methanogenesis is limited by a shortage of labile carbon. However, the very low fluxes measured in our study suggest that only a small proportion of the methane emitted from most wetlands can be due to root exudation. This in turn suggests that plants have a much greater potential to influence emissions by transporting methane internally or by altering redox conditions in the soil. Finally, our results indicate that management practices such as removing topsoil and mowing the vegetation could be an effective way of reducing methane emissions from wetlands. Further research would be needed, however, to determine whether any benefits would exceed the 'carbon debt' associated with removing the top soil.

Acknowledgments

We thank Matthias Suter and Dieter Ramseier for allowing us to use the study site and for providing historic data on hydrology and management of the site; and Matthew Scarborough for helping with conduct of the experiment.

Author Contributions

Conceived and designed the experiments: GSB PJE HOV. Performed the experiments: GSB. Analyzed the data: GSB. Contributed reagents/ materials/analysis tools: PJE HOV. Wrote the paper: GSB PJE HOV.

References

1. Aselmann I, Crutzen PJ (1989) Global distribution of natural freshwater wetlands and rice paddies, their net primary productivity, seasonality and possible methane emissions. J Atmos Chem 8: 307–358.
2. Cao M, Gregson K, Marshall S (1998) Global methane emission from wetlands and its sensitivity to climate change. Atmos Environ 32: 3293–3299.
3. IPCC (2007) Climate Change 2007; Solomon S, Qin D, Manning M, Chen Z, Marquis M, et al., editors. Cambridge, UK and New York, USA: Cambridge University Press.
4. LeMer J, Roger P (2001) Production, oxidation, emission and consumption of methane by soils: A review. Eur J Soil Biol 37: 25–50.
5. Rothman E, Bouchard V (2007) Regulation of carbon processes by macrophyte species in a Great Lakes coastal wetland. Wetlands 27: 1134–1143.
6. Lai DYF (2009) Methane Dynamics in Northern Peatlands: A Review. Pedosphere 19: 409–421.
7. Ding WX, Cai ZC, Tsuruta H, Li XP (2003) Key factors affecting spatial variation of methane emissions from freshwater marshes. Chemosphere 51: 167–173.
8. Stanley E, Ward A (2010) Effects of Vascular Plants on Seasonal Pore Water Carbon Dynamics in a Lotic Wetland. Wetlands 30: 889–900.
9. Joabsson A, Christensen TR, Wallen B (1999) Influence of vascular plant photosynthetic rate on CH4 emission from peat monoliths from southern boreal Sweden. Polar Res 18: 215–220.
10. Koelbener A, Ström L, Edwards PJ, Olde Venterink H (2010) Plant species from mesotrophic wetlands cause relatively high methane emissions from peat soil. Plant Soil 326: 147–158.
11. Ding WX, Cai ZC, Tsuruta H (2005) Plant species effects on methane emissions from freshwater marshes. Atmos Environ 39: 3199–3207.
12. Joabsson A, Christensen TR (2001) Methane emissions from wetlands and their relationship with vascular plants: an Arctic example. Glob chang Biol 7: 919–932.
13. Bhullar GS, Edwards PJ, Venterink HO (2013) Variation in the plant-mediated methane transport and its importance for methane emission from intact wetland peat mesocosms. J Plant Ecol 6: 298–304.

14. Bhullar GS, Iravani M, Edwards PJ, Olde Venterink H (2013) Methane transport and emissions from soil as affected by water table and vascular plants. BMC Ecol 13: 32.
15. Sebacher DI, Harriss RC, Bartlett KB (1985) Methane emissions to the atmosphere through aquatic plants. J Environ Qual14: 40–46.
16. Schimel JP (1995) Plant-transport and methane production as controls on methane flux from arctic wet meadow tundra. Biogeochemistry 28: 183–200.
17. Rice AL, Butenhoff CL, Shearer MJ, Teama D, Rosenstiel TN, et al. (2010) Emissions of anaerobically produced methane by trees. Geophys Res Lett 37: L03807.
18. Ward SE, Bardgett RD, McNamara NP, Ostle NJ (2009) Plant functional group identity influences short-term peatland ecosystem carbon flux: evidence from a plant removal experiment. Func Ecol 23: 454–462.
19. Dias ATC, Hoorens B, Van Logtestijn RSP, Vermaat JE, Aerts R (2010) Plant Species Composition Can Be Used as a Proxy to Predict Methane Emissions in Peatland Ecosystems After Land-Use Changes. Ecosystems 13: 526–538.
20. Christensen TR, Ekberg A, Strom L, Mastepanov M, Panikov N, et al. (2003) Factors controlling large scale variations in methane emissions from wetlands. Geophys Res Lett 30: 67 (61–64).
21. Waddington JM, Roulet NT (1996) Atmosphere-wetland carbon exchanges: Scale dependency of CO2 and CH4 exchange on the developmental topography of a peatland. Global Biogeoch Cycles 10: 233–245.
22. Dinsmore KJ, Skiba UM, Billett MF, Rees RM (2009) Effect of water table on greenhouse gas emissions from peatland mesocosms. Plant Soil 318: 229–242.
23. Inubushi K, Otake S, Furukawa Y, Shibasaki N, Ali M, et al. (2005) Factors influencing methane emission from peat soils: Comparison of tropical and temperate wetlands. Nutr Cycl Agroecosys 71: 93–99.
24. Chanton JP, Whiting GJ, Blair NE, Lindau CW, Bollich PK (1997) Methane emission from rice: Stable isotopes, diurnal variations and CO2 exchange. Global Biogeoch Cycles 11: 15–27.
25. Whiting GJ, Chanton JP (1993) Primary production control of methane emissions from wetlands. Nature 364: 794–795.
26. Ström L, Mastepanov M, Christensen TR (2005) Species-specific effects of vascular plants on carbon turnover and methane emissions from wetlands. Biogeochemistry 75: 65–82.
27. Bouchard V, Frey SD, Gilbert JM, Reed SE (2007) Effects of macrophyte functional group richness on emergent freshwater wetland functions. Ecology 88: 2903–2914.
28. Bhullar GS (2011) The role of plants in the production and transport of methane from wetland soils [PhD Dissertation]. Zurich: Swiss Federal Institute of Technology (ETH), Zurich. 95 p.
29. Bubier JL (1995) The relationship of vegetation to methane emission and hydrochemical gradients in northern peatlands. J Ecol 83: 403–420.
30. Moore T, Roulet N, Knowles R (1990) Spatial and temporal variations of methane flux from subarctic/northern boreal fens. Global Biogeoch Cycles 4: 29–46.
31. Long KD, Flanagan L.B., Cai T (2010) Diurnal and seasonal variation in methane emissions in a northern Canadian peatland measured by eddy covariance. Glob Chang Biol 16: 2420–2435.
32. Schutz H, Holzapfelpschorn A, Conrad R, Rennenberg H, Seiler W (1989) A 3-year continuous record on the influence of daytime, season, and fertilizer treatment on methane emission rates from an Italian rice paddy. Journal of J Geophys Res-Atmos 94: 16405–16416.
33. Maynard DG, Kalra YP, Crumbaugh JA (2008) Nitrate and exchangeable ammonium nitrate. In: Carter MR, Geregorich EG, editors. Soil sampling and methods of analysis. 2 ed: Canadian society of soil scientists. 71–80.
34. R Development Core Team (2008) R: A Language and Environment for Statistical Computing. R Foundation for Statistical Computing, Vienna, Austria: http://www.r-project.org/.
35. Morrissey LA, Livingston GP (1992) Methane emissions from Alaska arctic tundra- an assessment of local spatial variability. J Geophys Res-Atmos 97: 16661–16670.
36. Bubier JL, Moore TR, Bellisario L, Comer NT, Crill PM (1995) Ecological controls on methane emissions from a northern peatland complex in the zone of discontinuous permafrost, Manitoba, Canada. Global Biogeoch Cycles 9: 455–470.
37. Moore TR, Knowles R (1990) Methane Emissions from Fen, Bog and Swamp Peatlands in Quebec. Biogeochemistry 11: 45–61.
38. Saarnio S, Wittenmayer L, Merbach W (2004) Rhizospheric exudation of Eriophorum vaginatum L. – Potential link to methanogenesis. Plant Soil 267: 343–355.
39. Hirota M, Tang YH, Hu QW, Hirata S, Kato T, et al. (2004) Methane emissions from different vegetation zones in a Qinghai-Tibetan Plateau wetland. Soil Biol & Biochem 36: 737–748.
40. Torn MS, Chapin FS (1993) Environmental and biotic controls over methane flux from arctic tundra. Chemosphere 26: 357–368.
41. Chanton JP, Bauer JE, Glaser PA, Siegel DI, Kelley CA, et al. (1995) Radiocarbon Evidence for the Substrates Supporting Methane Formation within Northern Minnesota Peatlands. Geochim Cosmochim Acta 59: 3663–3668.
42. Ellenberg H, Weber HE, Düll R, Wirth V, Werner W, et al. (1991) Zeigerwerte von Pflanzen in Mitteleuropa. Scripta Geobotanica 18: 1–248.
43. Bakkenes MD, De Zwart D, Alkemade JRM (2002) MOVE nationaal Model voor de Vegetatie versie 3.2 Achtergronden en analyse van modelvarianten. IVM, Bilthoven – Nederland: 1–124.

3

Climatic Control on Plant and Soil $\delta^{13}C$ along an Altitudinal Transect of Lushan Mountain in Subtropical China: Characteristics and Interpretation of Soil Carbon Dynamics

Baoming Du[1], Chunjiang Liu[1,2]*, Hongzhang Kang[1,2], Penghua Zhu[2], Shan Yin[1,2], Guangrong Shen[1,2], Jingli Hou[3], Hannu Ilvesniemi[4]

1 School of Agriculture and Biology and Research Center for Low-Carbon Agriculture, Shanghai Jiao Tong University, Shanghai, China, **2** Key Laboratory of Urban Agriculture (South), Ministry of Agriculture, People's Republic of China, Shanghai, China, **3** Instrumental Analysis Center of SJTU, Shanghai Jiao Tong University, Shanghai, China, **4** Finnish Forest Research Institute, Vantaa, Finland

Abstract

Decreasing temperature and increasing precipitation along altitude gradients are typical mountain climate in subtropical China. In such a climate regime, identifying the patterns of the C stable isotope composition ($\delta^{13}C$) in plants and soils and their relations to the context of climate change is essential. In this study, the patterns of $\delta^{13}C$ variation were investigated for tree leaves, litters, and soils in the natural secondary forests at four altitudes (219, 405, 780, and 1268 m a.s.l.) in Lushan Mountain, central subtropical China. For the dominant trees, both leaf and leaf-litter $\delta^{13}C$ decreased as altitude increased from low to high altitude, whereas surface soil $\delta^{13}C$ increased. The lower leaf $\delta^{13}C$ at high altitudes was associated with the high moisture-related discrimination, while the high soil $\delta^{13}C$ is attributed to the low temperature-induced decay. At each altitude, soil $\delta^{13}C$ became enriched with soil depth. Soil $\delta^{13}C$ increased with soil C concentrations and altitude, but decreased with soil depth. A negative relationship was also found between O-alkyl C and $\delta^{13}C$ in litter and soil, whereas a positive relationship was observed between aromatic C and $\delta^{13}C$. Lower temperature and higher moisture at high altitudes are the predominant control factors of $\delta^{13}C$ variation in plants and soils. These results help understand C dynamics in the context of global warming.

Editor: Shuijin Hu, North Carolina State University, United States of America

Funding: This work was supported by the State Key Basic Research and Development Plan of China (2011CB403201) and the CFERN&GENE Award Funds on ecological paper. The funders had no role in study design, data collection and analysis, decision to publish, or preparation of the manuscript.

Competing Interests: The authors have declared that no competing interests exist.

* E-mail: chjliu@sjtu.edu.cn

Introduction

In terrestrial plant ecosystems, ^{13}C enrichment is generally greater in the soils than in vegetation due to the different fractionation of ^{13}C and ^{12}C in the biogeochemical processes [1,2]. In the last three decades, the C isotope technique has been a useful approach for understanding the effects of water deficit on plants [3,4], dynamics of soil organic C (SOC) [5,6,7], and local climate history [8]. $\delta^{13}C$ has been used as an index to study ecosystem response to climate and as a surrogate variable in modeling C fluxes in terrestrial ecosystems [9,10].

At a site scale, soil $\delta^{13}C$ is closely associated with local vegetation and climate history, and varies with the soil depth owing to its longer-term fractionation in deeper soil [11]. The impact of vegetation on soil $\delta^{13}C$ could vary with plant species and tissues. For instance, C_3 (Calvin cycle photosynthetic pathway) plants have $\delta^{13}C$ values of –22‰ to –35‰ (average –26.5‰), compared to C_4 plants whose $\delta^{13}C$ values are in the range of –8‰ to –16‰ (average –12.5‰) [8,12]. This difference help determine the relative proportion each vegetation type in soil organic C [13,14]. Among plants tissues, $\delta^{13}C$ is typically enriched by 1‰ to 2‰ in celluose and hemicelluloses, but depleted in ^{13}C by 2‰ to 6‰ in lignin, in comparison with whole-plant material [6,15].

At a broader scale, soil ^{13}C varies with spatial gradients by altitude, latitude, longitude due to changes of vegetation and climate [16]. Bird et al. (1996) reported that the average $\delta^{13}C$ was –28.3±0.6‰ in low latitude (0°N to 20°N or S), –27.7±0.6‰ in mid latitude soils (20° to 40°), and –27.3±0.7‰ in high latitude soils (40° to 90°) [17]. The overall increase of SOC $\delta^{13}C$ from low to high-latitude forests was −1‰. In mountain areas, climate varies with altitude resulting in different vegetation, litter input into soils, and litter decomposition rate, and therefore soil $\delta^{13}C$. Wei and Jia (2009) reported that soil $\delta^{13}C$ first decreased and then increased as altitude increased from 1000 m to 3800 m in Mount Gongga, southwestern China [18]. Consequently, ecosystems in mountain areas are very sensitive to changes in climate [19,20].

Lushan Mountain is located in the middle-lower plain of the Yangtze River in central subtropical China, with altitude from 30 m a.s.l. to 1470 m a.s.l. [21]. Along with the change of elevation is the opposite trend of heat and water (OHW), i.e., temperature decreases and precipitation increases from low to high elevations. Correspondingly, vegetation changes from evergreen broadleaf forests at the foothill to deciduous forests at the top of mountain. The OHW is a common mountain climate in subtropical China and has unique community composition, litterfall decomposition ratio, soil C biochemical processes for exploring the variations of plant and soil $\delta^{13}C$ with altitude, particularly in relation to climate change.

In this study, we examined the variation of $\delta^{13}C$ in fresh plant leaves, litter, and semi-decomposed and soil humic substances in order to characterize soil carbon dynamics at a local or regional scale in subtropical forest ecosystems in Lushan Mountain, China. The specific objectives were to: 1) determine the patterns of $\delta^{13}C$ variation in dominant plant species and soil with altitude, 2) investigate how soil $\delta^{13}C$ varies with soil depth at different altitudes, and 3) examine the effects of climatic factors on plant and soil $\delta^{13}C$ along the elevation gradient in Lushan Mountain.

Materials and Methods

Study Area

Study area is located in the Lushan Nature Reserve (29°31′–29°41′N, 115°51′– 116°07′E), the south of Jiujiang City, Jiangxi Province, China (Fig. 1). Permission was granted by the Nature Reserve of Lushan Jiangxi. Lushan is an isolated mountain body situated in the center of the vast plain of the middle and lower reaches of the Yangtze River, covering an area of about 300 km^2 along a altitude range from 50 m to 1474 m. This area has a subtropical monsoon climate. The mean annual precipitation (MAP) ranges from 1308 mm to 2068 mm, and the mean annual temperature (MAT) from 17.1°C to 11.6°C [21].

Owing to the variations in geology, climate, and vegetation along elevation, dominant soil types change from Ferric alisols at low elevations to Haplic alisols at high elevations [21] according to the FAO soil texture classification. The corresponding vegetation types are evergreen forests dominated by several Fagaceae tree species including *Castanopsis sclerophylla*, *Castanopsis eyrei* and *Lithocarpus glaber*, and some evergreen woodland species and shrubs at low altitudes of approximately 50 m to 600 m, evergreen broadleaf forests and some deciduous trees are mid altitudes of 600 m and 1000 m, and *Lindera obtusiloba* forest consisting of *Cerasus serrulata*, *Castanea seguinii*, *Tilia breviradiata*, and a few shrubs at about 1200 m.s.l. (Table S1). Some *Cryptomeria japonica* plantations were established about 50 years ago at mid altitudes.

Sample Stands and Collection

Permission was granted by the Nature Reserve of Lushan Jiangxi (29°31′–29°41′N, 115°51′–116°07′E) to carry out our study. Sample stands were chosen at an irregular altitude interval owing to natural forest fragmentation in mountain. In total, four study sites were established at altitudes of 219 m, 405 m, 780 m, and 1268 m (Fig. 1, Table 1). These study sites were not associated with endangered or protected species.At each site, three plots (each 20×20 m) were randomly delineated. In each plot, four 4 m^2 subplots were randomly chosen for shrub layer and four 1 m^2 subplots for herbaceous layer. In each plot, 100 leaves were collected from dominant tree (Table S1). Sample were collected from L (Litter) and LF (Semi-decomposition litter) horizons and mineral soil layers at 0–10 cm, 10–20 cm, 20–30 cm, 30–40 cm, 40–50 cm, and 50–60 cm depths. Five soil cores were randomly collected within each plot using a 2 cm-diameter stainless steel borer and bulked to make one composite sample by soil depth.

Soil samples were air dried, ground, and passed through a 2 mm sieve to remove coarse living roots and gravel before being ground and passed through a 0.149 mm mesh sieve prior to chemical analysis. Leaf and litter samples were oven dried (65°C) for a week to constant weight and ground to fine powder using a Tecator sample mill (Subang, Shanghai, China) prior to the chemical and isotopic analyses.

Chemical Analysis

The C isotope ratio ($\delta^{13}C$) of leaf, litter, and soil samples was determined using an elemental analysis–stable isotope ratio mass spectrometer (VarioElIII/Isoprime, Elementar, Hanau, Germany) operated at the Instrumental Analysis Center of Shanghai Jiao Tong University (SJTU). The results are reported as parts per thousand (‰) deviations from the Vienna–Pee Dee Belemnite (PDB) standard (uncertainty of ±0.1‰ uncertainty), which The is expressed as follows:

$$\delta^{13}C = \left({}^{\delta_{sample}}\!/_{\delta_{standard}} - 1 \right) \times 10^3$$

Where δ_{sample} is the $^{13}C/^{12}C$ ratio of the samples and $\delta_{standard}$ is the $^{13}C/^{12}C$ ratio of the reference standard (PDB) [22].

To express the absolute variation of soil $\delta^{13}C$ enrichment relative to litter, we define an absolute enrichment factor F_A as

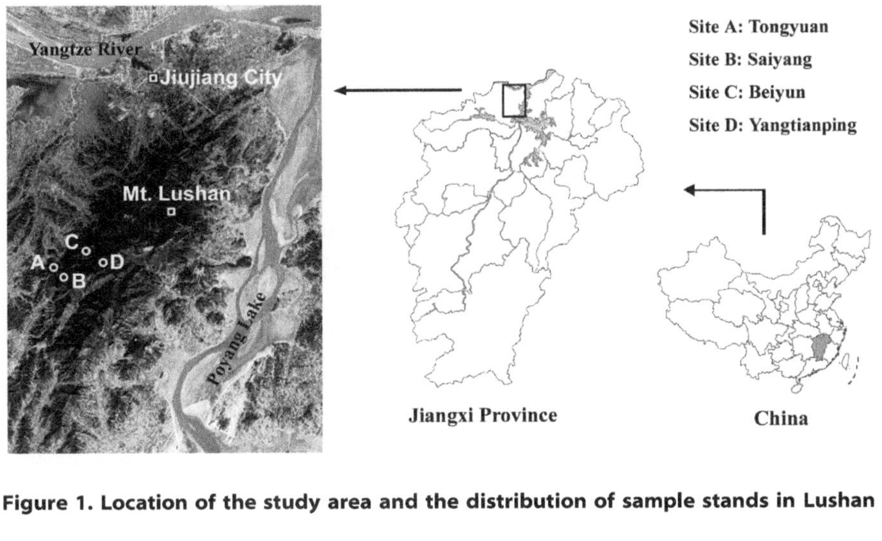

Figure 1. Location of the study area and the distribution of sample stands in Lushan Mountain, subtropical China.

Table 1. Features of climate and vegetation at different altitudes in Lushan Mountain.

Location	Altitude (ma.s.l.)	MAP[a] (mm)	MAT (°C)	TCM[b] (°C)	THM[c] (°C)	Growing season (Days)	Vegetation types[d]
Tongyuan	219	1429	16.2	3.8	28.5	262	EBF
Saiyang	405	1549	15.3	3.2	27.4	253	EBF
Beiyun	780	1794	13.6	1.9	25.1	234	EBMF
Yangtianping	1268	2112	11.3	0.3	22.2	209	DBF

[a]The climatic data from years 1971 to 2000 were obtained from the Lushan Meteorological Bureau.
[b]Temperature of the coldest month.
[c]Temperature of the hottest month.
[d]EBF represents for evergreen broadleaf forest; EBMF for evergreen broadleaf and needle-leaf mixed forest; and DBFfor deciduous broadleaf forest.

follows:

$$F_A = \delta^{13}C_{soili} - \delta^{13}C_{litter}$$

Where $\delta^{13}C_{soil\ i}$ is the $\delta^{13}C$ at *i*th soil layer and $\delta^{13}C_{litter}$is the $\delta^{13}C$ at the litter layer.

The rate of soil $\delta^{13}C$ enrichment varies with soil depth [1,2]. To express the relative enrichment of adjacent soil layers, we define the relative soil $\delta^{13}C$ enrichment factor F_R as follows:

$$F_R = \delta^{13}C_{soili} - \delta^{13}C_{soili-l}$$

Where $\delta^{13}C_{soil\ i}$ is the $\delta^{13}C$ at *i*th soil layer and $\delta^{13}C_{soil\ i-1}$ is the $\delta^{13}C$ at *i-1*th soil layer.

To explore the relationship between soil $\delta^{13}C$ and SOC concentration, as well as soil C functional groups, two soil variables were determined. The variation patterns in SOC concentration and soil C functional groups along the altitude gradient in Lushan Mountain will be presented in another paper. The SOC concentrations of soil samples from different depths were measured using dichromate oxidation method [23].

The chemical compositions of C in litter and soil layers at 0–10 cm, 30–40 cm, and 50–60 cm depths were analyzed with solid-state ^{13}C cross-polarization magic angle spinning–nuclear magnetic resonance (CP/MAS–NMR). The litter samples were dried to constant weight at 65°C and ground in a Wiley mill. The soil samples were pretreated with 10% (v/v) hydrofluoric acid (HF) before the NMR spectroscopy [24] to reduce Fe^{3+} and Mn^{2+} [25] and concentrate organic C for more accurate signal-to-noise ratio [24]. About 10 g of the ground sample was shaken with 50 ml HF for 2 h. After centrifugation (5,000 rpm) for 10 min, the supernatant was removed. The procedure was repeated five times. The remaining sediment was washed five times with 50 ml deionized water to remove residual HF before freeze drying.

The solid-state ^{13}C CP/MAS–NMR spectra of litter and soil samples were obtained at a frequency of 100.64 MHz using a Bruker AVANCEIII400 NMR spectrometer (BrukerBiospin, Rheinstetten, Germany) operated at 75.42 MHz for ^{13}C. The contact time was 1.5 ms, with 1 s recycle delay and the magnetic angle spinning rate was 5 kHz [24]. About 12,000 scans were collected for soil samples and 10,000 scans for litter samples [26]. The chemical shift regions 0–45 ppm, 45–110 ppm, 110–160 ppm, and 160–220 ppm were assigned to alkyl C, O-alkyl C, aromatic C, and carboxylic C, respectively [24,27]. The sources of organic carbon are: Alkyl C is derived from lipids, fatty acides and plant aliphatic polymers, O-alkyl C primarily from cellulose and hemicelluloses, as well as starch, proteins and carbohydrates, aromatic C from lignin and tannins, and carboxyl C from lipids, aliphatic esters, and amide carboxyls [28,29]. The signal intensities in the respective chemical-shift regions were expressed as a percentage of the area of the total spectra. The relative contents of different chemical structures were therefore calculated [26].

Statistical Analysis

Arithmetic means and standard deviation were calculated. A *t* test (i.e., least significant difference) was conducted to compare the means with a probability level of 0.05 for detecting significant differences. Linear regression analyses were used to examine the relationships between soil $\delta^{13}C$ and MAT, MAP, and SOC concentrations. All analyses were performed through SigmaPlot 10.0 (Systat Software, Richmond, CA, USA) and SAS V8.1 (SAS Institute Inc., Cary, North Carolina).

Results

Variations of Leaf, Litter, and Soil $\delta^{13}C$ with Altitude

The deciduous tree leaf $\delta^{13}C$ at 1268 m a.s.l. was –28.29‰, significantly lower than evergreen trees at lower altitudes (–27.65‰ to −26.98‰) (Fig. 2). A similar $\delta^{13}C$-altitude relationship was also evident in leaf litter, but not in semi-decomposed litter. Comparatively, fresh leaves had higher $\delta^{13}C$ than leaf litter or

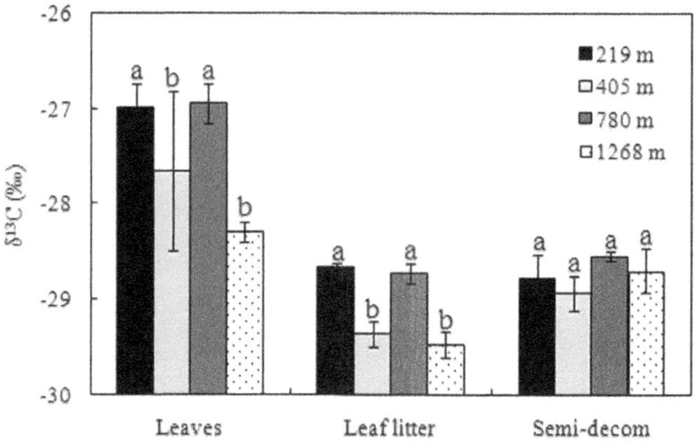

Figure 2. $\delta^{13}C$ values of leaves, leaf litter, and semi-decomposed litter in the natural secondary broadleaf stands at altitudes of 219, 405, 780, and 1268 m in the Lushan Mountain. The error bars are the standard errors (n = 3 for leaf litter and semi-decomposed litter, and n = 3 for 300 leaves). Different letters indicate significant differences among all classes at the different altitudes (P<0.05).

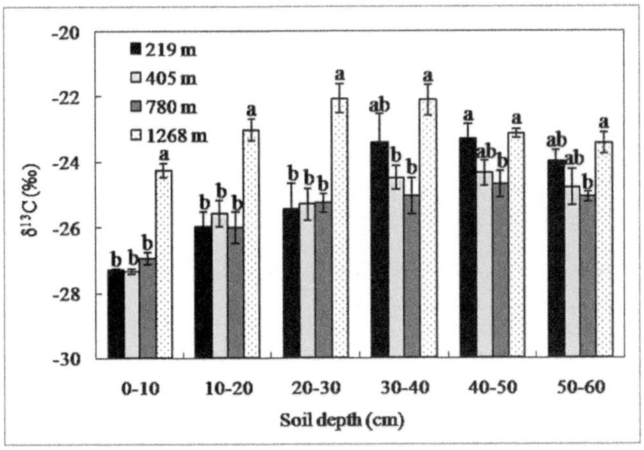

Figure 3. Soil δ^{13}C values of different layers in the natural secondary broadleaf stands at altitudes of 219, 405, 780, and 1268 m in the Lushan Mountain. The error bars represent standard errors means (n = 3). Different letters indicate significant differences among altitudes by soil depth ($P<0.05$).

semi-decomposed litter. For instance, δ^{13}C decreased from −26.98‰ in fresh leaves to −28.65‰ in leaf litter at 219 m a.s.l. (Fig. 2).

The surface layers of soils (0–10 cm, 10–20 cm, and 20–30 cm) at 1268 m was significantly higher than that at the other three lower altitudes. Ihe deeper soil layers (30–40 cm, 40–50 cm, and 50–60 cm), however, no significant difference by altitude occurred (Fig. 3).

Soil δ^{13}C Enrichment with Soil Depth

Soil δ^{13}C enrichment was greater with the increase of soil depth for all altitudes (Fig. S1). Absolute enrichment factor (F_A) generally increased from semi- decomposed layer, peaked at 30–40 cm, and remained stable at deeper soil (Fig. 4). The soil F_A at 1268 m were generally greater than that at the other three altitudes. For instance, the maximum F_A was 7.39‰ at 30–40 cm soil depth and 1268 m altitude, compared to the values of 3.66‰ to 5.23‰ at the same layer of three lower altitudes. The relative enrichment factor (F_R) was generally higher in the surface soil layers of 0–10 cm (1.49‰ to 4.43‰) and 10–20 cm (0.93‰ to 1.76‰) than in the deeper layers (Fig. 5).

Relationships of Soil δ^{13}C with SOC Concentration and Chemical Composition

Soil δ^{13}C increased (Fig. S1) with SOC concentration at each altitude, following a strong negative relationship ($p<0.01$) (Fig. 6). The relationship at the three sites of lower altitudes (219 m, 405 m and 780 m, all covered by evergreen forests) was different from that at the site of highest altitude (1268 m, covered by deciduous forests). Soil δ^{13}C was negatively correlated with O-alkyl C, but positively with aromatic C and carboxyl C (Fig. 7).

Relationships between Soil δ^{13}C and the Climatic Factors

In Lushan Mountain, temperature decreases with, whereas precipitation increases with the increase of elevation. The soil δ^{13}C decreased with MAT and increased with MAP in the three upper layers (0–10 cm, 10–20 cm, and 20–30 cm), whereas no clear trend existed for deeper layers (Fig. 8).

Discussion

Decoupled Patterns of Variations in Plant and Soil δ^{13}C along Elevation

In terrestrial ecosystems, plant functional types strongly affect site-level soil δ^{13}C through litter inputs [5,30,31]. For instance, Peri et al. (2012) reported that the soil δ^{13}C in *Nothofagus* forests was significantly associated with foliar δ^{13}C, both of which decreased with precipitation [28]. In the present study, however, fresh leaf and leaf litter δ^{13}C decreased with, whereas soil δ^{13}C increased with altitude (Figs. 2 and 3), a pattern that cannot be explained alone with the increasing precipitation by altitude in Lushan Mountain.

Different from some previous studies [33,34], the tree leaf δ^{13}C was significantly lower at the highest altitude, likely due to the special climate regime in Lushan Mountain where precipitation increases with and temperature decreases with altitude (Fig. 8). According to a general notion about plant ^{13}C discrimination [35], plants at moist sites tend to have high stomatal conductance (close to maximum), low water use efficiency, and high intercellular CO_2 concentration. This results in increasing discrimination against $^{13}CO_2$ during photosynthesis leading to low δ^{13}C values, in comparison with arid sites. In Lushan Mountain, plants experience greater drought stress at lower altitude sites owing to low precipitation and high temperature, resulting in high tissues δ^{13}C.

In the present study, soil δ^{13}C increased with altitude, consistent with the pattern found in previous studies [7,34,36,37,38]. For example, Townsend et al. (1995) reported an increase of soil δ^{13}C from −26.70 ‰ at 900 m a.s.l. to −25.90 ‰ at 1500 m a.s.l. in the island of Hawaii [32]. Similar results are also reported by Zimmermann et al. (2012) in a tropical forest in Peru where soil δ^{13}C values increased with elevation from −27.16‰ at 1700 m a.s.l. to −25.79‰ at 3030 m a.s.l. [38]. The major reason for the altitudinal variation of soil δ^{13}C in those studies is probably the influence of plant communities through the deposition of leaf litter, dead root material, and rhizodeposition [32]. In Lushan Mountain, however, increasing precipitation and decreasing temperature with altitude may have predominantly influence over soil δ^{13}C.

Relative to the bulk leaf δ^{13}C, sugars, starch, cellulose, protein, and organic aids are enriched, whereas lignin and lipids are depleted in δ^{13}C [6]. Therefore, organic matter with high concentrations of sugars, starch, and cellulose displays high δ^{13}C values. On the other hand, however, sugars, starch, cellulose, and protein, are more easily lost through litter decomposition than lignin [39,40]. This helps explain the high soil δ^{13}C of top soil layers (0–10 cm, 10–20 cm, and 20–30 cm) at the altitude of 1268 m (Fig. 3) where high moisture and low temperature not only reduces forest productivity and litter (organic matter) input to the soil, but also shows down decomposing activities of microbes.This may have led to accumulation of more less-decomposed organic matter in soils and therefore high soil δ^{13}C (accumulation of sugars, starch, cellulose, and protein).

Soil δ^{13}C Enrichment with Soil Depth by Altitude

In previous studies, enrichment factors (F_A in this study) were used to describe the variation in soil δ^{13}C enrichment relative to litter. However, the results of this study suggest that the relative enrichment factor (F_R) introduced in the present study better detect the difference of soil ^{13}C enrichment by depth than F_A used by previous studies (Figs. 4 and 5). For instance, F_R shows a more rapid change from semi-decomposed litter to soil layers (0–10 cm, 10–20 cm) than F_A (Fig. 5).

At all altitudes, soil δ^{13}C was enriched with soil depths from litter to O-layer and to mineral soil layers (Figs. 4 and S1). This

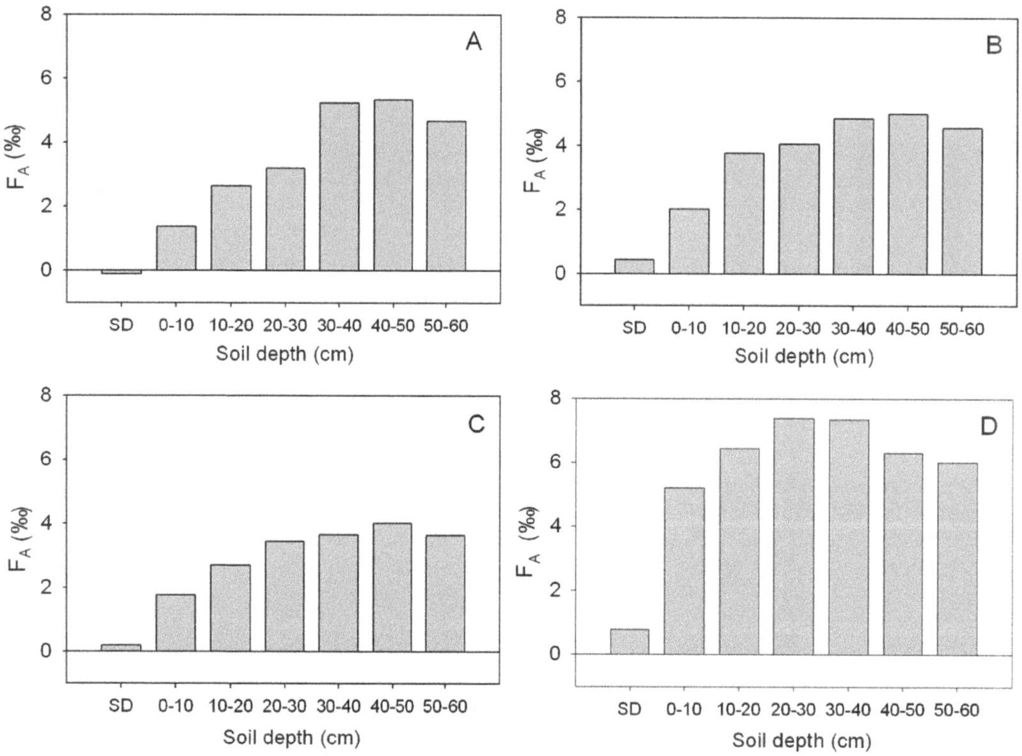

Figure 4. Absolute enrichment factor of soil δ[13] (F_A) by soil depth at altitudes of 219 (A), 405 (B), 780 (C), and 1268 m (D) in the Lushan Mountain. In the figures, SD represents the semi-decomposed litter layer.

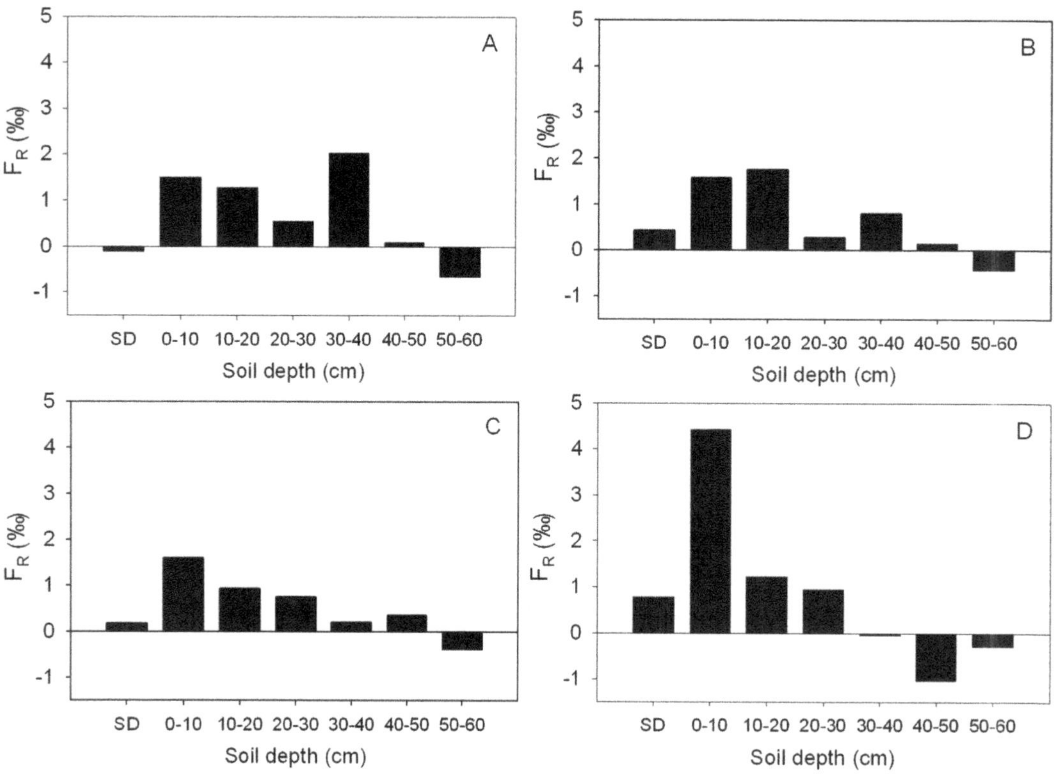

Figure 5. Relative enrichment factor of soil δ[13] (F_R) by soil depth at altitudes of 219 (A), 405 (B), 780, (C), and 1268 m (D) in the Lushan Mountain.

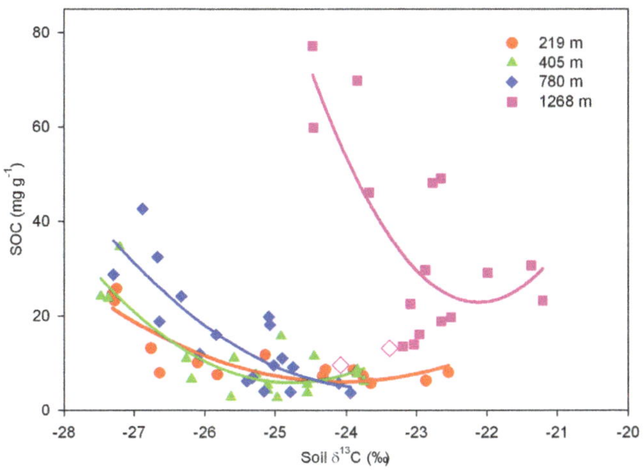

Figure 6. Relationships between SOC concentration (mg g^{-1}) and soil δ^{13}C by soil depth for four altitudes in the Lushan Mountain. The fitted models are: $y = 870.58 + 71.82x + 1.49x^2$, $r^2 = 0.76$, and $p = 0.0002$, for the 219 m site; $y = 1881.58 + 149.87x + 3.03x^2$, $r^2 = 0.72$, and $p < 0.0001$, for the 405 m site; $y = 1250.35 + 105.77x + 2.25x^2$, $r^2 = 0.68$, and $p = 0.0002$, for the 780 m site; and $y = 4265.44 + 383.58x + 8.67x^2$, $r^2 = 0.57$, and $p = 0.0041$, for the 1268 m site.

finding is consistent with the conclusion by previous studies [1,2,11,39]. In the present study, the absolute enrichment factor at 1268 m a.s.l. (F_A = 5‰ to 7.5‰) was greater than that at lower altitudes (F_A = 1.5‰ to 5.5‰). For example, The F_A of 0–10 cm at 1268 m a.s.l. was about 5‰ nearly twich of that of the same soil depth at all lower altitudes (Fig. S1). This result suggests that the climatic conditions (MAP = 2112 mm, MAT = 11.3°C) at 1268 m a.s.l. support a distinct fractionation compared to lower altitudes, particularly the sites at 219 m a.s.l. with MAP = 1429 and MAT = 16.2. Therefore, the climate regime lower temperature and higher moisture at high altitude strongly influences soil ^{13}C enrichment with soil depth [41,42].

Implications of Soil δ^{13}C in Ascertaining SOC Status

Soil C concentrations are strongly correlated with δ^{13}C in forest soil, following a negative relationship as demonstrated by previous studies [1,31,43]. In this study, soil δ^{13}C increased with soil depth, aromatic C and carbonyl C, but decreased with O-alkyl C (Figs. 3 and 7). The increase of ^{13}C and changes of SOC chemical composition with soil depth likely result from humification [44,45]. Microbial activities influence isotopic fractionation during SOC decomposition through differentiation use of substrate by different microbes [41,42,46] and isotopic effects on metabolic synthesis of secondary compounds (e.g., lipids, lignin, cellulose) [4,47]. However, lipids and lignin are degraded more slowly and tend to be ^{13}C depleted, whereas cellulose and carbohydrate degrade more rapidly and tend to be ^{13}C enriched [42,48]. Therefore, it is difficult to establish direct relationships between δ^{13}C and SOC chemical compositions with soil depth. The distinct high soil δ^{13}C at 1268 m a.s.l. is probably attributed to the accumulation of higher ^{13}C-based sugars, starch, cellulose, protein, and organic aids, resulting from slow litter decomposition in the surface soil layers (0–10 cm, 10–20 cm, and 20–30 cm) (Fig. 3). Therefore, the enrichment mechanisms of soil ^{13}C along the altitude gradient were different from those by soil depth in Lushan Mountain. The detail mechanisms need to be clarified in future research.

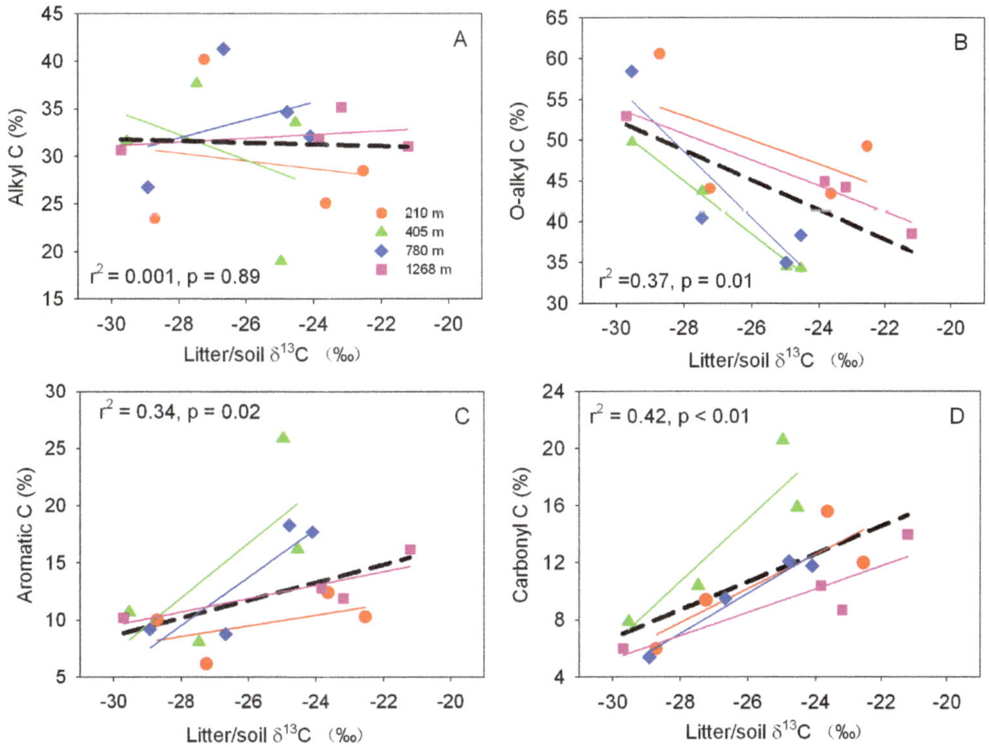

Figure 7. Relationships between soil δ^{13}C and Alkyl C (A), O-alkyl C (B), Aromatic C (C), and Carbonyl C (D) for the studied stands at the four latitudes in the Lushan Mountain. Dashed lines represent the general regression lines with all data, with significant level of $p < 0.05$, except for Alkyl C.

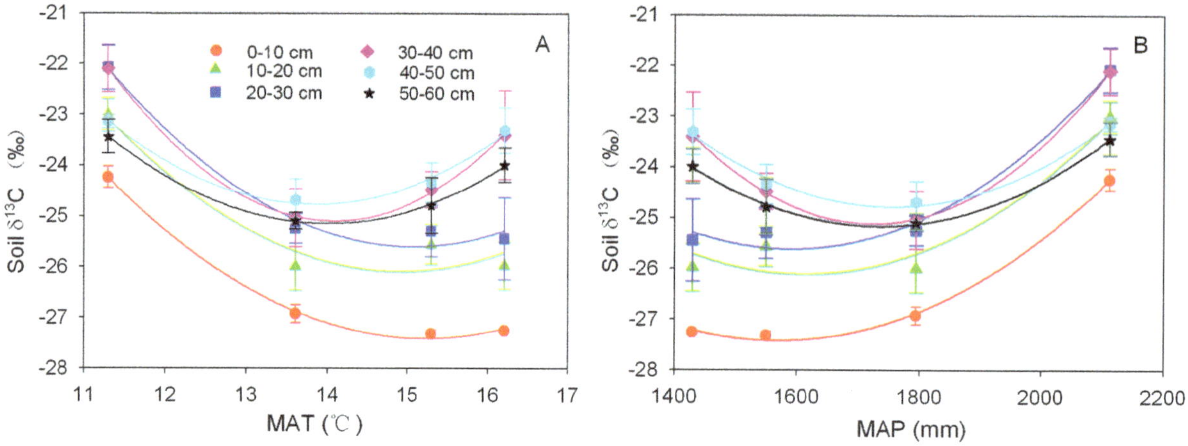

Figure 8. Variations of soil $\delta^{13}C$ with the mean annual temperature (MAT) (A) and mean annual precipitation (MAP) (B) by soil depth in the Lushan Mountain. The fitted model is $y=y_0+ax+bx^2$. The regression lines wtih MAT have $r^2=0.998$ and $p=0.0428$ for 0–10 cm layer, and $r^2=0.999$ and $p=0.0289$ for 30–40 cm layer (A), and those with MAP have $r^2=0.998$ and $p=0.0445$ for 0–10 cm layerand $r^2=0.999$ and $p=0.0175$ for 30–40 cm layer (B).

Concluding Remarks

The patterns of plant and soil $\delta^{13}C$ variations and their relation to the climate regime along an altitudinal gradient were studied in Lushan Mountain of central subtropical China. The results indicated that tree leaf $\delta^{13}C$ decreased and soil $\delta^{13}C$ increased with altitude. The decoupled pattern of plant and soil $\delta^{13}C$ was due to the climate regime of decreasing temperature and increasing precipitation with altitude in the study area, which result in decreased litter decomposition at high-latitude sites. These results have important implications for understanding C dynamics of subtropical forest ecosystems in the context of global warming.

Acknowledgments

We thank Li Zhang, Shi Xu, and Jieli Wu for their assistance in the sample analysis, and Jinhao Qian, Qin Zou, Kongfan Qian and Ming Du for their assistance with fieldwork. Dr. Rongzhou Man at Ontario Forest Research Institute, Canada, is gratefully acknowledged for his constructive comments and language checking. Chemical analyses were conducted in the Instrumental Analysis Center of SJTU, and the Nature Reserve of Lushan Jiangxi.

Author Contributions

Conceived and designed the experiments: CL HK. Performed the experiments: BD HK PZ. Analyzed the data: BD JH. Contributed reagents/materials/analysis tools: SY GS. Wrote the paper: BD CL HI.

References

1. Garten CT, Cooper LW, Post IIIWM, Hanson PJ (2000) Climate controls on forest soil C isotope ratios in the southern Appalachian mountains. Ecology 81: 1108–1119.
2. Boström B, Comstedt D, Ekblad A (2007) Isotope fractionation and ^{13}C enrichment in soil profiles during the decomposition of soil organic matter. Oecologia 153: 89–98.
3. Gouveia A, Freitas H (2009) Modulation of leaf attributes and water use efficiency in *Quercus suber* along a rainfall gradient. Trees 23: 267–275.
4. Hartman G, Danin A (2010) Isotopic values of plants in relation to water availability in the Eastern Mediterranean region. Oecologia 162: 837–852.
5. Ehleringer JR, Buchmann N, Flanagan LB (2000) Carbon isotope ratios in belowground carbon cycle processes. Ecological Applications 10: 412–422.
6. Bowling DR, Pataki DE, Randerson JT (2008) Carbon isotopes in terrestrial ecosystem pools and CO_2 fluxes. New Phytologist 178: 24–40.
7. Chen P, Wang G, Han J, Liu X, Liu M (2010) $\delta^{13}C$ diference between plants and soil organic matter along the eastern slope of Mount Gongga. Chinese Science Bulletin 1: 55–62.
8. Kohn MJ (2010) Carbon isotope compositions of terrestrial C_3 plants as indicators of (paleo) ecology and (paleo) climate. Proc Natl Acad Sci U S A 107: 19691–19695.
9. Ito A (2003) A global-scale simulation of the CO_2 exchange between the atmosphere and the terrestrial biosphere with a mechanistic model including stable carbon isotopes, 1953–1999. Tellus B 55: 596–612.
10. Kaplan JO, Prentice IC, Buchmann N (2002) The stable carbon isotope composition of the terrestrial biosphere: Modeling at scales from the leaf to the globe. Global Biogeochemical Cycles 16: 1060.
11. Dzurec RS, Boutton TW, Caldwell MM, Smith BN (1985) Carbon isotope ratios of soil organic matter and their use in assessing community composition changes in Curlew Valley, Utah. Oecologia 66: 17–24.
12. Ambrose SH, Sikes NE (1991) Soil carbon isotope evidence for Holocene habitat change in the Kenya Rift Valley. Science 253: 1402–1405.
13. Balesdent J, Mariotti A, Guillet B (1987) Natural ^{13}C abundance as a tracer for studies of soil organic matter dynamics. Soil Biology and Biochemistry 19: 25–30.
14. Skjemstad JO, Krull ES, Swift RS, Szarvas S (2008) Mechanisms of protection of soil organic matter under pasture following clearing of rainforest on an Oxisol. Geoderma 143: 231–242.
15. Benner R, Fogel ML, Sprague EK, Hodson RE (1987) Depletion of ^{13}C in lignin and its implications for stable carbon isotope studies. Nature 329: 708–710.
16. Li J, Ziegler SE, Lane CS, Billings SA (2013) Legacies of native climate regime govern responses of boreal soil microbes to litter stoichiometry and temperature. Soil Biology and Biochemistry 66: 204–213.
17. Bird MI, Chivas AR, Head J (1996) A latitudinal gradient in carbon turnover times in forest soils. Nature 381: 143–146.
18. Wei K, Jia G (2009) Soil n-alkane $\delta^{13}C$ along a mountain slope as an integrator of altitude effect on plant species $\delta^{13}C$. Geophysical Research Letters 36: L11401.
19. Löffler J, Anschlag K, Baker B, Finch OD, Diekkrüger B, et al. (2011) Mountain ecosystem response to global change. Erdkunde 65: 189–213.
20. Gottfried M, Pauli H, Futschik A, Akhalkatsi M, Barančok P, et al. (2012) Continent-wide response of mountain vegetation to climate change. Nature Climate Change 2: 111–115.

21. Liu X, Wang L (2009) Scientific survey and study of biodiversity on the Lushan nature reserve in Jiangxi province. Science Press.
22. Tu CL, Liu CQ, Lu XH, Yuan J, Lang YC (2011) Sources of dissolved organic carbon in forest soils: evidences from the differences of organic carbon concentration and isotope composition studies. Environmental Earth Sciences 63: 723–730.
23. Kalembasa SJ, Jenkinson DS (1973) A comparative study of titrimetric and gravimetric methods for the determination of organic carbon in soil. Journal of the Science of Food and Agriculture 24: 1085–1094.
24. Wang H, Liu SR, Mo JM, Wang JX, Makeschin F, Wolff M (2010) Soil organic carbon stock and chemical composition in four plantations of indigenous tree species in subtropical China. Ecological Research 25: 1071–1079.
25. Schmidt M, Knicker H, Hatcher PG, Kögel-Knabner I (1997) Improvement of ^{13}C and ^{15}N CPMAS NMR spectra of bulk soils, particle size fractions and organic material by treatment with 10% hydrofluoric acid. European Journal of soil Science 48: 319–328.
26. Jien SH, Chen TH, Chiu CY (2011) Effects of afforestation on soil organic matter characteristics under subtropical forests with low elevation. Journal of Forest Research 16: 275–283.
27. Rumpel C, Kögel-Knabner I, Bruhn F (2002) Vertical distribution, age, and chemical composition of organic carbon in two forest soils of different pedogenesis.Organic Geochemistry 33: 1131–1142.
28. Baldock JA, Oades JM, Nelson PN, Skene TM, Golchin A, et al. (1997) Assessing the extent of decomposition of natural organic materials using solid-state ^{13}CNMR spectroscopy. Australian Journal of Soil Research 35: 1061–1083.
29. García M, Faz CA (2012) Soil organic matter stocks and quality at high altitude grasslands of Apolobamba, Bolivia. Catena 94: 26–35.
30. Natelhoffer KJ, Fry B (1988) Controls on natural nitrogen-15 and carbon-13 abundances in forest soil organic matter. Soil Science Society of America Journal 52: 1633–1640.
31. Balesdent J, Girardin C, Mariotti A (1993) Site-related ^{13}C of tree leaves and soil organic matter in a temperate forest. Ecology 74: 1713–1721.
32. Peri PL, Ladd B, Pepper DA, Bonser SP, Laffan SW, et al. (2012) Carbon ($\delta^{13}C$) and nitrogen ($\delta^{15}N$) stable isotope composition in plant and soil in Southern Patagonia's native forests. Global Change Biology 18: 311–321.
33. Körner C, Farquhar GD, Roksandic Z (1988) A global survey of carbon isotope discrimination in plants from high altitude. Oecologia 74: 623–632.
34. Bird MI, Haberle SG, Chivas AR (1994) Effect of altitude on the carbon-isotope composition of forest and grassland soils from Papua New Guinea. Global Biogeochemical Cycles 8: 13–22.
35. Farquhar GD, Ehleringer JR, Hubick KT (1989) Carbonisotope discriminatio-nand photosynthesis. Annual Review of Plant Biology 40: 503–537.
36. Townsend AR, Vitousek PM, Trumbore SE (1995) Soil organic matter dynamics along gradients in temperature and land use on the island of Hawaii. Ecology 76: 721–733.
37. Powers JS, Schlesinger WH (2002) Geographic and vertical patterns of stable carbon isotopes in tropical rain forest soils of Costa Rica. Geoderma 109: 141–160.
38. Zimmermann M, Leifeld J, Conen F, Bird MI, Meir P (2012) Can composition and physical protection of soil organic matter explain soil respiration temperature sensitivity? Biogeochemistry 107: 1–14.
39. Melillo JM, Aber JD, Linkins AE, Ricca A, Fry B, et al. (1989) Carbon and nitrogen dynamics along the decay continuum: plant litter to soil organic matter. Plant and Soil 115: 189–198.
40. Coûteaux MM, Bottner P, Berg B (1995) Litter decomposition, climate and litter quality. Trends in Ecology and Evolution 10: 63–66.
41. Li J, Ziegler S, Lane CS, Billings SA (2012) Warming-enhanced preferential microbial mineralization of humified boreal forest soil organic matter: Interpretation of soil profiles along a climate transect using laboratory incubations. Journal of Geophysical Research: Biogeosciences 117: G02008.
42. Li J, Ziegler SE, Lane CS, Billings SA (2013) Legacies of native climate regime govern responses of boreal soil microbes to litter stoichiometry and temperature. Soil Biology and Biochemistry 66: 204–213.
43. Powers JS (2006) Spatial variation of soil organic carbon concentrations and stable isotopic composition in 1-ha plots of forest and pasture in Costa Rica: implications for the natural abundance technique. Biology and Fertility of Soils 42: 580–584.
44. Kramer MG, Sollins P, Sletten RS, Swart PK (2003) N isotope fractionation and measures of organic matter alteration during decomposition. Ecology 84: 2021–2025.
45. Quideau SA, Anderson MA, Graham RC, Chadwick OA, Trumbore SE (2000) Soil organic matter processes: characterization by ^{13}C NMR and ^{14}C measurements. Forest Ecology and Management 138: 19–27.
46. Ziegler SE, Billings SA, Lane CS, Li J, Fogel ML (2013) Warming alters routing of labile and slower-turnover carbon through distinct microbial groups in boreal forest organic soils. Soil Biology and Biochemistry 60: 23–32.
47. Cotrufo MF, Drake B, Ehleringer JR (2005) Palatability trials on hardwood leaf litter grown under elevated CO_2: a stable carbon isotope study. Soil Biology and Biochemistry 37: 1105–1112.
48. Ehleringer JR, Buchmann N, Flanagan LB (2000) Carbon isotope ratios in belowground carbon cycle processes. Ecological Applications 10: 412–422.

4

The Altitudinal Patterns of Leaf C:N:P Stoichiometry Are Regulated by Plant Growth Form, Climate and Soil on Changbai Mountain, China

Ning Zhao[1,2], Nianpeng He[1], Qiufeng Wang[1], Xinyu Zhang[1], Ruili Wang[1,2], Zhiwei Xu[1,2], Guirui Yu[1]*

1 Synthesis Research Center of Chinese Ecosystem Research Network, Key Laboratory of Ecosystem Network Observation and Modeling, Institute of Geographic Sciences and Natural Resources Research, Chinese Academy of Sciences, Beijing, China, 2 University of Chinese Academy of Sciences, Beijing, China

Abstract

Understanding the geographic patterns and potential drivers of leaf stoichiometry is critical for modelling the nutrient fluxes of ecosystems and to predict the responses of ecosystems to global changes. This study aimed to explore the altitudinal patterns and potential drivers of leaf C:N:P stoichiometry. We measured the concentrations of leaf C, N and P in 175 plant species as well as soil nutrient concentrations along an altitudinal transect (500–2300 m) on the northern slope of Changbai Mountain, China to explore the response of leaf C:N:P stoichiometry to plant growth form (PGF), climate and soil. Leaf C, N, P and C:N:P ratios showed significant altitudinal trends. In general, leaf C and C:N:P ratios increased while leaf N and P decreased with elevation. Woody and herbaceous species showed different responses to altitudinal gradients. Trees had the largest variation in leaf C, C:N and C:P ratios, while herbs showed the largest variation in leaf N, P and N:P ratio. PGF, climate and soil jointly regulated leaf stoichiometry, explaining 17.6% to 52.1% of the variation in the six leaf stoichiometric traits. PGF was more important in explaining leaf stoichiometry variation than soil and climate. Our findings will help to elucidate the altitudinal patterns of leaf stoichiometry and to model ecosystem nutrient cycling.

Editor: Han Y.H. Chen, Lakehead University, Canada

Funding: This study supported by the Major Program of the National Natural Science Foundation of China (Grant No. 31290220) (http://www.nsfc.gov.cn/Portal0/default152.htm). The funders had no role in study design, data collection and analysis, decision to publish, or preparation of the manuscript.

Competing Interests: The authors have declared that no competing interests exist.

* E-mail: yugr@igsnrr.ac.cn

Introduction

Understanding the spatial patterns and the controlling factors for leaf C:N:P stoichiometry is critical for elucidating the patterns of nutrient fluxes across ecological gradients and the response of vegetation to global change [1–4]. The latitudinal patterns of leaf stoichiometry have been widely investigated at regional [5–9] and global scales [2,10], and some general biogeographic patterns have been uncovered. However, variations in leaf stoichiometry along altitudinal gradients have met with ambiguous results. Some studies found that leaf C, N and P increase with elevation [11–14], but in other studies the opposite trend has also been found [15–20]. Previous studies on tropical, subtropical mountains and subarctic tundra suggest that leaf N and P declined with altitude [15–18]. In addition, non-linear relationships have been detected between leaf stoichiometric traits and altitude. For example, it is reported that leaf N and P at first increased and then decreased with increasing elevation in Mount Gongga, China and the Peruvian Andes [19–20]. Therefore, intensive studies are required to obtain a more general representation of the altitudinal patterns of leaf stoichiometry.

Plant growth form (PGF), climate and soil influence leaf C:N:P stoichiometry in a complex way [2,21–23]. A series of hypotheses has been developed to interpret the geographic patterns and their underlying mechanisms, including the Plant Physiological Hypothesis [2], the Biogeochemical Hypothesis [2], and the Growth Rate Hypothesis [24]. The Plant Physiological Hypothesis suggests that plants increase leaf N and P content to offset the deceleration of plant metabolic rates caused by the low activity of enzymes at low temperatures [2]. The Biogeochemical Hypothesis assumes that soil nutrient availability, which is influenced by temperature and precipitation through organic matter decomposition and leaching effects, respectively, has a significant effect on leaf nutrient concentrations [2,9,22,25–27]. The Growth Rate Hypothesis proposes that changes in plant growth rate bring a corresponding change in leaf stoichiometry. It claims that an elevated demand for P-rich ribosomal RNA under rapid growth drives variation in the P content and thus causes the corresponding change in the C:P and N:P ratios [24]. These hypotheses illustrate the factors driven by latitude that contribute to the biogeographic patterns of leaf stoichiometry. However, the montane landscape has its own unique features. For example, the environment and the vegetation type are known to change dramatically and rapidly with altitude, even over short distances. As we know, intrinsic to all mountainous areas are decreases in air temperature, soil nutriment availability and increases in relative humidity [28], which lead to more stressful environments for plants relative to lower altitudes. According to the Plant Physiological Hypothesis, plants should increase leaf N and P along elevation because of the reduced temperature. Whereas, to survival under low temperature stress and nutrient-limiting conditions plants tend to have lower growth rate [29]. So, it is more likely that leaf N and P decline while N:P increase along elevation according to the prediction of the

Biogeochemical Hypothesis and the Growth Rate Hypothesis. In fact, the discrepancy of previous results of altitudinal patterns suggests that the relationship between abiotic environmental conditions and leaf stoichiometry can vary among different regions or be impacted by different species composition, soil types or plant ecological strategies. Therefore, the relative effects of PGF, climate and soil on the altitudinal patterns of leaf stoichiometry need to be further understood.

Changbai Mountain is a volcanic mountain in northeastern China. The vegetation varies from broad-leaved forest at low elevation to alpine tundra at high elevation and has been deemed a mirror of the horizontal zonation of vegetation from temperate zones to frigid zones on the Eurasian continent [30–31]. In this study, we measured the concentrations of leaf C, N, and P in 175 plant species as well as the total and available N and P in the soil along an altitudinal transect (500–2300 m) on the northern slope of Changbai Mountain. The main objectives of this study were to (1) explore the altitudinal patterns of leaf stoichiometry; (2) compare the response of the leaf stoichiometric traits of different PGFs to the environmental gradient; and (3) determine the key factors controlling the altitudinal patterns of leaf stoichiometry.

Materials and Methods

Site description

Changbai Mountain (41°23'N–42°36'N, 126°55'E–129°00'E) is located in Jilin Province, northeastern China. It is the highest mountain in northeastern china and is the head of three large rivers (the Songhua River, the Yalu River and the Tumen River). The climate belongs to a temperate continental montane climate. As the elevation rises from 500 to 2744 m, the mean annual temperature (MAT) decreases from 3.5 to–7.4°C and the mean annual precipitation (MAP) increases from 720 to 1400 mm [32]. Changbai Mountain has obvious vertical vegetation zonation, including summer green broad-leaved forest (below 700 m), Korean pine and broad-leaved mixed forest (700–1100 m), dark-coniferous spruce and fir forest (1100–1800 m), sub-alpine birch forest (1800–2100 m), and alpine tundra (above 2100 m). Changbai Mountain is one of the few well-conserved natural ecosystems on Earth. All these factors make it an optimal site to investigate the altitudinal patterns of leaf stoichiometry.

Sampling and measurements

In early August 2012, we established six sampling sites (site A to site F) with the altitudinal gradient along the northern slope of Changbai Mountain. Locations and the main characteristics of sampling sites were showed in Fig. 1 and Table 1. Climatic data, such as mean annual temperature (MAT) and mean annual precipitation (MAP) were from literature (See Table 1 for more details) [33]. At each site, we set up four plots (30 m×40 m) where we collected leaves from the observed plant species and took soil samples. There was a total of 279 plant species collected across the six sites. Specifically, from site A to site F there was a distribution of 72, 91, 36, 38, 22, and 20 species, respectively (Table 1). If we consider the replication of plant species in the six sites, a total of 175 species belonging to 54 families were sampled from the six sites.

For each plant species, we collected sun-exposed and mature leaves (leaf blades for grasses) from five to ten individuals. Leaf samples were oven-dried at 60°C in the laboratory and were ground to a fine powder using a ball mill (MM400, Retsch, Germany) for chemical analysis. At each plot, soil samples were randomly collected from 30–50 points in the 0–10 cm and 10–30 cm layers using a soil sampler (diameter 6 cm), resulting in a mixed soil sampling (>5 kg) from each plot. The fresh soil samples were sieved through 2-mm meshes, and roots and visible organic debris were removed by hand. Approximately 100 g of each soil sample was air-dried in a ventilation room to analyse the soil properties (C and N, pH, and others). The remaining portion of each soil sample was stored at 4°C in refrigerators for available N and P analysis.

The total C and N concentrations of leaf and soil samples were determined by dry combustion using an elemental analyser (Vario MAX CN Elemental Analyzer, Elementar, Germany). Total P concentrations were measured by the ammonium molybdate method using a continuous-flow analyser (AutoAnalyzer3 Continuous-Flow Analyzer; Bran Luebbe, Germany) after H_2SO_4-$HClO_4$ digestion for leaves and H_2SO_4-H_2O_2-HF digestion for soil [34–35]. Soil inorganic N (NH_4^+-N and NO_3^--N) in the filtrates was extracted using 2 mol L^{-1} KCl and determined using a continuous-flow analyser (AutoAnalyzer 3 Continuous-Flow Analyzer; Bran Luebbe, Germany) [36]. To measure available P, fresh soil samples were extracted using 0.5 mol L^{-1} $NaHCO_3$, and the P concentration of the extract was determined by the ammonium molybdate method. Soil pH was determined with a pH meter using soil mixed with distilled water (ratio 1:2.5). For each soil variable, the value used here was the average for the two depths in each plot.

Ethics statement

Our field studies obtained special permission from Changbai Mountain National Reserve, Jilin Province, China. We have no commercial interests or conflicts of interest in performing this work.

Data analysis

The data were explored at the species level and the site-species level, respectively. At the species level, we statistically summarised the means and the coefficients of variation (CV) of the leaf C, N, P, and C:N:P ratios. Furthermore, all plant species were divided into three PGFs: herbs, shrubs, and trees. Differences in leaf stoichiometric traits among different PGFs were tested using analysis of variance (ANOVA) with the Duncan *post hoc* tests of significance.

At the site-species level, the relationships between leaf stoichiometric traits and altitude were explored by linear regressions after the leaf C, N, P, and C:N:P ratios were log-transformed to normalise the distributions. Furthermore, we compared the relationship between the leaf C, N, P and C:N:P ratios and altitude among PGFs in order to examine the responses of different PGFs to environmental gradients (Table S1).

The effects of PGF, climate and soil on the leaf C, N, P and C:N:P ratios were tested using the General Linear Model (GLM). As a first step, within climate (MAP and MAT) and soil variables (soil total carbon, STC; soil total nitrogen, STN; soil total phosphorus, STP; soil available nitrogen, SAN; soil available phosphorus, SAP and pH), stepwise selection of variables was performed to exclude variables that did not contribute significantly ($P<0.01$) to the explained variation (For full details see Table S2, S3, S4). Then, the partial General Liner Model (partial GLM) separated the variance explained by different factors into the independent effects of each individual factor and the interactive effects between factors (Table S5) (for details, see Heikkinen et al. 2005) [37].

All statistical analyses were performed with R 2.15.2 [38].

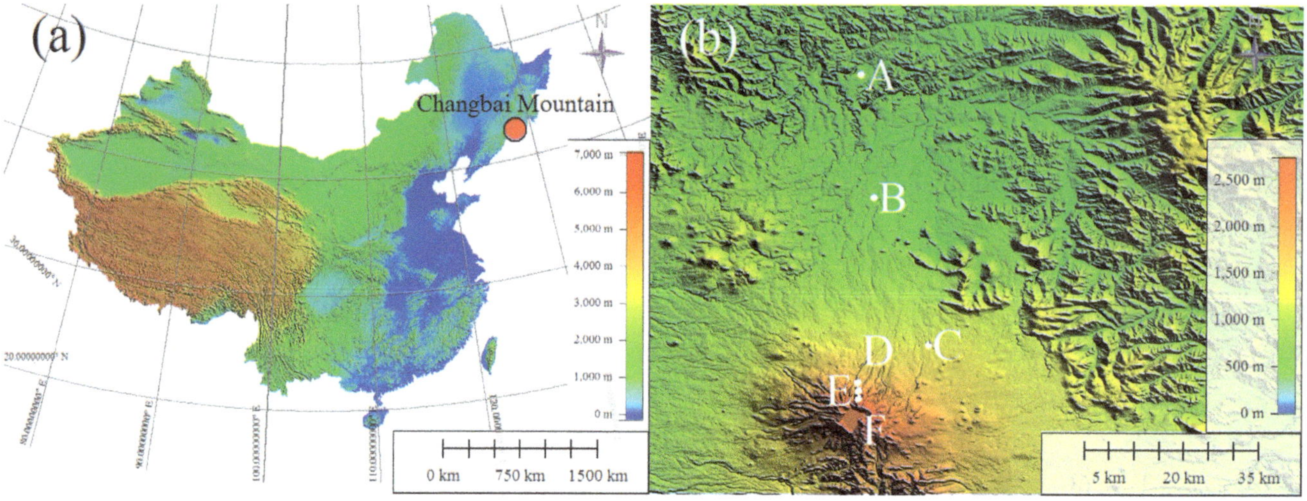

Figure 1. Experimental area and sampling sites along the northern slope of Changbai Mountain, China.

Results

Statistics of leaf C, N, P and C:N:P ratios

Leaf C, N, P and the C:N:P ratios of the plants on Changbai Mountain varied greatly with a range of 246.5–549.2 mg g^{-1} for C, 10.2–46.9 mg g^{-1} for N, and 0.8–5.5 mg g^{-1} for P. The C:N:P ratios ranged from 8.0 to 45.7 for C:N, 79.8 to 645.3 for C:P, and 3.4 to 24.8 for N:P. Leaf C, N, P and the C:N:P ratios varied 2 to 8 fold across species. Among leaf C, N, and P concentrations, leaf P had the greatest variation with a CV of 0.35, whereas leaf C had the smallest variation (CV = 0.08), which resulted in C:P having the greatest variation (CV = 0.43) and N:P having the smallest (CV = 0.28) (Table 2).

Leaf C, N, P and the C:N:P ratios differed significantly among different PGFs (Table 2). Leaf N and P were remarkably higher in herbaceous species than in woody species, whereas leaf C and C:N:P ratios were higher in trees and shrubs than in herbs (Table 2).

Patterns of leaf C, N, P and C:N:P ratios along altitudinal gradients

For all plant species, leaf C, N, P and the C:N:P ratios were significantly correlated with altitude ($P<0.001$) (Fig. 2). Leaf C increased significantly along elevation ($r^2 = 0.18$, $P<0.001$) (Fig. 2a), but leaf N and P decreased with increased altitude (Fig. 2c, e). Meanwhile, the leaf C:N:P ratios were positively correlated with altitude (Fig. 2b, d, f).

The leaf stoichiometric traits of different PGFs showed different responses to altitudinal gradients. Leaf C exhibited a remarkable increase along elevation for trees ($r^2 = 0.41$, $P<0.001$), shrubs ($r^2 = 0.50$, $P<0.001$), and herbs ($r^2 = 0.20$, $P<0.001$) (Fig. 2a). Leaf N was negatively correlated with altitude for herbs ($r^2 = 0.07$, $P<0.001$) (Fig. 2c) but showed no significant change for trees and shrubs. For shrubs and herbs, leaf P decreased significantly with increasing altitude (Fig. 2e).

Effects of PGF, climate and soil on leaf stoichiometric traits

The general linear model (GLM) showed that PGF, climate and soil together explained 17.6% to 52.1% of the variation in the six leaf stoichiometric traits (Fig. 3). The independent effect of PGF (*a*) showed the largest contribution (3.4–30.7%). The interactive effect of soil and climate (*bc*) also had substantial contributions (9.0–22.2%), although the independent effect of each factor was small. The total effect of PGF (*a+ab+ac+abc*) accounted for the largest contribution to the variations in leaf C and P concentrations and the C:P ratio (explained 25.7%, 20.4% and 25.2%, respectively), while the total effect of soil (*c+ac+bc+abc*) most contributed to leaf N and the C:N and N:P ratios (explained 14.0%, 18.4% and 17.4%, respectively).

Discussion

Variations of leaf C, N, P and C:N:P ratios

The average of leaf N, P and the C:N:P ratios of the 175 plant species on Changbai Mountain were different from those reported at regional scale, whereas leaf C showed no remarkable distinction. Our results showed that the mean of leaf C on Changbai Mountain was 438.56 mg g^{-1}, which is nearly equivalent with that of 213 species across the Chinese grassland (438 mg g^{-1}), as reported by He et al. (2006). However, the means of leaf N and P were 20.9% and 57.5% higher than those of the Chinese flora (19.96 and 1.41 mg g^{-1} for leaf N and P, respectively), while the N:P ratio was 19.0% lower (the N:P ratio was 14.2 for the Chinese flora) [23]. The higher concentrations of leaf N and P on Changbai Mountain may be attributed to the soil fertility being higher than the mean level of N and P storage in China [22,39], with the STN at 3.63 mg g^{-1} and the SAN at 0.06 mg g^{-1}. Similarly, the STP of Changbai Mountain (0.86±0.03 mg g^{-1}) was higher than the average of China (0.56 mg g^{-1}) [7].

We found that leaf P was more variable than leaf C and N, which is consistent with previous results at different spatial scales [8,40–42]. Our findings support the Stability of Limiting Elements Hypothesis to some extent, which suggests that nutrients required in a high concentration in plants and considered most frequently limiting in environment should show a small variation in their concentration and lower sensitivity to the environmental factors [22,43]. It implies that leaf P is less stable and has weaker stoichiometric homeostasis than leaf N. When comparing C:N:P ratios, the variation was largest for the C:P ratio and smallest for the N:P ratio (Table 2). The larger variation in the C:P ratio was mainly due to the large variation in P. In contrast, the N:P ratio

Table 1. Site descriptions of vegetation and soil properties.

Site	Vegetation type	Altitude (m)	Latitude	Longitude	Soil type	MAT (°C)	MAP (mm)	STC (mg g^{-1})	STN (mg g^{-1})	STP (mg g^{-1})	SAN (mg g^{-1})	SAP (mg g^{-1})	pH	No. of Species
Site A	Broad-leaved forest	540	42°37′	128°4′	Albi-Boric Argosols	2.9	632	88.54	7.39	1.54	0.08	0.02	5.31	72
Site B	Mixed coniferous broad-leaved forest	753	42°24′	128°5′	Albi-Boric Argosols	2.6	691	60.62	4.92	1.36	0.07	0.01	5.01	91
Site C	Dark-coniferous spruce-fir forest	1286	42°8′	128°11′	Bori-UdicCambosols	0.3	811	15.79	0.78	0.42	0.04	0.02	5.23	36
Site D	Ermans birch forest	1812	42°4′	128°4′	Umbri-GelicCambosols	−2.3	967	54.11	3.79	0.93	0.07	0.01	5.03	38
Site E	Alpine tundra	2008	42°3′	128°3′	Permi-GelicCambosols	−3.3	1038	43.52	2.71	0.51	0.06	0.01	5.02	22
Site F	Alpine tundra	2357	42°2′	128°3′	Permafrost cold Cambisols	−4.8	1154	31.37	2.20	0.40	0.05	0.01	5.14	20

MAT, mean annual temperature; MAP, mean annual precipitation; STC, soil total carbon; STN, soil total nitrogen; STP, soil total phosphorus; SAN, soil available nitrogen; SAP, soil available phosphorus. MAT, MAP and soil type are derived from Shen et al. (2013) [33].

Table 2. Leaf C, N, P and C:N:P ratios for plant species on the Changbai Mountain, northeast China.

		C (mg g^{-1})		N (mg g^{-1})		P (mg g^{-1})		C:N ratio		C:P ratio		N:P ratio	
	n	Mean	CV	Mean	CV	Mean	CV	Mean	CV	Mean	CV	Mean	CV
Herbs	105	422.98^{a†‡}	0.07	25.13^{a}	0.27	2.49^{a}	0.30	18.42^{a}	0.29	187.24^{a}	0.34	10.54^{a}	0.24
Shrubs	37	457.07^{b}	0.07	21.98^{b}	0.25	1.84^{b}	0.44	22.53^{b}	0.27	286.58^{b}	0.41	13.01^{b}	0.31
Trees	33	466.88^{b}	0.06	22.92^{b}	0.26	1.79^{b}	0.24	22.29^{b}	0.34	283.17^{b}	0.35	13.11^{b}	0.23
All species	175	438.56	0.08	24.13	0.27	2.22	0.35	19.71	0.32	226.33	0.43	11.54	0.28

†Mean values and the coefficient of variation (CV) are reported, along with the number of samples (n).
‡Differences among PGFs were tested using ANOVA with Duncan post hoc tests; different superscript letters (a and b) in each column indicate significant differences in the mean values at $P<0.05$.

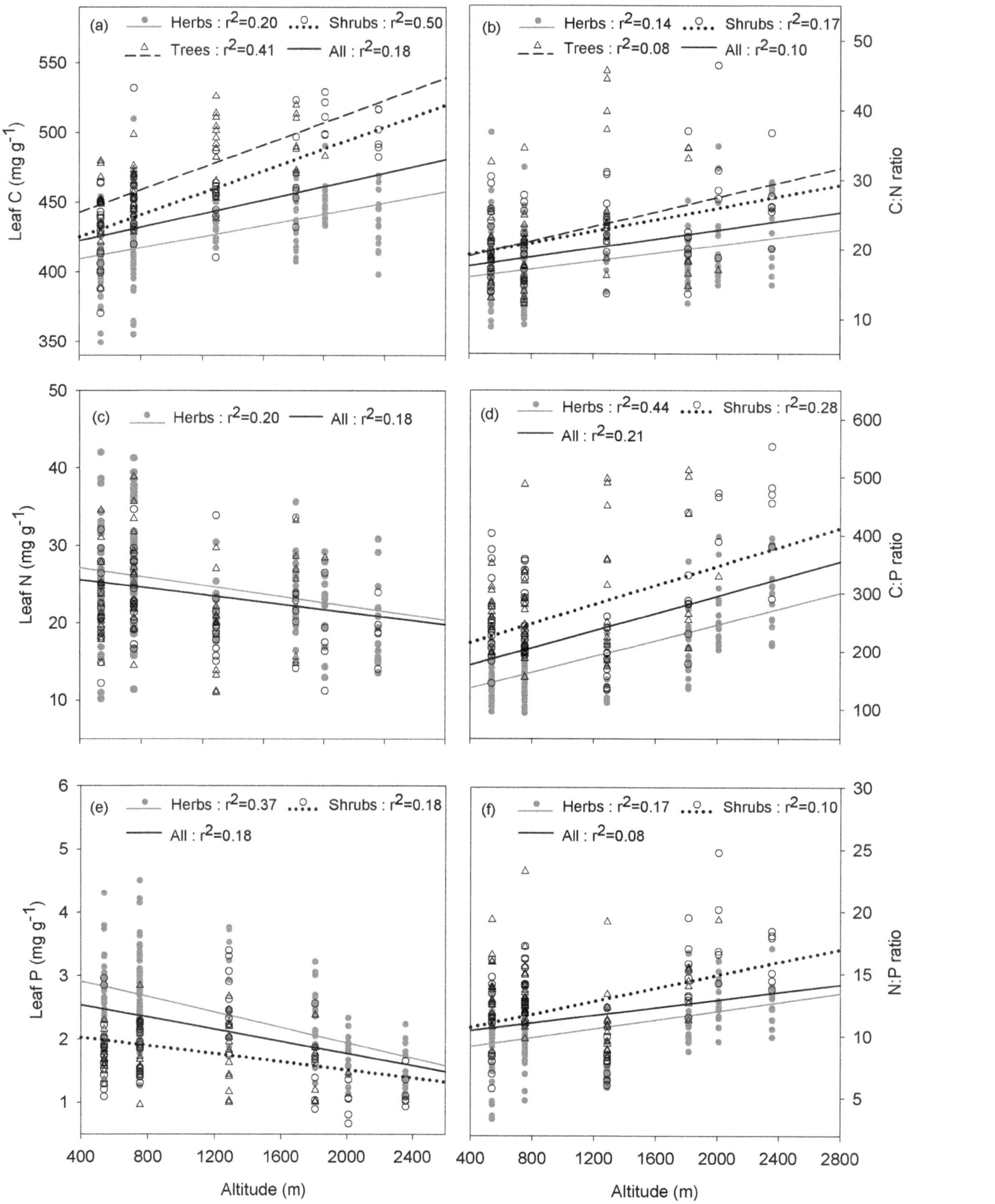

Figure 2. Changes in leaf C, N, P and C:N:P ratios with the altitudinal gradient. Lines are plotted if regressions were significant at $P<0.05$. Note log scale used on y-axis.

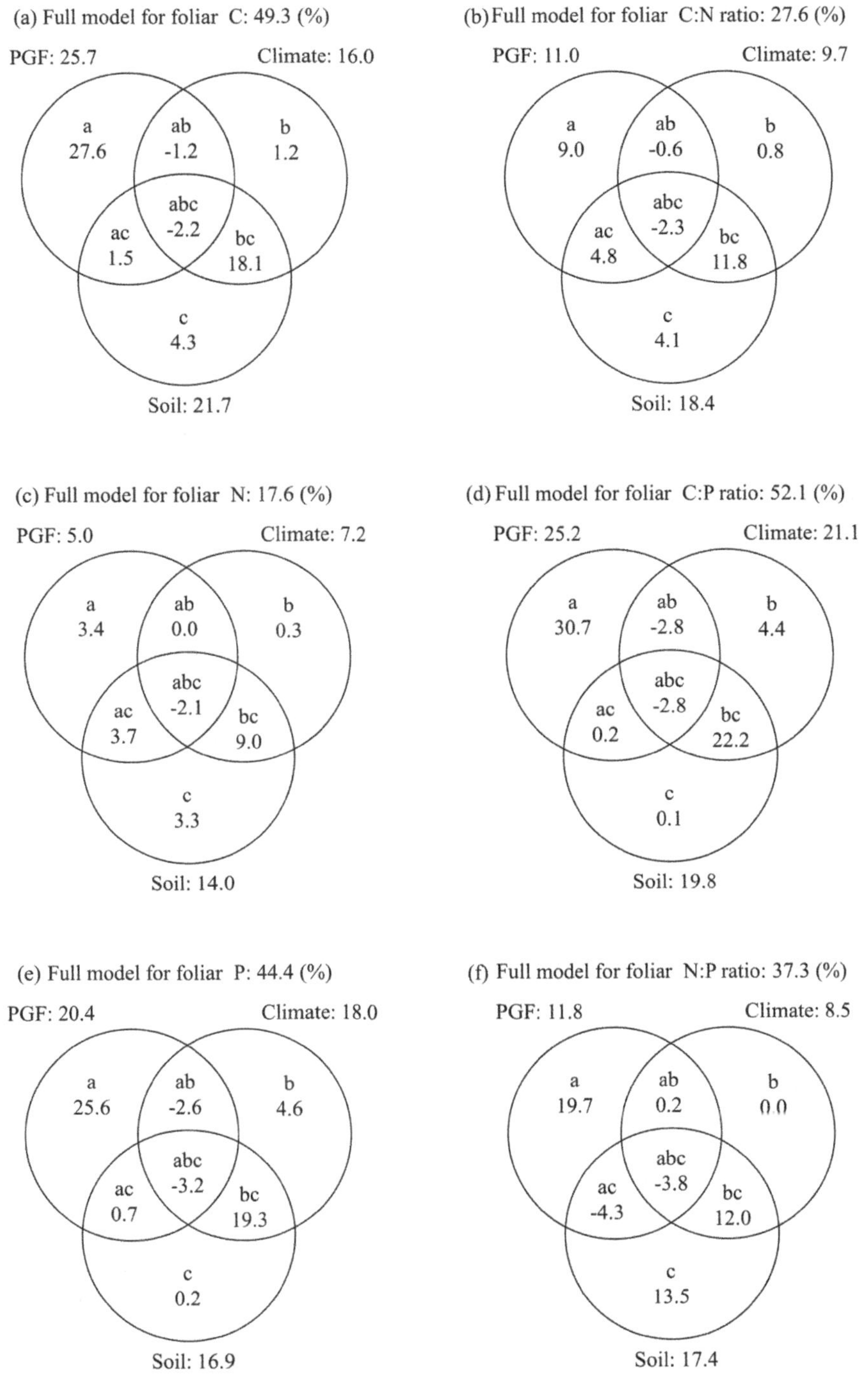

Figure 3. Variation partitioning (R^2, %) of PGF, climate and soil in accounting for the variations in leaf C, N, P and C:N:P ratios. *a*, *b*, and *c* denote the independent effect of plant growth form (PGF), climate and soil, respectively; *ab*, *ac*, and *bc* are the interactive effect between PGF and climate, PGF and soil, climate and soil, respectively; *abc* denotes the interactive effect among the three factors.

varied less than N or P alone because of the close relationship in biochemical reaction between N and P.

Altitudinal patterns of leaf C, N, P and C:N:P ratios

Leaf C, N, P and C:N:P ratios exhibited remarkable altitudinal patterns on Changbai Mountain. Leaf C increased significantly with altitude, in agreement with the results of a meta-analysis conducted on global mountain ranges [13]. The increased leaf C concentration was most likely caused by an increase in non-structural C (NSC), including starch, low molecular weight sugars and storage lipids. Plant species at higher elevations experience more stressed environments than do those at lower elevations; thus, higher NSC concentrations accumulated to balance the osmotic pressure of cells and to resist freezing [13,44–45].

Our results support the pattern that leaf N and P declined with elevation, consisting with results of some previous studies [15–18]. Meanwhile, the increased leaf N:P ratio was also found in studies conducted on subarctic tundra in northern Sweden and north-western Himalaya [14,18], although other studies found different tendency [12,17,20]. The deceleration of plant metabolic rates and the limitation of nutrient availability could interact to shape the altitudinal patterns of leaf N, P and N:P ratio. Some studies have demonstrated that low temperatures can limit soil microbe activity, resulting in low decomposition of organic matter and available nutrients in soil, and thus depress nutrient uptake by roots [2,25]. Furthermore, aggravated soil leaching caused by the increased precipitation at higher elevations deteriorates nutrient availability [46]. In fact, a soil fertility decrease along the altitudinal gradient was clearly observed on Changbai Mountain (Fig. S1). Therefore, declined metabolic rates and nutrient limitation together shaped the pattern that leaf N and P declined with elevation. The increased N:P ratio suggest that N concentrations decrease slower than P concentrations. Moreover, to survival under low temperature stress, plants tend to have lower growth rate which might lead to higher N:P ratio. As was mentioned above, the altitudinal patterns are consistent with the prediction of the Biogeochemical Hypothesis and the Growth Rate Hypothesis. It is possible to assume that, these hypotheses which based on latitudinal patterns can also well describe the drivers of altitudinal patterns.

The response of different plant growth forms

Different plant growth forms respond differently to environmental gradients for leaf stoichiometric traits. All the observed species were divided into three PGFs (trees, shrubs and herbs), and linear regressions of leaf stoichiometric traits against altitude were then performed to compare the response of different PGFs. The regression slopes indicate the degree of variation for leaf stoichiometric traits of different PGFs. A steeper slope indicates greater variation of leaf stoichiometric traits along elevation. Among the three PGFs, trees showed the steepest slope for leaf C, corresponding with the largest variation for the C:N and C:P ratios (Fig. 2, Table S1), while leaf N and P showed no significant change with increasing altitude. In contrast, herbs showed the largest variation in leaf N, P and the N:P ratio (Fig. 2, Table S1). The different responses may be influenced by the distinct life strategies exhibited by different PGFs. Slow-growing woody species are likely to be competitors or stress tolerators (*C* or *S* selected) with relatively conservative stoichiometric traits [47]. However, fast-growing herbaceous species, which adopt a ruderal strategy (*R*-selected), exhibit more flexibility for leaf stoichiometric traits. The more dramatic increase in leaf C in trees might be due to the tree experiencing critically low temperatures at lower elevation and latitude than smaller plants [48]. Thus, as we mentioned before, increased leaf C would help trees to resist freezing [13].

The relative effects of PGF, climate and soil

PGF, climate and soil jointly influenced the altitudinal patterns of leaf stoichiometry. GLM models were able to determine the explanatory capacity of PGF (*a*), climate (*b*) and soil (*c*) for different stoichiometric traits along the elevation gradient on Changbai Mountain. PGF (*a+ab+ac+abc*) accounted for the largest variations in leaf C, P, and the C:P ratio, while soil (*c+ac+bc+abc*) was better able to explain the variation of leaf N and the C:N and N:P ratios (Fig. 3, Table. S5). However, the collinearity among PGF, climate and soil may obscure their true explanatory capacity. The partial GLM provided a better way to divide the total effect into independent and interactive effects [37]. Our results showed that the independent effect of PGF (*a*) was the largest contributor to the explained variations in leaf C, N, P and C:N:P ratios. In contrast, the independent effects of climate (*b*) and soil (*c*) were small (Fig. 3, Table S5). Our results are different from recent studies carried out on forest and shrub biomes at regional scale. Chen et al. (2011) and Liu et al. (2012) reported that the independent effect of PGF had the largest contribution to leaf N variation, while environmental factors and PGF were both important for leaf P and the N:P ratio. A potential cause for such an inconsistency may be the different nutrient limitations in the different regions [8,49]. Plants are most likely P-limited in most regions of China [7], whereas the vegetation of Changbai Mountain is mainly limited by N (Table S6). The difference in nutrient limitations in different regions may obscure the roles of other factors. Overall, these results suggest that the relationships between leaf C:N:P stoichiometry and environment, depending on the most limiting factor, should be different in different regions.

The effects of environmental factors, PGF and interactions between them on stoichiometric traits have received much less attention in previous studies. In spite of the small independent effect, the interactive effect of climate and soil (*bc*) was the largest explanatory factor among the interactions for all six stoichiometric traits (Fig. 3, Table S5). This strong interaction of climate and soil may be the cause of the obvious change in soil fertility along the altitudinal gradient (Fig. S1). Note that the interactive effects were negative when the relationship between two factors is mainly suppressive rather than additive [22,37].

Concluding Remarks

This study comprehensively characterised the relative effects of PGF, climate and soil on the altitudinal patterns of leaf C:N:P stoichiometry on Changbai Mountain, China. The altitudinal patterns of leaf stoichiometry were influenced by PGF, climate, soil and their interactions. In general, PGF has a stronger influence than soil and climate. Overall, the relationships between leaf C:N:P stoichiometry and environment, depending on the most limiting factor, should be different in different regions.

Supporting Information

Figure S1 Changes in soil nutrient and pH value with the altitudinal gradient on Changbai Mountain. STC, soil total carbon; STN, soil total nitrogen; STP, soil total phosphorus; SAN, soil available nitrogen; SAP, soil available phosphorus. Error bars mean standard errors (SE) of variables.

Table S1 Linear regressions of leaf stoichiometric traits on altitude for different plant growth forms. "***" denotes $P<0.001$, "*" denotes $P<0.05$. Note log scale used on y-axis.

Table S2 Correlations between soil and environmental variables. MAT, mean annual temperature; MAP, mean annual precipitation; STC, soil total carbon; STN, soil total nitrogen; STP, soil total phosphorus; SAN, soil available nitrogen; SAP, soil available phosphorus. Pearson coefficients in bold and with an asterisk indicate the correlation is significant at $P<0.05$.

Table S3 Model summary for the stepwise multiple regressions of leaf stoichiometric traits on MAP and MAT. The variable that do not contribute significantly ($P<0.01$) to the explained variation will be excluded from the partial General Linear Models (partial GLM). MAT, mean annual temperature; MAP, mean annual precipitation.

Table S4 Model summary for the stepwise multiple regressions of leaf stoichiometric traits on soil variables. The variables that do not contribute significantly ($P<0.01$) to the explained variation will be excluded from the partial General Linear Models (partial GLM). STC, soil total carbon; STN, soil total nitrogen; STP, soil total phosphorus; SAN, soil available nitrogen; SAP, soil available phosphorus.

Table S5 Summary of the partial General Linear Models (partial GLM) for the effects of PGF, climate and soil nutrient on leaf stoichiometric traits. a, b, and c denote the independent effect of plant growth form (PGF), climate and soil, respectively; ab, ac, and bc are respectively the interactive effect between PGF and climate, PGF and soil, climate and soil; abc denotes the interactive effect among the three factors.

Table S6 Nutrition limitations of different vegetation types on Changbai Mountain. Nutrient limitations is diagnosed following criteria proposed by Koerselman (1996): If N:P ratio >16, plant growth is limited by P availability. If N:P ratio <14, plant growth is limited by N availability. If N:P ratio between 14 and 16, plant growth is co-limited by N and P together. "N" indicates N limitation; "N and P" indicates N and P co-limitation.

Acknowledgments

We thank teachers and students of Beihua University for field sampling assistance and the staff of Physical and Chemical Analysis Laboratory, Key Laboratory of Ecosystem Network Observation and Modelling of IGSNRR, CAS for laboratory analysis.

Author Contributions

Conceived and designed the experiments: NZ GY NH. Performed the experiments: NZ NH RW ZX XZ QW. Analyzed the data: NZ. Contributed reagents/materials/analysis tools: NZ GY. Wrote the paper: NZ GY NH XZ QW.

References

1. Elser JJ, Sterner RW, Gorokhova E, Fagan WF, Markow TA, et al. (2000) Biological stoichiometry from genes to ecosystems. Ecology Letters 3: 540–550. Available: http://onlinelibrary.wiley.com/doi/10.1111/j.1461-0248.2000.00185.x/abstract. Accessed 30 March 2014.
2. Reich PB, Oleksyn J (2004) Global patterns of plant leaf N and P in relation to temperature and latitude. Proceedings of the National Academy of Sciences of the United States of America 101: 11001–11006. Available: http://www.pnas.org/content/101/30/11001. Accessed 30 March 2014.
3. Ordoñez JC, Van Bodegom PM, Witte JPM, Wright IJ, Reich PB, et al. (2009) A global study of relationships between leaf traits, climate and soil measures of nutrient fertility. Global Ecology and Biogeography 18: 137–149. Available: http://onlinelibrary.wiley.com/doi/10.1111/j.1466-8238.2008.00441.x/abstract. Accessed 30 March 2014.
4. Liu J, Huang W, Zhou G, Zhang D, Liu S, et al. (2013) Nitrogen to phosphorus ratios of tree species in response to elevated carbon dioxide and nitrogen addition in subtropical forests. Global Change Biology 19: 208–216. Available: http://onlinelibrary.wiley.com/doi/10.1111/gcb.12022/abstract. Accessed 30 March 2014.
5. He JS, Fang J, Wang Z, Guo D, Flynn DFB, et al. (2006) Stoichiometry and large-scale patterns of leaf carbon and nitrogen in the grassland biomes of China. Oecologia 149: 115–122. Available: http://link.springer.com/article/10.1007%2Fs00442-006-0425-0. Accessed 30 March 2014.
6. Wu TG, Yu MK, Wang GG, Dong Y, Cheng XR (2012) Leaf nitrogen and phosphorus stoichiometry across forty-two woody species in Southeast China. Biochemical Systematics and Ecology 44: 255–263. Available: http://www.sciencedirect.com/science/article/pii/S0305197812001147. Accessed 30 March 2014.
7. Han WX, Fang JY, Guo DL, Zhang Y (2005) Leaf nitrogen and phosphorus stoichiometry across 753 terrestrial plant species in China. New Phytologist 168: 377–385. Available: http://onlinelibrary.wiley.com/doi/10.1111/j.1469-8137.2005.01530.x/abstract. Accessed 30 March 2014.
8. Chen YH, Han WX, Tang LY, Tang ZR, Fang JY (2011) Leaf nitrogen and phosphorus concentrations of woody plants differ in responses to climate, soil and plant growth form. Ecography 36: 178–184. Available: http://onlinelibrary.wiley.com/doi/10.1111/j.1600-0587.2011.06833.x/abstract. Accessed 30 March 2014.
9. Sardans J, Rivas-Ubach A, Peñuelas J (2011) Factors affecting nutrient concentration and stoichiometry of forest trees in Catalonia (NE Spain). Forest Ecology and Management 262: 2024–2034. Available: http://www.sciencedirect.com/science/article/pii/S0378112711005111. Accessed 30 March 2014.
10. Mcgroddy ME, Daufresne T, Hedin LO (2004) Scaling of C: N: P stoichiometry in forests worldwide: implications of terrestrial Redfield-type ratios. Ecology 85: 2390–2401. Available: http://www.esajournals.org/doi/abs/10.1890/03-0351. Accessed 30 March 2014.
11. Körner C (1989) The nutritional status of plants from high altitudes. Oecologia 81: 379–391. Available: http://link.springer.com/article/10.1007/BF00377088. Accessed 30 March 2014.
12. Soethe N, Lehmann J, Engels C (2008) Nutrient availability at different altitudes in a tropical montane forest in Ecuador. Journal of Tropical Ecology 24: 397–406. Available: http://journals.cambridge.org/action/displayAbstract?fromPage=online&aid=1924472. Accessed 30 March 2014.
13. Hoch G, Körner C (2012) Global patterns of mobile carbon stores in trees at the high-elevation tree line. Global Ecology and Biogeography 21: 861–871. Available: http://onlinelibrary.wiley.com/doi/10.1111/j.1466-8238.2011.00731.x/abstract. Accessed 30 March 2014.
14. Macek P, Klimeš L, Adamec L, Doležal J, Chlumská Z, et al. (2012). Plant Nutrient Content Does Not Simply Increase with Elevation under the Extreme Environmental Conditions of Ladakh, NW Himalaya. Arctic, Antarctic, and Alpine Research 44: 62–66. Available: http://www.bioone.org/doi/abs/10.1657/1938-4246-44.1.62. Accessed 30 March 2014.
15. Köhler L, Gieger T, Leuschner C (2006) Altitudinal change in soil and foliar nutrient concentrations and in microclimate across the tree line on the subtropical island mountain Mt. Teide (Canary Islands). Flora-Morphology, Distribution, Functional Ecology of Plants 201: 202–214. Available: http://www.sciencedirect.com/science/article/pii/S0367253005001283. Accessed 30 March 2014.
16. Macek P, Macková J, Bello F (2009) Morphological and ecophysiological traits shaping altitudinal distribution of three Polylepis treeline species in the dry tropical Andes. Acta Oecologica 35: 778–785. Available: http://www.sciencedirect.com/science/article/pii/S1146609X09001076. Accessed 30 March 2014.
17. Van de Weg MJ, Meir P, Grace J, Atkin O (2009) Altitudinal variation in leaf mass per unit area, leaf tissue density and foliar nitrogen and phosphorus content along an Andes-Amazon gradient in Peru. Plant Ecology & Diversity 2: 243–254. Available: http://www.tandfonline.com/doi/abs/10.1080/17550870903518045. Accessed 30 March 2014.
18. Sundqvist MK, Giesler R, Wardle DA (2011) Within-and across-species responses of plant traits and litter decomposition to elevation across contrasting vegetation types in subarctic tundra. Available: http://www.plosone.org/article/info%3Adoi%2F10.1371%2Fjournal.pone.0027056. Accessed 30 March 2014.
19. Shi W, Wang G, Han W (2012) Altitudinal Variation in Leaf Nitrogen Concentration on the Eastern Slope of Mount Gongga on the Tibetan Plateau, China.PLoS One 7: e44628. Available: http://www.plosone.org/article/info%3Adoi%2F10.1371%2Fjournal.pone.0044628. Accessed 30 March 2014.
20. Fisher JB, Malhi Y, Torres IC, Metcalfe DB, Weg MJ, et al. (2013) Nutrient limitation in rainforests and cloud forests along a 3,000-m elevation gradient in the Peruvian Andes. Oecologia 172: 889–902. Available: http://link.springer.com/article/10.1007%2Fs00442-012-2522-6. Accessed 30 March 2014.
21. Hou HY (1982) The Vegetation Geography of China with Reference to the Chemical Composition of Dominant Plants. Science Press, Beijing
22. Han WX, Fang JY, Reich PB, Woodward FI, Wang ZH (2011) Biogeography and variability of eleven mineral elements in plant leaves across gradients of climate, soil and plant functional type in China. Ecology Letters 14: 788–796. Available: http://onlinelibrary.wiley.com/doi/10.1111/j.1461-0248.2011.01641.x/abstract. Accessed 30 March 2014.
23. Zhang SB, Zhang JL, Slik JWF, Cao KF (2012) Leaf element concentrations of terrestrial plants across China are influenced by taxonomy and the environment. Global Ecology and Biogeography 21: 809–818. Available: http://onlinelibrary.wiley.com/doi/10.1111/j.1466-8238.2011.00729.x/abstract. Accessed 30 March 2014.
24. Elser JJ, Acharya K, Kyle M, Cotner J, Makino W, et al. (2003) Growth rate–stoichiometry couplings in diverse biota. Ecology Letters 6: 936–943. Available: http://onlinelibrary.wiley.com/doi/10.1046/j.1461-0248.2003.00518.x/abstract. Accessed 30 March 2014.
25. Aerts R, Chapin III FS (1999) The mineral nutrition of wild plants revisited: a re-evaluation of processes and patterns. Advances in Ecological Research 30: 1–

67. Available: http://www.sciencedirect.com/science/article/pii/S0065250408600161. Accessed 30 March 2014.
26. Hobbie SE, Gough L (2002) Foliar and soil nutrients in tundra on glacial landscapes of contrasting ages in northern Alaska. Oecologia 131: 453–462. Available: http://link.springer.com/article/10.1007%2Fs00442-002-0892-x. Accessed 30 March 2014.
27. Güsewell S (2004) N:P ratios in terrestrial plants: variation and functional significance. New Phytologist 164: 243–266. Available: http://onlinelibrary.wiley.com/doi/10.1111/j.1469-8137.2004.01192.x/abstract. Accessed 30 March 2014.
28. Barry RG (1981) Mountain weather and climate. 2nd ed. Routledge, London
29. Wright IJ, Reich PB, Westoby M, Ackerly DD, Baruch Z, et al. (2004). The worldwide leaf economics spectrum. Nature 428: 821–827. Available: http://www.nature.com/nature/journal/v428/n6985/full/nature02403.html. Accessed 30 March 2014.
30. Shao G, Shugart HH, Zhao G, Zhao S, Wang S, et al. (1996) Forest cover types derived from Landsat Thematic Mapper imagery for Changbai Mountain area of China. Canadian Journal of Forest Research 26: 206–216. Available: http://www.nrcresearchpress.com/doi/abs/10.1139/x26-024#.UzcDKaiSwSs. Accessed 30 March 2014.
31. Zhang M, Zhang XK, Liang WJ, Jiang Y, Dai GH, et al. (2011) Distribution of soil organic carbon fractions along the altitudinal gradient in Changbai Mountain, China. Pedosphere 21: 615–620. Available: http://www.sciencedirect.com/science/article/pii/S100201601160163X. Accessed 30 March 2014.
32. Zhu B, Wang XP, Fang JY, Piao SL, Shen HH, et al. (2010) Altitudinal changes in carbon storage of temperate forests on Mt Changbai, Northeast China. Journal of Plant Research 123: 439–452. Available: http://link.springer.com/article/10.1007%2Fs10265-009-0301-1. Accessed 30 March 2014.
33. Shen C, Xiong J, Zhang H, Feng Y, Lin X, et al.(2013) Soil pH drives the spatial distribution of bacterial communities along elevation on Changbai Mountain. Soil Biology and Biochemistry 57: 204–211. Available: http://www.sciencedirect.com/science/article/pii/S003807171200288X. Accessed 30 March 2014.
34. Kuo S (1996) Phosphorus. In: Sparks DL et al, Methods of soil analysis. Part 3. Chemical methods. Soil Science Society of America and American Society of Agronomy, Madison pp: 869–919. Available: https://dl.sciencesocieties.org/publications/books/abstracts/sssabookseries/methodsofsoilan3/869. Accessed 30 March 2014.
35. Bowman RA (1988) A rapid method to determine total phosphorus in soils. Soil Science Society of America Journal 52: 1301–1304. Available: https://www.soils.org/publications/sssaj/abstracts/52/5/SS0520051301. Accessed 30 March 2014.
36. Baumann F, He JS, Schmidt K, Khn P, Scholten T (2009) Pedogenesis, permafrost, and soil moisture as controlling factors for soil nitrogen and carbon contents across the Tibetan Plateau. Global Change Biology 15: 3001–3017. Available: http://onlinelibrary.wiley.com/doi/10.1111/j.1365-2486.2009.01953.x/abstract. Accessed 30 March 2014.
37. Heikkinen RK, Luoto M, Kuussaari M, Pöyry J (2005) New insights into butterfly–environment relationships using partitioning methods. Proceedings of the Royal Society B: Biological Sciences 272: 2203–2210. Available: http://rspb.royalsocietypublishing.org/content/272/1577/2203. Accessed 30 March 2014.
38. R Development Core Team (2012) R: A Language and Environment for Statistical Computing. R Foundation for Statistical Computing, Vienna, Austria http://www.r-project.org/
39. Yang YH, Ma WH, Mohammat A, Fang JY (2007) Storage, patterns and controls of soil nitrogen in China. Pedosphere 17: 776–785. Available: http://www.sciencedirect.com/science/article/pii/S1002016007600939. Accessed 30 March 2014.
40. Güsewell S, Koerselman W (2002) Variation in nitrogen and phosphorus concentrations of wetland plants. Perspectives in Plant Ecology, Evolution and Systematics 5: 37–61. Available: http://www.sciencedirect.com/science/article/pii/S1433831904700475. Accessed 30 March 2014.
41. Kerkhoff AJ, Enquist BJ, Elser JJ, Fagan WF (2005) Plant allometry, stoichiometry and the temperature-dependence of primary productivity. Global Ecology and Biogeography 14: 585–598. Available: http://onlinelibrary.wiley.com/doi/10.1111/j.1466-822X.2005.00187.x/abstract. Accessed 30 March 2014.
42. He JS, Wang L, Flynn DFB, Wang XP, Ma WH, et al. (2008) Leaf nitrogen: phosphorus stoichiometry across Chinese grassland biomes. Oecologia 155: 301–310. Available: http://link.springer.com/article/10.1007%2Fs00442-007-0912-y. Accessed 30 March 2014.
43. Sterner RW, Elser JJ (2002) Ecological stoichiometry: the biology of elements from molecules to the biosphere. Princeton University Press, Princeton. Available: http://library.wur.nl/WebQuery/clc/1680032. Accessed 30 March 2014.
44. Millard P, Sommerkorn M, Grelet GA (2007) Environmental change and carbon limitation in trees: a biochemical, ecophysiological and ecosystem appraisal. New Phytologist 175: 11–28. Available: http://onlinelibrary.wiley.com/doi/10.1111/j.1469-8137.2007.02079.x/abstract. Accessed 30 March 2014.
45. Hoch G, Popp M, Körner C (2002) Altitudinal increase of mobile carbon pools in Pinus cembra suggests sink limitation of growth at the Swiss treeline. Oikos 98: 361–374. Available: http://onlinelibrary.wiley.com/doi/10.1034/j.1600-0706.2002.980301.x/abstract. Accessed 30 March 2014.
46. Hedin LO, Vitousek PM, Matson PA (2003) Nutrient losses over four million years of tropical forest development. Ecology 84: 2231–2255. Available: http://www.esajournals.org/doi/abs/10.1890/02-4066. Accessed 30 March 2014.
47. Grime JP (1977) Evidence for the existence of three primary strategies in plants and its relevance to ecological and evolutionary theory. American Naturalist 111: 1169–1194. Available: http://www.jstor.org/discover/10.2307/2460262?uid=3737800&uid=2&uid=4&sid=21103576585301. Accessed 30 March 2014.
48. Körner C (2012) Treelines will be understood once the functional difference between a tree and a shrub is. Ambio 41: 197–206. Available: http://link.springer.com/article/10.1007%2Fs13280-012-0313-2. Accessed 30 March 2014.
49. Liu C, Wang XP, Wu X, Dai S, He JS, et al. (2012) Relative effects of phylogeny, biological characters and environments on leaf traits in shrub biomes across central Inner Mongolia, China. Journal of Plant Ecology 6: 220–231. Available: http://jpe.oxfordjournals.org/content/6/3/220. Accessed 30 March 2014.

5

Green Manure Addition to Soil Increases Grain Zinc Concentration in Bread Wheat

Forough Aghili[1]*, Hannes A. Gamper[1], Jost Eikenberg[2], Amir H. Khoshgoftarmanesh[3], Majid Afyuni[3], Rainer Schulin[4], Jan Jansa[5], Emmanuel Frossard[1]

1 Institute of Agricultural Sciences, Department of Environmental Systems Science, Swiss Federal Institute of Technology (ETH) Zürich, Switzerland, **2** Paul Scherrer Institute (PSI), Radioanalytics Laboratory, Villigen, Switzerland, **3** College of Agriculture, Department of Soil Sciences, Isfahan University of Technology, Isfahan, Iran, **4** Institute of Terrestrial Ecosystems, Department of Environmental Systems Science, Swiss Federal Institute of Technology (ETH) Zürich, Switzerland, **5** Institute of Microbiology, Academy of Sciences of the Czech Republic, Prague, Czech Republic

Abstract

Zinc (Zn) deficiency is a major problem for many people living on wheat-based diets. Here, we explored whether addition of green manure of red clover and sunflower to a calcareous soil or inoculating a non-indigenous arbuscular mycorrhizal fungal (AMF) strain may increase grain Zn concentration in bread wheat. For this purpose we performed a multifactorial pot experiment, in which the effects of two green manures (red clover, sunflower), $ZnSO_4$ application, soil γ-irradiation (elimination of naturally occurring AMF), and AMF inoculation were tested. Both green manures were labeled with ^{65}Zn radiotracer to record the Zn recoveries in the aboveground plant biomass. Application of $ZnSO_4$ fertilizer increased grain Zn concentration from 20 to 39 mg Zn kg^{-1} and sole addition of green manure of sunflower to soil raised grain Zn concentration to 31 mg Zn kg^{-1}. Adding the two together to soil increased grain Zn concentration even further to 54 mg Zn kg^{-1}. Mixing green manure of sunflower to soil mobilized additional 48 µg Zn (kg soil)$^{-1}$ for transfer to the aboveground plant biomass, compared to the total of 132 µg Zn (kg soil)$^{-1}$ taken up from plain soil when neither green manure nor $ZnSO_4$ were applied. Green manure amendments to soil also raised the DTPA-extractable Zn in soil. Inoculating a non-indigenous AMF did not increase plant Zn uptake. The study thus showed that organic matter amendments to soil can contribute to a better utilization of naturally stocked soil micronutrients, and thereby reduce any need for major external inputs.

Editor: Raffaella Balestrini, Institute for Sustainable Plant Protection - C.N.R., Italy

Funding: This study was supported by the grant 123920 of the Swiss National Science Foundation (SNF) and the Swiss Agency for Development and Cooperation (SDC) in the framework of Research Partnerships with Developing Countries. The funding body and host institutions had no role in study-design, data collection, analysis, and manuscript preparation. JJ was further supported by the J.E. Purkyně Fellowship and the long-term development program RVO61388971.

Competing Interests: The authors have declared that no competing interests exist.

* Email: forough.aghili@usys.ethz.ch

Introduction

Wheat is the third most-grown cereal after maize and rice, draws on the largest cropping area, and ranks first in global cereal trade [1]. It is the crop contributing most calories to large segments of the global human population. Yet, wheat-based diets do not provide sufficient zinc (Zn), which is a leading cause of widespread Zn deficiency [2]. Low grain Zn concentration is cause of low plant-availability of Zn in many soils on which wheat is produced [3,4]. Efforts have been made to breed wheat cultivars efficient in acquiring Zn from soil and effective in (re-) translocating Zn to grains [5,6]. Zinc supply from soil has, however, to be increased for such elite crop cultivars to become fully effective, which can be achieved via appropriate soil fertility management, such as application of mineral Zn fertilizer [7], addition of green manure [8], and/or a strengthening of nutrient acquisition traits, including symbioses of roots with naturally occurring and newly introduced arbuscular mycorrhizal fungi (AMF) [9–12].

Plants and probably also AMF take up Zn from the soil solution in form of free Zn^{2+} and $ZnOH^{+}$ ions, or as Zn-organic ligand complexes [13,14]. The depleted soil solution is replenished with Zn via desorption or dissolution from the soil matrix, followed by dispersion by diffusion [15]. Desorption from the soil matrix is controlled by soil pH, salinity, and clay, carbonate, and organic matter contents [15]. To promote Zn dissolution plants and microorganisms exude soil-acidifying protons and Zn-chelating low-molecular-weight organic compounds [16,17]. These Zn mobilizing activities of roots and microorganisms are promoted under iron (Fe), Zn, and other nutrient deficiencies [18]. The most well known classes of exuded organic Zn ligands are carboxylates, amides, and non-proteinaceous amino acids (= siderophores) [19,20]. A recent study on Zn uptake by wheat in the field [8] suggests that residues from the previous crop in crop rotations and thus possibly also addition of green manure to soil can raise the plant-available Zn in soil and promote Zn uptake in crop plants, mainly by contributing Zn-chelating dissolved organic carbon and amino acids.

Arbuscular mycorrhizal fungi, phylum Glomeromycota), because of their ubiquity and crucial involvement in plant nutrient acquisition and health, are a plant production-relevant group of root symbionts [21,22]. Their hyphae form effective, large-distance conduits for the transport of temporarily and spatially heterogeneously available nutrients of low mobility in soil, such as phosphate (P) and Zn [23]. Crop residue or green manure addition to soil and sporadic rain and irrigation may create nutrient flushes that crop plants with root systems extended by extraradical AMF hyphae may much more effectively utilize [24]. Arbuscular mycorrhizal fungi, whether indigenous or deliberately introduced as inoculants, were repeatedly reported to benefit Zn uptake in crop plants [9,10]. While the phenomenon of increased plant Zn uptake of mycorrhized as opposed to non-mycorrhized plants has already often been demonstrated [10], we still lack information about controlling factors [8] and the possible reinforcing effect of combined enrichment of soils with organic matter and mineral Zn fertilizer.

Here we report findings of a cross-factorial pot experiment on Zn nutrition of bread wheat involving the enrichment of calcareous soil with green manure of red clover and sunflower and inoculating plants with a non-indigenous AMF strain. For comparison and to study its interactions with the other experimental factors also a water-soluble mineral Zn fertilizer ($ZnSO_4$) was applied. To compare the effect of all naturally occurring AMF and possibly associated other soil microorganisms, the soil of half of the experimental units was γ-irradiated and re-inoculated with an AMF-free microbial filtrate of a water suspension of non-sterilized soil. An AMF inoculation treatment was further included in an attempt to assess the potential of this rapidly growing technology and for controlling against non-AMF mediated biotic effects by the soil γ-irradiation and microbial re-inoculation procedure. We predicted that 1) green manure addition to soil improves Zn uptake by wheat, 2) green manure addition improves Zn uptake from mineral Zn fertilizer and 3) inoculation of wheat plants with a foreign AMF strain improves Zn uptake, particularly from soil whose indigenous AMF had been killed by γ-irradiation.

Materials and methods

Experimental design

A complete cross-factorial experiment with the factors green manure (3 levels: none, red clover, sunflower), mineral Zn fertilization (2 levels: none, $ZnSO_4$), γ-irradiation (2 levels: non-irradiated; γ-irradiated and re-inoculated with soil microbes, except AMF), and AMF inoculation (2 levels: none, inoculated) was set up with each treatment replicated five times. The 120 experimental pots were arranged randomly on a glasshouse bench.

Soil origin and physicochemical soil analyses

The soil was collected in September 2009 to a depth of 30 cm from an arable field at the Rudasht research station (32, 29/N, 52, 10/E, 1560 m above sea level), located 65 km south-east of Isfahan in central Iran [permission for soil sampling was given by the Isfahan Agricultural and Natural Resource Research Center and for its import to Switzerland by the Swiss Federal Department of Agriculture]. The soil was air-dried and passed through a 5 mm sieve, before being shipped to Switzerland, where 5 kg were further passed through a 2 mm sieve for physicochemical analyses.

This experimental soil was a Typic Haplocambid, according to the USDA soil taxonomy [25]. The field where the soil was sampled had been under agricultural use with periodical irrigation from a nearby river. The crop rotation prior to soil sampling comprised barley, wheat, sugar beet, and wheat, interrupted by dry fallow periods. The climate in central Iran is semi-arid with a long-term mean annual temperature of 16.8°C, a mean annual rainfall lower than 100 mm, and a potential annual evaporation higher than 2000 mm [26].

The soil texture was a silty clay loam based on the measurements by the hydrometer method [27] and according to the USDA soil classification scheme [6]. The pH was 7.9 and the electrical conductivity (EC) 23 dSm^{-1}, as determined by the procedure of Rhoades [28], using an ISE pH meter 720A (Mettler Toledo LLC, Columbus, OH, USA) and an E- 518 conductometer (Metrohm Herisau, Switzerland), respectively. The soil was low in organic carbon [5 g $(kg\ soil)^{-1}$] and nitrogen [0.6 g $(kg\ soil)^{-1}$], as estimated by the methods of Walkley and Black [29] and on a Carlo Erba flash combustion CN-analyzer (NA1500, Carlo Erba, Milano, Italy), respectively. The soil contained 300 mg $CaCO_3$ $(kg\ soil)^{-1}$ as determined after neutralization with hydrochloric acid and back titration with sodium hydroxide [30].

The total Zn concentration of the soil was 80.2 mg Zn $(kg\ soil)^{-1}$, according to energy dispersive X-ray fluorescence spectrometry (XRF) measurements on a Spectra X-Lab 2000 instrument (SPECTRO Analytical Instruments GmbH, Kleve, Germany). Potentially plant-available Zn was extracted from 25 g of soil, using 50 ml of 0.05 M diethylene triamine-penta-acetic acid (DTPA), adjusted to pH 7.2 [31], and measured on an Agilent 7500 C inductively coupled plasma mass spectrometer (ICP-MS, Agilent Technologies, Santa Clara, California, USA). The amount of DTPA-extractable Zn was 0.45 mg Zn $(kg\ soil)^{-1}$.

Partial soil desalination, γ-irradiation and fertilization

Preliminary tests showed that wheat growth on the untreated soil was severely impaired by salinity stress (Forough Aghili et al., unpubl. data). Therefore, we flushed the soil in five rounds, with a volume ratio of 1:1 soil:deionized water each prior to the experiment. This treatment can be considered comparable to repeated flood irrigation, used by farmers at the field site prior to sowing. Soil washing lowered the EC from 23 to 6.7 dS m^{-1}, which is close to the threshold value of 5.9 dS m^{-1} for salinity-induced growth depression in wheat [32]. Flushing with water also slightly decreased the soil pH from 7.9 to 7.7. Analyses of the leachate indicated that large amounts of sodium, chlorine and sulfur had been removed, but virtually none of the total soil Zn (data not shown). The DTPA-extractable concentration of Zn in the soil even increased slightly from 0.45 to 0.47 mg Zn $(kg\ soil)^{-1}$. No spores of AMF could be retrieved from the leachate. The partially desalinated soil was again air-dried and sieved to a particle size of ≤5 mm.

Half of the soil was γ-irradiated with a dose of 25–75 kGy using a ^{60}Co source (Studer-Hard, Däniken, Switzerland, http://www.leoni-irradiation-services.com) in order to kill the indigenous soil microbes. To reintroduce soil microbes, but not AMF, 50 ml of a filtrate of a 2.5% (w:v) soil water suspension was added per kilogram soil. This microbial filtrate was prepared by passing the soil suspension twice through Whatman No 1 filter papers, which size-excluded AMF propagules. This procedure was recommended by Thompson [33] for studying the effect of AMF on Zn nutrition and is also utilized in many comparable studies to analyze the relative effect of different soil microorganisms on plant nutrition [34,35]. Indeed, soil γ-irradiation is currently the only way to create a control treatment, free of naturally occurring AMF, since mycorrhiza-defective mutants of wheat are not available [36].

The flushed soils (non-irradiated and γ-irradiated) received 400 mg K, 200 mg P, and 80 mg Fe per kilogram soil in the form of finely ground K_2SO_4, $CaHPO_4$ and $FeSO_4\ 7H_2O$ salts.

Preparation and application of ^{65}Zn-labeled green manure

We chose red clover (*Trifolium pratense L.*) and sunflower (*Helianthus annuus* L.) for the production of experimental green manures, because they are also often grown before wheat in the region of Isfahan [8]. Furthermore, Soltani et al. [8] found that compared to other possible green manure plants, red clover and sunflower introduced significant amounts of dissolved organic carbon into the soil solution, which was probably important for Zn solubilization.

Seeds of clover and sunflower were surface sterilized in 15% H_2O_2 for 15 min, rinsed with distilled water and germinated on moistened filter papers for two days. The most vigorously growing seedlings were transferred to a combined sand bed-hydroponics system, which consisted of compartmented seedling trays filled with silica sand of a particle size of 0.7–1.2 mm, partially immersed in a full-strength Hoagland nutrient solution [37]. After 28 days, the nutrient solution was exchanged for one that contained radioactive ^{65}Zn for an additional 26 days of vegetative plant development. While 30 liters of Hoagland nutrient solution were labeled with 5.55 MBq of ^{65}Zn for about 400 red clover plantlets, 40 l of Hoagland nutrient solution were labeled with 9.25 MBq of ^{65}Zn for about 140 sunflower plantlets. The ^{65}Zn radiotracer was added as carrier-free $ZnCl_2$, dissolved in weak HCl (Amersham plc, GE Healthcare, Little Chalfont, UK). The photoperiod in the glasshouse was set to 14 h, the minimal light intensity to 12 kLux, the temperature to 22°C during the day and 16°C at night, and the relative air humidity was allowed to vary between 30 and 40%.

At harvest, the entire shoots of red clover and only the leaf blades of sunflower, without petioles and stems, were collected, air-dried and finely ground to powder. The total tissue C and N concentrations were determined in 4 mg tissue samples using the above-mentioned CN-analyzer. The total P and Zn concentrations were measured in 100 mg subsamples of each of the green manures, using the above-mentioned ICP-MS after incineration at 550°C for 6 h and extracted with 3 ml of hot 14.4 M nitric acid (HNO_3). The concentration of radioactive ^{65}Zn in the green manures was measured by γ-spectrometry in 300 mg subsamples using IGC2 high purity germanium detectors (ORTEC Advanced Measurements Technologies Inc, Oak Ridge, Tennessee, USA). Selected characteristics of both green manures are presented in Table 1.

The green manures (Table 1) were homogenously mixed into the soils at a dosage of 4 g dry matter per kilogram of dry soil. This dosage corresponds approximately to an addition of 18 tons of dry matter per hectare considering a soil bulk density of 1.5 t m^{-3} and incorporation to a soil depth of 30 cm.

Mineral Zn fertilization

Half of the soil in each green manure × irradiation treatment combination was fertilized with water soluble Zn in the form of $ZnSO_4$ at a dosage of 4.7 mg Zn (kg soil)$^{-1}$, corresponding approximately to 19 kg Zn ha^{-1}.

AMF inoculation

Finally, half of all pots in each combination of the above treatments were inoculated with approximately 250 spores of a pot culture of the Swiss isolate BEG155 of the AMF species *Claroideoglomus claroideum* [38]. The spores of the inoculum were extracted by wet-sieving and decanting [39] and directly pipetted onto the roots of the transplanted wheat seedlings. This careful spore isolation and inoculation procedure prevented co-inoculation of many other microbes and introduction of organic matter in the form of root pieces, which would have been the case with so-called whole inoculum {= growth substrate, containing fungal spores, hyphae and colonized root fragments [40]}.

Data on AMF root colonization were not systematically collected in this experiment, since microscopic and molecular genetic quantification in a previous experiment with the same wheat cultivar and soil clearly showed a high infection pressure by the naturally occurring AMF (Forough Aghili et al., unpublished data). However, successful colonization of the roots by the AMF isolate BEG155 and by naturally occurring AMF were confirmed in a subset of all experimental treatments by quantitative polymerase chain reaction as in the previous study and additionally by phylogenetic sequence analysis of a 1.6 kb-long ribosomal DNA amplicon of the AMF colonizing the roots of a subset of all experimental units (Anouk Guyer et al., unpublished data).

Planting, growth conditions, harvest and mineral nutrient analyses

Wheat kernels {*Triticum aestivum* cv. Kavir, a cultivar widely grown in Iran [41]} were surface-sterilized with 15% H_2O_2 for 15 min and pre-germinated on moistened filter paper for three days prior to transplantation. Four seedlings were planted into each pot with 550 g soil and later thinned to the most vigorously growing two individuals. Nitrogen in the form of NH_4NO_3 was applied weekly, six times, amounting to 600 mg N (kg soil)$^{-1}$ until tillering.

Table 1. Mineral nutrient concentrations of the two types of applied green manure prepared from red clover and sunflower.

	Red clover (Mean ± SE)	**Sunflower (Mean ± SE)**
N mg (g DM*)$^{-1}$	43.3±0.6	58.4±0.4
C/N (mass ratio)	9.1±0.1	6.3±0.1
P mg (g DM)$^{-1}$	5.0±0.1	7.3±0.1
Zn µg (g DM)$^{-1}$	45.6±0.6	104.4±2.4
^{65}Zn kBq¶ (g DM)$^{-1}$	6.9±0.02	15.8±0.1

* DM: dry matter.

¶Radioactivity at the time of harvest of the experiment.

The means and associated standard errors (SE) of three replicate subsamples of the nitrogen (N) concentration, carbon (C) to N mass ratio, phosphorus (P) and zinc (Zn) concentrations, and ^{65}Zn concentration are listed.

The climatic conditions in the glasshouse were set to a photoperiod of 14 h with a minimum light intensity of 10 kLux, a 22/17°C day/night air temperature regime, and 40–45% relative air humidity. Soil moisture was kept at about 70% soil water holding capacity by daily watering to weight.

Harvest took place at grain maturity, 125 days after planting. The total weight of the grains and straw of the two plants in each pot was determined after drying to constant weight at 60°C for 72 h. Subsamples of 200 mg of finely ground wheat straw and grain were incinerated in porcelain crucibles at 550°C for 6 h and the ashes dissolved in 2 ml of boiling 13 M HNO_3. Zinc concentrations were measured using ICP-MS. Measurement accuracy and precision were ensured by regular inclusion of a certified plant reference standard in the analyses (hay powder IAEA-V-10, 24 mg Zn kg^{-1}) and by running blank samples with each sample batch. The concentration of radioactive ^{65}Zn was measured in the grains and straw by γ-spectrometry as described for the green manure above. Zinc uptake into roots could not be determined because it was not possible to recover the roots quantitatively from the clay-rich soil. Soil pH and the concentrations of total N and DTPA-extractable soil Zn were measured at the end of the experiment, using the above-described analytical procedures.

Calculations and data analyses

Plant Zn uptake from green manure, soil and mineral Zn fertilizer. Plant Zn uptake was calculated as the product of Zn concentration and biomass for both the grain and straw fractions and summed to obtain the amount of Zn contained in the entire aboveground biomass. All radioactivity measurements were related to the harvest time, taking the radioactive decay into account (^{65}Zn half-life time: 243.9 d). The percentage of Zn in the aboveground biomass derived from the green manure [Zn_{dgm} (%)] was calculated as:

$$Zn_{dgm}(\%) = 100 \times SA_p \times (SA_{gm})^{-1} \quad (1)$$

where SA_p [Bq (mg Zn^{-1})] is the specific activity (SA) of Zn in the aboveground biomass and SA_{gm} the specific activity of Zn in the added green manure. The specific activity of Zn in the plant was calculated as:

$$SA = [^{65}Zn] \times [Zn]^{-1} \quad (2)$$

where [^{65}Zn] stands for the concentration of radioactivity per kilogram of plant material [Bq (kg DM)$^{-1}$] and [Zn] stands for the concentration of total Zn in the plant [mg Zn (kg DM)$^{-1}$].

Using the equations 1 and 2, the amount of Zn taken up to the aboveground biomass from green manure [Zn_{dgm}, μg Zn (kg soil)$^{-1}$] was calculated as:

$$Zn_{dgm} = (Zn_{dgm} \times Zn_{upt}) \times 10^{-2} \quad (3)$$

where Zn_{upt} is the amount of Zn taken up to the aboveground plant biomass per kilogram of soil [μg Zn (kg soil)$^{-1}$].

Knowing Zn_{dgm}, it is possible to calculate the percentage of Zn recovered by the aboveground plant biomass from the added green manure [Zn_{rec_gm} (%)]:

$$Zn_{rec\_gm} = 100 \times Zn_{dgm} \times Zn_{added\_gm}{}^{-1} \quad (4)$$

where Zn_{added_gm} is the amount of Zn added with the green manure [μg Zn (kg soil)$^{-1}$].

Knowing Zn_{dgm}, it is also possible to calculate the amount of Zn in the plant that is derived from non-labeled Zn sources in soil, i.e. in the non-fertilized soil from the native soil Zn pool and in the mineral Zn-fertilized soil from the native soil Zn pool as well as the applied $ZnSO_4$ [Zn_{dsoil}, μg Zn (kg soil)$^{-1}$]:

$$Zn_{dsoil} = Zn_{upt} - Zn_{dgm} \quad (5)$$

Statistical analyses. The results were analyzed by fixed factor four-way or three-way analyses of variance (ANOVA) with two levels (γ-irradiated and non-irradiated) of 'soil γ-irradiation', three levels (none, red clover and sunflower) of 'green manure addition', two levels [0 and. 4.7 mg Zn (kg soil)$^{-1}$] of 'mineral Zn fertilization', and two levels (with and without inoculation with *C. claroideum*) of 'AMF inoculation'. The interaction terms were sequentially removed from the initially complete ANOVA model when not significant (backward elimination). In addition, a one-way ANOVA was carried out on the saved residuals of a three-way ANOVA with the factors 'soil γ-irradiation', 'mineral Zn fertilization' and 'AMF inoculation' in order to specifically analyze the effect of green manure addition. Compliance of data dispersion with ANOVA model assumptions was verified using Shapiro-Wilcoxon's normality and Levene's equal variance tests. If necessary, data were $\log_{10}$ or squareroot transformed. All statistical analyses were carried out using the software package SPSS version 17 (SPSS Inc., Chicago, Ill inois, USA). Means of factor levels were compared by least-significant-difference (LSD) tests at the significance level of $p<0.05$. The figures were prepared in SigmaPlot version 12 [Systat Software Inc. (SSI), San Jose, California, USA].

Results

Plant biomass and grain nitrogen concentration

The grain yield of the experimental wheat plants was not significantly affected by any of the treatments (Figure 1A, Table 2), but the large variance may have concealed a similar γ-irradiation effect as on the total aboveground biomass (Figure 1B). The total aboveground biomass (= straw + grains) was significantly larger for the wheat plants grown in γ-irradiated than non-irradiated soil ($F_{1,96} = 10.45$, $p = 0.002$). Inoculation with *C. claroideum* had no significant effect on aboveground biomass production (Table 2).

Both types of green manure significantly increased grain N concentration ($F_{2,70} = 31.41$, $p<0.001$, Table 2), which ranged between 30 and 45 g N (kg DM)$^{-1}$ in the plants grown on γ-irradiated soil and between 25 and 37 g N (kg DM)$^{-1}$ in the plants grown on non-irradiated soil (Figure 2). Soil γ-irradiation also led to a significant increase in grain N concentrations ($F_{1,70} = 10.04$, $p = 0.003$, Table 2). Inoculation with *C. claroideum* and $ZnSO_4$ addition had no significant effects on grain N concentration (Table 2).

Grain Zn concentration and Zn uptake to grains and aboveground biomass

Grain Zn concentration increased significantly with application of mineral Zn fertilizer ($F_{1,70} = 50.96$, $p<0.001$) and green manure ($F_{2,70} = 13.48$, $p<0.001$), but was not affected by inoculation of *C. claroideum* and soil γ-irradiation (Figure 3A, Table 2). The lowest grain Zn concentration [20±1.64 mg Zn (kg DM)$^{-1}$] was measured in the treatment with neither mineral Zn fertilization, nor addition of green manure, while the highest grain Zn

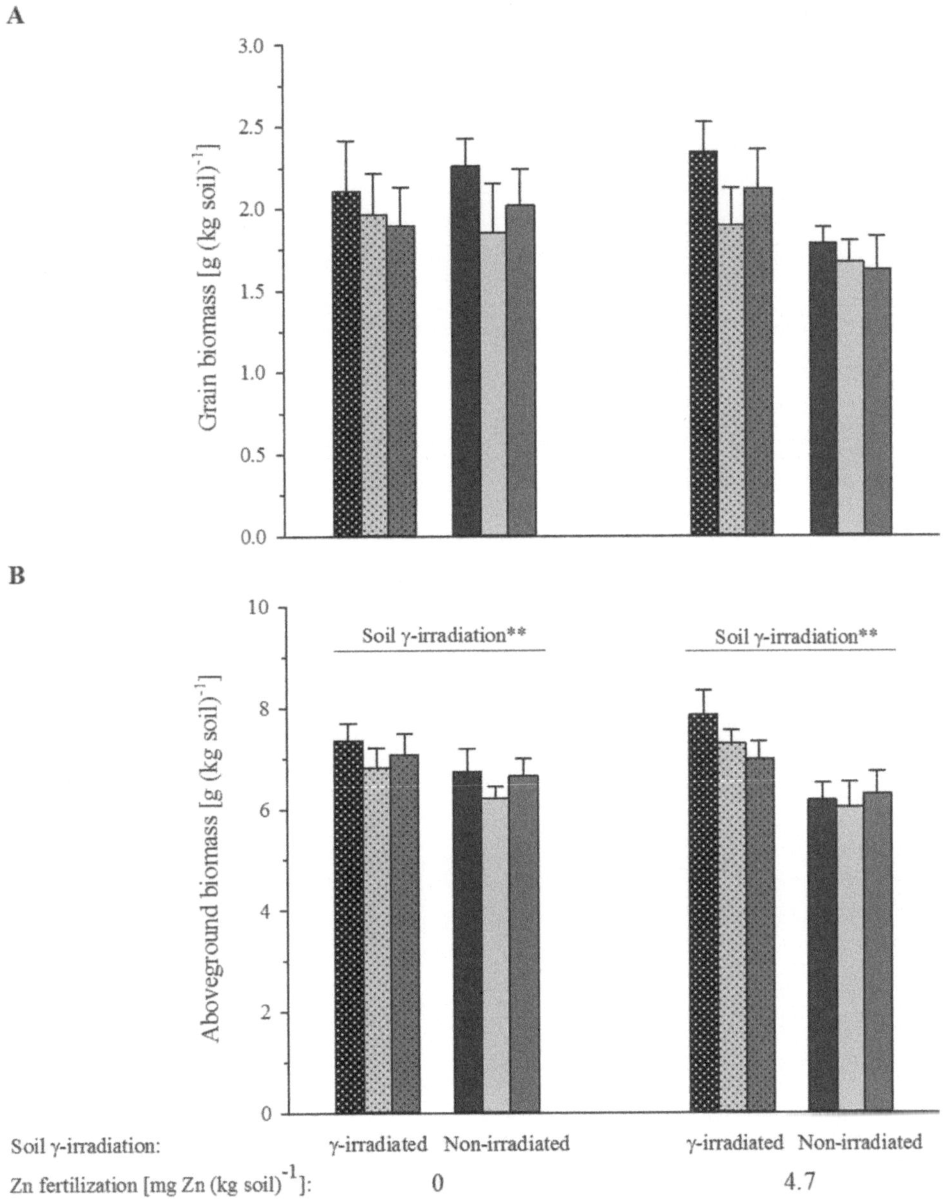

Figure 1. Grain yield (A) and aboveground biomass (B) of bread wheat grown in a calcareous soil. Black bar filling denotes the treatments without any added green manure, light grey bar filling the treatments in which the soil was amended with green manure of red clover, and dark grey bar filling those treatments in which green manure of sunflower was added to the soil. Bar hatching highlights the treatment missing the naturally occurring arbuscular mycorrhizal fungi after γ-irradiating the soil. Bars show mean values and associated standard errors of five experimental replicates. The significances (**, p<0.01) of soil γ-irradiation from a four-factorial analysis of variance are indicated. For full statistical details see Table 2.

concentration [54±2.32 mg Zn $(kg\ DM)^{-1}$] was found in the treatment with combined application of $ZnSO_4$ fertilizer and green manure of sunflower (Figure 3A).

The highest Zn concentration in the aboveground biomass [46.1±1.43 mg Zn $(kg\ DM)^{-1}$] was found in plants grown on non-irradiated soil amended with green manure of sunflower in combination with $ZnSO_4$ (Figure 3B). The lowest aboveground Zn concentration was observed in plants grown on γ-irradiated soil that had neither been fertilized with mineral Zn, nor amended with green manure (Figure 3B). Application of mineral Zn fertilizer reduced the differences between the aboveground biomass Zn concentrations of those plants that had received green manure and those that had not, resulting in a statistically significant interaction between green manure addition and mineral Zn fertilization ($F_{2,96} = 6.24$, $p < 0.01$; Table 2). Inoculation with *C. claroideum* had no significant effect on the concentration of Zn in the aboveground biomass (Table 2).

The amount of Zn taken up from the native soil Zn pool and mineral Zn fertilizer that was allocated to grains ranged between 46 and 88 μg Zn $(kg\ soil)^{-1}$, while the total amount of Zn accumulated in the aboveground biomass ranged between 117 and 367 μg Zn $(kg\ soil)^{-1}$ (Table 3). The lowest values were again observed in the absence of mineral Zn fertilization and green manure addition, while the highest values were observed after

Table 2. Results of fixed factor four-way analyses of variance (ANOVA) on plant and soil parameters.

ANOVA model	Grain biomass		Aboveground biomass		Grain N concentration		Grain Zn concentration		Aboveground Zn concentration		Zn uptake into grains	
Source of variance	**F value**	**df**	**F value**	**df**	**F value**	**df**	**F value**	**df**	**F value**	**df**	**F value**	**df**
Full model	0.29^{ns}	23	2.47^{*}	23	4.45^{***}	23	5.92^{***}	23	8.91^{***}	23	2.78^{**}	23
Green manure (A)	0.33^{ns}	2	0.26^{ns}	2	31.41^{***}	2	13.48^{***}	2	13.71^{***}	2	1.98^{ns}	2
Zn fertilization (B)	0.19^{ns}	1	0.12^{ns}	1	0.09	1	50.96^{***}	1	125.23^{***}	1	40.52^{***}	1
Soil γ-irradiation (C)	0.20^{ns}	1	10.45^{**}	1	10.04	1	2.99^{ns}	1	9.15^{**}	1	0.17^{ns}	1
Inoculation (D)	0.76^{ns}	1	0.13^{ns}	1	2.97	1	1.95^{ns}	1	1.04^{ns}	1	0.95^{ns}	1
A×B	r	2	r	2	R	2	12.71^{**}	2	6.24^{**}	2	12.71^{***}	2
A×C	r	2	0.61^{ns}	2	0.98^{ns}	2	3.36^{*}	2	4.41^{*}	2	1.22^{ns}	2
ANOVA model	**Zn uptake into aboveground biomass**		Zn_{dgm}		Zn_{rec_gm}		Zn_{dsoil}		**DTPA-extractable soil Zn**			
Source of variance	**F value**	**df**	**F value**	**df**	**F value**	**df**	**F value**	**df**	**F value**	**df**		
Full model	4.26^{***}	23	5.95^{***}	15	8.61^{***}	15	6.15^{***}	23	16.40^{***}	23		
Green manure (A)	4.13^{*}	2	39.92^{***}	2	72.25^{***}	2	3.50^{*}	2	85.22^{***}	2		
Zn fertilization (B)	53.49^{***}	1	21.23^{***}	1	22.31^{***}	1	77.13^{***}	1	168.84^{***}	1		
Soil γ-irradiation (C)	0.01^{ns}	1	5.21^{*}	1	8.85^{**}	1	1.14^{ns}	1	0.64^{ns}	1		
Inoculation (D)	2.34^{ns}	1	8.25^{**}	1	4.82^{*}	1	1.26^{ns}	1	2.10^{ns}	1		
A×B	0.49^{ns}	2	2.12^{ns}	2	0.13^{ns}	2	2.70^{ns}	2	2.17^{ns}	2		
A×C	0.44^{ns}	2	0.63^{ns}	2	4.64^{*}	2	2.95^{ns}	2	0.54^{ns}	2		

DTPA: diethylene triamine-penta-acetic acid.

na: not applied, r: removed from the statistical model, ns: not significant, *, $p<0.05$; **, $p<0.01$; ***, $p<0.001$.

The measurements of the plant parameters {grain biomass, total above ground biomass, grain N concentration, grain Zn concentration, aboveground Zn concentration, Zn uptake to grains, Zn uptake to the aboveground biomass, Zn derived from green manure in the aboveground biomass [Zn_{dgm}], proportion of the Zn recovered in the above ground biomass [Zn_{rec_gm}] of the Zn added with the green manure Zn derived from the soil and fertilizer in the aboveground biomass [Zn_{dsoil}]}and soil parameter (DTPA-extractable Zn from soil) were made at grain maturity. Effect sizes of the experimental factors and their significant interactions are indicated as F-values alongside with the statistical significance level. The interactions of the factors A×D, B×C, B×D, C×D, A×B×C, A×B×D, A×D×C, B×C×D, A×B×C×D were not significant in any of the 11 ANOVAs and thus these results are not listed.

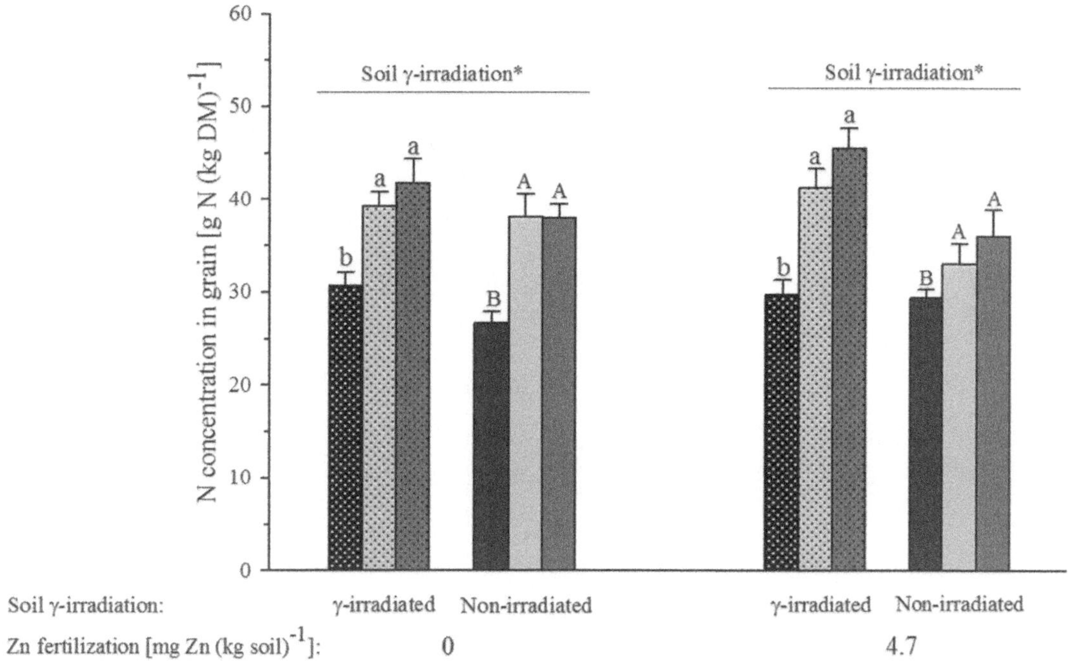

Figure 2. Grain nitrogen (N) concentration of wheat grown in a calcareous soil. Bar filling denoting the different green manure addition treatments is the same as in Figure 1. Bars show mean values and associated standard errors of five experimental replicates. The significances (*, $p<0.05$) of the effect of soil γ-irradiation from a four-factorial analysis of variance are shown. Different letters indicate statistical differences of separate least significant difference tests at $p<0.05$ for the different green manure addition treatments within the combinations of soil γ-irradiation and mineral Zn fertilization. For full statistical details see Table 2.

addition of mineral Zn fertilizer and green manure of sunflower. Adding green manure to the soil tended to increase Zn accumulation in the grains, when no mineral Zn fertilizer had been applied, but this effect was not statistically significant (Table 2). Inoculation with *C. claroideum* had no significant effect on Zn uptake to the grains (Table 2).

Zn derived from green manure and both, soil and mineral Zn fertilizer, in the aboveground biomass

The Zn from the added green manure recovered in the aboveground biomass (Zn_{rec_gm}) ranged between 2.5% and 4.1% for red clover, and between 1.4% and 2.7% for sunflower (Figure 4). These Zn recoveries from the two types of green manures differed significantly across all treatments ($F_{1,79}=72.25$, $p<0.001$). Mineral Zn fertilization lowered Zn_{rec_gm} for both forms of green manure ($F_{1,64}=22.31$, $p<0.001$, Figure 4 & Table 2), as did soil γ-irradiation ($F_{1,64}=8.85$, $p<0.01$). The recovered Zn fraction from sunflower residues was also reduced by the inoculation of *C. claroideum* when also mineral Zn fertilizer was applied ($F_{1,18}=10.84$, $p<0.01$; Figure 4).

The amount of Zn in the aboveground biomass derived from green manure (Zn_{dgm}) ranged between 4.5 μg Zn (kg soil)$^{-1}$ in the treatment of combined soil γ-irradiation, mineral Zn fertilization, and addition of green manure of red clover, and 11 μg Zn (kg soil)$^{-1}$ in the treatment with green manure of sunflower but without mineral Zn fertilization, soil γ-irradiation, and AMF inoculation (Table 3). The amount of plant Zn derived from green manure of sunflower (Zn_{dgm}) was consistently higher than that derived from red clover ($F_{1,64}=39.92$, $p<0.001$; Tables 2 & 3).

The amount of Zn derived from the native soil Zn pool (Zn_{dsoil}) was significantly increased with green manure addition, when no $ZnSO_4$ was applied ($F_{2,96}=3.50$, $p=0.034$, Table 2 & Figure 5). This apparent soil Zn mobilization was more pronounced for green manure produced from sunflower than from red clover (Figure 5). Under conditions of co-application of $ZnSO_4$, the effect of green manure addition was, however, variable and weak and statistically not significant. There was also no significant effect of γ-irradiation, nor any effect by the non-indigenous *C. claroideum* inoculant on the total amount of Zn derived from soil and mineral Zn fertilizer, either (Table 2).

Changes in DTPA-extractable Zn from soil, total N and pH in soil

The concentration of DTPA-extractable Zn from soil measured after plant harvest was significantly higher in the treatments with green manure than without green manure ($F_{1,96}=169$, $p<0.001$). Application of mineral Zn fertilizer also significantly increased DTPA-extractable Zn ($F_{2,96}=85.2$, $p<0.001$; Tables 2 & 3). There was a highly significant correlation between the total Zn taken up into the aboveground biomass (Y) and the DTPA-extractable Zn concentration in the soil (X) according to a double reciprocal regression model [$Y=1/(0.004+2.05/X)$] with $r^2=0.56$ and $p<0.0001$.

Green manure addition slightly decreased soil pH, independently of the soil γ-irradiation treatment ($F_{2,96}=139.4$, $p<0.001$; Table 4). The total soil N content was higher at harvest in the treatments that received green manure than in those that did not ($F_{2,96}=321.7$, $p<0.001$, Table 2 & 4).

Discussion

The results confirmed our first prediction, because green manure addition to soil increased Zn uptake by bread wheat. The findings, however, did not unequivocally support our second

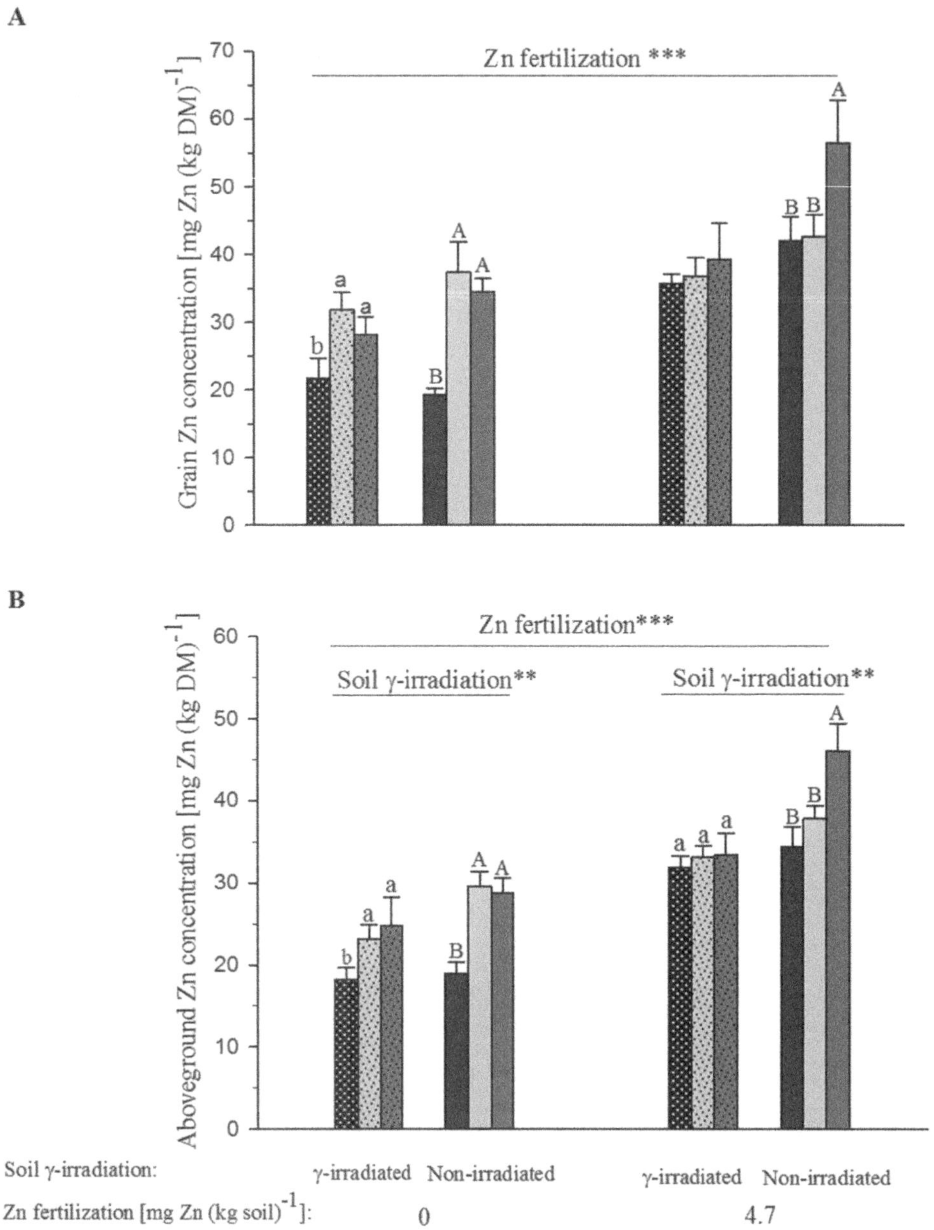

Figure 3. Grain (A) and aboveground (B) zinc (Zn) concentrations of wheat grown in a calcareous soil. Bar filling denoting the different green manure addition treatments is the same as in Figure 1. Bars show mean values and associated standard errors of five experimental replicates. The significances (**, $p<0.01$; ***, $p<0.001$) of the effect of mineral Zn fertilization, soil γ-irradiation of fixed factor four-way analyses of variance are shown. Different letters indicate statistical differences of separate least significant difference tests at $p<0.05$ for the different green manure addition treatments within the combinations of soil γ-irradiation and mineral Zn fertilization. For full statistical details see Table 2.

prediction, since utilization of the applied mineral Zn fertilizer by the plants did not improve in response to green manure addition. Also our last prediction that inoculation of a non-indigenous AMF strain would improve Zn transfer from soil to plants was not confirmed, not even in AMF- and thus competitor-free soil after removal of naturally occurring AMF by soil γ-irradiation. These two later findings appear to be related to the fact that Zn availability in the studied soil was neither limiting plant growth, nor yield.

In agreement with previous studies [43,44], we observed a clear increase in grain Zn concentration after application of mineral Zn fertilizer. However, application of mineral Zn fertilizer did neither translate into increased plant growth, nor higher grain yields, showing that Zn availability was not the primary plant growth-limiting factor in this soil. Sadeghzadeh [42] reports DTPA-extractable Zn levels below 0.6 mg Zn (kg soil)$^{-1}$ to limit wheat growth in calcareous soil. Since our soil contained 0.47 mg Zn (kg soil)$^{-1}$ and application of mineral Zn fertilizer did not promote wheat growth, it appears that the threshold for growth-limitation of soil Zn concentration may even be lower, although Zn uptake can obviously be raised by application of mineral Zn fertilizer.

Table 3. Effects of green manure addition and $ZnSO_4$ application to soil, soil γ-irradiation and inoculation of a non-indigenous arbuscular mycorrhizal fungus (AMF) on plant and soil parameters.

Green manure addition	Soil γ-irradiation	Zn added with $ZnSO_4$	Zn added with green manure	AMF inoculation	Zn uptake into aboveground biomass	Zn_{dgm}	Zn_{dsoil}	DTPA-extractable soil Zn
		µg Zn (kg soil)$^{-1}$			µg Zn (kg soil)$^{-1}$	µg Zn (kg soil)$^{-1}$		µg Zn (kg soil)$^{-1}$
None	**γ-irradiated**	0	0	no	159.3±12.4		159.3±12.4	450± 20
	γ-irradiated	0	0	yes	127.6± 6.9		127.6± 6.9	440± 32
	Non-irradiated	0	0	no	117.4± 13.2		117.4± 13.2	460± 35
	Non-irradiated	0	0	yes	127.2± 10.4		127.2± 10.4	480± 42
	γ-irradiated	4700	0	no	251.6± 16.6		251.6± 16.6	2150± 100
	γ-irradiated	4700	0	yes	227.3± 14.3		227.3± 14.3	1960± 93
	Non-irradiated	4700	0	no	206.6± 13.2		206.6± 13.2	2180± 110
	Non-irradiated	4700	0	yes	228.4± 27.5		228.4± 27.5	1990± 160
Red clover	**γ-irradiated**	0	183	no	152.9± 21.4	5.56± 0.59	147.4± 21.3	1220± 100
	γ-irradiated	0	183	yes	156.9± 25.3	5.86± 1.03	151.1± 19.6	1500± 120
	Non-irradiated	0	183	yes	205.0± 18.6	8.52± 0.35	196.5± 18.4	2010± 70
	Non-irradiated	0	183	yes	176.0± 23.8	6.10± 0.42	169.9± 20.2	2430± 110
	γ-irradiated	4700	183	no	224.0± 34.1	4.51± 0.39	219.5± 34.1	5050± 120
	γ-irradiated	4700	183	yes	243.9± 10.6	4.93± 0.38	239.0± 12.6	3550± 74
	Non-irradiated	4700	183	no	202.9± 25.2	5.32± 0.52	197.6± 23.8	4430± 32
	Non-irradiated	4700	183	yes	226.7± 22.4	6.11± 0.54	220.6± 21.5	3950± 92
Sunflower	**γ-irradiated**	0	418	no	227.9± 39.6	10.32± 1.39	217.6± 31.4	3310± 66
	γ-irradiated	0	418	yes	128.9± 11.4	8.38± 0.96	120.6± 22.6	1750± 120
	Non-irradiated	0	418	no	189.9± 14.3	11.05± 0.87	178.9± 16.4	3230± 86
	Non-irradiated	0	418	yes	174.9± 30.2	9.45± 1.61	165.5± 20.7	1950± 33
	γ-irradiated	4700	418	no	258.9± 25.8	8.44± 0.64	250.5± 41.3	3850± 82
	γ-irradiated	4700	418	yes	205.8± 20.4	5.94± 0.75	200.0± 19.5	4990± 36
	Non-irradiated	4700	418	no	367.0± 40.2	8.5± 0.64	358.5± 37.4	4540± 120
	Non-irradiated	4700	418	yes	227.9± 24.5	5.97± 0.80	222.0± 26.2	4680± 130

DTPA: diethylene-triamine-penta-acetic acid
Mean values and associated standard errors of five experimental replicates of Zn uptake to the aboveground biomass, Zn in the aboveground biomass derived from green manure (Zn_{dgm}), Zn derived from soil and fertilizer (Zn_{dsoil}) and DTPA-extractable Zn from soil at grain maturity are listed.

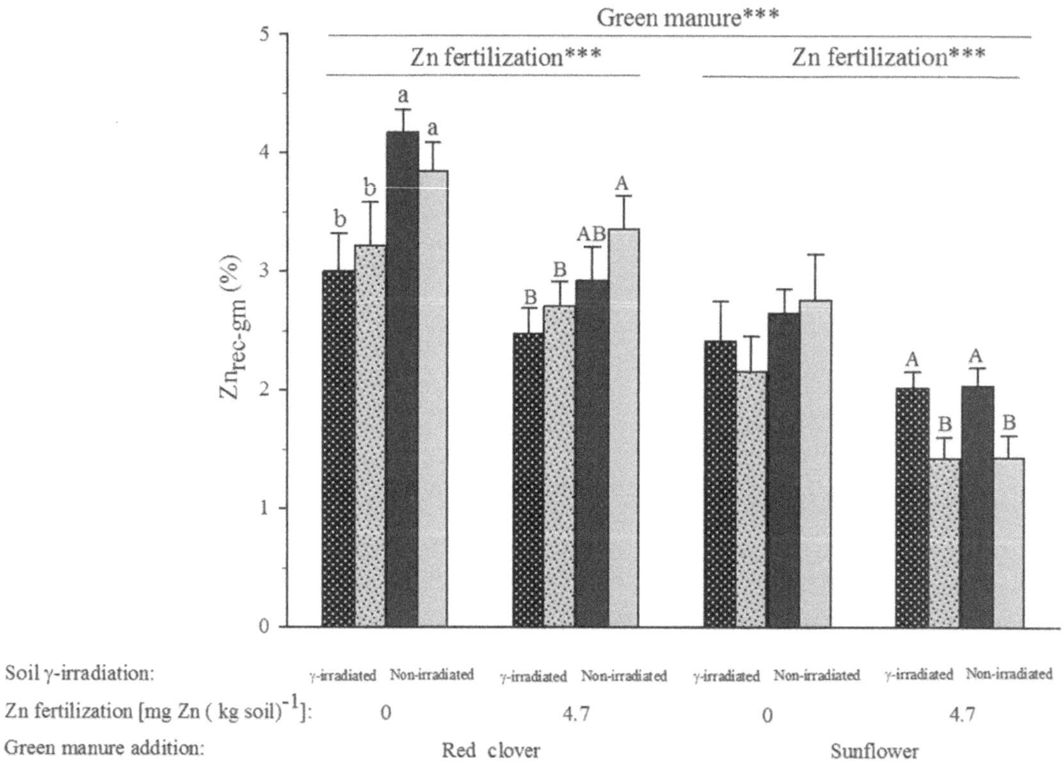

Figure 4. Percentage of zinc (Zn) recovered from green manure ($Zn_{rec\text{-}gm}$) in the aboveground wheat biomass. Black bar filling denotes treatments with arbuscular mycorrhizal fungus (AMF) inoculation and light grey filling treatments without AMF inoculation. Bar hatching highlights treatments in which the naturally occurring AMF were missing after γ-irradiating the soil. Bars show mean values and associated standard errors of five experimental replicates. The significances (***, $p<0.001$) of the effect of green manure addition and mineral Zn fertilization from a fixed factor four-way analysis of variance are indicated above the bars. For full statistical details see Table 2.

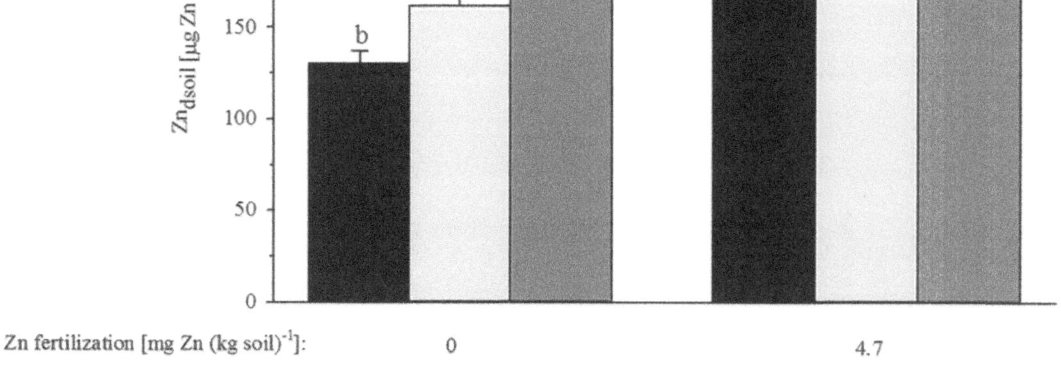

Figure 5. Zinc (Zn) uptake from soil to the aboveground biomass of bread wheat grown in a calcareous soil after addition of two types of green manure. The Zn taken up from soil either originated from the native soil Zn pool or from soil as well as applied $ZnSO_4$ fertilizer. Bar filling denoting the different green manure addition treatments is the same as in Figure 1. Bars show mean values and associated standard errors of five experimental replicates. The significance (***, $p<0.001$) of the effect of mineral Zn fertilization from a fixed factorial four-way analysis of variance is indicated above the bars. Different letters indicate statistical differences of separate least significant difference tests at $p<0.05$ for the different green manure addition treatments within the different Zn fertilization levels. For full statistical details see Table 2.

Table 4. Total soil nitrogen (N) concentration and pH in water at wheat harvest.

Soil	Green manure	Total N (g kg^{-1})	pH (H_2O)
		(Mean ± SE)	(Mean ± SE)
Non-irradiated	None	0.55 ± 0.02^{B}	7.75 ± 0.03^{A}
	+ Red clover	0.87 ± 0.02^{A}	7.50 ± 0.02^{B}
	+ Sunflower	0.88 ± 0.02^{A}	7.53 ± 0.01^{B}
γ-irradiated	None	0.63 ± 0.01^{b}	7.64 ± 0.02^{a}
	+ Red clover	0.89 ± 0.02^{a}	7.46 ± 0.01^{b}
	+ Sunflower	0.96 ± 0.02^{a}	7.48 ± 0.01^{b}

DTPA: diethylene-triamine-penta-acetic acid
The values represent the means and associated standard errors (SE) of 20 experimental units. Different superscript letters of the same type following the SE indicate statistical difference among the means of the three green manure treatments within each soil γ-irradiation treatment at $p<0.05$, according to least significance difference tests.

Mixing green manure to soil increased Zn uptake by wheat via Zn mobilizing effects

The observed increase in Zn uptake from the native soil Zn pool after green manure addition could originate from changes to the soil conditions, root growth, and/or plant physiology. Only addition of green manure to the soil increased the DTPA-extractable Zn from 0.47 to 0.5 or 2 mg Zn (kg soil)$^{-1}$, when green manure of red clover and sunflower were added, respectively. Combining $ZnSO_4$ application with green manure addition to soil increased the DTPA-extractable Zn concentration even further, namely up to 3.8 mg Zn (kg soil)$^{-1}$ when red clover was used and up to 5.0 mg Zn (kg soil)$^{-1}$ when sunflower was used, respectively (Table 3). We thus can conclude that the lack of a plant growth response to application of mineral Zn fertilizer (Figure 1A&B) must be explained by an alternative plant growth-limiting factor and not the >0.6 mg Zn (kg soil)$^{-1}$ of DTPA-extractable Zn [42].

The larger increase in DTPA-extractable soil Zn in response to addition of green manure of sunflower than such of red clover (Table 3) could have been due to the Zn-richer green manure of sunflower than the green manure of red clover. For production of the green manure of sunflower, but not that of red clover, only leaf laminae were used, which could explain the higher mineral nutrient concentrations of it (Table 1). Moreover, the relatively young age (54 day) and bigger sunflower than red clover seeds with own nutrient reserves may have further contributed to nutrient concentration differences of the two types of green manure (Table 1), translating into differences in plant Zn uptake.

There may have been further co-nutritional or root- and microbial growth-mediated Zn uptake stimulation, because the green manure produced of sunflower was also richer in N than that of red clover (Table 1). The green manure of the leaf laminae of the well N-supplied sunflower plants could have released N during its decomposition and stimulated root growth and possibly the synthesis of Zn facilitating N-rich plant and microbial compounds. Elevated root and microbial exudation and necro-mass in response to green manure of sunflower may also explain the higher DTPA-extractable Zn concentrations of the soil and increased plant Zn uptake. Root- and microbe-derived organic acids, amides and (phyto-) siderophores may have chelated and thereby mobilized Zn for uptake by the wheat plants. More N released from the decomposing green manure of more N-rich sunflower than red clover material (Table1) may have supported the N-demanding biosynthesis of Zn chelating compounds [8,14,45].

Elevated root and microbial activity in response to green manure addition may further have lowered soil pH (Table 4) and thereby mobilized soil Zn. Similarly, stimulation of nitrification upon higher N inputs with the green manure of sunflower than that of red clover and thus release of protons into the soil solution may have raised Zn bioavailability in soil for uptake by the mycorrhized wheat plants.

All in all, the effectiveness of green manure addition to soil to release Zn from the native soil Zn pool for plant uptake was remarkable (Figure 5). This stimulatory effect on plant Zn uptake diminished, however, when additionally $ZnSO_4$ fertilizer was applied. This is probably, because the applied $ZnSO_4$ fertilizer introduced already a high amount of Zn, compared to the amount solubilized by chelating compounds released upon addition of green manure, reducing Zn mobilization form soil.

Inoculation of a non-indigenous AMF strain did not increase Zn uptake in wheat

A previous experiment (Forough Aghili, unpublished) with the same soil and wheat cultivar and additional analyses on samples of this experiment (Anouk Guyer, unpublished data) showed that the roots of the wheat plants were colonized by naturally occurring AMF of this soil. 1.6 kb-long ribosomal DNA sequences confirmed root colonization by the inoculated Swiss strain of *C. claroideum* of the wheat plants raised in γ-irradiated soil. We suspect that maladaptation of the AMF inoculant to the living conditions in the experimental soil [46] must have compromised symbiotic Zn acquisition by wheat. The inoculated AMF was originally isolated from Swiss arable soil, not calcareous, and with a loamy texture, a lower, near neutral pH, and much richer in organic matter and bioavailable nutrients [47] than the study soil here. The improved Zn recovery from green manure of sunflower after application of mineral Zn fertilizer by the plants inoculated with the foreign AMF strain points at possibly complex soil-root-microbe interactions beneficial to mycorrhized plants. Lack, or just minor positive effects by the non-indigenous AMF inoculant should draw our attention to the fact that AMF inoculants can not be *a priori* considered effective in assisting crop plants to take up Zn from soil when introduced to soils of different physicchemical and biological properties.

γ-Irradiation has been shown to modify the physicochemical soil properties to some degree [48,49]. However, we believe that this did not invalidate our results. While the increased biomass (Figure 1) and N uptake (Figure 2) of the wheat plants grown in γ-irradiated soil can probably be related to released N from decomposing microbial necromass [49], the magnitude of this effect was certainly much smaller than that by the N added with the green manures (Figure 2, Table 2). Likewise, the levels of dissolved organic compounds [48] must have risen in response to γ-irradiation, but again this effect must have been considerably smaller than that caused by mixing 0. 4% (w/w) finely ground green manure to soil. Therefore, following Thompson [33], and until a mycorrhiza symbiosis-defective wheat mutant becomes available, we see no better approach than using γ-irradiation for the study of the influence of entire natural AMF assemblages on plant Zn uptake from real arable soil.

The finding of higher grain and aboveground Zn concentrations in plants raised on native than γ-irradiated soil (Figure 3) points at no Zn uptake stimulatory effect by elevated levels of dissolved organic compounds [48] and N after soil γ-irradiation [49]. The generally increased Zn tissue concentration after addition of green manure to soil (Figure 3) and higher Zn recovery from less N- and Zn-rich green manure of red clover than more N- and Zn-rich green manure of sunflower point at considerable influences by C- rather than N-limited saprotrophic soil microbes on Zn uptake of mycorrhized bread wheat plants. Being obligate biotrophs, AMF rely on saprotrophs for remobilization of mineral nutrients from decomposing green manure that they may as soon as available take up and then possibly deliver to their host plants as observed for the Zn from the green manure of sunflower (Figure 4). Much higher Zn recovery from the Zn- and N-poorer green manure of red clover than the Zn-richer and N-poorer green manure of sunflower (Figure 4) points at faster decomposition of the earlier than latter and thus also different speeds in nutrient recycling from different green manures. However, the two types of green manure appear not to have differed in Zn mobilization from soil (Figure 5).

Native as opposed to γ-irradiated soil after back-addition of an AMF propagule-free microbial filtrate of a water suspension of living soil does certainly not only differ in AMF occurrence. Besides the physicochemical changes discussed above [49], largely stochastic epidemic population growth of saprophytic organisms must occur [35]. Therefore, importance of fungi and microbes conferring similar nutritional benefits to plants as AMF [50] may increase in absence of otherwise naturally occurring AMF and much reduced population sizes of most of other just re-introduced soil microbes. Microbial nutrient competition with plants [34] may be particularly harsh when microbial population growth after inoculation coincides with release of nutrients from new microbial necromass and freshly applied mineral fertilizer, as it must have occurred in the γ-irradiated control soil of this experiment. The re-introduced soil microbes in the filtrate of the soil suspension and the newly inoculated foreign AMF inoculant may thus have mainly sequestered Zn and other nutrients in their newly formed biomass in γ-irradiated soil [34]. This would further explain why the inoculated foreign AMF strain did not assist much in plant Zn uptake.

In summary, this study advanced our knowledge on possible agronomic practices to biofortify cereal grains with Zn. i) It showed that addition of readily decomposable N-rich green manure to calcareous soil can raise grain Zn concentration of bread wheat to levels approaching those of only $ZnSO_4$ fertilizer-fed wheat plants; ii) The study showed that combined addition of N-rich green manure and $ZnSO_4$ fertilizer can considerably raise grain Zn concentration; iii) Mixing green manure to the soil showed that the native soil Zn pool can contribute considerably to total Zn accumulation in the aboveground plant biomass even when no additional mineral Zn fertilizer had been applied to the soil; iv) Lastly, the study emphasizes that care should be taken when trying to utilize AMF inoculants for supporting crop plants in their Zn uptake, when the fungi are not adapted to the prevailing soil conditions.

Acknowledgments

The authors would like to thank Dr. Laurie P. Mauclaire Schönholzer, Thomas Flura and Max Ruethi for their assistance in running the analytical equipment and Simon Ineichen for practical help in sample processing. We acknowledge critical comments by two anonymous reviewers that helped to clarify the main message of this study.

Author Contributions

Conceived and designed the experiments: JJ FA EF. Performed the experiments: FA. Analyzed the data: FA EF HAG. Wrote the paper: FA HAG EF RS. Provided the AM fungal inoculant: JJ. Provided the plant seeds and the soil: AHK MA. Enabled gamma-countingL JE. Approved the final manuscript version: JJ FA EF HAG AHK MA JE RS.

References

1. Curtis BC (2002) Wheat in the world. In B. C. . Curtis, S. . Rajaram, H. Gomez Macpherson (eds.), Bread wheat: Improvement and production. Food and Agriculture Organization of the United Nations, Rome, Italy.
2. Black RE, Victora CG, Walker SP, Bhutta ZA, Christian P, et al. (2013) Maternal and child undernutrition and overweight in low-income and middle-income countries. Lancet 382: 427–451.
3. Welch RM, Graham RD (2004) Breeding for micronutrients in staple food crops from a human nutrition perspective. J Exp Bot 55: 353–364.
4. Cakmak I, Kalayci M, Kaya Y, Torun AA, Aydin N, et al. (2010) Biofortification and localization of zinc in wheat grain. J Agr Food Chem 58: 9092–9102.
5. Cakmak I (2008) Enrichment of cereal grains with zinc: Agronomic or genetic biofortification? Plant Soil 302: 1–17.
6. Khoshgoftar AH, Shariatmadari H, Karimian N, Khajehpour MR (2006) Responses of wheat genotypes to zinc fertilization under saline soil conditions. J Plant Nutr 29: 1543–1556.
7. White PJ, Broadley MR (2011) Physiological limits to zinc biofortification of edible crops. Front Plant Sci 2: 80.
8. Soltani S, Khoshgoftarmanesh AH, Afyuni M, Shrivani M, Schulin R (2013) The effect of preceding crop on wheat grain zinc concentration and its relationship to total amino acids and dissolved organic carbon in rhizosphere soil solution. Biol Fert Soils 50: 239–247.
9. Cavagnaro TR (2008) The role of arbuscular mycorrhizas in improving plant zinc nutrition under low soil zinc concentrations: a review. Plant Soil 304: 315–325.
10. Lehmann A, Veresoglou SD, Leifheit EF, Rillig MC (2014) Arbuscular mycorrhizal influence on zinc nutrition in crop plants – A meta-analysis. Soil Biol Biochem 69: 123–131.
11. Ryan MH, Angus JF (2003) Arbuscular mycorrhizae in wheat and field pea crops on a low P soil: increased Zn-uptake but no increase in P-uptake or yield. Plant Soil 250: 225–239.
12. Ryan MH, Derrick JW, Dann PR (2004) Grain mineral concentrations and yield of wheat grown under organic and conventional management. J Sci Food Agr 84: 207–216.
13. Diesing WE, Sinaj S, Sarret G, Manceau A, Flura T, et al. (2008) Zinc speciation and isotopic exchangeability in soils polluted with heavy metals. Eur J Soil Sci 59: 716–729.
14. Gramlich A, Tandy S, Frossard E, Eikenberg J, Schulin R (2013) Availability of zinc and the ligands citrate and histidine to wheat: does uptake of entire complexes play a role? J Agric Food Chem 61: 10409–10417.
15. Alloway BJ (2009) Soil factors associated with zinc deficiency in crops and humans. Environ Geochem Hlth 31: 537–548.
16. Gao X, Zhang F, Hoffland E (2009) Malate exudation by six aerobic rice genotypes varying in zinc uptake efficiency. J Environ Qual 38: 2315–2321.
17. Daneshbakhsh B, Khoshgoftarmanesh AH, Shariatmadari H, Cakmak I (2013) Phytosiderophore release by wheat genotypes differing in zinc deficiency tolerance grown with Zn-free nutrient solution as affected by salinity. J Plant Physiol 170: 41–46.
18. Dakora FD, Phillips DA (2002) Root exudates as mediators of mineral acquisition in low-nutrient environments. Plant Soil 245: 35–47.
19. Hoffland E, Wei CZ, Wissuwa M (2006) Organic anion exudation by lowland rice (*Oryza sativa L.*) at zinc and phosphorus deficiency. Plant Soil 283: 155–162.
20. Ueno D, Rombola AD, Iwashita T, Nomoto K, Ma JF (2007) Identification of two novel phytosiderophores secreted by perennial grasses. New Phytol 174: 304–310.

21. Parniske M (2008) Arbuscular mycorrhiza: the mother of plant root endosymbioses. Nat Rev Microbiol 6: 763–775.
22. Bonfante P, Genre A (2010) Mechanisms underlying beneficial plant-fungus interactions in mycorrhizal symbiosis. Nat Commun 1: 48.
23. Fitter AH (1991) Costs and benefits of mycorrhizas - implications for functioning under natural conditions. Experientia 47: 350–355.
24. Joner EJ, Jakobsen I (1995) Growth and extracellular phosphatase-activity of arbuscular mycorrhizal hyphae as influenced by soil organic-matter. Soil Biol Biochem 27: 1153–1159.
25. Anonymous (2010). Keys to soil taxonomy. Natural Resources Conservation Service, United States Department of Agriculture, Washington, D.C., 338 p.
26. Mostafazadeh-Fard B, Mansouri H, Mousavi SF, Feizi M (2009) Effects of different levels of irrigation water salinity and leaching on yield and yield components of wheat in an arid region. J Irrig Drain E-Asce 135: 32–38.
27. Gee GW, Bauder JW (1986) Particle size analysis. In: A Klute, editor editors. Methods of soil analysis. Madison, Wisconsin, USA: American Society of Agronomy. pp. 383–411.
28. Rhoades JD (1996) Salinity: Electrical conductivity and total dissolved salts. Methods of soil analysis. Madison, Wisconsin, USA: Soil Science Society of America. pp. 417–435.
29. Walkey A, Black IA (1934) An examination of the Degtjareff method for determinig soil organic matter and proposed modification of the chromic acid titration method. Soil Sci 37: 29–38.
30. Burt R (2004) Soil survey laboratory methods manual: soil survey investigations report Nebraska: United States Department of Agriculture. Natural Resources Conservation Servic.
31. Lindsay WL, Norvell WA (1978) Development of a DTPA soil test for zinc, iron, manganese, and copper. Soil Sci Soc Am J 42: 421–428.
32. Maas EV (1986) Salt tolerance of plants. Appl Agric Res 1: 12–25.
33. Thompson JP (1990) Soil sterilization methods to show VA-mycorrhizae aid P and Zn nutrition of wheat in vertisols. Soil Biol Biochem 22: 229–240.
34. Nazeri N, Lambers H, Tibbett M, Ryan M (2013) Do arbuscular mycorrhizas or heterotrophic soil microbes contribute toward plant acquisition of a pulse of mineral phosphate? Plant Soil 373: 699–710.
35. van de Voorde TFJ, van der Putten WH, Bezemer TM (2012) Soil inoculation method determines the strength of plant–soil interactions. Soil Biol Biochem 55: 1–6.
36. Barker SJ, Stummer B, Gao L, Dispain I, O'Connor PJ, et al. (1998) A mutant in *Lycopersicon esculentum* Mill. with highly reduced VA mycorrhizal colonization, isolation and preliminary characterisation. Plant J 15: 791–797.
37. Hogland DR, Arnon OD (1938) The water-culture method of growing plants without soil. University of California college of agriculture, agricultural experiment station Berkeley, California. USA.
38. Schenck NC, Smith GS (1982) Additional new and unreported species of mycorrhizal fungi (Endogonaceae) from Florida. Mycologia 74: 77–92.
39. Gerdemann JW, Nicolson TH (1963) Spores of mycorrhizal *Endogone* species extracted from soil by wet sieving and decanting. T Brit Mycol Soc 46: 235–244.
40. Cavagnaro TR, Smith FA, Hay G, Carne-Cavagnaro VL, Smith SE (2004) Inoculum type does not affect overall resistance of an arbuscular mycorrhiza-defective tomato mutant to colonisation but inoculation does change competitive interactions with wild-type tomato. *New Phytol* 161: 485–494.
41. Karami M, Afyuni M, Khoshgoftarmanesh AH, Papritz A, Schulin R (2009) Grain zinc, iron, and copper concentrations of wheat grown in central Iran and their relationships with soil and climate variables. J Agr Food Chem 57: 10876–10882.
42. Sadeghzadeh B (2013) A review of zinc nutrition and plant breeding. J Soil Sci Plant Nutr 13: 905–927.
43. Kalayci M, Torun B, Eker S, Aydin M, Ozturk L, et al. (1999) Grain yield, zinc efficiency and zinc concentration of wheat cultivars grown in a zinc-deficient calcareous soil in field and greenhouse. Field Crop Res 63: 87–98.
44. Rengel Z (1999) Zinc deficiency in wheat genotypes grown in conventional and chelator-buffered nutrient solutions. Plant Sci 143: 221–230.
45. Neilands JB (1995) Siderophores: Structure and function of microbial iron transport compounds. J. Biol. Chem 270: 26723–26726.
46. Oliveira RS, Vosatka M, Dodd JC, Castro PML (2005) Studies on the diversity of arbuscular mycorrhizal fungi and the efficacy of two native isolates in a highly alkaline anthropogenic sediment. Mycorrhiza 16: 23–31.
47. Jansa J, Mozafar A, Anken T, Ruh R, Sanders IR, et al. (2002) Diversity and structure of AMF communities as affected by tillage in a temperate soil. Mycorrhiza 12: 225–234.
48. Marschner B, Bredow A (2002) Temperature effects on release and ecological properties of dissolved organic matter (DOM) in sterilized and biologically active soil samples. Soil Biol Biochem 34: 459–466.
49. McNamara NP, Black HIJ, Beresford NA, Parekh NR (2003) Effects of acute gamma irradiation on chemical, physical and biological properties of soils. Appl Soil Ecol 24: 117–132.
50. Kariman K, Barker SJ, Jost R, Finnegan PM, Tibbett M (2014) A novel plant–fungus symbiosis benefits the host without forming mycorrhizal structures. New Phytol 201: 1413–1422.

6

Availability and Temporal Heterogeneity of Water Supply Affect the Vertical Distribution and Mortality of a Belowground Herbivore and Consequently Plant Growth

Tomonori Tsunoda*, Naoki Kachi, Jun-Ichirou Suzuki

Department of Biological Sciences, Tokyo Metropolitan University, Hachioji, Tokyo, Japan

Abstract

We examined how the volume and temporal heterogeneity of water supply changed the vertical distribution and mortality of a belowground herbivore, and consequently affected plant biomass. *Plantago lanceolata* (Plantaginaceae) seedlings were grown at one per pot under different combinations of water volume (large or small volume) and heterogeneity (homogeneous water conditions, watered every day; heterogeneous conditions, watered every 4 days) in the presence or absence of a larva of the belowground herbivorous insect, *Anomala cuprea* (Coleoptera: Scarabaeidae). The larva was confined in different vertical distributions to top feeding zone (top treatment), middle feeding zone (middle treatment), or bottom feeding zone (bottom treatment); alternatively no larva was introduced (control treatment) or larval movement was not confined (free treatment). Three-way interaction between water volume, heterogeneity, and the herbivore significantly affected plant biomass. With a large water volume, plant biomass was lower in free treatment than in control treatment regardless of heterogeneity. Plant biomass in free treatment was as low as in top treatment. With a small water volume and in free treatment, plant biomass was low (similar to that under top treatment) under homogeneous water conditions but high under heterogeneous ones (similar to that under middle or bottom treatment). Therefore, there was little effect of belowground herbivory on plant growth under heterogeneous water conditions. In other watering regimes, herbivores would be distributed in the shallow soil and reduced root biomass. Herbivore mortality was high with homogeneous application of a large volume or heterogeneous application of a small water volume. Under the large water volume, plant biomass was high in pots in which the herbivore had died. Thus, the combinations of water volume and heterogeneity affected plant growth via the change of a belowground herbivore.

Editor: Mari Moora, University of Tartu, Estonia

Funding: This research was supported partly by the Japan Society for the Promotion of Science (11J06132 to TT). The funder had no role in study design, data collection and analysis, decision to publish, or preparation to the manuscript. No additional external funding received for this study.

Competing Interests: The authors have declared that no competing interests exist.

* E-mail: ttsunoda@tmu.ac.jp

Introduction

Water availability and its temporal variability (hereafter, 'water heterogeneity') affect plant biomass growth [1,2]. Plant responses to water frequency vary depending on nutrient availability and soil water content [3–6]. Soil water status also affects soil biota [7], and interacts with soil biota to affect plant growth. Empirical studies are necessary to clarify the interactive effects of water heterogeneity on soil biota and hence on plant growth because of the important influence of soil biota on plant growth and community dynamics [7,8].

In light of the presence of herbivorous insects in the soil, it is important that we elucidate the effects of water availability and heterogeneity on plants [9], because these insects influence the abundance, species diversity, and succession of plants [10–14]. Erb and Lu [15] pointed out that heterogeneity of soil abiotic factors such as moisture and nutrient availability alters the effects of belowground herbivores on plants. The interactions between water availability, heterogeneity, and belowground herbivory are likely to play crucial roles in plant growth.

Water availability and frequency affect the vertical distribution of belowground herbivores, and thus plant growth, because soil insects move vertically in response to changes in soil water status [16,17]. Grubs (Coleoptera: Scarabaeidae) are distributed deep in the soil in response to drought and shallow in the soil in response to irrigation [18]. The carrot-fly larva *Psila rosae* feeds on roots 15 cm below the ground in semi-dry soil, whereas in moist soil it feeds 1 cm from the soil surface [19]. Wireworms (Coleoptera: Elateridae) are distributed deep in the soil in summer but in shallow soil after heavy rain [20]. These changes in the vertical distribution of belowground herbivores affect plant mortality and growth [21]. Therefore, the amount and heterogeneity of water supply that determine soil moisture levels are likely to affect the vertical distribution of belowground herbivores and thus plant growth.

Soil moisture is one of the most important factors affecting the survival and abundance of belowground herbivorous insects [16,17,22]. Soil dryness increases mortality of belowground herbivores [23–26], whereas in moist soil mortality either does not change [24,27,28] or increases [23,29]. These findings suggest that the mortality of belowground herbivores in response to extreme water events has the potential to affect plant growth in various ways.

Table 1. Number of replicates, dead plants, and dead belowground herbivores, and the total sample size in biomass analyses of combinations of experimental treatments.

Water volume	Water heterogeneity	Belowground herbivore	Number of replicates	Number of dead plant	Number of dead belowground herbivore	Sample size in biomass analyses
Small	Homogeneous	Control	9	0	NA†	9
		Bottom	9	0	3	6
		Middle	9	0	0	9
		Top	9	2	1	6
		Free	9	0	1	8
	Heterogeneous	Control	9	0	NA†	9
		Bottom	9	0	4	5
		Middle	9	0	2	7
		Top	9	0	1	8
		Free	9	0	4	5
Large	Homogeneous	Control	9	0	NA†	9
		Bottom	9	0	2	7
		Middle	9	0	6	3
		Top	9	0	4	5
		Free	9	2	3	4
	Heterogeneous	Control	9	0	NA†	9
		Bottom	9	0	2	7
		Middle	9	0	2	7
		Top	9	1	2	6
		Free	9	1	3	5

†NA means not available due to the control treatment.

We conducted a growth experiment to test the hypothesis that changes in the amount and heterogeneity of water supply alter the vertical distribution and mortality of belowground herbivores and thus affect plant growth. Under heterogeneous supply of small amount water, belowground herbivores will become distributed deep in the soil to avoid the dry surface soil and feed on the fine root tips, which will hardly affect plant growth. In contrast, under homogeneous supply of large volume of water, herbivores will become distributed shallow in the soil and detach root connection by grazing, which has impact on plant growth. Soil water status resulting from changes in water supply amount and heterogeneity will also determine the fate of belowground herbivores. Unless belowground herbivores are present, plant growth will be no longer restricted by belowground herbivory.

Materials and Methods

Study species

One seedling of *Plantago lanceolata* L. (Plantaginaceae) was grown in each pot, to which we added a third-instar larva, or grub, of the belowground herbivorous insect *Anomala cuprea* Hope (Coleoptera: Scarabaeidae). The short-lived perennial forb, *P. lanceolata*, is cosmopolitan and has a rosette growth form. Seeds of *P. lanceolata* were collected from a population of more than 30 plants on a floodplain of the Tama River in Tokyo (35°38′N, 139°23′E). *P. lanceolata* is not endangered or protected species, and no specific permissions were required for this location to collect seeds of *P. lanceolata*. Larvae of *A. cuprea* feed on various herbaceous species [30,31], including *P. lanceolata* (T.Tsunoda, personal observation). Grubs were grown from eggs laid in humus by adult *A. cuprea* collected from a floodplain of the Tama River (35°38′N, 139°23′E) in June and July 2012. *A. cuprea* is not endangered or protected species, and no specific permissions were required for this location to collect insects.

Experimental design

The growth experiment was conducted from September to October 2012 in a plastic film greenhouse under natural sunlight in the experimental garden of Tokyo Metropolitan University (Hachioji, Tokyo; 35°37′N, 139°23′E). The mean annual precipitation in Hachioji has been 1602.3 mm year^{-1} over the past 30 years [32]. Seeds of *P. lanceolata* were sown in a tray of peat moss in a growth chamber (Koitotron PC-02, Koito Industries, Ltd., Kanagawa, Japan) at 25°C. Seven days after sowing, one seedling with cotyledons was transplanted into each plastic pot (20 × 20 × 18 cm deep). Each pot was filled with 4.8 L of a mix of granular red clay and black soil (ratio 1:1 v/v) with 16 g of slow-release fertilizer [Magamp K, 6:40:6:15 (N-P-K-Mg), Hyponex Japan, Osaka, Japan].

The experiment had a three-way factorial randomized block design with nine replications. The factors were water volume with two levels (large, 200 mL of water per day, and small, 100 mL of water per day), water supply frequency with two levels (homogeneous, daily watering, and heterogeneous, watering every 4 days), and belowground herbivore. The certain amount of water for any single watering was determined not to exceed the capacity of the pot. The total volume of water received over the period was the

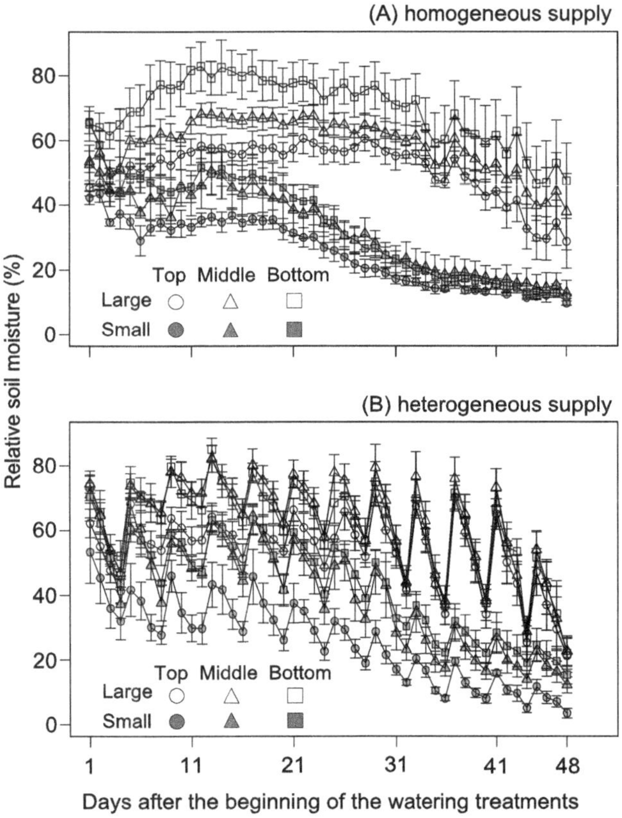

Figure 1. Mean relative soil moisture (± SE). (A) heterogeneous supply treatments; (B) homogeneous supply treatments; Top, middle, bottom: top, middle, and bottom zone treatments, respectively. Large, large water volume supply; Small, small volume supply. The soil moisture content was larger with a large water volume than a small volume. The soil moisture content in the bottom zone was the largest, and that in the top zone the smallest. The temporal variability in soil moisture content under heterogeneous water-supply conditions was larger than that under homogeneous conditions.

same between the homogeneous and heterogeneous supplies for each water volume regime.

There were five types of belowground herbivore treatment: in the top treatment, one grub was placed in the top zone; in the middle treatment, it was placed in the middle zone; and in the bottom treatment, it was placed in the bottom zone. In the control treatment, no grub was added to the pot. In the free treatment, no screens were added to restrict the grub movement. Two 20 × 20 cm stainless-steel wire-mesh screens (0.8-mm wire diameter, 5.5-mm mesh) were inserted into the pot soil to divide the soil evenly into three zones in all treatments except the free treatment. In the top, middle, and bottom treatments, one grub was introduced to the relevant zone through a hole (diameter 12 mm) on the side wall of the pot. The hole was closed with plastic film after the grub had been added. In the free treatment in which the grub was able to move freely around the pot, four screens of 20 × 10-cm stainless-steel wire-mesh were inserted, one along each wall of the pot, which assumed perfect mobility of the grub without any other differences from other three treatments. Because the insertion of the wire-mesh did not affect the plant growth in our previous experiment [33], no treatment without wire-mesh was included in this experiment.

For the first week after transplantation of the seedlings, 150 mL water was supplied to all pots every day. The treatments that combined different water volumes with different watering frequencies began at the start of the second week after the transplant. On day 20 after the beginning of the watering treatment, a grub was added to the relevant zone of the pot. In the free treatment, the grub was placed on the soil surface near the centre of the pot and left to burrow underground.

A plant survival was recorded every day after addition of the grub. If the roots became completely detached from the shoot and the leaves wilted, the plant was considered to die. When a plant died, we recorded survival of the grub.

On day 28 after addition of the grub, the plants were harvested and divided into shoots and roots. The shoots and roots were dried at 72°C for 3 days and weighed. At harvest, grub mortalities were also recorded.

Soil moisture measurement

Soil moisture (water content by volume) was measured with a soil moisture probe (ECH_2O, Decagon Devices, Inc., Pullman, Washington, USA) in an additional four replications without grubs of each watering treatment combination and soil zone during the experimental period. Measurements were taken every day, before the watering. Relative soil moisture content was calculated as the difference between the measured value and the minimum value during the experimental period, divided by the range between the maximum and minimum values during the experimental period [34]. The 4-day moving variance of the relative soil moisture content for the current day and the next 3 days was calculated to quantify the variability in water availability in each 4-day watering cycle. The coefficient of variation (CV) and temporal mean of the relative soil moisture content during each watering treatment were calculated. The CV was used as an index of temporal variability in soil moisture content during each watering treatment [34].

Table 2. Effects of water volume (V), water heterogeneity (H), and soil profile (P) on mean relative soil moisture content.

Source	χ^2	df	P
V	38.788	1	<0.001
H	0.108	1	0.742
P	6.048	2	0.049
V × H	0.822	1	0.365
V × P	0.660	2	0.719
H × P	0.551	2	0.759
V × H × P	0.599	2	0.741

Table 3. Effects of water volume (V), water heterogeneity (H), and soil profile (P) on mean CV of the relative soil moisture.

Source	χ^2	df	P
V	0.333	1	0.564
H	6.425	1	0.011
P	0.252	2	0.882
V × H	0.156	1	0.693
V × P	0.230	2	0.891
H × P	0.081	2	0.960
V × H × P	0.040	2	0.981

Data analysis

Relative soil moisture content and temporal variability in soil moisture were analyzed by using generalised linear mixed models (GLMMs) under the assumption of a Gaussian error distribution. In this model, the response variable was the relative soil moisture content or the CV; the explanatory variables were water volume, watering frequency, soil zone, and their interactions: the random factors were measuring date and block, with block as nested random effect within date.

Mortalities of plants and grubs were analyzed by using GLMMs with a binomial error distribution and logit-link function. In these models, the response variable was grub or plant mortality; the explanatory variables were water volume, watering frequency, belowground herbivore, and their interactions. The random factor was block.

Plant biomass as the sum of shoot and root biomass of each plant, was analyzed with a GLMM assuming a Gaussian error distribution. In this model, the response variable was plant biomass; the explanatory variables were water volume, watering frequency, belowground herbivore, and their interactions. The random factor was block. For each water volume, plant biomass was analyzed with the model lacking water volume as an explanatory variable.

In the analyses for continuous variables, if the assumption of homogeneity of variance was satisfied with Bartlett's test, then identity-link function was applied. If not, then log-link function was applied. Data from pots in which belowground herbivores died were treated as missing values. Number of samples in the biomass analyses was presented in Table 1.

Because several grubs were found dead at harvest, plant biomass was analyzed with generalised linear models (GLMs) for each water volume to evaluate changes in plant biomass due to herbivore mortality. In these models, the response variable was plant biomass; the explanatory variables were watering frequency, belowground herbivore, the survival of the belowground herbivore, and their interactions.

All analyses were performed with the statistical software R version 2.15.1 [35]. The lme4 package was used to calculate GLMMs by using maximum likelihood estimation. To determine the effects of the fixed factors (i.e., to calculate the in *P*-values), we used a likelihood ratio test to compare models with and without the variable of interest using a chi-squared test statistic [36]. The data were analyzed by GLMM framework because our data contain binary and continuous variables [37].

Results

Soil moisture

The mean relative soil moisture content was larger under a large water volume than a small volume (Figure 1, Table 2). The mean relative soil moisture content in the bottom zone was the largest, and that in the top zone was the smallest (Figure 1; Table 2).

The temporal variability in relative soil moisture content under heterogeneous water-supply conditions was larger than that under homogeneous conditions (Figure 1). The mean CV (%) of the relative soil moisture content was larger under heterogeneous water-supply conditions than under homogeneous conditions (Table 3).

Plant mortality

Plant mortality changed significantly with differences in the feeding zones of the belowground herbivore (Table 4). No plants died in the middle and bottom treatments, but three died in the top treatment and three in the free treatment. No plants died when a small volume of water was heterogeneously applied (Table 1).

Table 4. Effects of water volume (V), water heterogeneity (H), and belowground herbivore (B) on plant mortality.

Source	χ^2	df	P
V	0.753	1	0.386
H	0.763	1	0.382
B	8.621	3	0.035
V × H	6.234	3	0.101
V × B	2.156	1	0.142
H × B	0.000	3	1.000
V × H × B	0.000	3	1.000

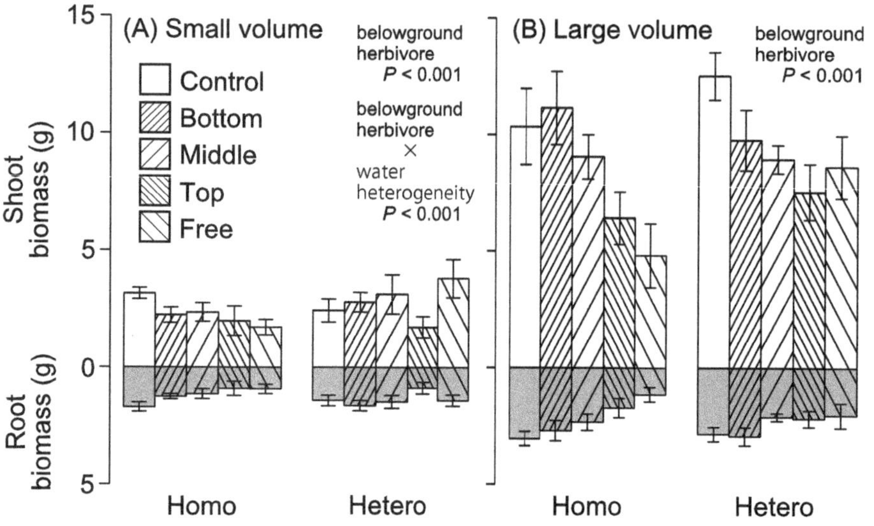

Figure 2. Mean of the shoot and root biomass (± SE). (A) Small water volume. (B) Large water volume. Control, no herbivore present; Bottom, Middle, Top; herbivores placed in the bottom, middle, and top zones, respectively, of the pot; Free, herbivore present and free to move in all zones of the pot. Homo, homogeneous water supply; Hetero, heterogeneous water supply. Biomass in the top treatment was the smallest between the middle, bottom, and control treatments in every watering treatment. Biomass in the free treatment was similar to that in the top treatment in every watering treatment, except for the heterogeneously-applied treatment with small volume of water.

Roots of the survived plants reached the bottom zone at the harvest.

Plant biomass

The three-way interaction between water volume, watering heterogeneity and belowground herbivory, significantly affected mean plant biomass (Figure 2, Table 5A). The herbivory effects on plant biomass differed between the combinations of water-supply volume and heterogeneity: belowground herbivory interacted with heterogeneous conditions and a small water volume (Figure 2A; Table 5B). When a small water volume was supplied, the plant biomass was larger in the control treatment than in the free treatment under homogeneous conditions and smaller than in the free treatment under heterogeneous conditions (Figure 2A). Under both homogeneous and heterogeneous conditions, plant biomass in the top treatment was the smallest between the three zone

Table 5. Effects of water volume (V), water heterogeneity (H), and belowground herbivore (B) on plant biomass.

(A) Three-way analysis			
Source	χ^2	df	P
A	1858.5	1	<0.001
H	23.367	1	<0.001
B	512.59	4	<0.001
A × H	1.084	1	0.298
A × B	32.951	4	<0.001
H × B	31.641	4	<0.001
A × H × B	27.610	4	<0.001
(B) Two-way analysis at small water volume			
Source	χ^2	df	P
H	4.443	1	0.035
B	28.933	4	<0.001
H × B	32.501	4	<0.001
(C) Two-way analysis at large water volume			
Source	χ^2	df	P
H	2.096	1	0.148
B	20.968	4	<0.001
H × B	1.304	4	0.861

treatments. Plant biomass in the free treatment was nearly the same as that in the top treatment with a homogeneous water supply and as large as that in the middle or bottom treatment with a heterogeneous water supply.

Under water supply of large volume, plant biomass significantly differed depending on the vertical distribution of a belowground herbivore (Figure 2B; Table 5C). In the treatment with herbivore in the top zone, plant biomass was remarkably small under both water heterogeneity conditions. Plant biomass in the free treatment was the smallest in all treatments with a homogeneous water supply, and plant biomass in the free treatment was slightly larger than in the top treatment with a heterogeneous water supply.

Mortality of belowground herbivores and its effect on plant biomass

The interaction between water volume and water supply heterogeneity significantly affected mortality of belowground herbivores ($\chi^2 = 5.249$, df $= 1$, $P = 0.022$). With a large water volume, 15 belowground herbivores in the homogeneous treatment died, as did 9 in the heterogeneous treatment (Table 1). With a small water volume, 5 died in the homogeneous treatment and 11 in the heterogeneous one (Table 1). Mortality of belowground herbivores did not differ between the feeding zones ($\chi^2 = 0.895$, df $= 3$, $P = 0.827$).

Mortality of the belowground herbivore significantly affected plant biomass with a large water volume (GLM, $F = 6.793$, df $= 1$, $P = 0.012$) but not with a small water volume (GLM, $F = 0.321$, df $= 1$, $P = 0.573$). Plant biomass in the pots with herbivore mortality was larger than those without mortality.

Discussion

Plant mortality occurred only in the treatments in which the herbivore was shallow in the soil to sever the aerial shoot from its root system. This plant mortality caused by belowground herbivory is consistent with our previous study [33]. With heterogeneous supply of small volume water, no plant was grazed at the root base and any plants did not die, probably because low moisture in the surface soil was not suitable for grubs to graze there.

The interaction between available volume and heterogeneity of water supply would change the vertical distribution of the grub in the free treatment. With a small volume of water under a heterogeneous supply, the mean plant biomass in the free treatment that was almost equivalent to those in the bottom and middle treatments, suggest the grubs occurred in the bottom and middle zones due to the low soil moisture levels and large moisture variability. Consequently, the herbivory effects on plant growth varied depending on the availability of water supply.

The vertical distribution of the belowground herbivore was consistent with the previous studies in which belowground herbivores were distributed deep in the soil under dry conditions [18–20]. The belowground herbivores may graze fine roots deep in the soil, which may promotes root turnover and enhances the absorption of resources [38]. Belowground herbivory can thus affect plant growth in either a negative or a positive way, depending on the heterogeneity of the water supply. In contrary, with the large amount of available water regardless of supply patterns, belowground herbivory occurred in shallow soil in the free treatment due to enough soil moisture. Therefore, plant biomass was significantly lower than that in the control treatment.

Mortality of belowground herbivores varied depending on available water, which altered plant growth. When the grubs died with enough volume of available water, plant biomass was large because of the negligible loss of roots. When the available water volume was small, herbivore mortality was higher under a heterogeneous water supply than a homogeneous one, which is consistent with previous studies [23–26]. However, the effects of herbivore mortality on plant biomass were not significant under a small water volume in this experiment because only few herbivores died.

To our knowledge, this is the first experiment to evaluate the simultaneous effects of available water volume, heterogeneity of water supply, and belowground herbivory on plant growth. The results are consistent with our hypothesis: the availability and heterogeneity of water supply changed the vertical distribution and mortality of belowground herbivores, and consequently plant growth. Therefore, the heterogeneity of soil water supply should be considered in root herbivory studies [15].

Severe climatic events attributable to climate change are already having serious consequences for plants and their herbivores [9,39–42]. We observed high mortality and changes in the vertical distribution of belowground herbivores under the most extreme conditions of water supply (i.e. small volume and heterogeneous supply); these consequently affected plant growth. Therefore, changes in the effects of belowground herbivores on plant growth are likely to occur under the severe weather conditions resulting from climate change.

Acknowledgments

We thank A Takaoka, K Ogata, and S Uchida of Tokyo Metropolitan University for their help with the experiment. We are grateful to Dr. Elly Morrien, an anonymous reviewer, and the academic editor for their constructive comments on our manuscript.

Author Contributions

Conceived and designed the experiments: TT JIS. Performed the experiments: TT. Analyzed the data: TT. Contributed reagents/materials/analysis tools: TT JIS. Wrote the paper: TT JIS. Provided editorial advice: NK.

References

1. Novoplansky A, Goldberg DE (2001) Effects of water pulsing on individual performance and competitive hierarchies in plants. Journal of Vegetation Science 12: 199–208. doi:10.2307/3236604.
2. Hagiwara Y, Kachi N, Suzuki J-I (2008) Effects of temporal heterogeneity of watering on size of an annual forb, *Perilla frutescens* (Lamiaceae), depend on soil nutrient levels. Botany 86: 1111–1116. doi:10.1139/B08-064.
3. Maestre FT, Reynolds JF (2007) Amount or pattern? Grassland responses to the heterogeneity and availability of two key resources. Ecology 88: 501–511. doi:10.1890/06-0421.
4. Heisler-White JL, Knapp AK, Kelly EF (2008) Increasing precipitation event size increases aboveground net primary productivity in a semi-arid grassland. Oecologia 158: 129–140. doi:10.1007/s00442-008-1116-9.
5. Knapp AK, Beier C, Briske DD, Classen AT, Luo Y, et al. (2008) Consequences of more extreme precipitation regimes for terrestrial ecosystems. Bioscience 58: 811–821.
6. Hagiwara Y, Kachi N, Suzuki J-I (2010) Effects of temporal heterogeneity of water supply on the growth of *Perilla frutescens* depend on plant density. Annals of Botany 106: 173–181. doi:10.1093/aob/mcq096.
7. Bardgett RD (2005) The Biology of Soil: A Community and Ecosystem Approach. Oxford University Press, Oxford.
8. Bardgett RD, Wardle DA (2010) Aboveground-Belowground Linkages: Biotic Interactions, Ecosystem Processes, and Global Change. Oxford Series in Ecology and Evolution, Oxford University Press, Oxford.

9. Staley JT, Johnson SN (2008) Climate change impacts on root herbivores, pp. 192–213, in SN Johnson and PJ Murray (eds.), Root Feeders – An Ecosystem Perspective. CABI, Wallingford.
10. Brown VK, Gange AC (1992) Secondary plant succession: how is it modified by insect herbivory? Vegetatio 101: 3–13.
11. De Deyn GB, Raaijmakers CE, Zoomer HR, Berg MP, De Ruiter PC, et al. (2003) Soil invertebrate fauna enhances grassland succession and diversity. Nature 422: 711–713.
12. Wardle DA, Bardgett RD, Klironomos JN, Setälä H, van der Putten WH, et al. (2004) Ecological linkages between aboveground and belowground biota. Science 304: 1629–1633.
13. van der Putten WH, Bardgett RD, de Ruiter PC, Hol WHG, Meyer KM, et al. (2009) Empirical and theoretical challenges in aboveground- belowground ecology. Oecologia 161: 1–14. doi 10.1007/s00442-009-1312-2.
14. Stein C, Unsicker SB, Kahmen A, Wagner M, Audorff V, et al. (2010) Impact of invertebrate herbivory in grasslands depends on plant species diversity. Ecology 91: 1639–1650.
15. Erb M, Lu J (2013) Soil abiotic factors influence interactions between belowground herbivores and plant roots. Journal of Experimental Botany 64: 1295–1303. doi:10.1093/jxb/ert007.
16. Villani MG, Wright RJ (1990) Environmental influences on soil macroarthropod behavior in agricultural systems. Annuan Review of Entomology 35: 249–269.
17. Barnett K, Johnson SN (2013) Living in the soil matrix. Advances in Insect Physiology 45: 1–52.
18. Villani MG, Wright RJ (1988) Use of radiography in behavioral studies of turfgrass-infesting scarab grub species (Coleoptera: Scarabaeidae). Bulletin of Entomological Society of America 34: 132–144.
19. Jones OT (1979) Responses of carrot fly larvae, *Psila rosae*, to components of their physical environment. Ecological Entomology 4: 327–334.
20. Lafrance J (1968) The seasonal movements of wireworms (Coleoptera: Elateridae) in relation to soil moisture and temperature in the organic soils of southwestern Quebec. The Canadian Entomologist 100: 801–807.
21. Davidson RL, Roberts RJ (1969) Scarab damage to grass and clover as influenced by depth of feeding. Bulletin of Entomological Research 58: 559–565.
22. Brown VK, Gange AC (1990) Insect herbivory below ground. Advances in Ecological Research 20: 1–58.
23. Campbell RE (1937) Temperature and moisture preferences of wireworms. Ecology 18: 479–489.
24. Moran NA, Whitham TG (1988) Population fluctuations in complex life cycles: An example from *Pemphigus* aphids. Ecology 69: 1214–1218.
25. Briones MJI, Ineson P, Piearce TG (1997) Effects of climate change on soil fauna; responses of enchytraeids, Diptera larvae and tardigrades in a transplant experiment. Applied Soil Ecology 6: 117–134.
26. Riis K, Esbjerg P (1998) Season and soil moisture effect on movement, survival, and distribution of *Cyrtomenus bergi* (Hemiptera: Cydnidae) within the soil profile. Environmental Entomology 27: 1182–1189.
27. Ladd TL, Buriff CR (1979) Japanese beetle: Influence of larval feeding on bluegrass yields at two levels of soil moisture. Journal of Economic Entomology 32: 341–360.
28. Régnière J, Rabb RL, Stinner RE (1981) *Popillia japonica*: Effect of soil moisture and texture on survival and development of eggs and first instar grubs. Environmental Entomology 10: 654–660.
29. Godfrey LD, Yeargan KV (1985) Influence of soil moisture and weed density on clover root curculio *Sitona hispidulus*, larval stress to alfalfa (*Medicago sativa*). Journal of Agricultural Entomology 2: 370–377.
30. Okuno T, Tanaka Y, Kimura Y, Yoneyama S (1978) Diseases and Pests of Flowers and Vegetables in Colour. Hoikusha Publishing CO., LTD. Osaka, Japan (in Japanese).
31. Sakai K, Fujioka M (2007) Atlas of Japanese Scarabaeoidea Vol. 2 Phytophagous group I. Roppon-Ashi Entomological Books. Tokyo, Japan (in Japanese).
32. Japan Meteorological Agency. *AMeDAS, Automated Meteorological Data Acquisition System*, URL http://www.data.jma.go.jp/obd/stats/etrn/index.php (in Japanese).
33. Tsunoda T, Kachi N, Suzuki J-I (2014) Effects of belowground vertical distribution of a herbivore on plant biomass and survival in *Lolium perenne*. Ecological Research 29: 351–355. doi:10.1007/s11284-014-1133-6.
34. James SE, Pärtel M, Wilson SD, Peltzer DA (2003) Temporal heterogeneity of soil moisture in grassland and forest. Journal of Ecology 91: 234–239. doi: 10.1046/j.1365-2745.2003.00758.x.
35. R Development Core Team (2012) *R: A language and environment for statistical computing*. R Foundation for Statistical Computing, Vienna, Austria. ISBN 3–900051–07–0, URL http://www.R-project.org/.
36. Crawley MJ (2007) The R Book. John Wiley and Sons, New York.
37. Bolker BM, Brooks ME, Clark CJ, Geange SW, Poulsen JR, et al. (2008) Generalized linear mixed models: a practical guide for ecology and evolution. Trends in Ecology and Evolution 24: 127–135. doi: 10.1016/j.tree.2008.10.008.
38. Ramsell J, Malloch AJC, Whittaker JB (1993) When grazed by *Tipula paludosa*, *Lolium perenne* is a stronger competitor of *Rumex obtusifolius*. Journal of Ecology 81: 777–786.
39. Staley JT, Hodgson CJ, Mortimer SR, Morecroft MD, Masters GJ, et al. (2007) Effects of summer rainfall manipulations on the abundance and vertical distribution of herbivorous soil macro-invertebrates. European Journal of Soil Biology 43: 189–198. doi:10.1016/j.ejsobi.2007.02.010.
40. Easterling DR, Meehl GA, Parmesan C, Changnon SA, Karl TR, et al. (2000) Climate extremes: observations, modeling, and impacts. Science 289: 2068–2074.
41. Smith MD (2011) An ecological perspective on extreme climatic events: a synthetic definition and framework to guide future research. Journal of Ecology 99: 656–663.
42. Reyer CPO, Leuzinger S, Rammig A, Wolf A, Bartholomeus RP, et al. (2013) A plant's perspective of extremes: terrestrial plant responses to changing climatic variability. Global Change Biology 19: 75–89.

7

Cultivation-Independent Screening Revealed Hot Spots of IncP-1, IncP-7 and IncP-9 Plasmid Occurrence in Different Environmental Habitats

Simone Dealtry[1], Guo-Chun Ding[1], Viola Weichelt[1], Vincent Dunon[2], Andreas Schlüter[3], María Carla Martini[4], María Florencia Del Papa[4], Antonio Lagares[4], Gregory Charles Auton Amos[5], Elizabeth Margaret Helen Wellington[5], William Hugo Gaze[5], Detmer Sipkema[6], Sara Sjöling[7], Dirk Springael[2], Holger Heuer[1], Jan Dirk van Elsas[8], Christopher Thomas[9], Kornelia Smalla[1]*

1 Julius Kühn-Institut – Federal Research Centre for Cultivated Plants (JKI), Institute for Epidemiology and Pathogen Diagnostics, Braunschweig, Germany, **2** Division of Soil and Water Management, KU Leuven, Heverlee, Belgium, **3** Center for Biotechnology (CeBiTec), Institute for Genome Research and Systems Biology, Bielefeld University, Bielefeld, Germany, **4** IBBM (Instituto de Biotecnología y Biología Molecular), CCT-CONICET-La Plata, Departamento de Ciencias Biológicas, Facultad de Ciencias Exactas, Universidad Nacional de La Plata, La Plata, Argentina, **5** School of Life Sciences, University of Warwick, Warwick, United Kingdom, **6** Laboratory of Microbiology, Wageningen University, Wageningen, The Netherlands, **7** Södertörns högskola (Sodertorn University), Inst. för Naturvetenskap, Miljö och medieteknik (School of Natural Sciences, Environmental Studies and media tech), Huddinge, Sweden, **8** University of Groningen, Groningen, The Netherlands, **9** School of Biosciences, University of Birmingham, Edgbaston, Birmingham, Warwick, United Kingdom

Abstract

IncP-1, IncP-7 and IncP-9 plasmids often carry genes encoding enzymes involved in the degradation of man-made and natural contaminants, thus contributing to bacterial survival in polluted environments. However, the lack of suitable molecular tools often limits the detection of these plasmids in the environment. In this study, PCR followed by Southern blot hybridization detected the presence of plasmid-specific sequences in total community (TC-) DNA or fosmid DNA from samples originating from different environments and geographic regions. A novel primer system targeting IncP-9 plasmids was developed and applied along with established primers for IncP-1 and IncP-7. Screening TC-DNA from biopurification systems (BPS) which are used on farms for the purification of pesticide-contaminated water revealed high abundances of IncP-1 plasmids belonging to different subgroups as well as IncP-7 and IncP-9. The novel IncP-9 primer-system targeting the *rep* gene of nine IncP-9 subgroups allowed the detection of a high diversity of IncP-9 plasmid specific sequences in environments with different sources of pollution. Thus polluted sites are "hot spots" of plasmids potentially carrying catabolic genes.

Editor: Axel Cloeckaert, Institut National de la Recherche Agronomique, France

Funding: This study was funded by the EU 7th Framework Programme (MetaExplore 222625) and the Inter-University Attraction Pole (IUAP) "μ-manager" of the Belgian Science Policy (BELSPO, P7/25). The funders had no role in study design, data collection and analysis, decision to publish, or preparation of the manuscript.

Competing Interests: The authors have declared that no competing interests exist.

* E-mail: kornelia.smalla@jki.bund.de

Introduction

The search for novel enzymes able to degrade recalcitrant natural contaminants such as chitins and lignins and man-made pollutants such as halogenated aliphatic and aromatic compounds motivated metagenomic explorations of various environments. It has been observed that the microbial metagenomes of open ecosystems, including soils and aquatic habitats, clearly represent rich reservoirs of genes that determine the desired enzymatic reactions in which chitinases, ligninases and dehalogenases are involved [1,2]. By anthropogenic activities, recalcitrant compounds have also been released as environmental pollutants. Typical metagenomic approaches employ genetic or activity screens of cloned large DNA fragments from various environments [3]. However, the idea of capturing complete mobile genetic elements (MGE) into suitable recipients might be an alternative and complementary approach to access the genes coding for novel enzymes or even complete degradative pathways. Mobile genetic elements such as plasmids are often found to play an important role in the adaptation of bacterial communities to changing and, due to pollutants, often challenging environmental conditions. For example, partial or complete degradative pathways were previously reported to be localized on plasmids belonging to the IncP-1, IncP-7 or IncP-9 group [1]. The present study aimed to monitor various environments for the abundance of these plasmids by using a cultivation-independent total community (TC-) DNA based approach to select the most promising habitats for mining plasmids potentially carrying genes coding for novel enzymes. We hypothesized that the frequency of occurrence of genes encoding the desired enzymatic activities is increased in the MGE gene pool. In particular, plasmids belonging to the incompatibility groups (Inc) P-1, P-7 and P-9 often carry genes responsible for the degradation of xenobiotic (man-made) and natural organic pollutants, being essential players in the adaptation of bacterial communities to new toxic compounds released in the environment [1]. Therefore, selected natural or treated environments were analyzed for the prevalence of plasmids belonging to the IncP-1, IncP-7 and IncP-9 groups by a cultivation-independent approach. Some of these environments were enriched for the desired degradation function by adding the relevant substrates, i.e. chitin, lignin and/or organohalogens. The habitats sampled included a variety of soils (one soil sample amended with chitin, peat bogs), biopurification systems (BPS) for pesticide removal from contaminated water, biogas production plants, wastewater, as well as aquatic (river bank sediments, sponges) environments from a wide range of geographic regions. Total community DNA was analyzed for the presence of IncP-1, IncP-7 and IncP-9 plasmids by means of PCR and subsequent Southern blot hybridization. A novel primer system for the specific amplification of IncP-9 plasmids was developed and tested in the present study. Southern blot hybridization using probes derived from reference plasmids belonging to different subgroups of IncP-1 plasmids provided new insights into their environmental dissemination. Our results showed a particularly widespread dissemination of IncP-1 plasmid-specific sequences. Different hot spots of plasmid occurrence were identified.

Materials and Methods

Ethics Statement of Provided Samples

None of the samples used in the present work involved any endangered or protected species. The marine sponges were obtained under legal permits from competent authorities: *Halichondria panicea* was obtained under a permit from authorities given to Wageningen University, while *Corticium candelabrum* and *Petrosia ficiformis* marine sponges were sampled under a Spanish permit to CEAB-CSIC. The sediments and soil originated from UK were taken from a river bed accessed from a public right of way and therefore no permissions were needed. Landsort Deep was sampled from a national environmental monitoring site (BY31) in conjunction with the Baltic Sea monitoring programme. The sampling permission was provided by the Stockholm University marine research Center. The Askö samples were sampled at the Stockholm Marine Research Center (now the Stockholm University Marine Research Center. The sediment and soils from Argentina were obtained from public locations as part of fundamental studies performed through a collaborative project with the agreement of the Facultad de Ciencias Exactas, Universidad Nacional de La Plata, and did not require any specific permission. The biopurification systems (BPS) samples were obtained from private land with permission from the local farmers in Kortrijk, Leefdaal, Lierde and Koksijde, located in Belgium.

Extraction of Total Community DNA (TC-DNA) and Metagenomic DNA (Pooled Fosmid Library) from Different Environmental Samples

The TC-DNA and/or metagenomic DNA (metagenomic DNA represented by the metagenomic pooled fosmid library from Baltic Sea) from different environmental samples originating from various geographic regions were extracted using different methods. The protocols used for TC-DNA extraction of each sample type are given in Table 1.

16S rRNA Gene PCR Amplification and Quantification

16S rRNA gene PCR amplification reaction was done as previously described by Heuer *et al.* (2009) [5] (product size of 1506 bp). The quality of the PCR product was determined by electrophoresis in 1% agarose gel and visualized with ethidium bromide staining and under UV light by comparison with the 1-kb gene-rulerTM DNA ladder (Fermentas, St Leon-Rot, Germany). Quantitative PCR (qPCR) targeting the 16S rRNA gene was performed with the TaqMan system as described by Suzuki *et al.* (2000) [6]. The 16S rRNA gene qPCR standard was made from cloned 16S rRNA gene amplicons (1467 bp) of *E. coli.* and 10^9, 10^8, 10^7 16S rRNA gene copy numbers were used.

Southern Blot-PCR Based Detection of IncP-1 Plasmids

IncP-1 plasmids belonging to the α, β, γ, δ and ε subgroups were detected based on the amplification of the *trfA* region (product size of 281 bp) from TC-DNA and metagenomic DNA using the primers described by Bahl *et al.* (2009) [2]. Digoxygenin-labeled probes targeting different IncP-1 plasmids subgroups were generated from reference plasmids belonging to the IncP-1α, β, γ, δ and ε plasmids (Table 2). The IncP-1 mixed probe was prepared by mixing probes generated for the different subgroups. The random primed digoxigenin labeling of PCR amplicons excised from preparative agarose gels was done according to the Roche manufacturer's protocol (Roche Diagnostics Deutschland GmbH, Mannheim, Germany).

Southern Blot-PCR Based Detection of IncP-7

PCR amplification of the *rep* region of IncP-7 plasmids (product size of 524 bp) from TC-DNA was performed as previously described by Izmalkova *et al.* (2005) [4]. Southern blotted PCR amplicons were hybridized at medium stringency with the dig-labeled IncP-7 probe generated from the reference plasmid pCAR1 isolated from *Pseudomonas resinovorans* according to the manufacturer's instructions (QIAGEN® Plasmid Mini Kit) (Table 2). The randomly primed digoxigenin labeling of PCR amplicons was done as described above.

Analyzing the Diversity and Abundance of IncP-9 Plasmids by a Novel PCR System Targeting the *oriV-rep* Region

To study the abundance and diversity of IncP-9 plasmids, a novel PCR system targeting the *oriV-rep* regions was developed and applied to detect IncP-9 plasmids in TC- and metagenomic DNA from all samples analyzed (Table 1).

Multiple alignments of 28 sequences of *oriV* (EU499619-EU499641, AF078924, AB237655, AJ344068, AB257759 and AF491307) and *rep* (EU499644-EU499666, AF078924, AB237655, AJ344068, AB257760 and AF491307) were performed with Molecular Evolutionary Genetics Analysis (MEGA 4). Conserved regions of sequences belonging to nine IncP-9 subgroups [7] were used for the primer design. The selected primer system consists of 21-mer degenerate forward primer (5-GAG GGT TTG GAG ATC ATW AGA-3) and reverse primer (5-GGT CTG TAT CCA GTT RTG CTT-3). *In silico* analysis showed no mismatch for at least 12 bp at the 3′ end of each primer and 1–4 mismatches for each sequence type at the 5′ end (Fig. S1). The expected amplicon size is 610–637 bp. The primers were further tested with plasmid DNA from the reference plasmids summarized in Fig. S1. None of the plasmids belonging to other incompatibility groups was amplified while the reference plasmids were amplified. The reaction mixture (25 μl) contained 1 μl template DNA (1–5 ng), 1× Stoffel buffer (Applied Biosystems, Foster, CA), 0,2 mM dNTPs, 2,5 mM $MgCl_2$, 2 μg/μl bovine serum albumin, 0.2 μM of each primer, and 2.5 U TrueStartTaq DNA polymerase (Stoffel fragment, Applied Biosystems). Denaturation was carried out at 94°C for 5 min, followed by 35 cycles

Table 1. Description of environmental samples analyzed and TC-DNA extraction applied.

Samples	Description of samples	TC-DNA extraction method
A	Biogas production plant fermentation sample from Bielefeld, Germany	[24]
B.1	*Biopurification system (BPS) from Leefdaal, Belgium	[17]
B.2	BPS from Leefdaal, Belgium	[17]
B.3	BPS from Leefdaal, Belgium	[17]
C.1	*BPS from Belgium (Pcfruit)	[17]
C.2	BPS from Belgium (Pcfruit)	[17]
C.3	BPS from Belgium (Pcfruit)	[17]
C.4	BPS from Belgium (Pcfruit)	[17]
C.5	BPS from Belgium (Pcfruit)	[17]
C.6	BPS from Belgium (Pcfruit)	[17]
D.1	*BPS from Lierde, Belgium	[17]
D.2	BPS from Lierde, Belgium	[17]
D.3	BPS from Lierde, Belgium	[17]
E.1	*BPS from Kortrijk, Belgium	[17]
E.2	BPS from Kortrijk, Belgium	[17]
E.3	BPS from Kortrijk, Belgium	[17]
F.1	*BPS from Koksijde, Belgium	[17]
F.2	BPS from Koksijde, Belgium	[17]
F.3	BPS from Koksijde, Belgium	[17]
G.1	Soil from La Plata, Argentina polluted with industrial residues and petrol	[18]
G.2	Soil from La Plata, Argentina polluted with industrial residues and petrol	[18]
G.3	Soil from La Plata, Argentina polluted with industrial residues and petrol	[18]
H.1	Sediments from La Plata, Argentina polluted with pesticides and petrol	[18]
H.2	Bordering soil from a water channel in La Plata, Argentina polluted with pesticides, residues from paper industry	[18]
H.3	Bordering soil from a water channel in La Plata, Argentina polluted with pesticides, residues from paper industry	[18]
J	Marginal river forest soil from La Plata, Argentina polluted with industrial residues	[18]
L.1	Bordering soil from a water channel in Buenos Aires, Argentina polluted with industrial residues	[18]
L.2	Bordering soil from a water channel in Buenos Aires, Argentina polluted with industrial residues	[18]
L.3	Bordering soil from a water channel in Buenos Aires, Argentina polluted with industrial residues	[18]
M	*Halichondria panicea* (marine sponge) from Oosterschelde, Netherlands	[25]
N	*Corticium candelabrum* (marine sponge) from Punta Santa Anna (Blanes), Spain	[25]
O	*Petrosia ficiformis* (marine sponge) from Punta Santa Anna (Blanes), Spain	[25]
P.1	Askö sediment from Baltic Sea Sweden (bottom fraction - anoxic)	[26]
P.2	Askö sediment from Baltic Sea Sweden (middle fraction - mixed anoxic/oxic)	[26]
P.3	Askö sediment from Baltic Sea Sweden (top fraction - oxic)	[26]
Q	Pooled fosmid library, Askö sediment, Baltic Sea	[3]
R	Landsort in Sweden	[26]
S.1	Sediment from a river in Warwickshire, UK	[27]
S.2	Sediment from a river in Warwickshire, UK	[28]
T	Soil from UK amended with chitin (Test site 1)	[28]

*BPS samples received water contaminated with different types of pesticides from spillage and residue water collected when cleaning the spraying equipment such as ethofumesate, fenpropimorf, fluroxypyr, glyphosate, linuron, metamitron and S-metalochlor (information provided by the farmers).

of 1 min at 94°C, 1 min at 53°C (primer annealing) and 2 min at 72°C and a final extension of 10 min at 72°C.

PCR amplicons of *oriV-rep* regions of nine IncP-9 subgroups IncP-9 plasmids (Table 2) were gel-purified and digoxigenin-labeled as described above. Southern blot hybridization of *oriV-rep* amplicons from different environmental samples listed above was performed with a mixture of these probes under medium stringency following the manufacturer's instructions (Roche Diagnostics Deutschland GmbH, Mannheim, Germany). Clone libraries were generated for these three BPS to confirm primer specificity. *oriV-rep* amplicons were gel-purified, ligated into pGEM vectors, and transformed into *E. coli* JM109 competent cells according to the instructions of the manufacturer. Clones containing the correct inserts were selected for sequencing.

Table 2. Generation of probes for Southern blot hybridization.

Probe	Reference plasmid	Plasmids host strain	Primers
IncP-1α	RP4	*E. coli*	[2]
IncP-1β	R751	*E. coli CM544*	[2]
IncP-1γ	pQKH54	*E. coli DH10B*	[2]
IncP-1δ	pEST4011	*Alcaligenes xylosoxidans EST4002*	[2]
IncP-1ε	p3-408	*E. coli cv601-GFP*	[2]
IncP-7	pCAR1,	*Pseudomonas resinovorans CA10*	[4]
IncP-9 α	pM3	*Pseudomonas putida*	*This study*
IncP-9 β	pBS2	*Pseudomonas putida BS268*	*This study*
IncP-9 γ	pSN11	*Pseudomonas putida BS349*	*This study*
IncP-9 δ	pSN11	*Pseudomonas putida SN11*	*This study*
IncP-9 ε	pMG18	*Pseudomonas putida AC34*	*This study*
IncP-9 ζ	pNL60	*Pseudomonas* spp. *18d/1*	*This study*
IncP-9 η	pNL15	*E. coli C600*	*This study*
IncP-9 θ	pSVS15	*Pseudomonas fluorescens SVS15*	*This study*
IncP-9 ι	pNL22	*Pseudomonas* spp. *41a/2*	*This study*

BLAST-N analysis was used to identify *oriV-rep* sequences of IncP-9. All sequences analyzed share high similarity with IncP-9 *oriV* or *rep* sequences in NCBI. The sequences and those of known *oriV-rep* sequences in the data base were aligned and phylogenetic tree was calculated according to the neighbor-joining method and boot-strapping analysis using MEGA 4.

Nucleotide Sequence Accession Numbers of Cloned IncP-9 *oriV-rep* Gene Amplicons

Amplicon sequences have been submitted to NCBI SRA with IncP-9 *oriV-rep* gene amplicons under accession numbers KF706553 - KF706633.

Results

Determination of Bacterial 16S rRNA Gene Copies by qPCR

To estimate the bacterial density of the different environmental samples analyzed, 16S rRNA gene copies were determined by quantitative real-time PCR from the TC-DNA. Most of the samples (Table 3) showed a high abundance of bacterial populations ranging from 10^8 to 10^9 16S rRNA gene copy numbers per gram of material. For a few samples significantly lower 16S rRNA gene copy numbers per gram of material (Tukey's test $p>0.05$) were detected (Table 3).

Distribution of IncP-1 Plasmids in Different Environments

To investigate the presence of IncP-1 plasmids in different habitats a detection system based on Southern blot-PCR was applied. Using the IncP-1 mixed probe from PCR products hybridization signals of the expected size (251 bp) were detected in a very wide range of different habitats (Table 3), indicating that IncP-1 plasmids of different subgroups are widely distributed. By using probes specific for the five different IncP-1 different subgroups (α, β, γ, δ and ε), differences in the composition of IncP-1 plasmids according to the geographic area and sample type were observed. Strong hybridization signals of IncP-1α plasmids were only observed in one TC-DNA from Askö sediment (Sweden), in TC-DNA from a biogas production plant (Germany) and fosmid DNA from Baltic Sea sediments. Strong hybridization signals were observed using the IncP-1β specific probe in the TC-DNAs of all biopurification system (BPS) samples from Belgium (Table 3, Fig. 1) and most of the sediment samples from Argentina, indicating that in these environments bacterial populations carrying IncP-1β plasmids were highly abundant. The highest IncP-1γ hybridization signal was observed in the TC-DNA of the BPS located in Kortrijk. Less intense IncP-1γ hybridization signals were detected in the TC-DNAs of other BPS from Belgium and in TC-DNA of sediments from Argentina. In all TC-DNAs of BPS from Belgium, strong IncP-1δ hybridization signals were observed and a weaker hybridization signal, compared to BPS TC-DNA, was detected in TC-DNA from sediments in Argentina. Very strong IncP-1ε hybridization signals were again detected in all BPS TC-DNAs from Belgium (Table 3, Fig. 2) and most of the sediments from Argentina. Using IncP-1 mixed-probe, strong hybridization signals were detected in soils from Argentina and soil treated with chitin from the UK, indicating a high abundance of IncP-1 plasmids.

Distribution of IncP-7 Plasmids in Different Environments

To investigate the occurrence of IncP-7 plasmids in different environments, a PCR-based detection approach was applied in combination with Southern blot hybridization. Strong hybridization signals were observed in all TC-DNAs from BPS analyzed (Fig. 3), indicating a high abundance of bacterial populations carrying in BPS IncP-7 plasmid. Less intense hybridization signals were observed in the TC-DNAs of seven sediment river samples from Argentina and in the TC-DNA from soil amended with chitin from the UK. Hybridization signals using the amplicon probe specific for IncP-7 plasmids were not detected in any of the other environmental samples analyzed (Table 3).

IncP-9 Plasmid Occurrence and Diversity in Different Environmental Samples

In order to verify the occurrence and diversity of IncP-9 plasmids in different habitats the new IncP-9 primer system developed in the present work was applied. Very strong hybridization signals were detected in all TC-DNAs of BPS samples, indicating that BPS are reservoirs of bacteria carrying IncP-9 plasmids. Less intense hybridization signals were observed in the TC-DNA of sediment samples from Argentina. A weaker hybridization signal was detected in the soil amended with chitin from the UK (Table 3).

To verify primer specificity and to gain insights into the IncP-9 plasmid diversity from BPS samples (indicated as a "hot spot" of IncP-9 plasmids), a clone library was generated with amplicons from PCR using primers targeting the IncP-9 *oriV-rep* region in TC-DNA of three different BPS. Sequencing revealed the presence of different IncP-9 subgroups while phylogenetic analysis (Fig. 4) showed IncP-9 plasmid types similar to *oriV-rep* sequences of pWWO and pM3 as well as several sequences that could not be affiliated to previously known IncP-9 plasmid groups indicating an undiscovered diversity of this plasmid group.

Discussion

In the present study a PCR-based screening combined with Southern blot hybridization allowed the detection of IncP-1, IncP-7 and IncP-9 plasmids in a wide range of different geographic areas and sample types. The results indicated a high abundance of these plasmids in environments with different sources of pollution.

Table 3. Bacterial densities and PCR-Southern blot hybridization detection of plasmid replicon-specific sequences belonging to the five IncP-1 subgroups, IncP-7 and IncP-9.

Sample	Description of samples	P-1	α	β	ε	γ	δ	P-7	P-9	16S log10/g
A	Biogas production plant from Bielefeld, Germany	+++	++	++	++	–	++	–	–	9,34
B.1	Biopurification system (BPS) from Leefdaal, Belgium	+++	–	++	+++	++	+++	+++	+++	9,32
B.2	BPS from Leefdaal, Belgium	+++	–	++	+++	++	+++	+++	+++	9,25
B.3	BPS from Leefdaal, Belgium	+++	–	++	+++	++	++	+++	+++	8,43
C.1	BPS from Belgium (Pcfruit)	+++	+	+++	+++	++	++	+++	+++	9,32
C.2	BPS from Belgium (Pcfruit)	+++	+	+++	++	++	+++	+++	+++	8,28
C.3	BPS from Belgium (Pcfruit)	++	–	++	++	–	+++	++	+	8,36
C.4	BPS from Belgium (Pcfruit)	+	–	++	+	+	+++	+++	+++	8,54
C.5	BPS from Belgium (Pcfruit)	+++	(+)	+++	+++	++	+++	+++	+++	8,66
C.6	BPS from Belgium (Pcfruit)	+	–	++	++	+	++	++	–	8,15
D.1	BPS from Lierde, Belgium	+++	–	++	–	–	+++	++	++	8,61
D.2	BPS from Lierde, Belgium	+++	–	++	++	–	+++	++	++	8,59
D.3	BPS from Lierde, Belgium	+++	–	++	+++	++	+++	++	++	8,31
E.1	BPS from Kortrijk, Belgium	+++	–	+++	+++	+++	+	+++	++	9,2
E.2	BPS from Kortrijk, Belgium	+++	–	+++	+++	+++	++	+++	+++	9,03
E.3	BPS from Kortrijk, Belgium	+++	–	++	++	+++	–	+++	+++	9,11
F.1	BPS from Koksijde, Belgium	++	(+)	++	+	–	+++	(+)	+++	9,01
F.2	BPS from Koksijde, Belgium	++	+++	++	–	–	++	–	+++	8,9
F.3	BPS from Koksijde, Belgium	++	(+)	++	+	–	+++	–	+++	8,95
G.1	Soil from La Plata, Argentina	+++	(+)	+	+	–	+++	+++	+++	8,55
G.2	Soil from La Plata, Argentina	+++	–	(+)	+	++	+	+++	+++	8,53
G.3	Soil from La Plata, Argentina	+++	–	++	–	++	+	–	(+)	8,22
H.1	Sediments from La Plata, Argentina	+++	++	+++	++	++	+++	+++	+++	8,96
H.2	Bordering soil from a water channel in La Plata, Argentina	+++	++	++	++	++	+++	+	+++	8,49
Sample	**Description of samples**	**P-1**	**α**	**β**	**ε**	**γ**	**δ**	**P-7**	**P-9**	**16S log10/g**
H.3	Bordering soil from a water channel in La Plata, Argentina	+++	+	++	++	++	+++	–	+++	8,7
I	Sweet-water soil from a river in La Plata, Argentina	+++	+	++	++	++	++	+++	+++	7,91
J	Marginal river forest soil from La Plata, Argentina	++	–	–	–	–	–	–	–	8,32
L.1	Bordering soil from a water channel in Buenos Aires, Argentina	+++	–	++	+	–	++	–	–	8,29
L.2	Bordering soil from a water channel in Buenos Aires, Argentina	+++	+	+++	++	++	+++	–	(+)	8,6
L.3	Bordering soil from a water channel in Buenos Aires, Argentina	+++	–	++	++	++	+++	++	+++	7,66
M	*Halichondria panicea* (marine sponge) from Oosterschelde, Netherlands	++	–	++	–	–	+++	–	–	7,32
N	*Corticium candelabrum* (marine sponge) from Punta Santa Anna (Blanes), Spain	++	–	+	–	–	+	–	–	8,18
O	*Petrosia ficiformis* (marine sponge) from Punta Santa Anna (Blanes), Spain	++	–	++	–	–	+	–	–	8,4
P.1	Askö sediment from Baltic Sea Sweden (bottom fraction - anoxic)	++	–	–	–	–	+	–	–	8,34
P.2	Askö sediment from Baltic Sea Sweden (middle fraction mixed anoxic/oxic)	++	++	+++	++	–	+	–	–	8,43
P.3	Askö sediment from Baltic Sea Sweden (top fraction - oxic)	+++	–	++	+++	–	+	–	–	8,09
Q	Pooled fosmid library, Askö sediment, Baltic Sea	+++	++	+	+	–	+	–	–	5,01
R	Landsort in Sweden	+++	++	–	–	–	–	–	–	8,16
S.1	Sediment from a river in Warwick, UK	+++	/	/	/	/	/	–	(+)	5,78
S.2	Sediment from a river in Warwick, UK	+++	/	/	/	/	/	++	–	6,26
T	Soil from Cuba amended with chitin (Test site 1)	+++	/	/	/	/	/	+++	++	6,95
negative control		–	–	–	–	–	–	–	–	

Table 3. Cont.

Sample	Description of samples	P-1	α	β	ε	γ	δ	P-7	P-9	16S log10/g
RP4 (IncP-1α)		+++	+++							
R751 (IncP-1β)		+++		+++						
pKJK5 (IncP-1ε)		++			++					
pQKH54 (IncP-1γ)		+++				+++				
pEST4011 (IncP-1δ)		+++					+++			
pCAR1 (IncP-7)								+++		
pNF 142 (IncP-9)									+++	

Hybridization signal: (+++) very strong, with exposure time up to five minutes; (++) strong, with exposure time up to one hour; (+) weak, with exposure time up to three hours; (−) none, with exposure time of more than three hours; (/) not analyzed.

It is tempting to speculate that degradative genes localized on the plasmid of these groups might contribute to the bacterial degradation of a variety of pollutants such as pesticides, due to the "metabolic complementation" resulting from the combination of different genes brought together by different plasmids. While IncP-1 plasmids typically host genes associated with the degradation of man-made pollutants (xenobiotics) [8], IncP-7 and IncP-9 plasmids often carry genes responsible for degradation of natural contaminants, such as polyaromatic hydrocarbons [9]. Screening TC-DNA revealed that IncP-1, IncP-7 and IncP-9 specific sequences vary according to sample type and degree of pollution. IncP-1 plasmid specific sequences were detected in a wide range of environments: marine sponges, soils and sediments, Baltic Sea sediment fosmid library, biogas production plant, river sediments, chitin-treated soils and BPS contaminated with pesticides. Very strong hybridization signals for all different IncP-1 subgroups tested except for IncP-1α plasmids were especially observed in the BPS samples heavily contaminated with pesticides, indicating an

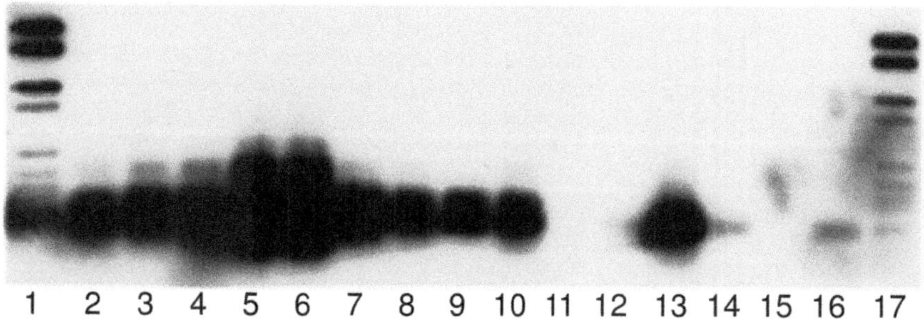

Figure 1. Biopurification systems (BPS). Hybridization of Southern-blotted PCR products obtained with *trfA* primer system from TC-DNA of BPS (IncP-1β specific group). Lanes: 1 and 17, dig ladder; lanes 2 to 4, BPS from Lierde, Belgium; lanes 5 to 7, BPS from Kortrijk, Belgium; lanes 8–10, BPS from Koksijde, Belgium; lane 11, negative control; lanes 12–16, IncP-1 positive controls RP4 (α), R751 (β), pKJK5 (ε), pQKH54 (γ) and pEST4011 (δ). Exposure time of 5 min.

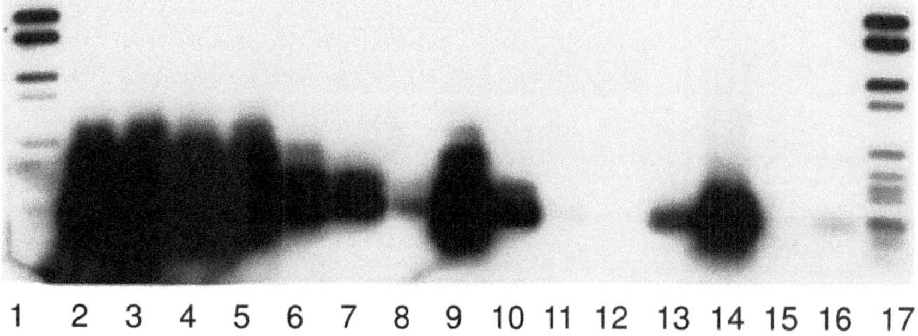

Figure 2. Biopurification systems (BPS). Hybridization of Southern-blotted PCR products obtained with *trfA* primer system from TC-DNA of BPS with the IncP-1ε specific probe. Lanes: 1 and 17, dig ladder; lanes 2 to 4 BPS from Lierde, Belgium; lanes 5 to 7, BPS from Kortrijk, Belgium; lanes 8 to 10, BPS from Koksijde, Belgium; lane 11, negative control; lanes 12 to 15, IncP-1 positive controls RP4 (α), R751 (β), pKJK5 (ε), pQKH54 (γ) and pEST4011 (δ). Exposure time of 5 min.

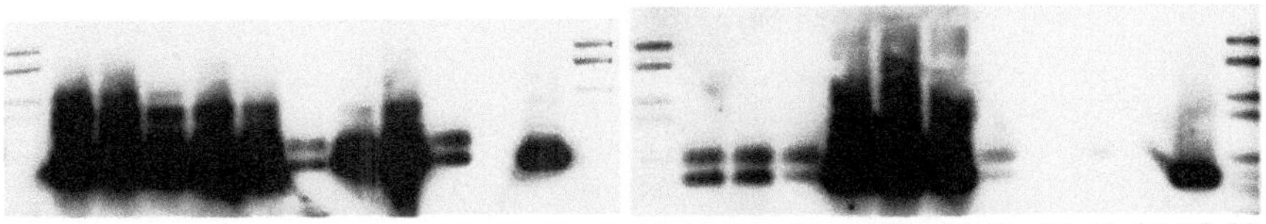

Figure 3. Biopurification systems (BPS). Hybridization of Southern-blotted PCR products obtained with *rep* primer system from TC-DNA of BPS with the IncP-7 probe generated from pCAR1. Lanes: 1, 13 and 26, dig ladder; lanes 2 to 4, BPS from Leefdaal, Belgium; lanes 5 to 10, BPS from Belgium (Pcfruit); lanes 15 to 17, BPS from Lierde, Belgium; lanes 18 to 20, BPS from Kortrijk, Belgium; lanes 21 to 23, BPS from Koksijde, Belgium; lanes 11 and 24, negative control; lanes 12 and 25 IncP-7 positive control pCAR-1. Exposure time of 5 min.

unusual high abundance of bacterial populations carrying IncP-1 plasmids. Indeed, the use of BPS, defined as a pollution control technique employing microorganisms to degrade pesticides through biodegradation processes [10], in on-farm treatment of water contaminated with pesticides has substantially increased and enhanced the degradation rates [11]. Strong hybridization signals of IncP-1β and IncP-1ε plasmids observed in all BPS samples and in some sediments from Argentina contaminated with oil, suggested that IncP-1β and IncP-1ε plasmids might be important in the local adaptation of bacteria to changing environmental conditions [12,13]. Strong IncP-1 plasmid hybridization signals observed in sediments from different regions: Warwick (UK), La Plata (Argentina) and sediments from Sweden indicated that IncP-1 plasmids might also have an important ecological role in the adaptation and biodegradation processes in sediments as previously reported already for mercury-contaminated sediments in Kazachstan [14]. The apparently high abundance of IncP-1 plasmids in soils from different regions contaminated with different pollutants, such as soils from Argentina polluted with oil and soils from the UK enriched with chitin, also suggested that IncP-1 plasmids might substantially contribute to the adaptation and survival of the soil bacterial communities in response to wide range of environmental pollutants [8,15–17]. The results from several studies suggested a correlation between IncP-1 plasmid abundance and pollution as hypothesized by Smalla *et al.* (2006) and confirm

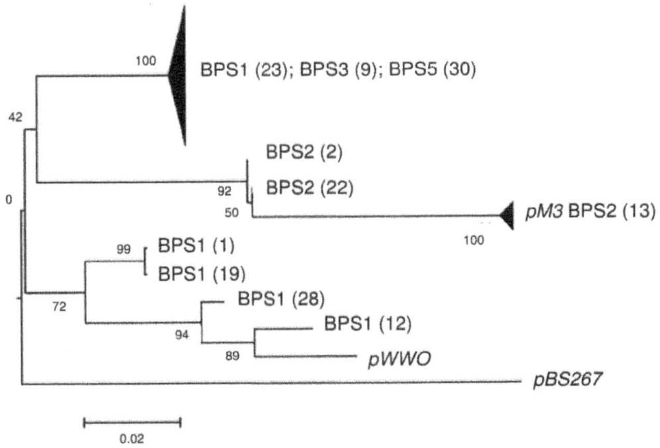

Figure 4. Neighbor-Joining phylogenetic tree based on the multiple alignment of cloned amplicon sequences of the *oriV-rep* IncP-9 gene. Sequences from known IncP-9 plasmids have been included as references. Value at each node is percent bootstrap support of 1,000 replicates. BPS1; BPS2 and BPS5 correspond to three different biopurification systems (BPS), located in Belgium. Numbers in brackets correspond to number of clones and numbers without brackets correspond to the clone designation.

previously published quantitative data on the abundance of IncP-1 plasmids in BPS samples from one BPS site by means of a qPCR targeting the *korB* gene. Obviously, the relative abundance of IncP-1 plasmids can only be precisely quantified by quantitative real-time PCR. However, the recently developed *korB* quantitative PCR system [18] cannot indicate the relative abundance of the different IncP-1 subgroups which was achieved with specific probes for different IncP-1 groups used in the present study in a semi-quantitative manner.

The study by Sevastsyanovich *et al.* (2008) already showed that IncP-9 plasmid diversity is much broader than previously imagined. In view of this huge plasmid diversity, a novel IncP-9 primer system was developed and established in the present work. Typically, IncP-9 plasmids are related to the degradation of natural pollutants as polyaromatic hydrocarbons [19]. However, the detection of very strong IncP-9 hybridization signals mainly in BPS indicated that populations carrying IncP-9 plasmids are also important players in the degradation of man-made pollutants or wood-derived aromatic compounds. IncP-9 plasmids often possess different aromatic-ring degradation genes. BPS typically contain wood chips but also various aromatic ring-containing pesticides such as bentazon, epoxiconazol and diflufencian [20], which could explain the high abundance of IncP-9 plasmids observed in BPS. Cloning and sequencing of amplicons obtained with the novel IncP-9 primers from BPS TC-DNA confirmed not only the specificity of the primers but also showed the presence of plasmids with high similarity to pWWO, that were previously reported to carry degradative genes (Fig. 4) [21,22]. The presence of several sequences with high similarity to the *oriV-rep* sequence of pM3, an antibiotic resistance plasmid belonging to the IncP-9α subgroup, in BPS 2 might be caused by manure addition in the beginning of every year (on March) by the farmers as a C-source in BPS material. Therefore, the addition of manure in BPS as nutrient source for the microorganisms might be reconsidered and replaced for an alternative one.

The indication of high abundance of IncP-9 plasmids in soils from Argentina contaminated with oil is not too surprising. IncP-9 plasmids are important vehicles for the dissemination of genes coding for enzymes involved in the degradation of polycyclic aromatic hydrocarbons (PAH) and are very often found in environments polluted with oil [23] (Flocco *et al.*, unpublished).

PCR-Southern blot hybridization results showed that bacteria hosting IncP-7 plasmids were also highly abundant in BPS, indicating a role of these plasmids in the degradation of man-made pollutants such as pesticides. It can be concluded that PCR-Southern blot hybridization detection of plasmid-specific sequences from TC-DNA is a suitable and specific but semi-quantitative approach to investigate the occurrence of plasmid-specific sequences in different environments and in a large number of samples. The detection of plasmids was possible independently of the cultivation of their original hosts [1] and indicated "hot spots" of IncP-1, IncP-7 and IncP-9 plasmids, such as BPS.

Author Contributions

Performed the experiments: SD VW. Analyzed the data: SD GCD HH KS. Contributed reagents/materials/analysis tools: HH CT. Wrote the paper: SD. Contributed sending samples: VD AS MCM MFP AL GCA EMHW WHG D Sipkema SS D Springael JDE. Corrected the writing: KS.

References

1. Heuer H, Smalla K (2012) Plasmids foster diversification and adaptation of bacterial populations in soil. FEMS Microbiology Reviews 36: 1083–1104.
2. Bahl MI, Burmølle M, Meisner A, Hansen LH, Sørensen SJ (2009) All IncP-1 plasmid subgroups, including the novel ε subgroup, are prevalent in the influent of a Danish wastewater treatment plant. Plasmid 62: 134–139.
3. Hardeman F, Sjoling S (2007) Metagenomic approach for the isolation of a novel low-temperature-active lipase from uncultured bacteria of marine sediment. FEMS Microbiology Ecology 59: 524–534.
4. Izmalkova TY, Sazonova OI, Sokolov SL, Kosheleva IA, Boronin AM (2005) The P-7 incompatibility group plasmids responsible for biodegradation of naphthalene and salicylate in fluorescent pseudomonads. Microbiology 74: 290–295.
5. Schauss K, Focks A, Heuer H, Kotzerke A, Schmitt H, et al. (2009) Analysis, fate and effects of the antibiotic sulfadiazine in soil ecosystems. Trac-Trends in Analytical Chemistry 28: 612–618.
6. Suzuki MT, Taylor LT, DeLong EF (2000) Quantitative analysis of small-subunit rRNA genes in mixed microbial populations via 5′-nuclease assays. Applied and Environmental Microbiology 66: 4605–4614.
7. Sevastsyanovich YR, Krasowiak R, Bingle LEH, Haines AS, Sokolov SL, et al. (2008) Diversity of IncP-9 plasmids of *Pseudomonas*. Microbiology-Sgm 154: 2929–2941.
8. Krol JE, Penrod JT, McCaslin H, Rogers LM, Yano H, et al. (2012) Role of IncP-1 beta plasmids pWDL7::rfp and pNB8c in chloroaniline catabolism as determined by genomic and functional analyses. Applied and Environmental Microbiology 78: 828–838.
9. Jutkina JHE, Vedler E, Juhanson J, Heinaru A (2011) Occurrence of plasmids in the aromatic degrading Bacterioplankton of the Baltic Sea. Genes 2: 853–868.
10. Castillo MdP, Torstensson L, Stenström J (2008) Biobeds for environmental protection from pesticide use - a review. Journal of Agricultural and Food Chemistry 56: 6206–6219.
11. Omirou M, Dalias P, Costa C, Papastefanou C, Dados A, et al. (2012) Exploring the potential of biobeds for the depuration of pesticide-contaminated wastewaters from the citrus production chain: Laboratory, column and field studies. Environmental Pollution 166: 31–39.
12. Trefault N, De la Iglesia R, Molina AM, Manzano M, Ledger T, et al. (2004) Genetic organization of the catabolic plasmid pJP4 from *Ralstonia eutropha* JMP134 (pJP4) reveals mechanisms of adaptation to chloroaromatic pollutants and evolution of specialized chloroaromatic degradation pathways. Environmental Microbiology 6: 655–668.
13. Oliveira CS, Lazaro B, Azevedo JSN, Henriques I, Almeida A, et al. (2012) New molecular variants of epsilon and beta IncP-1 plasmids are present in estuarine waters. Plasmid 67: 252–258.
14. Smalla K, Haines AS, Jones K, Krögerrecklenfort E, Heuer H, et al. (2006) Increased abundance of IncP-1 beta plasmids and mercury resistance genes in mercury-polluted river sediments: First discovery of IncP-1 beta plasmids with a complex mer transposon as the sole accessory element. Applied and Environmental Microbiology 72: 7253–7259.
15. Top EM, Springael D (2003) The role of mobile genetic elements in bacterial adaptation to xenobiotic organic compounds. Current Opinion in Biotechnology 14: 262–269.
16. Sen DY, Brown CJ, Top EM, Sullivan J (2013) Inferring the Evolutionary History of IncP-1 Plasmids Despite Incongruence among Backbone Gene Trees. Molecular Biology and Evolution 30: 154–166.
17. Dunon VSK, Bers K, Lavigne R, Smalla K, Springael D (2013) High prevalence of IncP-1 plasmids and *IS1071* insertion sequences in on-farm biopurification systems and other pesticide polluted environments. FEMS Microbiology Ecology.
18. Jechalke S, Dealtry S, Smalla K, Heuer H (2013) Quantification of IncP-1 plasmid prevalence in environmental samples. Applied and Environmental Microbiology 79: 1410–1413.
19. Gomes NCM, Flocco CG, Costa R, Junca H, Vilchez R, et al. (2010) Mangrove microniches determine the structural and functional diversity of enriched petroleum hydrocarbon-degrading consortia. FEMS Microbiology Ecology 74: 276–290.
20. Fetzner S, Lingens F (1994) Bacterial Dehalogenases: Biochemistry, Genetics, and Biotechnological Applications. Microbiological Reviews 58: 641–685.
21. Greated A, Lambertsen L, Williams PA, Thomas CM (2002) Complete sequence of the IncP-9 TOL plasmid pWW0 from *Pseudomonas putida*. Environmental Microbiology 4: 856–871.
22. Sota M, Yano H, Ono A, Miyazaki R, Ishii H, et al. (2006) Genomic and functional analysis of the IncP-9 naphthalene-catabolic plasmid NAH7 and its transposon Tn4655 suggests catabolic gene spread by a tyrosine recombinase. Journal of Bacteriology 188: 4057–4067.
23. Izmalkova TY, Mavrodi DV, Sokolov SL, Kosheleva IA, Smalla K, et al. (2006) Molecular classification of IncP-9 naphthalene degradation plasmids. Plasmid 56: 1–10.
24. Zhou Y (1996) Two-phase anaerobic digestion of water hyacinth pretreated with dilute sulphuric acid. Huanjing Kexue 17: 13–92.
25. Sipkema D, Blanch HW (2010) Spatial distribution of bacteria associated with the marine sponge *Tethya californiana*. Marine Biology 157: 627–638.
26. Edlund A, Hardeman F, Jansson JK, Sjoling S (2008) Active bacterial community structure along vertical redox gradients in Baltic Sea sediment. Environmental Microbiology 10: 2051–2063.
27. Gaze WH, Zhang LH, Abdouslam NA, Hawkey PM, Calvo-Bado L, et al. (2011) Impacts of anthropogenic activity on the ecology of class 1 integrons and integron-associated genes in the environment. ISMEJournal 5: 1253–1261.
28. Byrne-Bailey KG, Gaze WH, Kay P, Boxall ABA, Hawkey PM, et al. (2009) Prevalence of sulfonamide resistance genes in bacterial isolates from manured agricultural soils and pig slurry in the United Kingdom. Antimicrobial Agents and Chemotherapy 53: 696–702.

8

Can Biochar and Phytoextractors Be Jointly Used for Cadmium Remediation?

Huanping Lu[1,2], Zhian Li[1]*, Shenglei Fu[1], Ana Méndez[3], Gabriel Gascó[4], Jorge Paz-Ferreiro[1,4]

1 Key Laboratory of Vegetation Restoration and Management of Degraded Ecosystems, South China Botanical Garden, Chinese Academy of Sciences, Guangzhou, China, **2** University of Chinese Academy of Sciences, Beijing, China, **3** Departamento de Ingeniería de Materiales, Universidad Politécnica de Madrid, Madrid, Spain, **4** Departamento de Edafología, Universidad Politécnica de Madrid, Madrid, Spain

Abstract

Phytoremediation of soils contaminated with cadmium was tested after liming (CaO) or biochar addition, using red amaranth (*Amaranthus tricolor* L.) as test plant species. Two biochars with contrasting characteristics were prepared from two feedstocks and added to the soil at a rate of 3% (w:w): Eucalyptus pyrolysed at 600°C (EB) and poultry litter at 400°C (PLB). Liming was carried out in two treatments (CaO1) and (CaO2) to the same pH as the treatments EB and PLB respectively. Total plant mass increased in soils amended with PLB and with a mixture of PLB and EB; however this was not sufficient to increase the efficiency of phytoextraction. Bioavailable and mobile fractions of Cd diminished after liming or biochar addition. Our study infers that, both the amount of Cd immobilized and the main mechanism responsible for this immobilization varies according to biochar properties.

Editor: Richard G. Haverkamp, Massey University, New Zealand

Funding: J. Paz-Ferreiro was sponsored by the Chinese Academy of Sciences (fellowship for young international scientists number 2012Y1SA0002). The authors acknowledge support from the National Natural Science Foundation of China (No. 40871221 and No. 41301571), and the Research Fund Program of Guangdong Provincial Key Laboratory of Environmental Pollution Control and Remediation Technology (2013K0008). The funders had no role in study design, data collection and analysis, decision to publish, or preparation of the manuscript.

Competing Interests: The authors have declared that no competing interests exist.

* E-mail: lizan@scbg.ac.cn

Introduction

Industrialisation and technical advances have led to an increase in the use of heavy metals. Anthropogenic activities, including smelting, mining, use of pesticides, fertilisers and sludge are responsible for the input of high levels of heavy metals into the soil. In the last years there is an increasing concern about soil heavy metal pollution. Soil heavy metal pollution poses a risk to the environment and to human health due to biomagnification (increases in metal concentration as the element passes from lower to higher trophic levels). An additional concern is that, contrary to organic substances, heavy metals are non-degradable and accumulate in the environment.

In particular, cadmium pollution is a concern in vast areas of China and some areas in Japan, Taiwan and other countries, rendering valuable farmland into non-cultivated areas. Cadmium is known to originate from lithogenic sources in areas with high abundance in zinc and lead ores. However, in the last years the agricultural use of fertilisers and the addition of sewage sludges to the soil have resulted in an increase in soil Cd concentrations. In general plants tolerate Cd better than human and animals and Cd toxicity in plants appears at higher concentrations than in animals or humans. In humans, excessive Cd intake can cause the "Itai-itai disease" and has also been related to some types of cancer or to kidney damage.

In the last years there has been an increasing interest in the use of biochar to tackle soil heavy metal pollution [1], [2], [3]. Biochar influences a number of biogeochemical processes and in general there has been reported a positive effect on plant productivity [4] while simultaneously mitigating greenhouse emissions (as CO_2 and N_2O) from soil. Several co-benefits may result from the use of biochar including positive effects on soil field capacity, nutrient availability, fertilizer use efficiency, pH ("liming effect"), cation exchange capacity (CEC) and soil biological properties [5]. Therefore, in the past few years, there has been a growing interest in the use of biochar as a soil amendment. The potential of biochar to immobilize toxic substances, in particular cadmium, from different types of soil is well documented [2], [3], however, more uncertainties prevail about the mechanisms involved in heavy metal immobilization and their relative importance.

Biochars using plant materials as feedstocks are considered a soil conditioner, due to their low leachable nutrient contents [6]. On the contrary, manure-derived biochar can act as soil fertilizer and conditioner due to NPK release [7].

On the other hand, phytoextraction consists in the use of hyperaccumulator plants to remove heavy metals from the environment. This is a relatively new technology developed from 1990 onwards. Because of its low costs, phytoremediation can be considered as a relatively attractive technique to restore or partially decontaminate a site compared to other options, e.g. [8], [9]. Other advantages of this technique would include its good perception among the general public. Plant species used for phytoextraction must not only accumulate high amounts of the target element but also have a high growth rate, tolerate toxic effects of heavy metals, be adapted to local soil and climate, be resistant to pathogen and pests, be easy to cultivate and repulse herbivores to avoid food chain contamination [9], [10]. Management of soil pH (liming) and soil nutrients have been utilised in the past in order to increase metal hyperaccumulation. In particular, liming is more effective when heavy metal concentrations in the soil are too elevated to reduce plant stress [11], [12] [13]. Organic matter amendments from several sources have been jointly used with phytoremediators and results showed that organic matter had either null or positive effect in enhancing phytoremediation [12], [14].

Of the more than 400 plants identified as phytoextractors most are Ni-hyperaccumulators and only a handful of plants accumulating Cd have been identified. Moreover, most Cd-accumulating plants present slow growth rates. Recently *Amaranthus tricolor* L. has been identified as a Cd-accumulator [15]. This species presents the main advantage of a rapid growth and a relative high yield compared to other Cd accumulating plants [15].

Simultaneous use of biochar and phytoremediation may seem contradictory, because until now most research results indicate that biochar reduce the bioavailability of heavy metals [2], [3] but plants need high concentrations of soluble metals to extract and to accumulate them. In fact, exploration of the efficiency in the joint use of biochar amendments and phytoremediation, has been attempted before [16], [17] for the case of Cd using *Brassica napus* L. as phytoextractor and Miscanthus biochar [16] or for the case of multicontaminated soils using different biochars and plant species [17].

Based on the above considerations we performed an experiment with the following aims: a) To explore the possibilities of biochars with contrasting characteristics on Cd remediation b) To explore the possibility of combining biochar with phytoremediation c) To deepen our mechanistic knowledge about how biochar contributes to heavy metal immobilization, using lime controls.

Material and Methods

Ethics statement

No specific permissions were required for collecting the soil, poultry litter or Eucalyptus used in this study.

Our study did not involve the use of endangered or protected species.

Soil sampling and characteristics

The soil was collected from the surface layer (0–20 cm) of a cropland area, near a waste landfill site in the suburb of Guangzhou, China (23° 07′ N and 113° 21′ E). Guangzhou is located in the subtropical humid area having an average annual temperature of 12.7°C and annual average precipitation of 1700 mm. According to FAO, the soil is classified as a fimic anthrosol.

Part of the soil was sieved to 2 mm to conduct general analyses, while the rest was sieved to 10 mm to conduct the experiment. Before the starting of the pot experiment, the soil had a total organic carbon content of 1.98%, total nitrogen content of 0.142%, pH of 6.00, total phosphorus of 690 mg kg^{-1}, available phosphorus of 126 mg kg^{-1} and total Cd 6.1 mg kg^{-1}.

Preparation and characterization of biochar

Two biochars were used in this experiment using poultry litter and eucalyptus as feedstocks. Poultry litter was collected at the Experimental Poultry Farm of South China Agricultural University (23° 09′ N and 113° 21′ E) located at Guangzhou, China. Chicken at this facility are organically bred and fed with a mixture of corn, wheat bran and soybean pulp. After collection, litter was dried in an oven (70°C) for 24 hours and sieved to 1 mm.

Eucalyptus (*Eucalyptus urophylla* S.T. Blake) wood was collected at Heshan Hilly Land Experiment Station of the Chinese Academy of Sciences in Heshan, China (22° 84′ N and 112° 54′ E). Wood was collected from branches of four different individuals, mixed together and brought to the South China Botanical Garden where it was ground to wood chips with a particle size smaller than 0.5 mm.

Biochars were prepared as described by [18] with samples being pyrolysed by increasing the temperature to 400°C (poultry litter biochar) or 600°C (Eucalyptus biochar) at a rate of 10°C min^{-1}. The final temperature was maintained for 2 h. Different temperatures were used in the preparation of the biochars as our main aim was to produce two biochars with contrasting characteristics.

Mesocosm experiments

On the 13th April of 2013, a mesocosm using a fully replicated randomised experiment was set up in a greenhouse in South China Botanical Garden. Each mesocosm consisted in a pot filled with 1.5 kg of soils in pots with plants and 500 g of soil in pots without crops. The reason for using different amounts of soil was to minimise the amount of biochar prepared, while allowing the crops to have enough soil to grow. The soils were watered to 60% of field capacity and watered daily to account for moisture losses.

The experiment studied two factors, namely the type of amendment and the presence/absence of phytoremediator, having 4 replications per treatment. The treatments in the amendment factor were: Control (C), poultry litter biochar (3% w/w) (PLB), eucalyptus biochar (3% w/w) (EB), poultry litter biochar (1.5% w/w) + eucalyptus biochar (1.5% w/w) (BB), liming with CaO as necessary to increase soil pH as much as EB (CaO1) and liming with CaO as necessary to increase soil pH as much as PLB (CaO2).

With respect to the factor involving presence/absence of phytoremediator, half of the mesocosm were planted with 3 seeds of red amaranth (Amaranth tricolor L.), while in the remainder half the vegetative cover was absent.

After 60 days soils and plant were collected for analyses.

Soil analysis

Soil pH was determined using a 1:2.5 soil:water ratio. Total carbon and total nitrogen were determined using using a vario ISOTOPE CUBE elemental analyzer (elementar, Germany). Plant-available metals were extracted from treated soil using diethylenetriaminepentaacetic acid–$CaCl_2$–triethanolamine (DTPA) as described elsewhere [19]. The mobile forms of heavy metals were extracted using 0.1 M $CaCl_2$ [20].

Plant analysis

The crops were harvested and separated into roots, stems and leaves. Then they were rinsed firstly with tap water followed by distilled water. Samples were oven-dried for 72 h at 65°C. Once dried, plant biomass was weighted and ground. For digestion, a microwave closed system (Multiwave3000, Anton Paar, Austria) was used. Samples (0.1 g) were digested with 5 mL of nitric acid and 1 mL of hydrogen peroxide in a microwave digestion system for 30 min and diluted with deionized water to 30 mL. Cadmium concentrations were analyzed using an ICP-MS spectrometer (Agilent 7700x, USA).

Biochar analysis

pH and TOC in the biochar were determined as described for the soil. Biochar nitrogen adsorption analysis to determine BET surface was carried out at 77 K in a Micromeritics Tristar 3000 (Instituto de Catálisis y Petroquímica, CSIC, Spain).

Total pore volume (V) was estimated from the amount of nitrogen adsorbed at a relative pressure. Average pore size (D) was estimated from V and BET surface using the following equation and assuming a cylindrical pore shape:

$D = 4V/S_{BET}$.

Proximate analysis was calculated by thermogravimetry using a Labsys Setaram equipment. Samples were heated up to 600°C under an N_2 atmosphere at a flux of 40 mL min^{-1} using a heat rate of 20°C min^{-1}. Moisture content (H) was determined as weight loss at 120°C and Volatile matter (VM) was determined as the weight loss from 120°C to 600°C. At 600°C, air flux was introduced until a constant weight was reached and ashes were determined as the final weight of the samples. Fixed carbon (FC) was calculated by difference.

Cadmium in the biochars was determined as described for the soil samples.

Phytoextraction indices

The amount of Cd transported from soil to shoots was calculated using the bioconcentration factor (BCF), [21].

Accumulation Factor = Cd concentration plant tissue/Cd concentration in soil.

The ability of each species to translocate Cd from the roots to the shoots was calculated by the translocation index (TI) [22].

TI (%) = Cd concentration in aerial parts/Cd concentration in roots x 100.

Statistical analysis

Statistical analyses (calculation of means and standard deviations, differences between treatments) were performed using SPSS 15.0 package. Differences of means were tested using a two-way ANOVA with the presence/absence of plant and the type of amendment (eucalyptus biochar, poultry litter biochar, CaO1, CaO2, mixture of biochar or none) as factors. Means were considered to be different when $P<0.05$ using the Tukey's test.

Results and Discussion

Biochar properties

The physical characteristics of both biochars were contrasted (Table 1). Eucalyptus biochar (EB) had a larger surface area and pore volume compared to the poultry litter biochar (PLB) due to its higher content in carbon. Biochar pore volume has been directed related before to surface area [23] and both, pore volume and surface area, are expected to have larger values in biochars prepared at higher temperatures [24]. Rouquerol et al [25] defined the internal diameter range of pores for porous solids, suggesting <2.0 nm as micropores, 2.0–50.0 nm as mesopores, and >50.0 nm as macropores. Our results showed clear evidence that the PLB was dominated by mesopores, while EB was dominated by micropores.

Biochar pHs are very basic according to other works that attributed this fact to the polymerisation/condensation reactions and acidic surface group releases during pyrolysis [26]. The total Cd content of biochars was negligible. Finally, the PLB presented a higher content in ash, VM and a lower content in FC than EB due to the different composition of the raw materials and pyrolysis conditions.

Soil properties

The addition of lime or biochar clearly modified soil pH (see Table 2). All treatments increased significantly the pH compared to the control ($P<0.001$). While the control soil had a pH of 6.03, this increased to 7.62 for the treatment PLB, 7.01 for EB and 7.45 for BB. The pH of PLB was significantly higher than that of EB.

Total carbon and total nitrogen increased in BB and EB compared to the control. However, in the case of PLB only an increase in total nitrogen with respect to the control was found.

DTPA extracted-Cd and $CaCl_2$ extracted-Cd can be considered respectively as the bioavailable and mobile forms of Cd present in soil. Our study showed that both fractions were affected by the type of amendment (in both cases $P<0.001$); however only $CaCl_2$ extracted-Cd was affected by the presence of red amaranth ($P<0.010$). Moreover, for $CaCl_2$ extracted-Cd there was an interaction phytoremediation x treatment ($P<0.010$). Thus, mobile forms of Cd were higher in the EB soil with amaranth with respect to the one without amaranth. This leads us to think that, at least to some extent, plants with a high affinity for heavy metals can mobilize part of the heavy metal stabilized by biochar.

Incorporation of biochar reduced $CaCl_2$ extracted Cd compared to the control by 97, 67 and 92% respectively for PLB, EB and BB (see Figure 1). The value for PLB was similar to the one found by [16] using Miscanthus biochar. These values were 40, 10 and 29% comparing the reduction with respect to the control in the case of DTPA extracted Cd (Figure 2).

In the last years a number of studies have shown that biochars have a potential for metal adsorption and immobilization, but there is no agreement about what drives biochar effects on soil heavy metals. While some authors postulate that this effect is

Table 2. General properties of the soils at the end of the experiment.

	pH	Carbon (%)	Nitrogen (%)
Control	6.03±0.03 a	1.76±0.03 a	0.123±0.005 a
PLB	7.62±0.03 d	2.35±0.12 a	0.171±0.006 b
EB	7.01±0.02 b	5.19±0.52 c	0.176±0.010 b
CaO1	6.81±0.06 b	1.79±0.23 a	0.128±0.012 a
CaO2	7.08±0.12 bc	1.73±0.12 a	0.122±0.009 a
BB	7.45±0.22 cd	3.80±0.40 b	0.172±0.020 b

Different letters in the same column indicate statistical significant differences ($P<0.05$).

Table 1. General characteristics of the biochars used in the experiment.

	pH	Carbon (%)	Nitrogen (%)	Cd (mg kg^{-1})	Surface area (m^2 g^{-1})	Average pore width (nm)	Pore volume (cm^3 g^{-1})	Ash (%)	VM (%)	FC (%)
PLB	10.02	16.77	1.37	n.a.	7.418	15.406	0.0286	74.95	7.66	17.39
EB	10.40	81.03	1.07	n.a.	334.560	1.928	0.1612	1.74	2.19	96.06

VM and FC stand for volatile matter and fixed carbon, respectively.

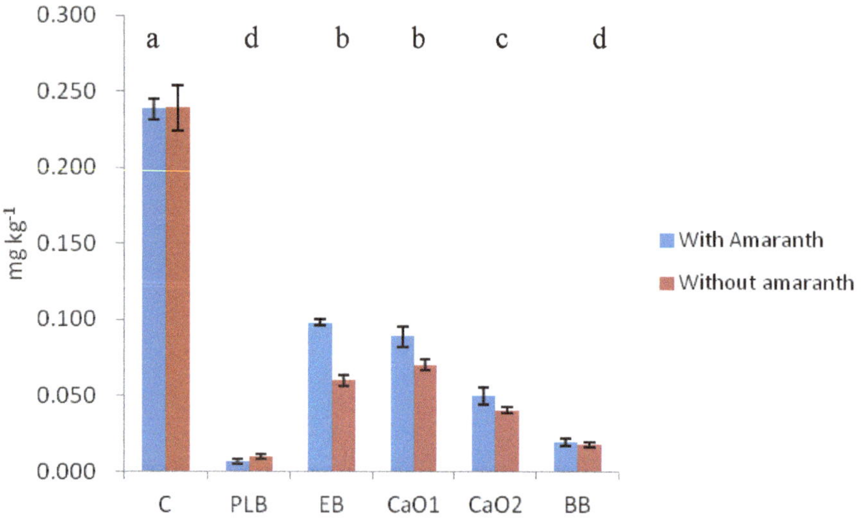

Figure 1. $CaCl_2$ extracted Cd for all the treatments at the end of the experiment. Different letters in the same indicate statistical significant differences for the factor "type of amendment" (P<0.05).

mainly pH mediated [27], other authors pointed out to high surface are and pore volume [23], [28], or the amount of oxygen functional groups [29]. Our study confirms the capacity of biochar to immobilize cadmium and points out to different mechanisms depending on biochar type. This experiment demonstrated that the EDTA and $CaCl_2$ extractable Cd in EB was similar to that of CaO1, pointing out to Cd immobilization as caused primarily by changes in soil pH for EB and pointing to negligible effects of biochar surface area, contrary to the suggestion of other authors [30]. However, this was not the case of PLB, which had a very different EDTA and $CaCl_2$ extractable Cd compared to CaO2. We would expect that Cd immobilisation in the PLB treatment was driven both by pH changes and by precipitation with mineral ash. Thus, PLB has a high content of ash, which must be enriched by mineral salts of K, Ca, Mg, C and P among other elements. Cadmium can precipitate as insoluble phosphate and carbonate salts, in particular and high pH values. Another mechanism that could explain Cd immobilization in the PLB treatment would be the presence of high oxygen-containing functional groups in the poultry litter biochar. These groups are particularly effective for heavy metal stabilization in soils with low organic carbon contents [29].

Plant growth and Cd uptake

Plant biomass increased as a consequence of liming and poultry litter biochar addition (Figure 3). This way, total plant biomass was 3.44 and 2.24 times higher under PLB and BB respectively than in the control. Treatments CaO1, CaO2 and EB showed no difference with the control. We would like to highlight that plant height was higher in all treatments with respect to the control for one month following the start of the experiment (data not shown), with the addition of two different biochars (BB) performing better than any other treatment. This suggests that at least in the initial state of growing there could be some synergistic effect between different types of biochar.

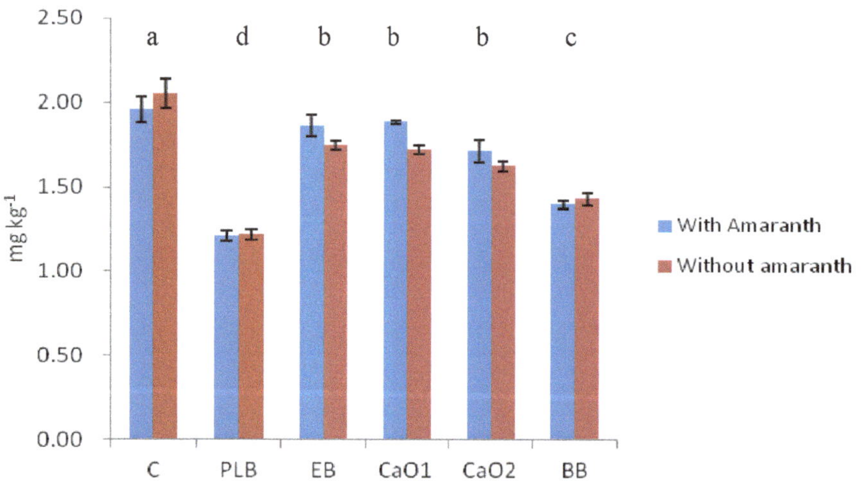

Figure 2. DTPA extracted Cd for all the treatments at the end of the experiment. Different letters in the same indicate statistical significant differences for the factor "type of amendment" (P<0.05).

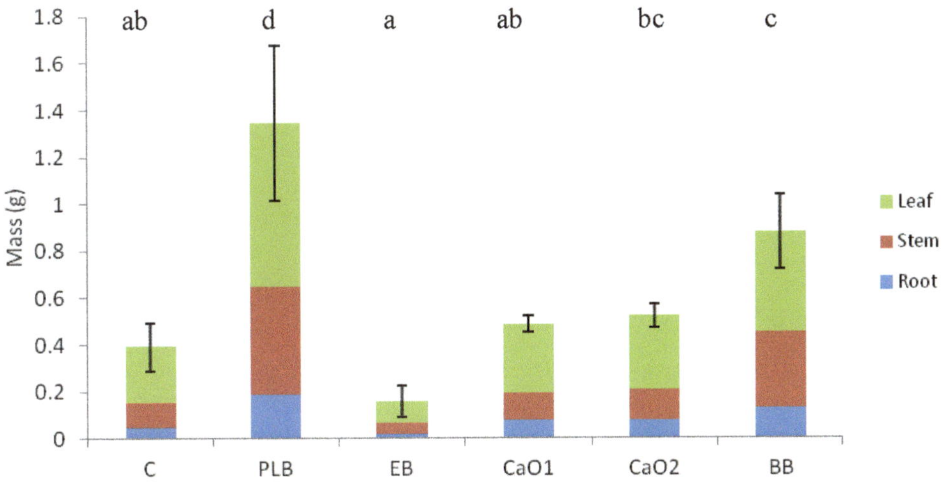

Figure 3. Total plant biomass at the end of the experiment. Different letters in the same indicate statistical significant differences for the factor "type of amendment" ($P<0.05$).

Cadmium concentration in the different plant tissues was reduced by every treatment. Concentrations diminished more for PLB and for BB (with both treatments having similar values) than in EB. The concentrations of Cd in the root, shoots and leaves of the plants were in all cases over the range of concentrations (see Table 3) that is usually considered as excessive or toxic (5–30 mg kg^{-1}, [31]) for humans, but we did not observe any signs of Cd toxicity in the plants.

Quantification of phytoextraction efficiency

All treatments extracted significantly less Cd from the soil than the control ($P<0.001$) as shown in Figure 4. EB treatment removed only 22% of the Cd accumulated by Amaranthus in the control, while this percentage was 49% in the case of PLB.

The translocation index (TI) was 120.5 in the control soil and dropped to 71.0 for PLB (see Table 3). The bioconcentration factor (BCF) varied from 14.76 in the control soil to 2.17 in PLB. The treatments EB, CaO1, CaO2 and BB also had a lower transfer factor than the control soil, being 8.12 5.67, 4.05 and 2.71 respectively.

The values of TI for EB and the control soil were similar probing that EB did not inhibit heavy metal transportation from roots to shoots, but did from soil to shoots (see BCF values). On the other hand, liming and PLB inhibited both, transportation from soil to roots and from roots to the aerial parts of the plant.

Both indices, TI and BCF are important in screening hyperaccumulators species. Yoon et al. [32] recommends the use of plants that have a TI of more than 100% and BCF values higher than 1. While all our treatments had BCF values higher than 1, the low values of TI would not make realistic to combine amaranth with biochar for phytoextraction purposes due to the excessive amount of time required. However, the increase of biomass observed with PLB and BB, could have some useful implications that will be next addressed.

Significance of our study

Based on our study, it seems that using biochar alone as a soil amendment is a promising technique to immobilize Cd in polluted areas, but a careful choice of feedstock and pyrolysis temperature should be made. A study on five sites concerning rice grown in polluted areas demonstrated that biochar can reduce rice Cd uptake through a reduction in Cd mobility [1]. However, PLB would have a greater potential than the wheat straw biochar used by [1] as its use resulted in greater heavy metal immobilization, as measured by $CaCl_2$ extraction.

Other experiments with Cd and biochar have shown that Cd mobility and plant availability can be reduced with increasing doses of biochar, at least up to an amendment of 10% biochar [16]. In addition to its use as a phytoremediator, amaranth has been used for as forage species or for human consumption. The values obtained in this study for Cd concentrations in the leaves of

Table 3. Phytoremediation parameters and Cd concentration in different plant tissues at the end of the experiment.

	BCF	TI	Root Cd concentration	Leaf Cd concentration	Stem Cd concentration
Control	12.76±1.90 a	120.27±5.42 a	76.40±4.65 a	108.57±7.87 a	52.15±7.87 a
PLB	2.17±0.42 d	70.74±20.57 b	17.79±0.58 d	15.65±1.78 d	7.31±1.78 d
EB	8.12±0.78 b	93.22±15.39 ab	53.62±4.90 b	57.77±2.47 b	30.33±2.47 b
CaO1	5.67±0.80 c	79.31±7.68 b	42.03±3.12 bc	39.27±3.12 c	17.57±3.12 c
CaO2	4.05±0.26 cd	80.21±9.89 b	29.91±1.45 cd	28.53±0.83 cd	12.07±0.83 cd
BB	2.71±0.20 d	71.80±10.27 b	22.05±1.67 d	21.64±0.73 d	7.77±0.73 d

Different letters in the same column indicate statistical significant differences ($P<0.05$).

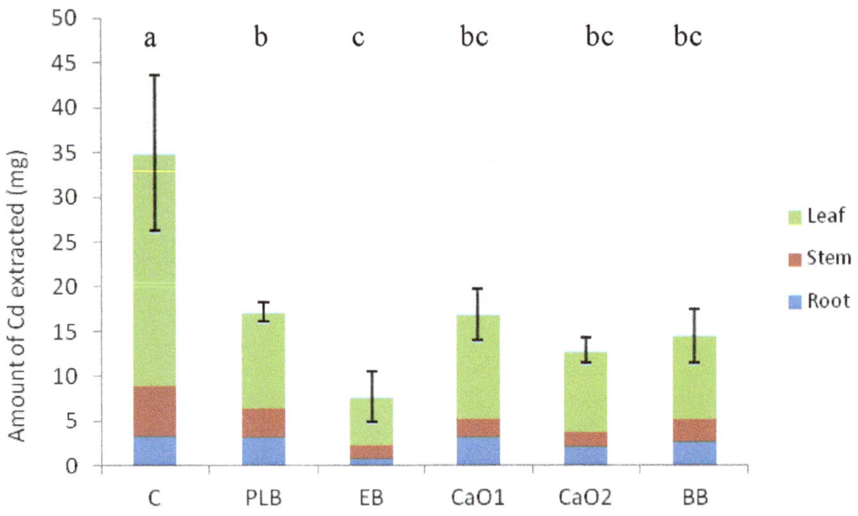

Figure 4. Total amount of Cd extracted by plants. Different letters in the same indicate statistical significant differences for the factor "type of amendment" ($P<0.05$).

plants (>10 mg kg^{-1}) would discourage the use of amaranth for this purpose, even in the treatments with lower leaf Cd concentration.

In the last years much research about biochar has been emerged and some unintended consequences have been identified including the potential presence of toxic compounds in biochars [33] and increasing difficulties for weed control due to pesticide adsoption on biochar surface [34]. Our study is, to our knowledge, pioneering (but see also [14], [17]) with respect to addressing a joint use of phytoremediation and biochar for the remediation of heavy metal polluted soils and raises awareness over a new unintended consequence (limitation on the use of phytoextractors), which can arise from biochar management. Our results point out that biochars could impede the transfer of heavy metals to phytoextractors and slow down remediation processes. However, these results should be carefully extrapolated to the field scale, as the yield metal extracted by phytoremediators can vary greatly among field and mesocosm studies [35].

Our study did not account for other potentially beneficial effects which are concomitant to the use of phytoremediation and are derived from the establishment of a plant cover. In this aspect, introduction of vegetation in a polluted area can help to prevent erosion or contaminant leaching. In addition, it could be expected that the vegetation will help the improvement of soil biological properties. Moreover, the harvested biomass of the phytoremediator could be used as feedstock for biochar production. This option would only be feasible if the feedstock biomass does not contain excessive levels of heavy metals. Indeed, Méndez et al. [2] have demonstrated that after pyrolysis of a sewage sludge containing heavy metals, the major part of the metal was immobilized in the biochar and as a consequence, addition of sewage sludge biochar did not result in increased mobile fractions of metals in soil. However, metal concentrations in a biochar prepared using a metallophyte as feedstock would presumable be higher than in the sewage sludge used by Méndez et al. [2]. Carbon sequestration is another benefit that could arise from the combined use of phytoextractors and biochar. Thus, we could devise some situations where it could be advantageous to combine phytoextraction with biochar amendment. This type of scenarios would include soils with low organic matter contents and with acidic pH values which would compromise the growth of phytoextractors.

There are two big interrogations regarding the future use of biochar and phytoremediation as joint strategies to address heavy metal pollution. Firstly, our experiment has only considered a single heavy metal (cadmium). According to other studies [3], [36] biochar has a moderate affinity for Cd which would make our results difficult to extrapolate to heavy metals that are not so likely to be immobilized by biochar or to multicontaminated soils. Secondly, the development of biochar-based fertilizers [37] could change the present paradigm of soil heavy metal remediation, as we could expect that with a careful choice of biochar lessened amounts would be needed to obtain a plant effect, resulting in fewer amounts of biochar added to the soil and consequently diminished pollutant immobilization. Another future line for this type of studies would be to use phytoremediators and biochars targeting at different heavy metals (or the biochar not to target any metal at all), as the biochar effect is strongly dependent on the chemical element immobilized.

Author Contributions

Conceived and designed the experiments: HPL ZAL SF AM GG JPF. Performed the experiments: HPL JPF. Analyzed the data: HPL ZAL SF AM GG JPF. Contributed reagents/materials/analysis tools: ZAL SF. Wrote the paper: HPL ZAL SF AM GG JPF.

References

1. Bian R, Chen D, Liu X, Cui L, Li L, et al. (2013) Biochar soil amendment as a solution to prevent Cd-tainted rice from China: Results from a cross-site field experiment. Ecol Eng 58: 378–383.
2. Méndez A, Gómez A, Paz-Ferreiro J, Gascó G (2012) Effects of biochar from sewage sludge pyrolysis on Mediterranean agricultural soils. Chemosphere 89: 1354–1359.
3. Uchimiya M, Lima IM, Klasson KT, Wartelle LH (2010) Contaminant immobilization and nutrient release by biochar soil amendment: Roles of natural organic matter. Chemosphere 80: 935–940.
4. Liu X, Zhang A, Ji C, Joseph S, Bian R, et al. (2013) Biochar's effect on crop productivity and the dependence on experimental conditions- a meta-analysis of literature data. Plant Soil 373: 583–594.

5. Paz-Ferreiro J, Fu S, Méndez A, Gascó G (2014) Interactive effect of biochar and the earthworm Pontoscolex corethrurus on plant productivity and soil enzyme activity. J Soil Sediment 14: 483–494.
6. Steiner C, Teixeira WG, Lehmann J, Nehls T, de Macedo JLV, et al. (2007) Long term effects of manure, charcoal and mineral fertilization on crop production and fertility on a highly weathered Central Amazonian upland soil. Plant Soil 291: 275–290.
7. Chan KY, Van Zwieten L, Meszaros I, Downie A, Joseph S (2008) Using poultry litter biochars as soil amendments Aust J Soil Res 46: 437–444.
8. Paz-Ferreiro J, Lu H, Fu S, Méndez A, Gascó G (2014) Use of phytoremediation and biochar to remediate heavy metal polluted soils: A review. Solid Earth 5: 65–75.
9. Ali H, Khan E, Sajad MA (2013) Phytoremediation of heavy metals-Concepts and applications. Chemosphere 91: 869–881.
10. McGrath SP, Zhao FJ (2003) Phytoextraction of metals and metalloids from contaminated soils. Curr Opin Biotechnol 14: 277–282.
11. Adriano DC, Wenzel WW, Vangrosveld J, Nolam NS (2004) Role of assisted natural remediation in environmental clean-up. Geoderma 122: 121–142.
12. de Abreu CA, Coscione AR, Pires AM, Paz-Ferreiro J (2012) Phytoremediation of a soil contaminated by heavy metals and boron using castor oil plants and organic matter amendments. J Geochem Explor 123: 3–7.
13. Komárek M, Vaněk A, Ettler V (2013) Chemical stabilization of metals and arsenic in contaminated soils using oxides-a review. Environ Pollut 172: 9–22.
14. Gabos MB, Casagrande G, Abreu CA, Paz-Ferreiro J (2011) Use of organic matter to mitigate multicontaminated soil and sunflower plants as phytoextractor. Rev Bras Eng Agr Amb 15: 1298–1306.
15. Watanabe T, Murata Y, Osaki M (2009) Amaranthus tricolor has the potential for phytoremediation of cadmium-contaminated soils. Commun Soil Sci Plant Anal 40: 3158–3169.
16. Houben D, Evrard L, Sonnet P (2013) Beneficial effects of biochar application to contaminated soils on the bioavailability of Cd, Pb and Zn and the biomass production of rapeseed (Brassica napus L.). Biomass Bioenerg 57: 196–204.
17. Fellet G, Marmiroli M, Marchiol L (2014) Elements uptake by metal accumulator species grown on mine tailings amended with three types of biochar. Sci Total Environ 468–469: 598–608.
18. Gascó G, Paz-Ferreiro J, Méndez A (2012) Thermal analysis of soil amended with sewage sludge and biochar from sewage sludge pyrolysis. J Thermal Anal Calorim 108: 769–775.
19. Lindsay WL, Norvell WA (1969) Development of a DTPA micronutrients soil test. Agron Abs 69: 84.
20. Ure AM, Davidson CM, Thomas RP (1995) Single and sequential extraction schemes for trace metal speciation in soil and sediment. Quality Assurance for Environmental Analysis, vol. 20. Elsevier Science B.V., pp. 505–523.
21. Zhuang P, Yang Q, Wang H, Shu W (2007) Phytoextraction of heavy metals by eight plant species in the field. Water Air Soil Pollut 184: 235–242.
22. Zacchini M, Pietrini F, Mugnozza GS, Iori V, Pietrosanti L, et al. (2009) Metal tolerance, accumulation and translocation in poplar and willow clones treated with cadmium in hydroponics. Water Air Soil Poll 197: 23–34.
23. Downie A, Crosky A, Munroe P (2009) Physical properties of biochar. J. Lehmann, S. Joseph (Eds.), Biochar for Environmental Management: Science and Technology, Earthscan, London, UK.
24. Kameyama K, Miyamoto T, Shiono T, Shinogi Y (2012) Influence of sugarcane bagasse-derived biochar application on nitrate leaching in calcaric dark red soil. J Environ Qual 41: 1131–1137.
25. Rouquerol F, Rouquerol I, Sing K (1999) Adsorption by Powders and Porous Solids, Academic Press, London, UK.
26. Méndez A, Gascó G, Freitas MMA, Siebielec G, Stuczynsky T, et al. (2005) Preparation of carbon-based adsorbents from pyrolysis and air activation of sewage sludges. Chem Eng J 108: 169–177.
27. Houben D, Evrard L, Sonnet P (2013) Mobility, bioavailability and pH-dependent leaching of cadmium, zinc and lead in a contaminated soil amended with biochar. Chemosphere 92: 1450–1457.
28. Mohan D, Pittman CU Jr, Brcika M, Smith F, Yancey B, et al. (2007) Sorption of arsenic, cadmium and lead by chars produced from fast pyrolysis of wood and bark during bio-oil production. J Colloid Interface Sci 310: 57–73.
29. Uchimiya M, Chang SC, Klasson KT (2011) Screening biochars for heavy metal retention in soil: Role of oxygen functional groups. J Hazard Mater 190: 432–444.
30. Park JH, Choppala GK, Bolan NS, Chung JW, Chuasavathi T (2011) Biochar reduces the bioavailability and phytotoxicity of heavy metals. Plant Soil 348: 439–451.
31. Kabata-Pendias A, Mukherjee AB (2007) Trace elements from soil to human. Berlin: Springer.
32. Yoon J, Cao X, Zhou Q, Ma LQ (2006) Accumulation of Pb, Cu, and Zn in native plants growing on a contaminated Florida site. Sci Total Environ 368: 456–464.
33. Kookana RS, Sarmah AK, Van Zwieten L, Krull E, Singh B (2011) Biochar application to Soil: Agronomic and environmental benefits and unintended consequences. Adv Agron 112: 103–144.
34. Yu XY, Ying GG, Kookana RS (2009) Reduced plant uptake of pesticides with biochar additions to soil. Chemosphere 76: 665–671.
35. Ji P, Sun T, Song Y, Ackland ML, Liu Y (2011) Strategies for enhancing the phytoremediation of cadmium-contaminated agricultural soils by *Solanum nigrum* L. Environ Pollut 159: 762–768.
36. Namgay T, Singh B, Singh BP (2010) Influence of biochar application to soil on the availability of As, Cd, Cu, Pb, and Zn to maize (*Zea mays* L.). Aust J Soil Res 48: 638–647.
37. Joseph SD, Graber ER, Chia C, Munroe P, Donne S, et al. (2013) Shifting paradigms Shifting paradigms: development of high-efficiency biochar fertilizers based on nano-structures and soluble components. Carbon Management 4: 323–343.

9

A 2-Year Field Study Shows Little Evidence That the Long-Term Planting of Transgenic Insect-Resistant Cotton Affects the Community Structure of Soil Nematodes

Xiaogang Li[1,2], Biao Liu[1]*

1 Nanjing Institute of Environmental Sciences, Ministry of Environmental Protection of China, Nanjing, Jiangsu Province, China, 2 Key Laboratory of Soil Environment and Pollution Remediation, Institute of Soil Science, Chinese Academy of Sciences, Nanjing, Jiangsu Province, China

Abstract

Transgenic insect-resistant cotton has been released into the environment for more than a decade in China to effectively control the cotton bollworm (*Helicoverpa armigera*) and other Lepidoptera. Because of concerns about undesirable ecological side-effects of transgenic crops, it is important to monitor the potential environmental impact of transgenic insect-resistant cotton after commercial release. Our 2-year study included 1 cotton field where non-transgenic cotton had been planted continuously and 2 other cotton fields where transgenic insect-resistant cotton had been planted for different lengths of time since 1997 and since 2002. In 2 consecutive years (2009 and 2010), we took soil samples from 3 cotton fields at 4 different growth stages (seedling, budding, boll-forming and boll-opening stages), collected soil nematodes from soil with the sugar flotation and centrifugation method and identified the soil nematodes to the genus level. The generic composition, individual densities and diversity indices of the soil nematodes did not differ significantly between the 2 transgenic cotton fields and the non-transgenic cotton field, but significant seasonal variation was found in the individual densities of the principal trophic groups and in the diversity indices of the nematodes in all 3 cotton fields. The study used a comparative perspective to monitor the impact of transgenic insect-resistant cotton grown in typical 'real world' conditions. The results of the study suggested that more than 10 years of cultivation of transgenic insect-resistant cotton had no significant effects–adverse or otherwise–on soil nematodes. This study provides a theoretical basis for ongoing environmental impact monitoring of transgenic plants.

Editor: Nicolas Desneux, French National Institute for Agricultural Research (INRA), France

Funding: This research was jointly supported by the National Special Transgenic Project (2011ZX08012-005; 2013ZX08012-005), the National Science and Technology Pillar Program (No. 2013BAD11B01), the Jiangsu Province Science Foundation for Youths (BK2012498). The funders had no role in study design, data collection and analysis, decision to publish, or preparation of the manuscript.

Competing Interests: The authors have declared that no competing interests exist.

* E-mail: 85287064@163.com

Introduction

Many crops have been transformed to provide enhanced resistance against pests and diseases. Crops expressing δ-endotoxins of *Bacillus thuringiensis* (*Bt*) active against Lepidopteran and Coleopteran insect pests are the most widely grown [1]. The global area of transgenic crops has increased from 1.7 million hectares in 1996 to 160 million hectares in 2011 [1]. Transgenic insect-resistant cotton expressing Cry1Ab/c and/or CpTI (Cowpea Trypsin Inhibitor) has been released into the environment for commercial cultivation for more than a decade in China. Its planted area currently represents 71.5% of the total cotton grown in China [1]. These lines effectively control cotton bollworm and other Lepidoptera, resulting in a significant reduction in the usage of chemical insecticides, thus protecting the environment and human health while yielding substantial socioeconomic benefits [2–4]. Nevertheless, as with any technology, there have been questions about the potential environmental risks associated with transgenic plants. One of the major ecological concerns about the environmental risks of transgenic insect-resistant plants is the potential effects of these plants on non-target organisms [5–7].

The effects of transgenic insect-resistant cotton on non-target pests and natural enemies have been extensively assessed [8,9]. Most studies have found no convincing and meaningful negative effects of transgenic insect-resistant cotton on the population density, abundance, species richness and diversity of non-target arthropod natural enemies [9–13]. Pollinators are also important non-target organisms, and the impacts of transgenic crops on these organisms have also been evaluated. Feeding tests, as well as field surveys, have been extensively performed to evaluate the safety of *Bt* plants for honey bees or pollinating beetles, and no significant adverse effects on longevity, feeding and learning behavior, the development of the hypopharyngeal glands or superoxide dimutase activity have been observed in these insects [14–21].

Transgenic insect-resistant cotton could also affect soil organisms. Transgenic proteins, such as Cry1Ab and Cry1Ac, can be released into the soil from cotton residues, root exudates and pollen during growth and after harvest [22,23]. Once in the soil, the toxins can be bound to clay and humus particles [24]. This state protects them from biodegradation and preserves their insecticidal activity [25], and it may pose a potential, inadvertent risk to soil-dwelling organisms [5,26]. Free-living soil nematodes are the most abundant and species-rich metazoan group in soils [27]. Nematodes are useful indicators of soil quality because of their great diversity and participation in many functions at different levels of the food webs in the soil. They are relatively simple to separate and enumerate. Their populations, in contrast to those of bacteria, are stable in response to changes in soil moisture and temperature [28]. Impacts on soil nematodes are, therefore, an important aspect of the environmental risk assessment and post-release monitoring of transgenic insect-resistant plants.

Certain *Bt* toxins, e.g., Cry5B, Cry6A, Cry14A and Cry21A, have been found to have direct toxic effects on some nematode species [29]. Nevertheless, Cry1Ac and CpTI expressed simultaneously by transgenic crops have not been evaluated for their effects on nematodes. To date, studies on the effects of transgenic insect-resistant plants, such as those expressing Cry1Ab, on soil nematodes have produced contrasting results. Negative effects of Cry1Ab protein on the growth, number of eggs and reproduction of soil nematodes have been detected in the rhizosphere soil of transgenic Cry1Ab maize [30]. However, the results of laboratory and field studies have generally shown no consistent effects of transgenic insect-resistant plants on soil nematodes [31–34]. For example, no significant differences in the numbers, communities and biodiversity of nematodes have been found in the soil of maize expressing the Cry1Ab protein relative to non-*Bt* maize [31,32]. A significant but transient decrease in the numbers of nematodes in soil under *Bt* maize expressing the Cry1Ab protein at 3 different field sites was found in a comparison with non-*Bt* maize, whereas studies conducted in a greenhouse showed no toxic effects of the Cry1Ab protein on populations of nematodes [35,36]. The reasons for the differences between the 2 studies are unclear, but they may have resulted from different environmental conditions in the greenhouse and the field, which could affect the interactions between plants and soil organisms.

As the results of these studies only reflect short histories of transgenic cultivation, we still face considerable gaps in our scientific understanding of longer-term community-level impacts on soil nematodes from the cultivation of transgenic insect-resistant crops [37]. The biosafety regulations for genetically modified organisms (GMOs) in many jurisdictions, including both China and the European Union (EU), require monitoring of environmental impacts after the environmental release and commercial cultivation of transgenic crops [38,39]. Transgenic insect-resistant cottons have now been planted for more than a decade in China, a nation prominent in pioneering the use of this new technology. The cotton fields of China therefore offer a valuable opportunity to address scientific questions of longer-term impacts and to fulfill the ongoing demands of biosafety regulations. The purpose of this study was to investigate the population density and community structure of soil nematodes over 2 years, comparing the soil nematodes of a conventional non-transgenic cotton plantation with those from soils that have been planted with transgenic cotton expressing Cry1Ab/c and CpTI for up to 10 years, and to provide a theoretical basis for environmental impact monitoring of transgenic plants.

Materials and Methods

Plant Material and Field Trial

This study was conducted in fields at a cotton farm in Baibi town, Anyang, Henan Province, China. The farm belongs to the Cotton Research Institute (CRI) of the Chinese Academy of Agricultural Sciences (CAAS). This field site is in the North Temperate Zone and has a continental monsoon climate. The annual mean temperature is 13.6 °C, and the annual mean precipitation is 606.1 mm. Three types of cotton fields were selected for this study (Table 1). Field T-1 was originally planted with non-transgenic cotton, then sown beginning in 2002 with the transgenic insect-resistant cotton line Zhong-41 expressing Cry1Ab/c and CpTI. Field T-2 was planted with 2 transgenic Cry1Ab/c cotton lines, Zhong-29 and Zhong-30, from 1999 to 2001 and subsequently planted with Zhong-41 beginning in 2002. Field CK has been planted with the conventional non-transgenic cotton line Zhong-35 since 1999. Zhong-29 and Zhong-30 were developed by CRI, CAAS, and approved by the Ministry of Agriculture (MOA) of China in 1998. Zhong-41 was developed jointly by CRI and the Biotechnology Research Institute of CAAS and approved by the MOA in 2002 [40]. The content of Cry1Ab/c expressed in the leaves of Zhong-41, Zhong-29 and Zhong-30 was determined, and the results indicated that the transgene expression remained stable during the survey period (Table 1).

The 3 cotton fields monitored were all located at 36° 7′ N and 116° 22′ E and were composed of a cambisol-type soil (FAO (1998) classification) with the following properties (on a dry mass basis): pH (soil: water ratio 1:2.5) 7.82, organic C 16.10 g•kg^{-1}, total *N* 0.84 g•kg^{-1}, total *P* 0.85 g•kg^{-1}, total *K* 7.61 g•kg^{-1}, available *P* 26.62 mg•kg^{-1}, available *K* 134.47 mg•kg^{-1} and soil clay (<0.002 mm) 9.29%. The fields CK, T-1 and T-2 were distributed side by side from south to north and separated by belts 50 m wide. The cotton growing season extended from April to November annually. The agricultural practice for the cotton in the 3 fields was the same as that used for conventional cottons. Fertilizer was applied at the seedling stage and at the budding stage. During the growing seasons, chemical pesticides were used for pest control as necessary (Fig. 1). In addition, all 3 fields lay fallow from November to the following April.

Table 1. Planting information for 3 cotton fields.

Treatments (Cotton fields)	Cotton lines	Transgenes	Contents of Cry1Ab/c (ng •g^{-1} fresh tissue)	Planting time
CK	Zhong-35	–	–	1999 to 2010
T-1	Zhong-41	Cry1Ab/c[#] & CpTI	150–600	2002 to 2010
T-2	Zhong-29 & Zhong-30	Cry1Ab/c	200–550	1999 to 2001
	Zhong-41	Cry1Ab/c & CpTI	150–600	2002 to 2010

[#]Cry1Ab/c represents a fusion gene of Cry1Ac and Cry1Ab.

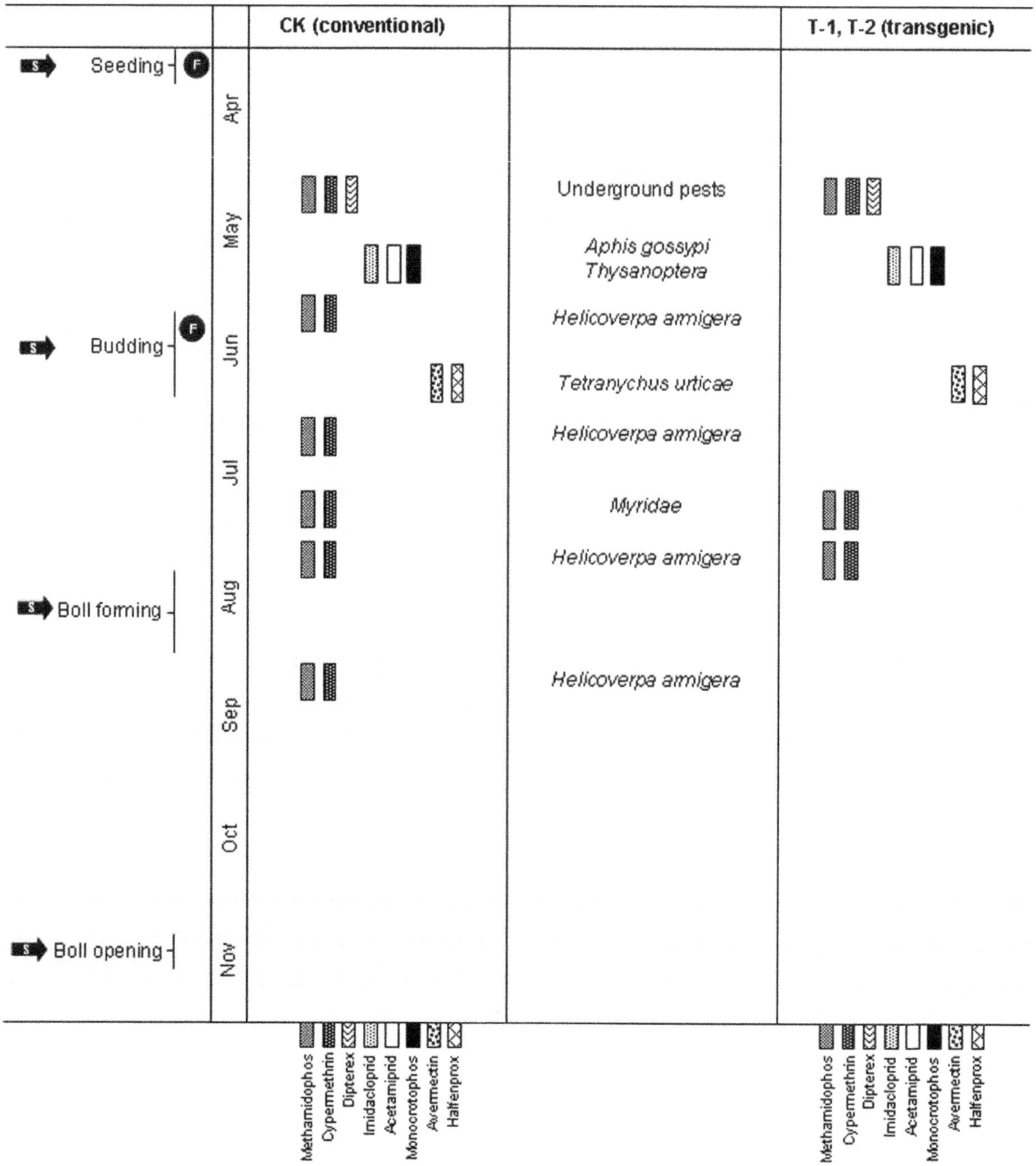

Figure 1. Schematic of agricultural practices applied to the 3 cotton fields during this study (2007–2010). Patterned bars represent different pesticides, with target pest listed in central column. Black circles with 'F' indicate fertilizer application, and black arrows with 'S' represent sampling times. Symbol placement is indicative of timing but is not precise.

Soil Sample Collection

Soil samples were collected 4 times per year from 2009 to 2010, inclusive, coinciding with the major growth stages of cotton, namely, seedling (April), budding (June), boll-forming (August) and boll-opening (November). Fifteen meters were left at both ends of every treatment to eliminate marginal field effects on soil sampling. Each type of treatment field was established in triplicate, and the plot size for each replication was 0.17 hectare. One soil sample was collected between 0 and 20 cm deep with 5 cores using a soil auger with a 4 cm diameter and then placed in a sterile plastic bag. Three soil samples were taken in each replication according to the checkerboard method [41]. The soil samples were immediately transported to the laboratory to isolate the soil nematode specimens. The data from 3 soil samples in each replication were pooled, and 3 replicates from each field were used in further statistical analyses.

Extraction and Identification of Nematodes

Nematodes were extracted from fresh soil equivalent to 100 g dried soil with a sieving process followed by sugar flotation [42].

Table 2. Composition and abundance (individual/100 g dry soil) of the soil nematodes extracted from different samples from 3 cotton fields during 2009–2010. P_i denotes the proportion of soil nematode individuals in each treatment.

Trophic groups	Genus	Treat-ment	2009				2010				Total	Percentage of total	Average abundance
			Seedling	Budding	Boll forming	Boll opening	Seedling	Budding	Boll forming	Boll opening			
Plant parasites	*Tylenchus*	CK	0.31	0.71	1.30	2.40	1.85	0.51	1.10	1.99	10.17	4.2%	
		T-1	0.00	0.34	2.02	2.10	1.47	0.29	1.39	1.54	9.15	3.7%	4.0%
		T-2	0.54	0.00	2.29	1.93	1.47	0.51	0.69	1.73	9.17	3.9%	
	Filenchus	CK	2.90	1.47	2.72	1.95	3.08	1.30	3.16	3.28	19.86	8.2%	
		T-1	2.43	1.41	1.55	2.99	2.93	1.67	3.22	3.12	19.32	7.9%	8.2%
		T-2	2.86	1.49	2.78	2.48	3.12	1.54	2.71	3.04	20.01	8.5%	
	Psilenchus	CK	0.75	0.51	0.00	0.00	0.00	0.00	0.00	0.00	1.26	0.5%	
		T-1	1.66	0.57	0.00	0.00	0.00	0.29	0.29	0.00	2.80	1.1%	1.1%
		T-2	0.00	0.31	0.59	0.96	0.98	0.00	0.29	0.85	3.97	1.7%	
	Helicotylenchus	CK	1.95	1.55	4.28	3.99	1.95	2.48	1.67	2.66	20.54	8.5%	
		T-1	2.39	2.08	4.22	4.12	1.99	2.37	2.04	2.30	21.51	8.8%	8.6%
		T-2	2.68	1.56	3.82	4.29	1.67	2.27	1.79	2.12	20.20	8.6%	
	Pratylenchus	CK	2.80	1.81	2.91	2.17	1.54	1.54	1.20	2.20	16.17	6.7%	
		T-1	2.25	1.20	2.77	2.21	1.39	1.39	1.39	2.20	14.79	6.0%	6.4%
		T-2	2.94	1.81	2.65	1.93	1.10	1.39	1.30	1.90	15.01	6.4%	
	Paratylenchus	CK	1.52	1.21	0.00	1.72	0.00	0.00	0.51	0.00	4.96	2.1%	
		T-1	2.67	1.41	1.13	1.99	0.00	0.00	0.69	0.00	7.90	3.2%	2.9%
		T-2	2.43	1.10	1.94	2.15	0.69	0.00	0.00	0.00	8.31	3.5%	
	Longidorella	CK	0.29	0.98	0.55	1.25	0.85	0.00	0.00	0.00	3.92	1.6%	
		T-1	0.00	0.93	1.05	1.48	0.69	0.00	0.51	0.29	4.95	2.0%	1.7%
		T-2	0.30	0.31	0.54	0.31	0.98	0.00	0.29	0.98	3.71	1.6%	
Bacterivores	*Mesorhabditis*	CK	1.66	1.47	2.74	2.61	0.85	0.85	0.29	0.85	11.31	4.7%	
		T-1	2.33	1.33	2.78	2.59	1.20	0.00	0.69	0.00	10.92	4.5%	4.9%
		T-2	2.18	1.47	2.76	2.88	1.30	0.29	0.51	1.47	12.86	5.5%	
	Protorhabditis	CK	2.09	1.00	1.89	3.32	1.61	1.47	2.20	1.95	15.53	6.4%	
		T-1	2.05	0.93	2.60	2.49	1.61	1.30	1.95	1.73	14.65	6.0%	6.2%
		T-2	1.80	1.12	1.70	2.74	1.90	1.85	1.39	1.67	14.16	6.0%	
	Eucephalobus	CK	0.55	0.98	1.09	1.17	1.20	0.85	1.47	0.69	8.01	3.3%	
		T-1	0.30	2.37	1.66	2.31	1.10	1.30	1.39	1.20	11.64	4.8%	3.6%
		T-2	0.00	0.72	0.35	0.41	0.69	1.39	1.20	1.67	6.43	2.7%	
	Heterocephalobus	CK	1.60	1.62	1.69	0.69	2.77	2.48	1.61	2.16	14.62	6.1%	
		T-1	2.25	1.83	0.42	0.69	2.54	2.46	1.67	2.16	14.04	5.7%	5.8%

Table 2. Cont.

Trophic groups	Genus	Treat-ment	2009				2010				Total	Percentage of total	Average abundance
			Seedling	Budding	Boll forming	Boll opening	Seedling	Budding	Boll forming	Boll opening			
		T-2	1.35	1.43	0.00	1.11	2.59	2.51	1.99	1.99	12.97	5.5%	
	Acrobeles	CK	1.94	0.87	1.30	0.81	0.29	0.51	0.00	0.00	5.72	2.4%	
		T-1	1.25	0.82	1.99	1.04	0.00	0.00	0.29	0.00	5.38	2.2%	2.0%
		T-2	1.44	0.29	0.35	0.00	0.51	0.00	0.51	0.00	3.10	1.3%	
	Acrobeloides	CK	2.85	2.83	2.35	2.47	2.75	2.40	2.98	3.28	21.92	9.1%	
		T-1	2.89	3.32	0.42	2.54	2.85	2.34	3.19	3.39	20.94	8.6%	9.0%
		T-2	2.68	2.54	1.95	2.52	2.83	2.30	3.26	3.50	21.58	9.2%	
	Panagrolaimus	CK	0.29	0.30	1.31	1.25	0.00	0.00	0.00	0.51	3.66	1.5%	
		T-1	0.53	0.34	1.44	0.69	0.00	0.00	0.29	0.29	3.58	1.5%	1.5%
		T-2	0.00	0.87	0.00	0.00	0.51	0.29	0.51	1.20	3.38	1.4%	
	Plectus	CK	0.72	0.87	1.09	0.00	0.00	0.29	0.00	0.00	2.97	1.2%	
		T-1	1.14	0.95	1.06	0.69	0.00	0.00	0.51	0.00	4.35	1.8%	1.5%
		T-2	1.16	0.87	0.59	0.00	0.51	0.00	0.29	0.00	3.42	1.5%	
	Chronogaster	CK	0.31	0.29	0.00	1.08	0.00	0.00	0.51	0.00	2.19	0.9%	
		T-1	0.00	0.31	1.06	0.69	0.29	0.85	0.69	0.00	3.89	1.6%	1.1%
		T-2	0.30	0.00	0.00	0.00	0.00	0.69	0.29	0.69	1.97	0.8%	
	Pseudoaulolaimus	CK	0.29	0.00	0.00	0.00	1.73	2.12	1.73	0.85	6.73	2.8%	
		T-1	0.30	0.00	0.86	0.00	0.98	1.85	1.39	0.29	5.66	2.3%	2.5%
		T-2	0.00	0.00	0.00	0.00	1.10	1.73	1.47	0.98	5.28	2.3%	
	Alaimus	CK	1.25	0.87	1.39	1.90	2.54	1.54	1.10	1.61	12.20	5.1%	
		T-1	1.43	1.20	1.05	1.92	2.54	1.47	1.30	0.98	11.90	4.9%	4.9%
		T-2	1.44	1.32	0.57	0.92	2.12	1.47	1.73	1.30	10.88	4.6%	
Fungivores	*Ditylenchus*	CK	0.74	1.21	1.09	0.00	0.51	0.69	0.85	0.00	5.10	2.1%	
		T-1	0.73	1.17	0.74	1.96	0.29	0.51	1.20	0.00	6.61	2.7%	2.4%
		T-2	0.54	0.00	0.99	2.18	0.51	0.00	1.20	0.29	5.71	2.4%	
	Aphelenchus	CK	1.14	1.39	1.96	1.79	0.00	0.00	0.00	0.00	6.27	2.6%	
		T-1	1.85	2.01	1.91	1.24	0.00	1.10	0.00	0.00	8.10	3.3%	3.3%
		T-2	2.05	2.74	2.18	1.63	0.00	0.85	0.00	0.00	9.46	4.0%	
	Aphelenchoides	CK	2.79	2.34	1.30	1.75	1.67	1.14	0.98	0.69	12.68	5.3%	
		T-1	2.60	2.84	1.28	1.55	1.47	0.29	0.29	0.29	10.59	4.3%	4.9%
		T-2	2.40	3.23	1.40	2.38	1.10	0.85	0.00	0.29	11.65	5.0%	
	Dorylaimoides	CK	0.87	0.00	0.80	0.69	0.29	0.00	0.00	0.00	2.64	1.1%	
		T-1	0.88	0.00	0.00	0.00	0.00	0.00	0.29	0.00	1.17	0.5%	1.0%

Table 2. Cont.

Trophic groups	Genus	Treat-ment	2009				2010				Total	Percentage of total	Average abundance
			Seedling	Budding	Boll forming	Boll opening	Seedling	Budding	Boll forming	Boll opening			
		T-2	1.79	0.69	0.81	0.00	0.00	0.00	0.00	0.00	3.30	1.4%	
Predators	*Mesodorylaimus*	CK	0.29	0.00	1.24	1.09	0.00	0.00	0.29	0.00	2.91	1.2%	
		T-1	0.72	0.00	1.31	0.00	0.29	0.29	0.98	0.00	3.59	1.5%	1.2%
		T-2	0.55	0.00	0.00	0.00	0.29	0.29	0.51	0.69	2.33	1.0%	
	Thonus	CK	0.31	1.54	1.54	1.38	0.85	0.29	0.98	1.47	8.36	3.5%	
		T-1	0.31	2.30	0.71	1.46	0.69	0.51	0.85	1.10	7.92	3.2%	3.3%
		T-2	0.72	1.42	1.72	1.09	0.29	0.98	1.10	0.51	7.83	3.3%	
Omnivores	*Eudorylaimus*	CK	0.31	0.00	1.09	0.85	0.00	0.00	0.00	0.00	2.25	0.9%	
		T-1	0.00	0.00	0.76	1.19	0.51	0.00	0.29	0.00	2.75	1.1%	0.9%
		T-2	0.31	0.29	0.54	0.00	0.29	0.00	0.00	0.29	1.72	0.7%	
	Epidorylaimus	CK	1.96	1.31	0.00	0.00	0.69	0.29	1.47	1.39	7.11	2.9%	
		T-1	1.52	0.34	0.00	0.00	0.29	1.10	1.30	0.98	5.52	2.3%	2.5%
		T-2	1.85	0.69	0.00	0.00	0.85	0.85	1.10	0.29	5.63	2.4%	
	Microdorylaimus	CK	0.72	0.86	1.54	0.00	2.04	1.10	1.39	1.67	9.32	3.9%	
		T-1	0.54	1.62	0.41	1.50	1.95	0.98	0.00	1.54	8.54	3.5%	3.5%
		T-2	0.72	1.12	0.59	0.00	1.73	0.98	0.69	1.20	7.05	3.0%	
	Aporcelaimellus	CK	0.26	0.14	0.49	0.42	6.06	0.36	0.20	0.69	0.60	1.1%	
		T-1	0.41	0.00	0.49	0.35	4.72	0.15	0.00	0.69	0.49	1.0%	1.2%
		T-2	0.29	0.42	0.68	0.35	7.35	0.01	0.60	0.43	0.09	1.5%	

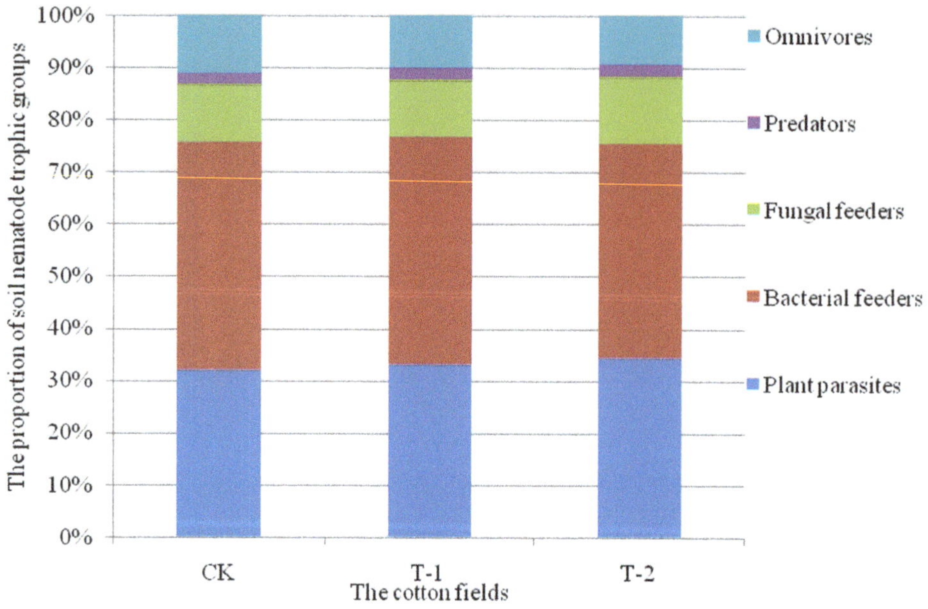

Figure 2. The soil nematode composition in 3 cotton fields during 2009–2010.

The nematodes were heat killed and fixed in 4% formaldehyde. They were then counted under a dissecting microscope at 25× magnification. A total of 100 specimens per sample were then randomly selected and identified to the genus level, as described in Liang et al. (2009), at 200× magnification using an inverted compound microscope [43].

Nematode Community Analysis

Nematode abundances were ln (x +1) transformed prior to statistical analysis and expressed as numbers per 100 g dry soil. Nematode biodiversity was measured with the Simpson and Shannon–Wiener diversity indices. Both indices are sensitive to the abundance of the most common/dominant species in a population [44]. Simpson's index is defined by the equation $C = 1 - \sum_{i=1} (P_i)^2$.

Table 3. Generalized Linear Mixed Model (GLMM) results for overall effects on the numbers of soil nematodes.

Effects	Degrees of freedom	F Value	*p* Value
Total soil nematodes			
Treatments	2;14	0.69	0.52
Sampling time	7;14	24.53	0.00
Filenchus			
Treatments	2;14	0.12	0.88
Sampling time	7;14	9.98	0.01
Helicotylenchus			
Treatments	2;14	1.00	0.39
Sampling time	7;14	49.47	0.00
Acrobeloides			
Treatments	2;14	0.18	0. 84
Sampling time	7;14	5.34	0.00

The Shannon–Wiener diversity index is defined by the equation $H = -\sum_{i=1} P_i \ln P_i$. In these equations, P_i denotes the proportion of soil nematode individuals in each treatment; $P_i = N_i/N$, where N_i is the abundance of the ith species and N is the overall total abundance in each treatment. $P_i \geq 10\%$ represents dominant groups, $10\% > P_i \geq 1\%$ represents common groups and $P_i < 1\%$ represents rare groups.

Nematode taxa were ranked along a colonizer–persister (c–p) scale of 1–5 according to Bongers and Ferris (1999) [45]. The Maturity Index (MI) was calculated using the equation of Bongers (1990) [46]:$MI = -\sum_{i=1}^{n} cp_i \times P_i$, where P_i is the frequency of the taxon in the sample and cp_i is the c-p value of taxon i.

Nematode taxa were classified into 5 main trophic groups: bacterial feeders, fungal feeders, plant parasites, omnivores and predators [47]. Based on the c–p and feeding type classification, nematode taxa were also categorized in functional guilds according to Ferris *et al.* (2001): Ba_n, Fu_n, Ca_n, and Om_n = bacterial feeders, fungal feeders, predators, and omnivores, respectively, with n = c–p value [48]. The following indices were calculated to describe the enrichment and structure conditions as well as the predominant decomposition channels in the soil food webs: Enrichment inde**x**: $EI = 100 \times \frac{e}{e+b}$, Structure index: $SI = 100 \times \frac{s}{s+b}$, Channel index: $CI = 100(0.8\ Fu_2)/(3.2\ Ba_1 + 0.8\ Fu_2)$, where $b = (Ba_2 + Fu_2) \times 0.8$, $e = Ba_1 \times 3.2 + Fu_2 \times 0.8$, and $s = Ca_2 \times 0.8 + (Ba_3 + Ca_3 + Fu_3 + Om_3) \times 1.8 + (Ba_4 + Ca_4 + Fu_4 + Om_4) \times 3.2 + (Ba_5 + Ca_5 + Fu_5 + Om_5) \times 5$.

The response of the soil nematode community to the factors 'treatment' and 'sampling time' was examined with a 2-way ANOVA (Proc GLM). Significance was measured at the alpha = 0.05 level. Principal component analysis (PCA), a repeated-measures multivariate ordination analysis, was performed to identify the influence of treatment and sampling time on community structure (SPSS 13.0 for Windows). The principal components whose eigenvalue exceeded 1 were selected for the analysis.

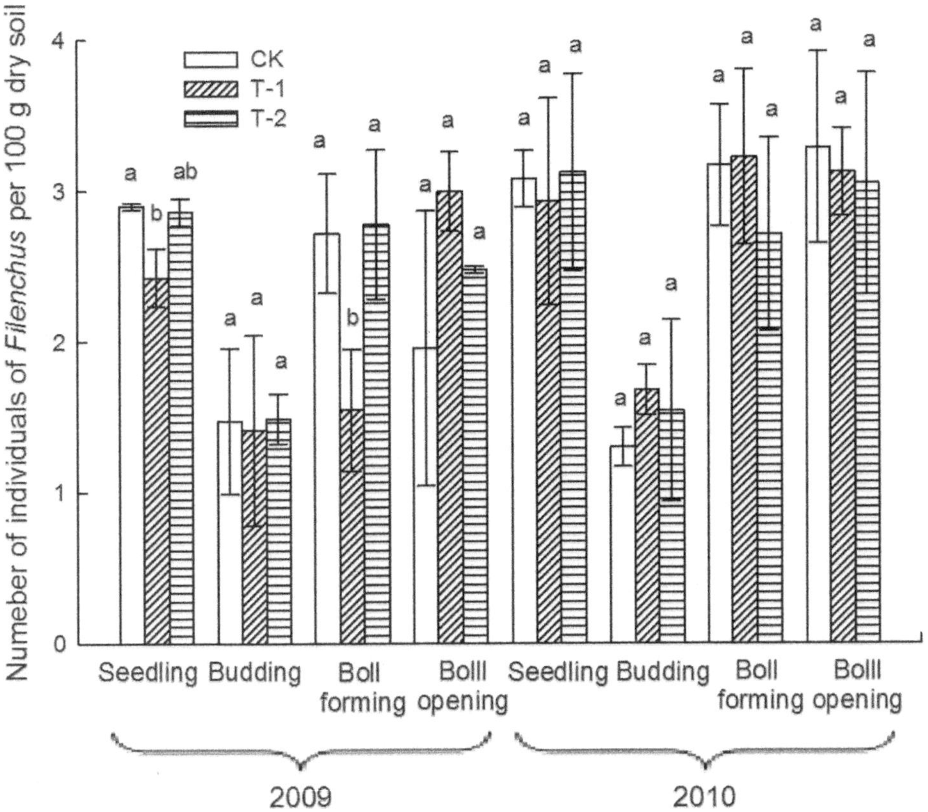

Figure 3. Number of *Filenchus* in 3 cotton fields at different sampling times. Error bars indicate standard errors (n = 3). Different letters above bars denote a statistically significant difference between the means of the fields.

Results

The Composition of the Soil Nematode Community in the 3 Cotton Fields

Twenty-eight genera of soil nematodes were identified in the 3 cotton fields during different cotton growth stages over 2 years (Table 2). The overall results showed that the most abundant common groups in CK, T-1 and T-2 were *Helicotylenchus*, *Filenchus* and *Acrobeloides*. Most of the genera collected from the soil, such as *Tylenchus*, *Pratylenchus*, *Paratylenchus*, *Mesorhabditis*, *Protorhabditis*, *Eucephalobus*, *Heterocephalobus*, *Acrobeles*, *Pseudoaulolaimus*, *Alaimus*, *Ditylenchus*, *Aphelenchus*, *Aphelenchoides*, *Thonus*, *Epidorylaimus* and *Microdorylaimus*, represented common groups. *Eudorylaimus* was a rare group. Over the 2-year field period, the composition of soil nematode communities was essentially uniform in the transgenic insect-resistant cotton fields and the non-transgenic cotton field (Fig. 2). In brief, the soil nematode communities in the 3 fields did not differ significantly.

Effect of Transgenic Insect-resistant Cotton on the Number of Soil Nematodes

Due to the particularly arid conditions occurring throughout 2010, the number of total soil nematodes for the 4 sampling times in 2010 was obviously less than that for the 2009 samples (Table 2). During the 2-year sampling period, the number of total soil nematodes and the most abundant nematodes, such as *Filenchus*, *Helicotylenchus* and *Acrobeloides*, in each cotton field varied significantly among different sampling times (i.e., different growth stages of the plants; $p<0.01$; Table 3) but did not differ significantly overall among the 3 cotton fields ($p>0.05$; Table 3). At the seedling and boll-forming stages in 2009, the number of *Filenchus* in T-1 was significantly lower than that in CK ($p<0.05$; Fig. 3). At the seedling stage in 2009, the number of *Helicotylenchus* in T-2 was significantly greater than that in CK ($p<0.05$; Fig. 4). At boll-forming stage in 2009, the number of *Acrobeloides* in T-1 was significantly lower than that in CK and T-2 ($p<0.05$; Fig. 5). For the other sampling times, the numbers of *Filenchus, Helicotylenchus* and *Acrobeloides* did not vary significantly among the 3 fields ($p>0.05$; Figs. 3–5).

Effect of Transgenic Insect-resistant Cotton on Soil Nematode Trophic Groups

In all, 26–57% of the nematodes were bacterial feeders, 24–47% plant parasites, 1–25% fungal feeders, 3–15% omnivores and 1–8% predators. The feeding-type composition was relatively constant throughout the 2-year survey, with average proportions of 41–44% bacterial feeders, 32–34% plant parasites, 11–13% fungal feeders, 9–11% omnivores and 2% predators (Table 4). These values showed no significant differences in feeding-type composition among the 3 cotton fields at different sampling times ($p>0.05$; one-way ANOVA). During the 2 years (2009 and 2010), the proportions of bacterial feeders, fungal feeders, plant parasites, omnivores and predators in each cotton field varied significantly among sampling times ($p<0.05$; Table 5) but did not differ significantly overall among the 3 cotton fields ($p>0.05$; Table 5).

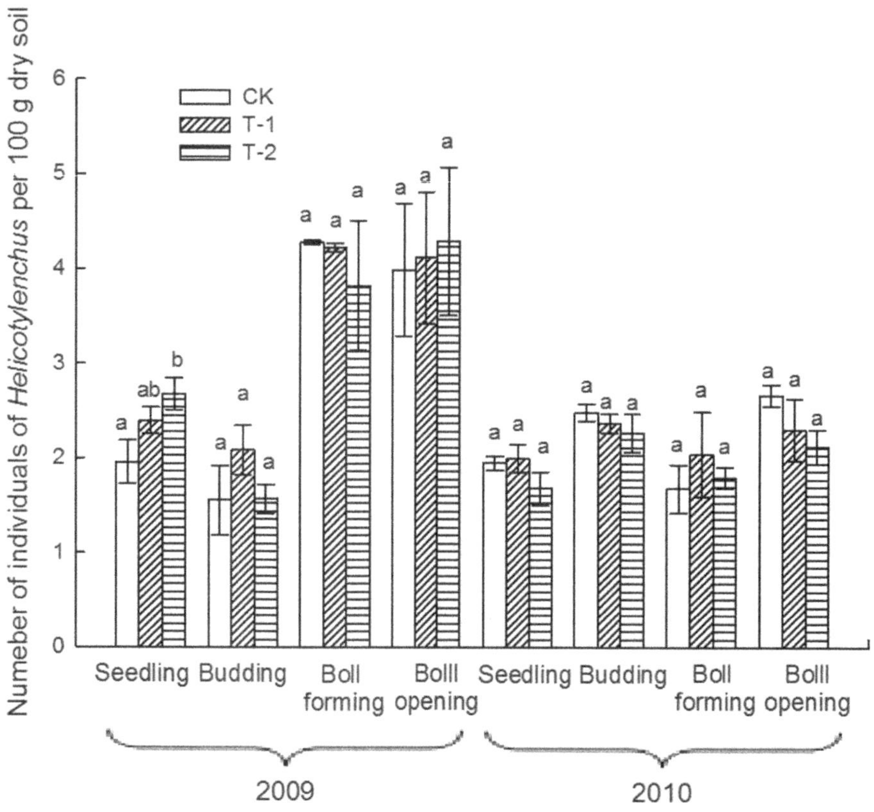

Figure 4. Number of *Helicotylenchus* in 3 cotton fields at different sampling times. Error bars indicate standard errors (n = 3). Different letters above bars denote a statistically significant difference between the means of the fields.

Effect of Transgenic Insect-resistant Cotton on the Ecological Indices of Soil Nematodes

The initial values of the Shannon–Wiener, Simpson, EI, SI, CI and MI ecological indices were obtained from the analysis of the nematode fauna at the seedling stage. Differences (Δ) from these initial values were obtained for each cotton field at the budding, boll-forming and boll-opening stages (Fig. 6). A univariate general linear model analysis suggested that these differences in the ecological indices varied significantly overall among different sampling times ($p<0.05$) but did not differ significantly overall among the 3 cotton fields ($p>0.05$). A further analysis used a one-way ANOVA with a priori contrasts to compare these ecological indices for the conventional cotton field with the corresponding indices for the 2 transgenic fields but detected no differences in the values of the ecological indices of the transgenic fields relative to the conventional cotton field at the 8 sampling times.

Principal Component Analysis of Soil Nematode Composition in Cotton Fields

Five principal components were selected for the analysis based on a cumulative contribution rate of 85% for the principal components extracted. The contribution rates of the first 2 principal components were 31.73% and 20.53%, respectively (Fig. 7). Different sampling times showed a distinct separation along the principal component axes, whereas different fields formed a cluster at the same sampling time. The first principal component axis clearly separated the budding, boll-forming and boll-opening stages. The second principal component axis clearly distinguished the seedling and budding stages (Fig. 7).

Discussion

With the cultivation of more varieties of transgenic insect-resistant plants and their large-scale planting, environmental impact monitoring after commercial release has attracted increasing attention from the scientific community [37,49–51]. High population densities and large numbers of species of nematodes occur in almost all soils. In this study, potential effects on soil nematodes at the community level were monitored during a 2-year survey to assess the environmental risks associated with transgenic fields planted with transgenic insect-resistant cotton for more than 10 years relative to the risks associated with a field planted with non-transgenic cotton. Based on the nematode communities examined at the genus level, the overall findings of the 2-year field study showed that the community structure of the soil nematodes was similar in the 3 cotton fields and that the most abundant common groups of soil nematodes all included *Helicotylenchus*, *Filenchus* and *Acrobeloides*. In soil planted with *Bt* maize and *Bt* eggplant expressing the Cry1Ac and Cry3Bb1 proteins, respectively, no effects were found on nematode community structure [32,52]. However, a distinct shift in community structure, i.e., a significantly higher proportion of mycophagous nematodes and a lower proportion of phytophagous nematodes, were observed in soil planted with *Bt* canola relative to soil planted with the respective non-*Bt* isoline [53]. In a study of the effects of Mon88017 cultivation on the community structure of the indigenous soil-inhabiting nematodes, no significant differences in the generic composition of the nematodes was found between a *Bt* maize plot and non-*Bt* maize plots, but a significant shift in generic composition was found at the final sampling date [48]. In a

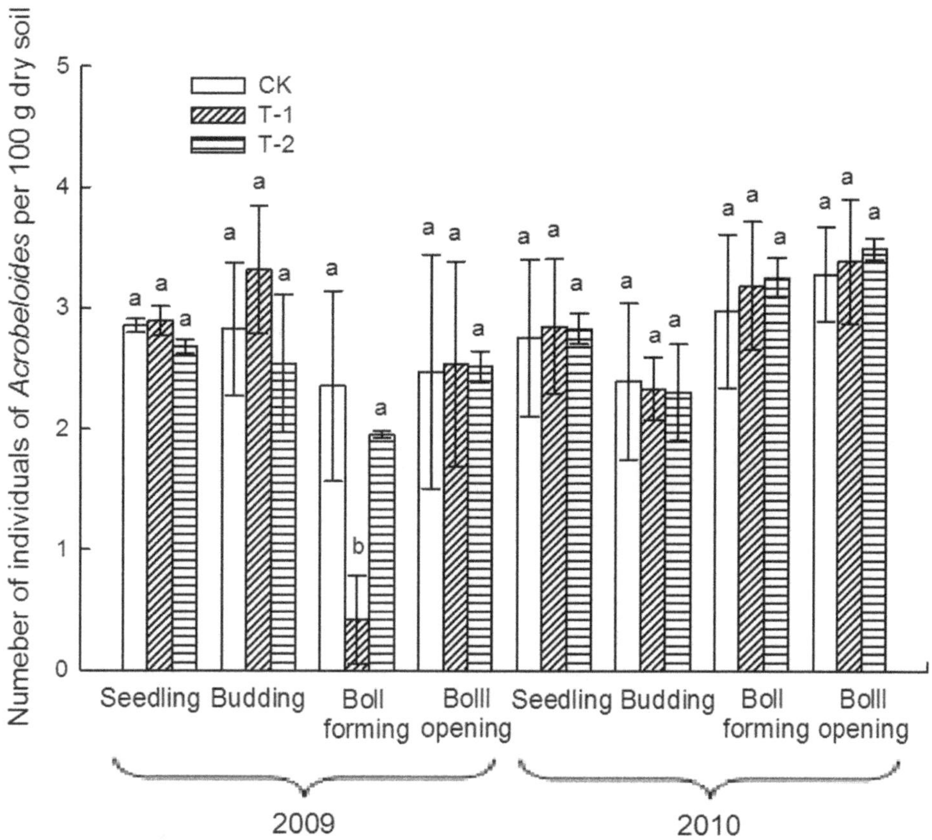

Figure 5. Number of *Acrobeloides* in 3 cotton fields at different sampling times. Error bars indicate standard errors (n = 3). Different letters above bars denote a statistically significant difference between the means of the fields.

Table 4. Changes in the composition of nematode trophic groups (%) among different cotton fields at different sampling times for the 2-year survey (mean ± SE).

Trophic groups	Treatment	2009				2010			
		Seedling	Budding	Boll forming	Boll opening	Seedling	Budding	Boll forming	Boll opening
Plant parasites	CK	31.6±0.8^a	29.5±2.1^a	31.6±4.7^a	37.1±2.9^a	31.3±2.6^a	26.4±3.7^a	28.9±2.4^a	36.1±6.7^a
	T-1	32.3±1.9^a	25.1±1.7^a	36.2±3.8^a	37.1±1.3^a	30.7±3.2^a	26.9±2.4^a	32.8±3.9^a	40.4±3.8^a
	T-2	33.2±0.2^a	24.0±2.3^a	46.4±9.7^a	43.5±7.6^a	34.0±4.5^a	23.9±4.3^a	27.4±2.8^a	36.7±4.2^a
Bacterial feeders	CK	40.8±4.9^a	39.6±11.7^a	40.0±6.3^a	42.1±5.8^a	46.5±4.6^a	56.5±7.6^a	44.9±1.2^a	42.3±4.7^a
	T-1	41.0±5.8^a	42.4±5.2^a	43.6±8.5^a	39.0±3.6^a	47.6±5.5^a	51.7±9.4^a	46.0±2.7^a	42.9±6.9^a
	T-2	34.9±10.0^a	38.8±9.6^a	26.3±13.7^a	32.7±7.4^a	47.8±7.3^a	52.5±3.7^a	51.0±4.5^a	50.0±5.6^a
Fungal feeders	CK	16.7±5.9^a	17.7±2.4^a	13.9±3.2^a	11.7±3.2^a	8.4±2.7^a	8.3±1.4^a	6.9±1.3^a	2.5±0.6^a
	T-1	17.1±4.4^a	19.0±2.3^a	11.2±3.1^a	11.8±2.3^a	6.4±3.7^a	8.5±1.7^a	6.1±2.7^a	1.2±0.7^a
	T-2	19.2±7.3^a	24.3±6.1^a	17.1±4.7^a	19.2±3.5^a	5.5±1.6^a	7.1±2.7^a	4.7±1.9^a	2.0±1.3^a
Predators	CK	1.8±0.3^a	5.5±1.5^a	7.5±1.9^a	6.8±2.1^a	2.9±1.2^a	1.3±0.9^a	4.8±2.1^a	5.2±1.8^a
	T-1	2.9±0.2^a	7.3±1.3^a	5.7±2.2^a	3.6±1.5^a	3.6±1.8^a	3.6±1.5^a	6.3±1.8^a	4.7±2.2^a
	T-2	3.6±1.2^a	5.2±2.8^a	5.5±1.9^a	3.4±1.7^a	2.0±0.5^a	5.3±2.8^a	6.2±2.7^a	4.2±1.7^a
Omnivores	CK	9.0±1.3^a	7.8±3.2^a	7.1±2.7^a	2.3±1.1^a	11.0±2.5^a	7.6±2.3^a	14.5±1.9^a	13.9±5.1^a
	T-1	6.7±3.5^a	6.2±1.8^a	3.3±1.9^a	8.4±3.7^b	11.8±1.3^a	9.3±3.8^a	8.8±2.1^a	10.8±2.3^a
	T-2	9.0±0.3^a	7.7±1.4^a	4.8±2.3^a	1.3±2.8^a	10.7±2.9^a	11.2±3.7^a	10.7±2.9^a	7.1±3.8^a

Mean values and standard error of 3 replicates are presented. Use of the same letter as a superscript indicates that variable means did not differ significantly among different treatments at each sampling date (p>0.05).

Table 5. Generalized Linear Mixed Model (GLMM) results for the overall effects on soil nematode trophic groups.

Effects	Degrees of freedom	F Value	*P* Value
Plant parasites			
Treatments	2;14	0.62	0.55
Sampling time	7;14	6.28	0.00
Bacterial feeders			
Treatments	2;14	0.85	0.45
Sampling time	7;14	4.56	0.01
Fungal feeders			
Treatments	2;14	2.07	0.16
Sampling time	7;14	15.77	0.00
Predators			
Treatments	2;14	0.12	0.89
Sampling time	7;14	4.24	0.01
Omnivores			
Treatments	2;14	0.78	0.48
Sampling time	7;14	1.99	0.13

few cases, the studies cited reported some differences in the nematode communities among the different treatments at certain sampling times, but no consistent significant differences were found over the entire sampling period, and these studies only covered a short sampling time period. The current study observed no significant differences in soil nematode communities between fields of transgenic insect-resistant cotton and a field of non-transgenic cotton, and the study indicated that long-term cultivation of transgenic insect-resistant cotton had no significant impact on the composition and community structure of soil nematodes in agricultural soils.

Laboratory and field studies have shown no consistent effects of transgenic insect-resistant maize on the number of soil nematodes relative to non-*Bt* maize [31,33,34]. The cultivation of *Bt* maize expressing the Cry1Ab protein significantly decreased the number of soil nematodes at 3 different field sites relative to non-*Bt* maize, whereas a study conducted in a greenhouse showed no toxic effects of the Cry1Ab protein on populations of nematodes. In contrast, both the sampling site and the time had greater significant influences on the population density of soil nematodes than that of the maize lines [35,36]. Based on a GLMM analysis of data collected over 2 years of sampling, our results indicated that the number of functional guilds in the 3 cotton fields showed significant seasonal variation (necessarily following the progression of different cotton growth stages). However, the effect of long-term cultivation of transgenic insect-resistant cotton (T-1 and T-2) on the number of total soil nematodes and on certain dominant groups, such as *Filenchus*, *Helicotylenchus* and *Acrobeloides*, was not significant relative to the values found for conventional cotton cultivation. In only a few instances in our study did we find significant differences in the numbers of soil nematodes between fields of different cultivars. No consistent trend was found over the 2 study years. This general finding is in accord with work by other teams in other regions and on a variety of crops [31,33,35,36,54,55].

Parasitic nematodes can cause considerable economic damage worldwide to many types of crops, including cotton in the major growing areas of certain countries [56,57]. Several plant parasitic nematodes, such as *Helicotylenchus*, *Tylenchorhynchus*, *Tylenchus*, *Pratylenchus* and *Filenchus*, have been detected in the cotton fields of north China [58–60]. Our study found that *Filenchus*, *Helicotylenchus*, *Tylenchus* and *Pratylenchus* were the principal genera of plant parasitic nematodes in the 3 cotton fields. These results were consistent with the findings of other studies conducted in China. Certain genera, e.g., *Meloidogyne*, *Rotylenchulus* and *Belonolaimus*, that are known to infest cotton were not detected in the soils of the 3 cotton fields. The principal explanation for this result might be that these nematodes are of concern in the United States, India, Pakistan, Egypt and Brazil but not in China [61] and are seldom separated from soils sampled in the cotton-growing region of north China [58–60]. Moreover, these nematodes primarily parasitize the root tissues of the plant and are very rare in the soil around the plant roots. Finally, outbreaks of cotton nematode diseases have not been recorded in the 3 cotton fields for more than 20 years. Therefore, the nematode communities collected from the soil samples in the study were typical of China.

Functional analyses and indices such as those used in ecological studies have proven relatively useful in detecting the true effects of the cultivation of transgenic plants [18,34,62]. Over 2 years of sampling, we observed strong and significant seasonal variations in the ecological indices of the soil nematodes collected in all 3 cotton fields, in parallel with the distinct growth stages of cotton. However, no statistically significant effects of the long-term cultivation of transgenic insect-resistant cotton (in fields T-1 and T-2) on the ecological indices of the soil nematodes were evident relative to the non-*Bt* cotton (field CK). This result agreed with those reported by Manachini and Lozzia (2002) and Höss et al. (2011), who found that *Bt* maize expressing the Cry1Ab and Cry3Bb1 toxins had no significant effects on the diversity of soil nematodes [32,34].

The accumulation of *Bt* protein in soil has long been posited as one of the main putative mechanisms for the effects of transgenic *Bt* plants on soil organisms [9,68–70]. However, many studies have demonstrated that Cry proteins degrade rapidly in soil under laboratory conditions [70–72] and, hence, are unlikely to accumulate or persist in fields where *Bt* crops have been planted for years [73–75]. For example, the content of Cry1Ab protein was above the detection limit of an ELISA test in only half of the soil samples obtained from transgenic plots, ranging from 0.19 to 1.31 ng g^{-1} dry weight [34]. To address this possible but unlikely occurrence at our study site, we determined the residual levels of Cry1Ac protein in the soils of T-1 and T-2 using a QualiplateTM kit for Cry1Ab/Cry1Ac (EnviroLogix, USA) and found extremely low levels of Cry1Ac protein, below the quantitative limit of the kit [76]. These results agreed with the findings of others [73–75], indicating that Cry1Ab/c protein did not accumulate in cambisol soil with prolonged planting of transgenic cotton.

There was also no indication that the community structure of the soil nematodes was influenced by indirect effects of transgenic proteins, such as Cry1Ab and Cry1Ac, via the food web. Microbial communities such as bacteria, actinomycetes and fungi and soil invertebrate communities such as Collembola, Opisthophora and Acarina, which were studied in the same cotton fields investigated in the present study, showed no significant differences in abundance and diversity between transgenic insect-resistant cotton and non-*Bt* cultivars [76,77]. These findings are consistent with the results of another study examining the effect of transgenic insect-resistant plants on microbial populations [75,78,79] and other soil organisms, such as Collembola, mites and earthworms [62,80–82]. Clearly, soil organisms are not impacted or are only

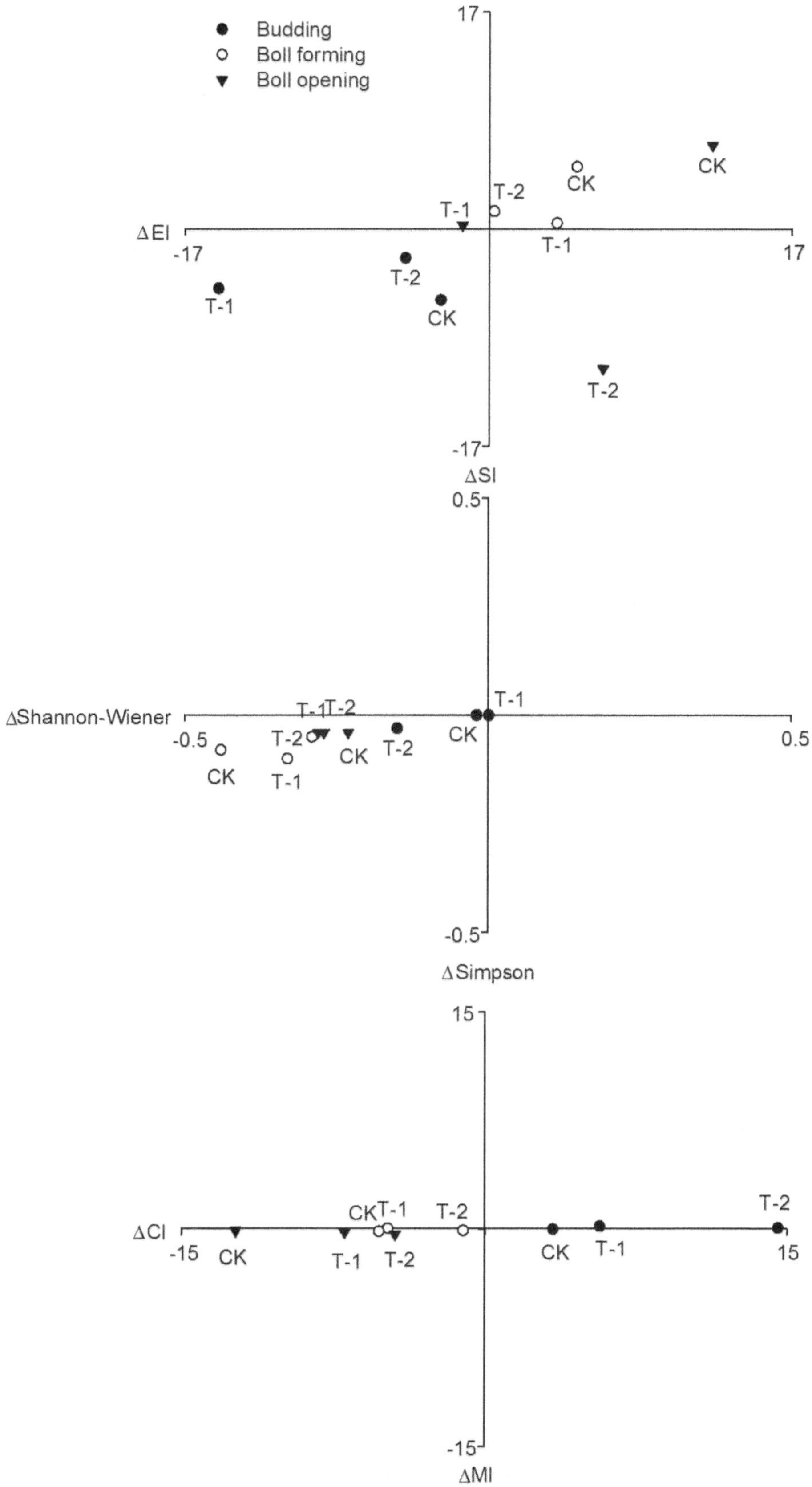

Figure 6. Changes in the values of ecological indices for soil nematode faunal analysis relative to seedling stage among different cotton fields.

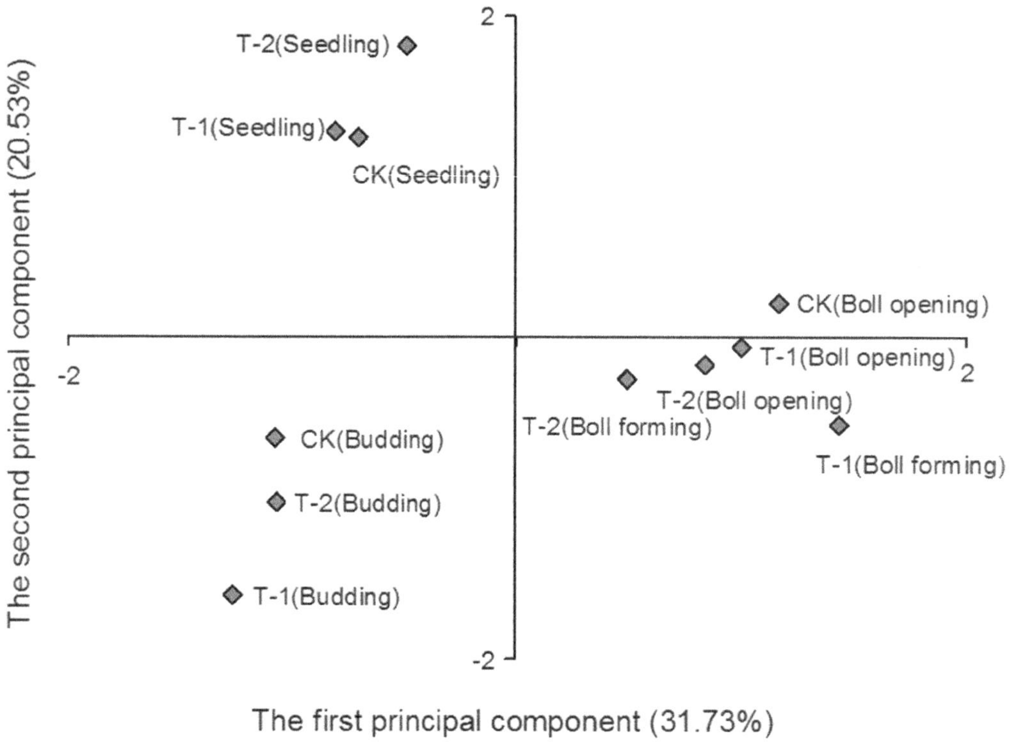

Figure 7. Factor loading graph of principal component analysis of soil nematode generic composition in cotton fields at different sampling times.

slightly impacted by the cultivation of transgenic insect-resistant cotton.

Many studies on the effects of pesticides (primarily nematicides and insecticides) on total nematode abundance and feeding groups under field or semi-field conditions have been conducted. The majority of these studies have indicated no negative effects of pesticides such as malathion, imicyafos and carbofuran on nematodes [63–66]. In one study, a decrease in abundance was observed as an effect of pesticides such as nemacur [67]. In the current study, the same spraying dose of chemical pesticides (i.e., avermectin and halfenprox) used to control insects with piercing-sucking mouthparts, such as the red spider, was applied to the 3 fields. However, the applications of chemical pesticides (i.e., methamidophos and cypermethrin) used to control Lepidoptera, such as the cotton bollworm, in the 2 transgenic cotton fields were fewer in number than those applied to the conventional cotton field (Fig. 1). Therefore, the differences in the spraying of chemical pesticides to control Lepidoptera in the 3 cotton fields might be another important influence on the soil nematode community, and the total pesticide applications could mask the effects of different cotton lines in the present study, or the effects of transgenic cotton lines on the soil nematode community might be smaller than those of the pesticide regimes.

Transgenic plants must be monitored for environmental risk after being commercially released [50,83]. The risks potentially posed by transgenic plants, especially *Bt* crops, to the environment have been extensively assessed worldwide over the past 10 years, and no scientific evidence has shown that the cultivation of *Bt* crops has caused sustained environmental harm to communities of soil organisms, such as nematodes, earthworms, collembolans or mites [33,35,50,62]. However, the soil environment is a very complex ecosystem in which many factors affect the soil biota. In field studies, high variability in biotic parameters is inherent and usually present, and this variability must be considered seriously if the ecological risks posed by transgenic plants are to be monitored. Moreover, the soil biota may be strongly stressed as a result of the influence of environmental factors (e.g., pH, salinity, redox potential, vegetation and water-holding capacity), which may cause higher or lower levels of sensitivity to transgenic plants [23,40,56]. Therefore, it is necessary to continue monitoring the effects of transgenic plants on the soil ecosystem in different environments and to define the ecological significance of the planting of transgenic crops.

Author Contributions

Conceived and designed the experiments: BL XGL. Performed the experiments: XGL. Analyzed the data: XGL. Contributed reagents/materials/analysis tools: XGL BL. Wrote the paper: XGL.

References

1. James C (2011) Global Status of Commercialized Biotech/GM Crops: 2011. ISAAA Brief No. 43. ISAAA, Ithaca, New York.
2. Lu YH, Wu KM, Jiang YY, Guo YY, Desneux N (2012) Widespread adoption of *Bt* cotton and insecticide decrease promotes biocontrol services. Nature 487: 362–365.
3. Choudhary B, Gaur K (2010) *Bt* cotton in india: a country profile. ISAAA Series of Biotech Crop Profiles. ISAAA, Ithaca, New York.
4. Hutchison WD, Burkness EC, Mitchell PD, Moon RD, Leslie TW, et al. (2010) Areawide suppression of European corn borer with *Bt* maize reaps savings to non-Bt maize growers. Science 330: 222–225.

5. Lilley AK, Bailey MJ, Cartwright C, Turner SL, Hirsch PR (2006) Life in earth: the impact of GM plants on soil ecology? Trends Biotech 24: 9–14.
6. Cortet J, Griffiths BS, Bohanec M, Demsar D, Andersen MN, et al. (2007) Evaluation of effects of transgenic *Bt* maize on microarthropods in a European multi-site experiment. Pedobiologia 51: 207–218.
7. Rose R, Dively GP (2007) Effects of insecticide treated and lepidopteran-active *Bt* transgenic sweet corn on the abundance and diversity of arthropods. Environ Entomol 36: 1254–1268.
8. Yu HL, Li YH, Wu KM (2011) Risk assessment and ecological effects of transgenic *Bacillus thuringiensis* crops on non-target organisms. J Integr Plant Biol 53: 520–538.
9. Duan JJ, Lundgren JG, Naranjo S, Marvier M (2010) Extrapolating non-target risk of *Bt* crops from laboratory to field. Biol Lett 6: 74–77.
10. Naranjo SE (2005) Long-term assessment of the effects of transgenic *Bt* cotton on the abundance of nontarget arthropod natural enemies. Environ Entomol 34: 1193–1210.
11. Marvier M, McCreedy C, Regetz J, Kareiva P (2007) A meta-analysis of effects of *Bt* cotton and maize on non-target invertebrates. Science 316: 1475–1477.
12. Li GP, Feng HQ, McNeil JN, Liu B, Chen PY, et al. (2011) Impacts of transgenic *Bt* cotton on a non-target pest, *Apolygus lucorum* (Meyer-Dür) (Hemiptera: Miridae), in northern China. Crop Prot 30: 1573–1578.
13. Xu Y, Wu KM, Li HB, Liu J, Ding RF, et al. (2012) Effects of transgenic *Bt*+CpTI cotton on field abundance of non-target pests and predators in Xinjiang, China. J Integr Agr 11: 1493–1499.
14. Liu B, Shu C, Xue K, Zhou KX, Li XG, et al. (2009) The oral toxicity of the transgenic *Bt* plus CpTI cotton pollen to honeybees (*Apis mellifera*). Ecotox Environ Safe 72: 1163–1169.
15. Han P, Niu CY, Lei CL, Cui JJ, Desneux N (2010) Quantification of toxins in a Cry1Ac+CpTI cotton cultivar and its potential effects on the honey bee *Apis mellifera* L. Ecotoxicology 19: 1452–1459.
16. Chen LZ, Cui JJ, Ma WH, Niu CY, Lei CL (2011) Pollen from Cry1Ac/CpTI-transgenic cotton does not affect the pollinating beetle *Haptoncus luteolus*. J Pest Sci 84: 9–14.
17. Han P, Niu CY, Biondi A, Desneux N (2012) Does transgenic Cry1Ac+CpTI cotton pollen affect hypopharyngeal gland development and midgut proteolytic enzyme activity in the honey bee *Apis mellifera* L. (Hymenoptera, Apidae)? Ecotoxicology 21: 2214–2221.
18. Wolfenbarger LL, Naranjo SE, Lundgren JG, Bitzer RJ, Watrud LS (2008) *Bt* crop effects on functional guilds of non-target arthropods: A meta-analysis. PloS ONE 3 (5): e2118. doi: 10.1371/journal.pone.0002118.
19. Duan JJ, Marvier M, Huesing J, Dively G, Huang ZY (2008) A meta-analysis of effects of *Bt* crops on honeybees (Hymenoptera: Apidae). PLoS ONE 3 (1): e1415. doi:10.1371/journal.pone.0001415.
20. Rose R, Dively GP, Pettis J (2007) Effects of *Bt* corn pollen on honey bees: emphasis on protocol development. Apidologie 38: 368–377.
21. Hofs JL, Schoeman AS, Pierre J (2008) Diversity and abundance of flower-visiting insects in *Bt* and non-*Bt* cotton fields of Maputaland (KwaZulu Natal Province, South Africa). Int J Trop Insect Sci 28: 211–219.
22. Gupta VVSR, Watson S (2004) Ecological impacts of GM cotton on soil biodiversity: Below-ground production of *Bt* by GM cotton and *Bt* cotton impacts on soil biological processes. Australian Government Department of the Environment and Heritage, CSIRO Land and Water.
23. Knox OGG, Gupta VVSR, Nehl DB, Stiller WN (2007) Constitutive expression of Cry proteins in roots and border cells of transgenic cotton. Euphytica 154: 83–90.
24. Tapp H, Stotzky G (1998) Persistence of the insecticidal toxin from *Bacillus thuringiensis* subsp. *kurstaki* in soil. Soil Biol Biochem 30: 471–476.
25. Koskella J, Stotzky G (1997) Microbial utilization of free and clay-bound insecticidal toxins from *Bacillus thuringiensis* and their retention of insecticidal activity after incubation with microbes. Appl Enviro Microbiol 63: 3561–3568.
26. Andrássy I (1992) A short census of free-living nematodes. Fund Appl Nematol 15: 187–188.
27. Blair JM, Bohlen PJ, Freckman DW (1996) Soil invertebrates as indicators of soil quality. In: Doran JW, Jones AJ, editors. Methods for Assessing Soil Quality. SSSA Special Publication No. 49. Soil Science Society of America, Madison, 283–301.
28. Sanvid O, Stark M, Romeis J, Bigler F (2006) Ecological impacts of genetically modified crops: experiences from ten years of experimental field research and commercial cultivation. Swiss Expert Committee for Biosafety. ART-Schriftenreihe 1: 1–84.
29. Wei JZ, Hale K, Carta L, Platzer E, Wong C, et al. (2003) *Bacillus thuringiensis* crystal proteins that target nematodes. PANS 100: 2760–2765.
30. Lang A, Arndt M, Beck R, Bauchhenss J, Pommer G (2006) Monitoring of the environmental effects of the *Bt* gene. Bayerische Landesanstalt für Landwirtschaft. Available: http://www.lfl-neu.bayern.de/publikationen/daten/schriftenreihe_url_1_43.pdf. Accessed 2012 Nov 3.
31. Saxena D, Stotzky G (2001) *Bacillus thuringiensis* (*Bt*) toxin released from root exudates and biomass of *Bt* corn has no apparent effect on earthworms, nematodes, protozoa, bacteria, and fungi in soil. Soil Biol Biochem 33: 1225–1230.
32. Manachini B, Lozzia GC (2002) First investigations into the effects of *Bt* corn crop on Nematofauna. Boll Zool Agraria e di Bachicoltura Serie II 34: 85–96.
33. Al-Deeb MA, Wilde GE, Blair JM, Todd TC (2003) Effect of *Bt* corn for corn rootworm control on nontarget soil microarthropods and nematodes. Environ Entomol 32: 859–865.
34. Höss S, Nguyen HT, Menzel R, Pagel-Wieder S, Miethling-Graf R, et al. (2011) Assessing the risk posed to free-living soil nematodes by a genetically modified maize expressing the insecticidal Cry3Bb1 protein. Sci Total Envir 409: 2674–2684.
35. Griffiths BS, Caul S, Thompson J, Birch ANE, Scrimgeour C, et al. (2005) A comparison of soil microbial community structure, protozoa, and nematodes in field plots of conventional and genetically modified maize expressing the *Bacillus thuringiensis* Cry1Ab toxin. Plant Soil 275: 135–146.
36. Griffiths BS, Caul S, Thompson J, Birch ANE, Scrimgeour C, et al. (2006) Soil microbial and faunal community responses to *Bt* maize and insecticide in two soils. J Environ Qual 35: 734–741.
37. Bartsch D, Devos Y, Hails R, Kiss J, Krogh PH, et al. (2010) Environmental impact of genetically modified maize expressing cry1 proteins. Genet Modif Plants Biotechnol Agr For 64: 575–614.
38. The States Council of China (2001) Regulations on safety of agricultural genetically modified organisms. The Central People's Government of the People's Republic of China. Available: http://www.gov.cn/flfg/2005-08/06/content_21003.htm. Accessed 2013 Feb 7.
39. EFSA (2011) Guidance on the post-market environmental monitoring (PMEM) of genetically modified plants. J EFSA 9(8): 2316. doi:10.2903/j.efsa.2011.2316.
40. Guo XM, Liu JS (2003) The guidance for seed introduction of cotton. The JinDun Publishing House, Beijing.
41. Committee of handbook of soil fauna research methods (1998) Handbook of soil fauna research methods. China Forestry Publishing House, Beijing.
42. Jenkins WR (1964) A rapid centrifugal-floatation technique for separating nematodes from soil. Plant Dis Rep 48: 692.
43. Liang WJ, Lou YL, Li Q, Zhong S, Zhang XK, et al. (2009) Nematode faunal response to long-term application of nitrogen fertilizer and organic manure in Northeast China. Soil Biol Biochem 41: 883–890.
44. Magurran AE (2004) Measuring biological diversity. Blackwell Science, Oxford.
45. Bongers T, Ferris H (1999) Nematode community structure as a bioindicator in environmental monitoring. Trend Ecol Evolut 14: 224–228.
46. Bongers T (1990) The maturity index: an ecological measure of environmental disturbance based on nematode species composition. Oecologia 83: 14–19.
47. Yeates GW, Bongers T, de Goede RGM, Freckman DW, Georgieva SS (1993) Feeding habits in soil nematode families and genera – an outline for soil ecologists. J Nematol 25: 315–31.
48. Ferris H, Bongers T, de Goede RGM (2001) A framework for soil food web diagnostics: extension of the nematode faunal analysis concept. Appl Soil Ecol 18: 13–29.
49. Graef F, Züghart W, Hommel B, Heinrich U, Stachow U, et al. (2005) Methodological scheme for designing the monitoring of genetically modified crops at the regional scale. Environ Monit Assess 111: 1–26.
50. Sanvido O, Aviron S, Romeis J, Bigler F (2007) Challenges and perspectives in decision-making during post-market environmental monitoring of genetically modified crops. J Für Verbraucherschutz 2: 37–40.
51. Züghart W, Benzler A, Berhorn F, Sukopp U, Graef F (2008) Determining indicators, methods and sites for monitoring potential adverse effects of genetically modified plants to the environment. the legal and conceptional framework for implementation. Euphytica 164: 845–852.
52. Manachini B, Fiore MC, Landi S, Arpaia S (2003) Nematode species assemblage in Bt-expressing transgenic eggplants and their isogenic control. Biodiversity Implications of Genetically Modified Plants. September 7–13, Ascona, Switzerland.
53. Manachini B, Landi S, Fiore MC, Festa M, Arpaia S (2004) First investigations on the effects of *Bt*-transgenic *Brassica napus* L. on the trophic structure of the nematofauna. IOBC/WPRS Bull 27: 103–108.
54. Powell JR, Levy-Booth DJ, Gulden RH, Asbil WL, Campbell RG, et al. (2009) Effects of genetically modified, herbicide-tolerant crops and their management on soil food web properties and crop litter decomposition. J Appl Ecol 46: 388–396.
55. Green J, Wang D, Lilley CJ, Urwin PE, Atkinson HJ (2012) Transgenic potatoes for potato cyst nematode control can replace pesticide use without impact on soil quality. PloS ONE 7 (2): e30973. doi:10.1371/journal.pone.0030973.
56. Karuri HW, Amata R, Amugune N, Waturu C (2010) Occurrence and distribution of soil nematodes in cotton (*Gossypium hirsutum* L.) production areas of Kenya. Afr J Agr Res 5: 1889–1896.
57. Lingaraju S, Sonavane P, Jamadar MM, Harlapur SI, Bhat RS, et al. (2012) Plant parasitic nematodes associated with *Bt* cotton. Curr Sci India 103: 926–932.
58. Wang RX, Yang ZW, Pang HL, Cheng P (1989) Studies on the effect of major groups of nematode in the cotton fields of Shaanxi province on *Fusarium* wilt of cotton. Acta Phytopathologica Sinica 19: 205–209.
59. Dong DF, Chen YF, Steinberger Y, Cao ZP (2008) Effects of different soil management practices on soil free-living nematode community structure, Eastern China. Can J Soil Sci 88: 115–127.
60. Li KM, Liang Z, Xu WL, Zhang XB, Tunuhe, etal. (2009) The species of parasitic nematodes in cotton field of Xinjiang. Acta Agri Boreali-occidentalis Sinica 18: 273–275.

61. Robinson AF (2007) Nematode management in cotton. In: Ciancio A, Mukerji KG, editors. Integrated management and biocontrol of vegetable and grain crops nematodes. Springer: Springer Netherlands. 149–182.
62. Priestley AL, Brownbridge M (2009) Field trials to evaluate effects of *Bt*-transgenic silage corn expressing the Cry1Ab insecticidal toxin on non-target soil arthropods in northern New England, USA. Transgenic Res 18: 425–443.
63. Moser T, Schallnaß HJ, Joses SE, Gestel CAMV, Koolhaas JEE, et al. (2004) Ring-testing and field-validation of a terrestrial model ecosystem (TME)–an instrument for testing potentially harmful substances: effects of carbendazim on nematodes. Ecotoxicology 13: 61–74.
64. Parmelee RW, Phillips CT, Checkai RT, Bohlen PJ (1997) Determining the effects of pollutants on soil faunal communities and trophic structure using a refined microcosm system. Environ Toxicol Chem 16: 1212–1217.
65. Wada S, Toyota K (2008) Effect of three organophosphorous nematicides on non-target nematodes and soil microbial community. Microbes Environ 23: 331–336.
66. Chelinho S, Sautter KD, Cachada A, Abrantes I, Brown G, et al. (2011) Carbofuran effects in soil nematode communities: Using trait and taxonomic based approaches. Ecotox Environ Safe 74: 2002–2012.
67. Pen-Mouratov S, Steinberger Y (2005) Responses of nematode community structure to pesticide treatments in an arid ecosystem of the Negev Desert. Nematology 7: 179–191.
68. Lee L, Saxena D, Stotzky G (2003) Activity of free and clay-bound insecticidal proteins from *Bacillus thuringiensis* subsp. Israelensis against the mosquito *Culex pipiens*. Appl Enviro Microbiol 69: 4111–4115.
69. Stotzky G (2004) Persistence and biological activity in soil of the insecticidal proteins from *Bacillus thuringiensis*, especially from transgenic plants. Plant Soil 266: 77–89.
70. Icoz I, Stotzky G (2008) Cry3Bb1 protein from *Bacillus thuringiensis* in root exudates and biomass of transgenic corn does not persist in soil. Transgenic Res 17: 609–620.
71. Sims SR, Ream JE (1997) Soil inactivation of the *Bacillus thuringiensis* subsp. *kurstaki* Cry IIA insecticidal protein within transgenic cotton tissue: laboratory microcosm and field studies. J Agr Food Chem 45: 1502–1505.
72. Hopkins DW, Gregorich EG (2003) Detection and decay of the *Bt* endotoxin in soil from a field trial with genetically modified maize. Eur J Soil Sci 54: 793–800.
73. Head G, Surber JB, Watson JA, Martin JW, Duan JJ (2002) No detection of Cry1Ac protein in soil after multiple years of transgenic *Bt* cotton (Bollgard) use. Environ Entomol 31: 30–36.
74. Ahmad A, Wilde GE, Zhu KY (2005) Detectability of coleopteran-specific Cry3Bb1 protein in soil and its effect on nontarget surface and below-ground arthropods. Environ Entomol 34: 385–394.
75. Icoz I, Saxena D, Andow DA, Zwahlen C, Stotzky G (2008) Microbial populations and enzyme activities in soil in situ under transgenic corn expressing Cry proteins from *Bacillus thuringiensis*. J Environ Qual 37: 647–662.
76. Li XG, Liu B, Wang XX, Han ZM, Cui JJ, et al. (2012) Field trials to evaluate effects of continuously planted transgenic insect-resistant cottons on soil invertebrates. J Environ Monit 14: 1055–1063.
77. Li XG, Liu B, Cui JJ, Liu DD, Ding S, et al. (2011) No evidence of persistent effects of continuously planted transgenic insect-resistant cotton on soil microorganisms. Plant Soil 339: 247–257.
78. Devare M, Londoño-R LM, Thies JE (2007) Neither transgenic *Bt* maize (MON863) nor tefluthrin insecticide adversely affect soil microbial activity or biomass: A 3-year field analysis. Soil Biol Biochem 39: 2038–2047.
79. Fließbach A, Messmer M, Nietlispach B, Infante V, Mäder P (2012) Effects of conventionally bred and *Bacillus thuringiensis* (*Bt*) maize varieties on soil microbial biomass and activity. Biol Fert Soils 48: 315–324.
80. Bitzer RJ, Rice ME, Pilcher CD, Pilcher CL, Lam WF (2005) Biodiversity and community structure of epedaphic and euedaphic springtails (Collembola) in transgenic rootworm *Bt* corn. Environ Entomol 34: 1346–1376.
81. Cortet J, Griffiths BS, Bohanec M, Demsar D, Andersen MN, et al. (2007) Evaluation of effects of transgenic *Bt* maize on microarthropods in a European multi-site experiment. Pedobiologia 51: 207–218.
82. Duc C, Nentwig W, Lindfeld A (2011) No adverse effect of genetically modified antifungal wheat on decomposition dynamics and the soil fauna community – a field study. PloS ONE 6(10): e25014. doi:10.1371/journal.pone.0025014.
83. Sanvido O, Widmer F, Winzeler M, Bigler F (2005) A conceptual framework for the design of environmental post-market monitoring of genetically modified plants. Environ Biosaf Res 4: 13–27.

10

Effect of Optimal Daily Fertigation on Migration of Water and Salt in Soil, Root Growth and Fruit Yield of Cucumber (*Cucumis sativus* L.) in Solar-Greenhouse

Xinshu Liang, Yinan Gao, Xiaoying Zhang, Yongqiang Tian, Zhenxian Zhang, Lihong Gao*

Beijing Key Laboratory of Growth and Developmental Regulation for Protected Vegetable Crops, Department of Vegetable Science, China Agricultural University, Beijing, P.R. China

Abstract

Inappropriate and excessive irrigation and fertilization have led to the predominant decline of crop yields, and water and fertilizer use efficiency in intensive vegetable production systems in China. For many vegetables, fertigation can be applied daily according to the actual water and nutrient requirement of crops. A greenhouse study was therefore conducted to investigate the effect of daily fertigation on migration of water and salt in soil, and root growth and fruit yield of cucumber. The treatments included conventional interval fertigation, optimal interval fertigation and optimal daily fertigation. Generally, although soil under the treatment optimal interval fertigation received much lower fertilizers than soil under conventional interval fertigation, the treatment optimal interval fertigation did not statistically decrease the economic yield and fruit nutrition quality of cucumber when compare to conventional interval fertigation. In addition, the treatment optimal interval fertigation effectively avoided inorganic nitrogen accumulation in soil and significantly ($P<0.05$) increased the partial factor productivity of applied nitrogen by 88% and 209% in the early-spring and autumn-winter seasons, respectively, when compared to conventional interval fertigation. Although soils under the treatments optimal interval fertigation and optimal daily fertigation received the same amount of fertilizers, the treatment optimal daily fertigation maintained the relatively stable water, electrical conductivity and mineral nitrogen levels in surface soils, promoted fine root (<1.5 mm diameter) growth of cucumber, and eventually increased cucumber economic yield by 6.2% and 8.3% and partial factor productivity of applied nitrogen by 55% and 75% in the early-spring and autumn-winter seasons, respectively, when compared to the treatment optimal interval fertigation. These results suggested that optimal daily fertigation is a beneficial practice for improving crop yield and the water and fertilizers use efficiency in solar greenhouse.

Editor: Jin-Song Zhang, Institute of Genetics and Developmental Biology, Chinese Academy of Sciences, China

Funding: This work was supported by the earmarked fund for Modern Agro-industry Technology Research System in China (CARS-25) and the Special Fund for Research in the Public Interest provided by the Chinese Ministry of Water Resources (201001061). The funders had no role in study design, data collection and analysis, decision to publish, or preparation of the manuscript.

Competing Interests: The authors have declared that no competing interests exist.

* E-mail: gaolh@cau.edu.cn

Introduction

Protected vegetable production systems have been rapidly developed in recent decades in China. Solar-greenhouse, an unheated plastic greenhouse using solar light energy to make sure that vegetables grow normally, plays more and more important roles in China's vegetable production and supplication during the winter [1]. However, excessive irrigation and fertilization are commonly practiced in the solar-greenhouse production systems. For instance, some investigations have revealed that irrigation water rate is 1000 mm each year and fertilizer N apparent recovery efficiency can be less than 10% using conventional management practices [2,3]. Consequently, redundant water and fertilizers affect environmental protection by nutrient accumulation and soil salinization [4–6].

It has become the focus of agricultural field that the efficient water and fertilizer management methods are applied in vegetable production in greenhouse. For example, Cabello et al. [7] demonstrated that under moderate deficit irrigation (90% evapotranspiration) condition, reducing the inputs of nitrogen fertilizer did not reduce yield of melon, but increased water and nitrogen fertilizer use efficiencies. Similarly, Mahajan and Singh [8] found that fertigation through reducing rates of water and nitrogen fertilizer promoted fruit yield and quality, and water and fertilizer use efficiencies in greenhouse tomato cultivation system. In addition, both subsurface drip irrigation and alternate furrow irrigation can significantly improve the root growth, and thus enhance the yield and water and fertilizer use efficiencies [9–12].

However, conventional interval fertigation is still common in intensive vegetable production systems in China. In general, the period between successive irrigation and fertilization events is 7 to 10 days, or even longer. Consequently, the nutrient concentration in the root-zone soil may be in excess of plant requirement for growth on the first day after fertigation, and then decreases gradually to reach deficit levels before the day preceding the next fertigation event, and eventually inhibit crop growth and development [13]. Therefore, there is a need for ensuring water and nutrient in root-zone soil stable to reduce adverse effects caused by their fluctuation on crop growth under intensive vegetable management practices.

Drip fertigation, a technique used to manage soil water and nutrition supplies according to the actual water and nutrient requirement of the plants, can be applied to vegetable production to maintain stable water and nutrition contents in the root zone of crops. It applies frequent and small amounts of soluble fertilizers along with water by exerting the soil buffer characteristic and by reducing the time interval between successive irrigations [14,15]. This technique is still rarely implemented in soil cultivation systems in developing countries, despite of the fact that it has been applied for many years in developed countries.

The efficiency of a fertigation treatment can be evaluated with a combination of the availability and distribution of soil water and nutrients, the growth and development of crop root, and the formation of crop yield. These parameters are relatively simple, rapid, easy to obtain and cost effective. Among these parameters, particular attention should be paid to plant root since the morphology and spatial configuration of root can significantly affect soil water and nutrient transformation, mobilization and use efficiency by plant and crop yield [16–18]. However, due to the complexity of root growth environment and the limitations of root research methods, little information is available regarding the vegetable root growth under daily fertigation based on crop requirement under greenhouse soil cultivation conditions.

Cucumber is one of the major greenhouse vegetables in China. However, inappropriate irrigation and fertilizing practices have caused soil nutrient imbalance that negatively affects cucumber growth and the reduction of soil water and fertilizer use efficiencies. It is important for farmers to manage soil water and nutrition supplies according to the actual water and nutrient requirement of the plants during greenhouse cucumber production seasons. Thus, the objectives of this study were to investigate how the migration of water and salt in soil, and root growth and fruit yield of cucumber are affected by the daily fertigation based upon the actual water and nutrient requirement of the plants.

Materials and Methods

Ethics statement

No specific permissions were required for the described field study.The experiment was carried out in 2011 in a typical solar greenhouse of Fangshan District Agricultural Science Research Institute, Beijing, China (39.7°N; 116.1°E), which is an experimental station of China Agricultural University. The field study did not involve endangered or protected species.

Site description and experimental design

The greenhouse was covered with polyethylene film (ground area 56 m×7 m) without supplementary lighting and heating. Daily average soil temperatures and air temperature in solar-greenhouse were shown in Fig. 1. Both the two temperatures in April gradually increased until each reached the highest value in July and August, and then gradually decreased. Furthermore, Daily average soil temperatures at 10 cm depth from conventional interval fertigation (CK), optimal interval fertigation (OIF) and optimal daily fertigation (ODF) were not significantly different. Field experiments were conducted on a sandy loam soil (36% sand, 48% silt and 16% clay). The topsoil (0–30 cm layer) had a pH (1:2.5 soil/water) value of 7.37, an electrical conductivity (EC) (1:5 soil/water) value of 1.17 dS m^{-1}, a field capacity of 20.8% and a bulk density of 1.36 g cm^{-3}, and contained 15.8 g kg^{-1} organic matter, 0.96 g kg^{-1} total nitrogen (N), 195 mg kg^{-1} inorganic N, 193 mg kg^{-1} available phosphorus (P) and 275 mg kg^{-1} available potassium (K).

The experimental period comprised two growth cycles including the early-spring (ES) (February 1-July 29) and autumn-winter (AW) (August 1-December 25) seasons. The varity of cucumber was Zhongnong No. 26. This varity was a high-quality and largely fruit-bearing cucumber hybrid. The average fruit length and width of this varity were 30 and 3.3 cm, respectively. Due to its high resistance to several diseases (including powdery mildew, downy mildew and gray mold) and tolerance to both low temperature and weak light, this varity was very suitable for cultivation in solar-greenhouse. Cucumber seedlings with two leaves were transplanted by hand, before which, soils were incorporated with organic fertilizer (basal fertilizer) at designed rates (Table 1), and were ploughed and harrowed to a depth of 30 cm. Plant pruning was performed as follows: all lateral branches were removed by hand, however, the axial shoot of cucumber was remained and it climbed upward along a vertical rope. Planting density of cucumber was 5 plants m^{-2}.

The experiment consisted of three treatments:

(1) Conventional interval fertigation (CK): Conventional organic manure (basal fertilizer) and chemical fertilizer were applied at rates based on average fertilization level used by greenhouse cucumber growers in the suburb of Beijing (Table 1). A detailed description of chemical fertilizer topdressing/irrigation rates under the CK was given in Fig. 2.

(2) Optimal interval fertigation (OIF): The organic manure (basal fertilizer) application rate was half the rate conventionally applied per hectare (Table 1). Based on the N requirement for cucumber growth [19–21] and N fertilizer recommendation, the total N rates applied by topdressing were 427.5 and 151.5 kg N ha^{-1} in the ES and AW seasons (Table 1), respectively. These total mineral N (N_{min}) application rates were calculated using the method of soil N balance, where expected yield of solar greenhouse cucumber were 120 and 60 t ha^{-1} in the ES and AW seasons, respectively. The equation [22] was as follows:

$$N_{recommend} = N_{crop} + N_{safty} + N_{loss} - N_{initial} - N_{manure} - N_{mineralization} \quad (1)$$

Where $N_{recommend}$ is Recommended fertilizer N; N_{crop} is crop N uptake; N_{safty} is soil N_{min} safety margin; N_{loss} is N loss; $N_{initial}$ is soil N_{min} in the root zone before transplanting; N_{manure} is N_{min} from the mineralization of organic fertilizer; $N_{mineralization}$ is N_{min} from the mineralization of organic nitrogen in soil.

Chemical fertilizer topdressing/irrigation frequencies and irrigation rates were the same as the CK, except fertilizer topdressing rates. A detailed description of chemical fertilizer topdressing/irrigation rates under the treatment OIF was given in Fig. 2.

(3) Optimal daily fertigation (ODF): The total amounts of organic manure (basal fertilizer) and chemical fertilizer were the same as treatment OIF (Table 1), however, chemical fertilizer was applied automatically daily according to the actual water and nutrient requirement of the plants. A detailed description of chemical fertilizer topdressing/irrigation rates under the treatment ODF was given in Fig. 2.

The total irrigation amount was the same for three treatments in whole growth period of cucumber. Fertigation was used under transparent thin film for all treatment. Irrigations were applied 16

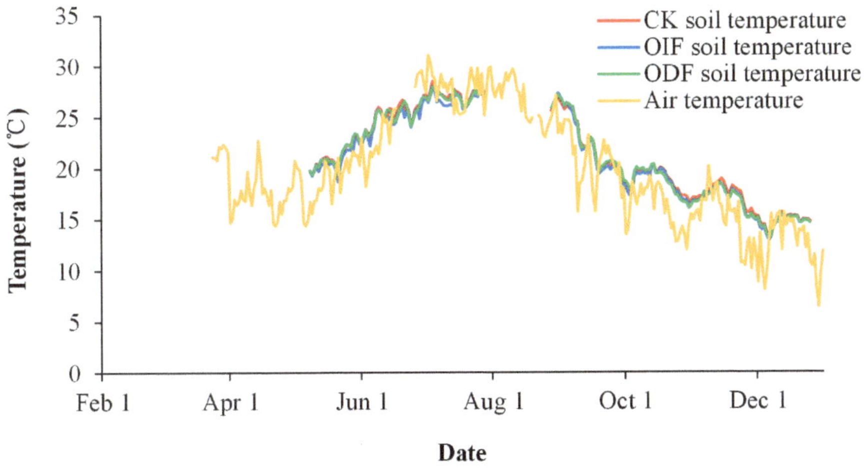

Figure 1. Daily average soil temperatures at 10 cm depth from conventional interval fertigation (CK), optimal interval fertigation (OIF) and optimal daily fertigation (ODF) and daily average air temperature at 150 cm height in cucumber greenhouse cropping system in 2011 at Fangshan, Beijing suburbs. These data were determined by RTH-1010 TPE rensin-shield sensor and RT-12 Thermo Recorder made in Japan. Time interval for data recording was set to 10 minutes.

and 7 times in the ES and AW seasons, respectively. The irrigation rate each time for CK and OIF teatments was 22.5–30 mm according to requirement rule of water for cucumber [23–25].

The chemical fertilizer used in this study was a special water-soluble fertilizer for cucumber called Shengdanshu (a local fertilizer). This fertilizer had three $N:P_2O_5:K_2O$ formulations, i.e., 20:20:20, 19:8:27 and 18:6:34, which were used at the early harvest stage (April 21 – May 12 during ES season, September 25 – October 11 during AW season), the middle harvest stage (May 13 – July 5 during ES season, October 12 – November 30 during AW season) and the late harvest stage (July 6 – July 29 during ES season, December 1 – December 25 during AW season), respectively. Fertilizers were applied 12 and 6 times in the ES and AW seasons, respectively. Except different fertigation methods related to different treatments, the same local management practices were applied in all treatments. The experiment was a randomized block design with four replications and the size of each replicate plot was 3.9 m×4.8 m. Each replicate plot had three cultivation furrows and was separated from the adjacent plots by plastic films buried at a depth of 50 cm.

Soil water content, EC and mineral N content

To evaluate the migration of soil water and salt under different fertigation treatments, soil samples from five cores per subplot were collected five times within a single fertigation cycle during the middle fruit harvest period when daily fruit production was very high in each cropping season. For the ES season, soils were sampled on May 27, May 28, May 31, June 2 and June 4. For the AW season, the corresponding sampling times were October 26, October 27, November 1, November 6 and November 12. Soil samples were taken at 0–15, 15–30 and 30–45 cm depth. Soil samples of each plot at each depth were mixed thoroughly and passed through a 2-mm sieve. Sub-samples of 20 g fresh soil were dried for 12 h at 105°C, and then soil water content was determined at the ratio of water and dry soil weight. Sub-samples of about 300 g fresh soil used to measure EC were air-dried passed through a 1-mm sieve. Soil EC was analyzed from a 1:5 (*w*/*v*) soil (air dried) to water ratio using an EC meter and combination glass electrodes (FE30, METTLER TOLEDO, Shanghai, China). Sub-samples of 12 g fresh soil used to estimate mineral N were submerged into 100 ml 0.01 M $CaCl_2$ solution and shaken for 1 hour to extract inorganic N. The extracts were filtered and

Table 1. Amounts of fertilizers used in the treatments CK (conventional interval fertigation), OIF (optimal interval fertigation) and ODF (optimal daily fertigation) in the ES (early-spring) and AW (autumn-winter) seasons.

Cropping season	Treatment	Organic manure[a]	Chemical fertilizer (kg ha^{-1})			Irrigation water
		(t ha^{-1})	N	P_2O_5	K_2O	(mm)
ES	CK	45.0	819.0	690.4	1006.2	349.2
	OIF	22.5	427.5	361.3	524.1	349.2
	ODF	22.5	427.5	361.3	524.1	349.2
AW	CK	60.0	465.0	312.0	593.4	147.2
	OIF	30.0	151.5	123.1	183.1	147.2
	ODF	30.0	151.5	123.1	183.1	147.2

[a]For the ES season, total N, total P and total K contents of the organic manure were 2.09% (N), 2.06% (P_2O_5) and 1.34% (K_2O), respectively. For the AW season, total N, total P and total K contents of the organic manure were 1.30% (N), 0.5% (P_2O_5) and 1.06% (K_2O), respectively.

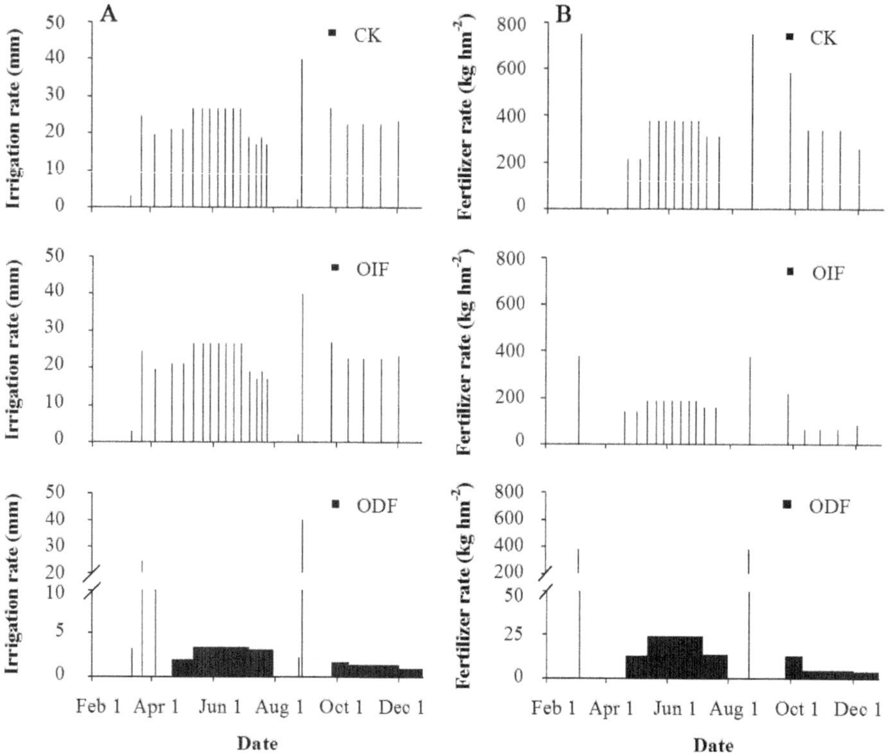

Figure 2. Irrigation scheduling (A) and fertilization scheduling (B) under three different fertigation ways in a solar greenhouse cucumber cultivation system. All fertilizer varity were compound fertilizer. The two $N:P_2O_5:K_2O$ formulations of basal fertilizer were respectively (18:46:0) and (15:15:15) during ES and AW seasons. The three $N:P_2O_5:K_2O$ formulations of topdressing fertilizer were respectively (20:20:20), (19:8:27) and (18:6:34) for early fruit stage, middle fruit stage, late fruit stage during ES and AW seasons.

analyzed using an continuous flowing analyzer (TRAACS2000, USA) to determine NO_3^--N and NH_4^+-N contents [26].

Cucumber root morphologic characters

Once the final harvest was completed, one typical cucumber root in each plot was collected with an Eijelkamp root auger (length = 0.15 m, diameter = 0.08 m) from the 0–15, 15–30 and 30–45 cm soil layers. For each layer, 9 holes were drilled around the cucumber main root in a shape of 3×3 cross square. The roots in the soil were carefully selected and washed to acquire the roots from different soil layers. Root morphology was analyzed by using fresh roots and a root scanner system (EPSON EXPRESSION 4990, Japan). Data were then analyzed with the WinRHIZO root analysis software (LC4800-II LA2400; Saint foy, Canada) to determine the root characteristics, including root length, root surface, root volume and average diameter. The root characteristics could be divided into four classification according to the root diameters aggregated into classes of 0.0–0.5, 0.5–1.0, 1.0–1.5 and ≥1.5 mm. The scanned roots were oven-dried at 65°C until weight constancy and weighed.

Cucumber fruit yield, irrigation water use efficiency and partial factor productivity of applied nitrogen

Economic yield was measured for whole cucumber growth cycles in each plot and translated into economic yield weight per hectare. The ratio of yield to water supply was referred to as irrigation water use efficiency (*IWUE*, kg mm^{-1}):

$$IWUE = Y/W \quad (2)$$

where Y and W represent the economic yield (kg ha^{-1}) and the amount of water (mm) applied to the cucumber during the growing cycle [7], respectively. The ratio of yield to N supply is referred to as partial factor productivity of applied N (PFP_N, kg kg^{-1}):

$$PFP_N = Y/F \quad (3)$$

where F is the amount of fertilizer N (kg) applied to the cucumber during the growing cycle [27].

Vitamin C, soluble sugar and nitrate contents in cucumber fruit

To estimate cucumber fruit quality under different fertigation treatments, three quality parameters which were concerned by local residents were made. Fresh fruit samples of cucumber in each plot were collected when daily fruit production was very high, and their appearance should be similar and marketable. Fresh fruits were washed, chopped and mixed in the lab to assess different quality indexes. Contents of Vitamin C and soluble sugar were determined by the methods of 2, 6-dichloro-indophenol titration [28] and anthrone ethyl acetate colorimetic [29], respectively. Content of nitrate was measured by the method of sulfuric acid-acid [30].

Statistical analysis

SPSS 17.0 was used to analyse data. Treatment means were separated using the least significant difference (LSD) test at $P<0.05$. Principal component analysis (PCA) was done to comprehensively determine the whole effect of different fertigation methods on several root morphologic characters (Root dry weight, length, surface area, average diameter and volume).

Results

Soil water content, EC and mineral N content

Soil water content under the treatments OIF and CK increased rapidly and then decreased gradually within an irrigation cycle (Fig. 3). However, it maintained a relatively stable level under the treatment ODF. No significant difference was found in soil water content between the treatments OIF and CK in all tested soil layers in both ES and AW seasons. The treatment ODF decreased the amplitude (i.e. the difference between maximum and minimum values) of soil water content in all tested soil layers during ES season and that in the 0–15 cm soil layer during AW season in an irrigation cycle, when compared to the treatments OIF and CK (Table 2). In addition, the relative soil water content under the treatment ODF was higher in the 0–15 cm soil layer than in the 15–30 and 30–45 cm soil layers, suggesting that the soil water was concentrated by the treatment ODF in the 0–15 cm soil layer.

The EC values in the 0–15 cm soil layer were higher and more variable than those in the 15–30 and 30–45 cm soil layers, suggesting the salts were concentrated in the 0–15 cm soil layer (Fig. 3). The CK showed the highest and most variable EC values in the 0–15 cm soil layer, resulting in the highest amplitude of soil EC value (Table 2). Generally, in the ES season, the treatments OIF and ODF did not show significant differences in EC values in the 0–15 cm soil layer. However, in the AW season, EC values under the treatment OIF were significantly higher and more variable than those under the treatment ODF in most sampling times. The soil EC values were not significantly different ($P>0.05$) between any two treatments in the 15–30 and 30–45 soil layers in both ES and AW seasons.

Similar to soil EC values, generally, soil mineral N (N_{min}) contents in the 0–15 cm soil layer were higher and more variable than those in the 15–30 and 30–45 cm soil layers (Fig. 3). In addition, the CK had the higher and more variable soil N_{min} contents in the 0–15 cm soil layer. However, the amplitude of soil N_{min} content was significantly higher under the CK than under the treatments OIF and ODF in the 15–30 and 30–45 soil layers in ES and AW seasons, respectively (Table 2).

Principal component analysis of root growth properties

Root dry weight, length, surface area, average diameter and volume were used to evaluate the plant root growth. Since there were statistical links among each other of these root properties (all $P<0.05$), we used the principal component analysis (PCA) to transform these correlated root variables (Table 3) into two principal components (i.e. PC_1 and PC_2) to obtain two princioal expressions as follow:

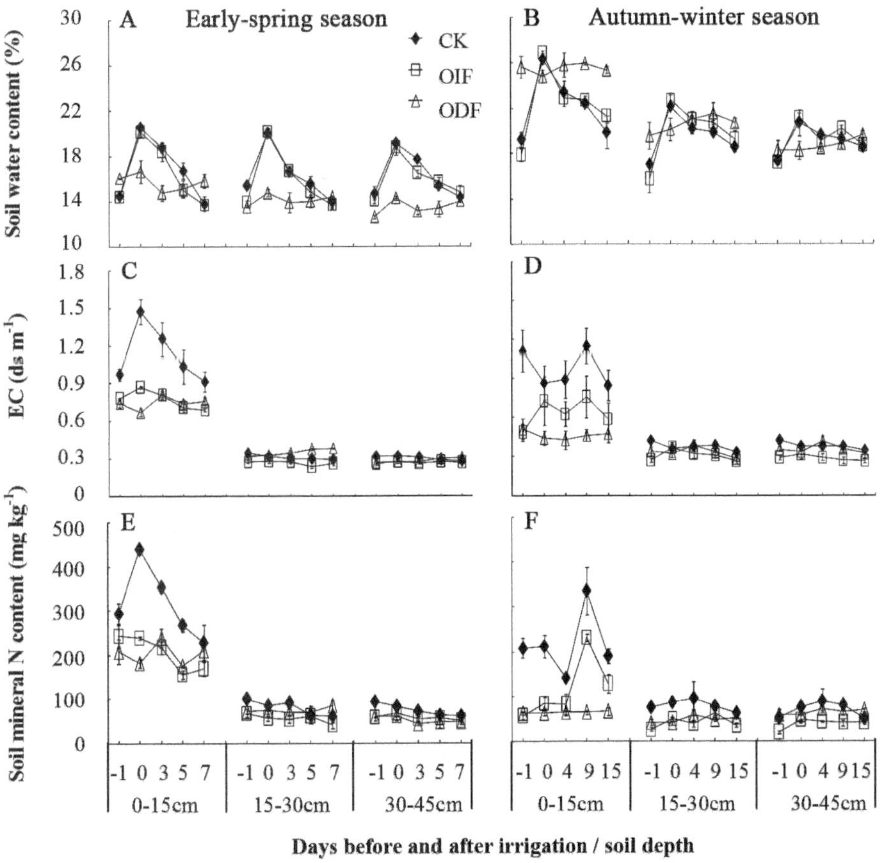

Figure 3. Effects of conventional interval fertigation (CK), optimal interval fertigation (OIF) and optimal daily fertigation (ODF) on changes of soil water content (A and B), EC value (C and D) and mineral N content (E and F) in the early-spring (ES) and autumn-winter (AW) seasons. The numbers on the abscissa represent the days before (negative value), during (zero) and after (positive value) irrigation. Bars represent standard errors.

Table 2. Effects of conventional interval fertigation (CK), optimal interval fertigation (OIF) and optimal daily fertigation (ODF) on the amplitudes (the difference between maximum and minimum values) of soil water content, EC value and mineral N content in the early-spring (ES) and autumn-winter (AW) seasons.

Cropping season	Treatment	Soil water (%)			Soil EC (mS cm^{-1})			Soil mineral N (mg kg^{-1})		
		0–15 cm[a]	15–30 cm	30–45 cm	0–15 cm	15–30 cm	30–45 cm	0–15 cm	15–30 cm	30–45 cm
ES	CK	7.0 a[b]	6.1 a	5.0 a	0.717 a	0.066 a	0.071 a	230.7 a	59 a	43.9 a
	OIF	6.5 a	6.6 a	5.0 a	0.204 b	0.062 a	0.033 a	114.5 b	39.9 b	35.6 a
	ODF	3.4 b	2.2 b	2.1 b	0.154 b	0.087 a	0.077 a	91.6 b	32.6 b	41.8 a
AW	CK	8.4 a	5.2 a	3.8 a	0.682 a	0.108 a	0.094 a	205.2 a	77.5 a	76.1 a
	OIF	9.0 a	7.0 a	4.0 a	0.495 a	0.130 a	0.076 a	177.6 a	38.7 a	36.4 b
	ODF	3.5 b	4.0 a	2.2 a	0.223 b	0.112 a	0.097 a	23.9 b	29.6 a	25.2 b

[a]Soil sampling layers
[b]The same letter in the same data column denotes no significant difference ($P \leq 0.05$) by LSD.

$$PC_1 = 0.452X_1 + 0.489X_2 + 0.544X_3 - 0.243X_4 + 0.449X_5 \quad (4)$$

$$PC_2 = 0.418X_1 - 0.357X_2 - 0.076X_3 + 0.705X_4 + 0.441X_5 \quad (5)$$

where X_1, X_2, X_3, X_4 and X_5 represent the root dry weight, length, surface area, average diameter and volume, respectively. The eigenvalue contribution rates of PC_1 and PC_2 were 65.95% and 31.60%, respectively.

Since the greater values of root dry weight, length, surface area and volume and the thinner root diameter, the root system is more powerful for absorption of water and nutrient, the comprehensive principal component (CPC) was received as follow:

$$CPC = 0.6774PC_1 - 0.3155PC_2 = 0.201X_1 + 0.444X_2 + 0.389X_3 - 0.349X_4 + 0.219X_5 \quad (6)$$

The treatment CK showed the lowest PC_1 and CPC values in the ES season, but the highest PC_2 values in the ES and AW seasons and PC_1 and CPC values in the AW season (Table 4). However, the treatment OIF showed the lowest PC_1 and CPC values in the AW season. The treatment ODF showed the lowest PC_2 value in the ES and AW seasons, but the highest PC_1 and CPC values in the ES season.

Table 4. Effects of conventional interval fertigation (CK), optimal interval fertigation (OIF) and optimal daily fertigation (ODF) on the first (PC1) and second (PC2) principal components and comprehensive principal component (CPC) values of root characteristic parameters in the early-spring (ES) and autumn-winter (AW) seasons.

Principal component	ES season			AW season		
	CK	OIF	ODF	CK	OIF	ODF
PC_1	−0.09	1.05	1.61	0.50	−2.62	−0.46
PC_2	1.28	1.26	−0.81	−0.22	−0.58	−0.93
CPC	−0.46	0.30	1.32	0.40	−0.93	−0.01

Root length and distribution in different soil layers

In general, root length of four diameter classes (i.e. 0.0–0.5, 0.5–1.0, 1.0–1.5 and >1.5 mm) decreased with rooting depth under all fertigation treatments (Fig. 4). However, while the fine roots (<0.5 mm diameter) mainly concentrated at 0–15 cm depth, the thicker roots (>0.5 mm diameter) were more evenly distributed

Table 3. Effects of conventional interval fertigation (CK), optimal interval fertigation (OIF) and optimal daily fertigation (ODF) on the root weight, length, surface area, average diameter and volume in the early-spring (ES) and autumn-winter (AW) seasons.

Cropping season	Treatment	X_1	X_2	X_3	X_4	X_5
		Root weight (g)	Root length (cm)	Root surface area (cm^2)	Root average diameter (mm)	Root volume (cm^3)
ES	CK	1.68 a[a]	4913.69 b	653.73 a	0.416 a	8.58 a
	OIF	1.86 a	5641.86 ab	732.78 a	0.407 a	9.31 a
	ODF	1.67 a	7368.76 a	781.54 a	0.370 b	8.47 a
AW	CK	1.51 a	6104.37 a	715.41 a	0.388 a	8.35 a
	OIF	1.09 b	3870.42 b	481.92 b	0.399 a	5.92 b
	ODF	1.31 a	5855.82 ab	659.25 a	0.384 a	7.08 b

[a]The same letter in the same data column denotes no significant difference ($P \leq 0.05$) by LSD.

with soil depth. Root length of 0.0–0.5 mm diameter was statistically affected by fertigation treatment within the uppermost 15 cm of the soil depth, however, root length of 0.5–1.5 mm diameter was statistically affected by fertigation treatment within the uppermost 30 cm of the soil depth ($P<0.05$; Fig. 4A,B). Root length of >1.5 mm diameter was only statistically affected by fertigation treatment in the 15–30 cm soil layer in the AW season ($P<0.05$; Fig. 4B). The treatment OIF significantly increased the root length of 0.5–1.0 mm diameter in the 0–15 cm soil layer in the ES season, however, the treatment ODF significantly increased the root length of 0.0–1.5 mm diameter in the 0–15 cm soil layer in the ES season and the root length of 1.0–1.5 mm diameter in the 0–15 cm soil layer in the AW season, when compared to the CK. Generally, the root lengths were not significantly different ($P>0.05$) between the treatment OIF and CK in the 15–30 and 30–45 soil layers in both ES and AW seasons. However, the treatment ODF significantly decreased the root length in the 15–30 and 30–45 soil layers in both ES and AW seasons, when compared to the CK.

Cucumber economic yield, IWUE and PFP_N

In both ES and AW seasons, cucumber economic yield, IWUE, PFP_N from the treatment ODF were significantly ($P<0.05$) higher than those from the CK (Table 5). No significant ($P>0.05$) difference was found in both cucumber economic yield and IWUE between the treatments OIF and CK, however, PFP_N from the treatment OIF was significantly ($P<0.05$) higher than that from the CK in both ES and AW seasons. The treatment ODF increased cucumber economic yield, IWUE and PFP_N by 6.2%, 6.1% and 103% in the ES season, and 8.3%, 8.4% and 232% in the AW season when compared to the CK, respectively. The treatment OIF increased PFP_N by 88% and 209% in the ES and AW seasons when compared to the CK, respectively.

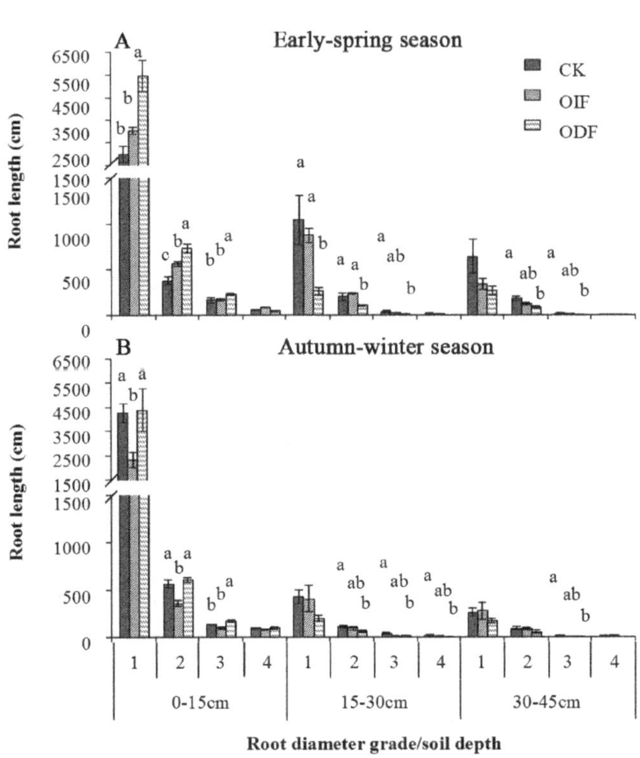

Figure 4. Effects of conventional interval fertigation (CK), optimal interval fertigation (OIF) and optimal daily fertigation (ODF) on the length and distribution of root with different diameter grades in different soil layers in the early-spring (A) and autumn-winter (B) seasons. The numbers 1, 2, 3 and 4 on the abscissa represent four root diameter scales, 0.0–0.5, 0.5–1.0, 1.0–1.5 and ≥1. 5 mm, respectively. Bars represent standard errors. The same letter in the same data column denotes no significant difference ($P\leq0.05$) by LSD.

Vitamin C, soluble sugar and nitrate in cucumber fruit

In general, there was no significant ($P>0.05$) difference in soluble sugar in cucumber fruit between the optimal fertigation treatments (i.e. OIF and ODF) and the CK (Table 6). However, the nitrate in cucumber fruit was significantly ($P<0.05$) decreased by the treatment OIF in the ES and AW seasons, and by the treatment ODF in the AW season, when compared to the CK. In addition, vitamin C in cucumber fruit was significantly ($P<0.05$) increased by the treatment ODF in both ES and AW seasons, when compared to the treatment OIF.

Discussion

Effects of different fertigation treatments on soil water and nutrients

The spatial and temporal distribution of water and nutrients is very heterogeneous in soil. Plant root growth not only can be induced by nutrient supply intensity on the whole, but also can be effected by spatiotemporal variation of water and nutrients [31,32]. Thus, it is important to maintain a relative stable nutrient supply in plant root-zone. Our results clearly showed that increasing the frequency of fertigation could decrease the amplitudes of water and nutrient contents in soils (Table 2). In general, optimal daily fertigation significantly decreased the amplitudes of water content, EC value and mineral N content in the 0–15 cm soil layer in the AW season, when compared to optimal interval fertigation (ODF *vs* OIF; Table 2). Thus, the water and nutrient limitation for plant root growth is probably not likely in soils under optimal daily fertigation. In addition, since the topsoil (0–15 cm) received more water and nutrient than the subsoil (15–45 cm) under fertigation conditions, optimal daily fertigation significantly promoted the root length of <1.5 mm diameter in the 0–15 cm soil layer when compared to optimal interval fertigation (ODF *vs* OIF; Fig. 4). These results are in accordance with previous researches [33–36] and further demonstrate that using the traditional interval fertigation may lead to larger fluctuation of water and nutrients in soil and inhibit plant root growth [37,13]. Furthermore, since soil relative water contents in the 0–15 cm soil layer generally exceeded 25.5% and were higher than soil field capacity (20.8%) under optimal daily fertigation in the AW season (Fig. 3), there is a possibility for our study to further reduce the irrigation rate.

Effects of different fertigation treatments on plant root growth

Plant roots are able to make different responses to different water and nutrient contents of soils. Although excessive fertilizer application can increase crop yield, it can inhibit plant root growth due to the high nutrient concentration in root-zone. Furthermore, unused nutrients will accumulate in the soils and finally enhance the potential threat to the environment. In contrast, root growth can be depressed when soil nutrient is in deficit [38,2,39,40]. In general, optimal fertilizer application can promote plant root growth and improve fertilizer use efficiency. In this study, however, optimal fertilizer application had no promoting effect on both cucumber root growth (Fig. 4B) and economic yield

Table 5. Effects of conventional interval fertigation (CK), optimal interval fertigation (OIF) and optimal daily fertigation (ODF) on cucumber economic yield, irrigation water use efficiency (IWUE) and partial factor productivity of applied nitrogen (PFP_N) in the early-spring (ES) and autumn-winter (AW) seasons.

Cropping season	Treatment	Economic yield (t ha^{-1})	IWUE (kg mm^{-1})	PFP_N (kg kg N^{-1})
ES	CK	132.6b[a]	379. 8b	161. 9c
	OIF	130.2b	372. 8b	304. 5b
	ODF	140.8a	403. 1a	329. 2a
AW	CK	41.2b	279. 8b	88. 6c
	OIF	41.5b	281. 7b	273. 7b
	ODF	44.6a	303. 2a	294. 5a

[a]The same letter in the same data column denotes no significant difference ($P \leq 0.05$) by LSD.

(Table 5) in the AW season under same fertilizer application ways (OIF *vs* CK). The explanation is that although the amount of inorganic fertilizer application was sufficient for plant root growth under the treatment OIF during middle fruit harvest period, it was insufficient for plant root growth during the late fruit harvest period, resulting in significant inhibition of root growth and no significant effect on the cucumber economic yield. For instance, contents of mineral nitrogen in soil laye of 0–30 cm of the treatments CK, OIF and ODF during the late fruit harvest period were respectively 93, 35 and 61 mg kg^{-1}. The contents of mineral nitrogen of the treatment OIF was obviousely less than 50 mg kg^{-1}, which is the lower limit of soil N_{min} content for depth 0–30 cm for normal growth and development of cucumber [41], leading to significantly inhibit root growth in the AW seasnon. The results presented here suggested that fertilizer should be applied according to the actual nutrient requirement of the crops at different growing stages.

Effects of different fertigation treatments on cucumber fruit yield

Root growth is closely coordinated with shoot growth [42]. Generally, spatial distribution of roots in soils, together with distribution of water and nutrients in root-zone and water and nutrients requirements for root and shoot growth, determine the water and nutrient use efficiency of crops and the economic yield formation [43]. The data on economic yield in this study showed that although the amount of fertilizer applied to soil was visibly lower under optimal interval fertigation than under conventional interval fertigation (Table 1), optimal interval fertigation still met the water and nutrient requirements of cucumber (Fig. 3) and maintained root growth (Table 3 and Fig. 4), resulting in the higher efficiency of roots as absorbing organs and no significant effect on cucumber economic yield (Table 5). Thus, excessive fertilizer applied to soil under conventional interval fertigation did not make much contribution to increasing crop yield. In contrast, it can inhibit plant root growth due to the high nutrient concentration in root-zone.

Although soils under the treatments OIF and ODF received the same amount of fertilizer (Table 1), the treatment ODF significantly increased economic yield when compared to the treatment OIF (Table 5). This can be explained by the fact that ODF maintained sufficient and relatively stable water and nutrients in soil (Table 2) and promoted proliferation of fine roots which increased root surface area (Table 3). In addition, ODF reduced the spatiotemporal variation of water in root-zone when compare to OIF (Fig. 3). It was also supported by two recent studies: Gao et al. [14] found that the technology of solution daily application based on the requirement of crops had a promoting effect on tomato yield in a plotted system; Yoshida [44] reported that this technology could increase both the yield of tomato fruit and the water and fertilizers use efficiency by plant.

Root function is related to the morphological and physiological characteristics of root. However, this study did not consider the root physiological function, hence, further work is required test the root physiological function to elucidate the mechanism of communication of root and shoot growth.

Table 6. Effects of conventional interval fertigation (CK), optimal interval fertigation (OIF) and optimal daily fertigation (ODF) on the contents of vitamin C, soluble sugar and nitrate in cucumber fruit in the early-spring (ES) and autumn-winter (AW) seasons.

Cropping season	Treatment	Vitamin C (mg 100 g^{-1})	Soluble sugar (mg g^{-1})	Nitrate (ug g^{-1})
ES	CK	6. 94ab[a]	20. 78a	92. 23a
	OIF	6. 71b	19. 81a	76. 97b
	ODF	7. 31a	21. 12a	84. 94ab
AW	CK	5. 51ab	32. 56a	137. 68a
	OIF	5. 33b	32. 31a	78. 65b
	ODF	5. 76a	31. 02a	63. 89b

[a]The same letter in the same data column denotes no significant difference ($P \leq 0.05$) by LSD.

Effects of different fertigation treatments on nitrate content in cucumber fruit

In general, vegetables can provide 80% of nitrate absorbed into the human body [45]. However, part of these nitrate can be reduced to nitrite by bacteria, which may induce lower oxygen-carrying capacity of blood and methemoglobinemia. In addition, nitrite can produce a reaction with secondary amines, such as amides and amino acids, possibly causing the formation of nitrosamines, which may induce cancer in the digestive system of human body [46]. Therefore, it is important to control and maintain low nitrate content in edible part of vegetables. In this study, both OIF and ODF treatments significantly decreased nitrite content, compared to CK treatment. The relative high nitrate content in fruit under CK can be explained by the previous research that surplus nitrate can be partly absorbed and stored by the plant than its need, in order to maintain normal growth requirement when nitrate supply is in deficit [47]. The results presented here suggested that reducing the fertilizer amounts, according to the actual nutrient requirement of the crops at different growing stages, is beneficial to fruit quality and human healthy without having a significant effect on cucumber fruit yield.

Conclusions

On the premise of optimal management of irrigation rate for cucumber, optimal interval fertilization based on nutrient target values, including soil mineral nutrient safety margin and crop uptake, and soil nutrients in root-zone had no significant effect on economic yield, but increased partial factor productivity of applied N. Optimal daily fertigation based on the actual requirement of crops maintained the relatively stable water and nutrients in soils, reduced the spatiotemporal variation of water and nutrients in root-zone and promoted cucumber fine root (<1.5 mm diameter) growth. In addition, optimal daily fertigation maintained significantly higher cucumber economic yield, irrigation water use efficiency and partial factor productivity of applied N than optimal interval fertigation.

Acknowledgments

The authors thank you very much to all the staff of Fangshan District Agricultural Science Research Institute.

Author Contributions

Conceived and designed the experiments: XL ZZ LG. Performed the experiments: XL XZ. Analyzed the data: XL YG. Wrote the paper: XL YT ZZ LG.

References

1. Gao LH, Qu M, Ren HZ, Sui XL, Chen QY, et al. (2010) Structure, function, application and ecological benefit of single-slope, energy-efficient solar greenhouse in China. Horttechnology 20: 626–631.
2. He F, Chen Q, Jiang R, Chen X, Zhang F (2007) Yield and nitrogen balance of greenhouse tomato (Lycopersicum esculentum Mill) with conventional and site-specific nitrogen management in Northern China. Nutrient Cycling in Agroecosystems 77: 1–14.
3. Hvistendahl M (2010) China's Push to Add by Subtracting Fertilizer.Sci. 327: 801–801.
4. Shi WM, Yao J, Yan F (2009) Vegetable cultivation under greenhouse conditions leads to rapid accumulation of nutrients, acidification and salinity of soils and groundwater contamination in South-Eastern China. Nutrient Cycling in Agroecosystems 83: 73–84.
5. Thompson R, Martinez-Gaitan C, Gallardo M, Gimenez C, Fernandez M (2007) Identification of irrigation and N management practices that contribute to nitrate leaching loss from an intensive vegetable production system by use of a comprehensive survey. Agricultural Water Management 89: 261–274.
6. Tian Y, Liu J, Zhang X, Gao L (2010) Effects of summer catch crop, residue management, soil temperature and water on the succeeding cucumber rhizosphere nitrogen mineralization in intensive production systems. Nutrient Cycling in Agroecosystems 88: 429–446.
7. Cabello MJ, Castellanos MT, Romojaro F, Martinez-Madrid C, Ribas F (2009) Yield and quality of melon grown under different irrigation and nitrogen rates. Agricultural Water Management 96: 866–874.
8. Mahajan G, Singh KG (2006) Response of Greenhouse tomato to irrigation and fertigation. Agricultural Water Management 84: 202–206.
9. Jokinen K, Sarkka LE, Nakkila J, Tahvonen R (2011) Split root fertigation enhances cucumber yield in both an open and a semi-closed greenhouse. Sci Hortic (Amsterdam) 130: 808–814.
10. Wang Y, Liu F, Andersen MN, Jensen CR (2010) Improved plant nitrogen nutrition contributes to higher water use efficiency in tomatoes under alternate partial root-zone irrigation. Funct Plant Biol 37: 175–182.
11. Zhang L, Gao L, Zhang L, Wang S, Sui X, et al.(2012) Alternate furrow irrigation and nitrogen level effects on migration of water and nitrate-nitrogen in soil and root growth of cucumber in solar-greenhouse. Sci Hortic (Amsterdam) 138: 43–49.
12. Zotarelli L, Scholberg JM, Dukes MD, Munoz-Carpena R, Iceman J (2009) Tomato yield, biomass accumulation, root distribution and irrigation water use efficiency on a sandy soil, as affected by nitrogen rate and irrigation scheduling. Agricultural Water Management 96: 23–34.
13. Matsuo N, Mochizuki T (2009) Assessment of three water-saving cultivations and different growth responses among six rice cultivars. Plant Prod Sci 12: 514–525.
14. Gao YM, Li JS, Cao YE (2006) Study on the fertigation of tomato soil culture by drip irrigation in greenhouse. Acta Agriculturae Boreali-Occidentalis Sinica 15: 121–126. (in Chinese with English abstract)
15. Zhao SM, Li BM (2001) Introduction to present status,development and new techniques of nutri-culture systems in Japan. Transactions of the Chinese Society of Agricultural Engineering 17: 171-173. (in Chinese with English abstract)
16. Lynch JP (1995) Root architecture and plant productivity. Plant Physiol 109: 7–13.
17. Lynch JP (2007) Roots of the second green revolution. Australian Journal of Botany 55: 493–512.
18. Lynch JP, Brown KM (2012) New roots for agriculture: exploiting the root phenome. Philos Trans R Soc Lond B Biol Sci 367: 1598–1604.
19. Pei XB, Zhang FM, Wang L (2002) Effect of light and temperature on uptake and distribution of nitrogen,phosphorus and potassium of solar greenhouse cucumber. Zhongguo Nong Ye Ke Xue 35: 1510–1513. (in Chinese with English abstract)
20. Wang JF, Xing SZ, Chen SY, Yu QY, Li XL (2005) Dry matter accumulation in and NPK uptake by non-polluted cucumber in protected cultivation. Chinese Journal of Soil Science 36: 708–711. (in Chinese with English abstract)
21. Yu SF, Gao XB, Lu LP (2000) The study on nutrient absorption of cucumber in chinese lean-to solar greenhouse. China Vegetables: 10–11. (in Chinese with English abstract)
22. Chen Q, Zhang FS, Li XL (2005) Application of N management strategies in vegatable production. China Vegetables: 57–63. (in Chinese with English abstract)
23. An SW, Wang YQ, Li HL, Wang SJ, Gao LH (2010) Effects of different irrigation quantities on growth, yield and fruit qual ity of tomato in solar greenhouse. Acta Agriculturae Boreali-Occidentalis Sinica: 188–192. (in Chinese with English abstract)
24. Sun LP, Wen YG, Wang SZ, Zhao JW, Gao LH (2010) Effects of irrigation on water distribution of cucumber in solar greenhouse. Acta Agriculturae Boreali-Occidentalis Sinica: 173–178. (in Chinese with English abstract)
25. Wei Y, Sun L, Wang S, Wang Y, Zhang Z, et al. (2010) Effects of different irrigation methods on water distribution and nitrate nitrogen transport of cucumber in greenhouse. Transactions of the Chinese Society of Agricultural Engineering 26: 67–72. (in Chinese with English abstract)
26. Houba VJG, Temminghoff EJM., Gaikhorst GH (2000) Soil analysis procedures using 0.01M calcium chloride as extraction regent. Commun Soil Sci Plant Anal 31: 1299–1396.
27. Zhang FS,Wang JQ, Zhang WF, Cui ZL, Ma WQ, et al. (2008) Nutrient use efficiencies of major cereal crops in China and measures for improvement. Tu Rang Xue Bao 45: 915–924.
28. Li H (2000) Principles and techniques of plant physiological biochemical experiment. Beijing: Higher Education Press. 246–248 p.
29. Li H (2000) Principles and techniques of plant physiological biochemical experiment. Beijing: Higher Education Press. 195–197 p.
30. Cataldo D, Maroon M, Schrader L, Youngs V (1975) Rapid colorimetric determination of nitrate in plant tissue by nitration of salicylic acid 1. Commun Soil Sci Plan Anal 6: 71–80.
31. De Kroon H, Huber H, Stuefer J F, Van Groenendael J M (2005) A modular concept of phenotypic plasticity in plants. New Phytol 166: 73–82.
32. Hodge A (2006) Plastic plants and patchy soils. J Exp Bot 57: 401–411.
33. Rajput TBS, Patel N (2006) Water and nitrate movement in drip-irrigated onion under fertigation and irrigation treatments. Agricultural Water Management 79: 293–311.

34. Silber A, Xu G, Levkovitch I, Soriano S, Bilu A, et al. (2003) High fertigation frequency: the effects on uptake of nutrients, water and plant growth. Plant Soil 253: 467–477.
35. Silber A, Bruner M, Kenig E, Reshef G, Zohar H, et al. (2005) High fertigation frequency and phosphorus level: Effects on summer-grown bell pepper growth and blossom-end rot incidence. Plant Soil 270: 135–146.
36. Xu G, Levkovitch I, Soriano S, Wallach R, Silber A (2004) Integrated effect of irrigation frequency and phosphorus level on lettuce: P uptake, root growth and yield. Plant Soil 263: 297–309.
37. Belder P, Spiertz J, Bouman B, Lu G, Tuong T (2005) Nitrogen economy and water productivity of lowland rice under water-saving irrigation. Field Crop Res 93: 169–185.
38. Boomsma C, Santini J, Tollenaar M, Vyn T (2009) Maize per-plant and canopy-level morpho-physiological responses to the simultaneous stresses of intense crowding and low nitrogen availability. Agron J 101: 1426–1452.
39. Mi GH, Chen FJ, Wu QP, Lai NW, Yuan LX, et al. (2010) Ideotype root architecture for efficient nitrogen acquisition by maize in intensive cropping systems. Sci China Life Sci 53: 1369–1373.
40. Shen J, Li C, Mi G, Li L, Yuan L, et al. (2012) Maximizing root/rhizosphere efficiency to improve crop productivity and nutrient use efficiency in intensive agriculture of China. J Exp Bot.
41. Guo R, Li X, Christie P, Chen Q, Zhang F (2008) Seasonal temperatures have more influence than nitrogen fertilizer rates on cucumber yield and nitrogen uptake in a double cropping system. Environ pollut 151: 443–451.
42. Wang H, Inukai Y, Yamauchi A (2006) Root development and nutrient uptake. CRC Crit Rev Plant Sci 25: 279–301.
43. Garnett T, Conn V, Kaiser B N (2009) Root based approaches to improving nitrogen use efficiency in plants. Plant Cell Environ 32: 1272–1283.
44. Yoshida C, Iwasaki Y, Makino A, Ikeda H (2011) Effects of Irrigation Management on the Growth and Fruit Yield of Tomato under Drip Fertigation. Horticultural Research (Japan) 10: 325–331.
45. Jackson T, Westermann D, Moore D (1966) The effect of chloride and lime on the manganese uptake by bush beans and sweet corn. Soil Science Society of America Journal 30: 70–73.
46. Wang XZ, Cheng BS, Zhang GZ (1991) Nitrate and its influence factors in vegetables. Chinese Bulletin of Botany 8: 34–37. (in Chinese with English abstract).
47. Koch G, Schulze ED, Percival F, Mooney H, Chu C (1988) The nitrogen balance of Raphanus sativus x raphanistrum plants. II. Growth, nitrogen redistribution and photosynthesis under NO^{3-} deprivation. Plant Cell Environ 11: 755–767.

11

How Do Earthworms, Soil Texture and Plant Composition Affect Infiltration along an Experimental Plant Diversity Gradient in Grassland?

Christine Fischer[1,2]*, Christiane Roscher[3], Britta Jensen[4], Nico Eisenhauer[5], Jussi Baade[6], Sabine Attinger[1,7], Stefan Scheu[4], Wolfgang W. Weisser[8], Jens Schumacher[9], Anke Hildebrandt[1,2]

1 Institute of Geosciences, Friedrich-Schiller-University Jena, Jena, Germany, **2** Max Planck Institute for Biogeochemistry, Jena, Germany, **3** UFZ, Helmholtz Centre for Environmental Research, Department of Community Ecology, Halle, Germany, **4** J. F. Blumenbach Institute of Zoology and Anthropology, Georg August University of Göttingen, Göttingen, Germany, **5** Institute of Ecology, Friedrich Schiller University Jena, Jena, Germany, **6** Department of Geography, Friedrich Schiller University Jena, Jena, Germany, **7** UFZ, Helmholtz Centre of Environmental Research, Department of Computational Hydrosystems, Leipzig, Germany, **8** Technical University Munich, Terrestrial Ecology Research Group, Department of Ecology and Ecosystem Management, Center for Food and Life Sciences Weihenstephan, Freising, Germany, **9** Institute of Stochastics, Friedrich Schiller University Jena, Jena, Germany

Abstract

Background: Infiltration is a key process in determining the water balance, but so far effects of earthworms, soil texture, plant species diversity and their interaction on infiltration capacity have not been studied.

Methodology/Principal Findings: We measured infiltration capacity in subplots with ambient and reduced earthworm density nested in plots of different plant species (1, 4, and 16 species) and plant functional group richness and composition (1 to 4 groups; legumes, grasses, small herbs, tall herbs). In summer, earthworm presence significantly increased infiltration, whereas in fall effects of grasses and legumes on infiltration were due to plant-mediated changes in earthworm biomass. Effects of grasses and legumes on infiltration even reversed effects of texture. We propose two pathways: (i) direct, probably by modifying the pore spectrum and (ii) indirect, by enhancing or suppressing earthworm biomass, which in turn influenced infiltration capacity due to change in burrowing activity of earthworms.

Conclusions/Significance: Overall, the results suggest that spatial and temporal variations in soil hydraulic properties can be explained by biotic processes, especially the presence of certain plant functional groups affecting earthworm biomass, while soil texture had no significant effect. Therefore biotic parameters should be taken into account in hydrological applications.

Editor: Andrew Hector, University of Oxford, United Kingdom

Funding: The Jena Experiment was funded by the German Science Foundation (DFG, FOR456/1451, www.dfg.de <http://www.dfg.de>), with additional support from the Friedrich Schiller University Jena and the Max Planck Society. This specific research project was further supported by the Helmholtz Impulse and Networking Fund through the Helmholtz Interdisciplinary Graduate School for Environmental Research (HIGRADE) and AquaDiv@Jena, a project funded by the initiative "ProExzellenz" of the German Federal State of Thuringia. The funders had no role in study design, data collection and analysis, decision to publish, or preparation of the manuscript.

Competing Interests: The authors have declared that no competing interests exist.

* E-mail: Fischer.christine@uni-jena.de

Introduction

The water balance of soils is determined by the interaction of water supply and water removal due to processes such as precipitation, infiltration, run-off, percolation and evapotranspiration. For efficient soil and water management, knowledge on soil hydraulic properties, including soil hydraulic conductivity and infiltration characteristics, is necessary to understand how rainwater moves from the soil surface to the groundwater. Hydraulic conductivity describes the capacity of a porous medium to transmit water. It depends on total pore space, pore size distribution and tortuosity [1]. Soil pores are of various origin. The smallest ones (micropores) are related to the grain size distribution and constitute the largest fraction of the total pore volume [2–4]. Larger pores (often referred to as meso- and macropores) make up a characteristic property of the soil structure [5], [6]. Soil structure is determined by aggregates of different sizes, divided into intra-aggregate and inter-aggregate pore structures [7]. Intra-aggregate pores include micro- and mesopores, whereas inter-aggregate pores include meso- and macropores [8].

Traditionally predictions of hydraulic conductivity are based on soil texture, bulk density or organic matter content (mainly intra-aggregate pores) [9–11], implying a decrease in hydraulic conductivity with increasing fraction of fine grains. However, this relationship can be weakened by soil structuring processes forming larger inter-aggregate pores such as interpedal voids and biopores [3], [12]. Macropores constitute a comparatively small fraction of the total pore space, but can contribute substantially to total flow through the porous medium, mainly during high intensity rainfall events [13–15].

Both biotic and abiotic processes contribute to shape soil structure and aggregation of macropores. Drying and freezing causes fissures and cracks in the soil, which are prone to conduct water [16], [17]. Macropores created by both earthworms and plant roots also play a major role for preferential flow [5], [14], [18]. For example, Weiler and Naef [19] observed in grassland that the flow rate through vertically oriented macropores formed by earthworms or plant roots was higher than through the soil matrix. On the other hand, root growth can have opposite effects by clogging of pore space and thus decrease hydraulic conductivity [20].

Macropores formed by earthworms range between 2 to 11 mm in diameter [21] depending on the ecological group of earthworms, i.e. endogeic, epigeic and anecic [22]. Endogeic and epigeic earthworms that live in upper mineral soil or at the soil surface mainly form small and tortuous pores ranging between 2 and 5 mm in diameter [23]. In contrast, anecic species form pores larger than 5 mm in diameter, which may reach as deep as 2 m into the soil [24] and thus enhance infiltration into deep soil layers [25]. As a consequence of the different burrowing behaviors the impact on water flow through soil varies among the different ecological earthworm groups [18]. Further, roots form voids of different size, but the majority of pores stemming from root growth are smaller (0.1–0.6 mm) than those from earthworms [26]. However, root induced pores differ with plant species and can be much larger. For example, most of the root channels formed by the legume alfalfa were between 0.5 and 2.5 mm in diameter [27]. Besides the formation of macropores mentioned above, biotic processes are also involved in forming and stabilizing soil structure [28], [29]. Soil aggregates are more stable in biologically active soil with high carbon content, which therefore is associated with low soil bulk density and high porosity [30].

The processes contributing to structure the soil and shape its hydraulic properties are closely interlinked. For example, earthworm activity depends on a number of factors which influence soil structure and hydraulic properties, such as soil type [31] and texture [24], management practices [32] and vegetation cover [33]. Earthworms also alter above- and belowground plant productivity by forming macropores with the effect varying with plant species and functional group diversity [34], [35]. Experimental studies have shown that above- and belowground biomass production change with increasing plant diversity [36–38]. Furthermore, larger and longer macropores have been shown to correlate with increased plant biomass production and earthworm abundance [39]. Plant species richness does not only affect rooting density, but also improves soil stability, accumulation of organic matter and promotes the activity for soil biota [5], [40]. In addition, the presence of certain plant functional groups, such as legumes and grasses, has been shown to affect the abundance and activity of soil organisms [41], [42]. Understanding how plant species diversity, plant functional group composition and earthworms influence soil water fluxes and the resulting soil moisture distribution is important to improve predictions on how water fluxes will change in taxonomically simplified ecosystems.

Our measurements were conducted in the frame of the Jena Experiment [43], a long-term grassland biodiversity experiment with plots varying in plant species diversity and including experimental blocks differing in soil structure and texture. The design of the experiment provides the opportunity to disentangle the relative importance of soil physical and biological processes for infiltration capacity.

Materials and Methods

Ethics Statement

No specific permission was required for the described field studies. The field site of the Jena Experiment is a former arable land leased by the research consortium from an agricultural collective for the duration of the research grant. The land owner gave the permission for this study and field work including soil sampling and other experimental manipulations. The experiments did not involve endangered or protected species.

Experimental Design

The study was performed on the field site of the Jena Experiment which is located in the floodplain of the Saale river near Jena (Thuringia, Germany; 50°55′N, 11°35′E, 130 m above sea level). Mean annual air temperature is 9.3°C and mean annual precipitation is 587 mm [44]. Before the establishment in 2002, the experimental field site was an arable land and highly fertilized over the last decades. After the last harvest in autumn 2000 the field was ploughed and kept fallow throughout 2001, and the experimental plant communities were established in spring 2002 [43]. The soil of the experimental site is an Eutric Fluvisol [45] developed from up to 2 m thick loamy fluvial sediments [43]. The soil texture in the upper 10 cm of the soil profile changes with increasing distance from the river gradually from sandy loam to silt clay. The sand content decreases from 40% near the river to 11% at distance, while the silt content increases proportionally from 44% to 66%. The clay content (16–23%) shows no significant spatial trend (Table S1). Plots were assembled on a 10 ha area into four blocks, arranged parallel to the river, thus accounting for changes in soil texture.

A pool of 60 native plant species common of Central European mesophilic grasslands was used to establish a gradient of plant species richness (1, 2, 4, 8, 16 and 60) and functional group richness (1, 2, 3 and 4) on 80 plots each of 20×20 m size. To account for differences in morphology and physiology, species were assigned to four functional groups: grasses (16 species), small herbs (12 species), tall herbs (20 species) and legumes (12 species). In addition to test for effects of plant species richness and functional group richness, the experimental design allows for tests caused by the presence and absence of certain functional groups and texture. Plant species richness and functional groups varied in a near-orthogonal design, because the lowest plant species richness level cannot be combined with the highest functional group number and at the 16 plant species level it was not possible to create pure legume and small-herb species mixtures. The plots were mown twice a year, and the mown material was removed from the plots shortly after cutting. All plots were weeded regularly to maintain the target species composition. More details on the experiment and management are given in Roscher et al. [43].

Earthworm Density Manipulations

Earthworm abundance was observed and manipulated on subplots of the main experimental plots with species-richness levels of 1 (12 plots), 4 (16 plots) and 16 (14 plots) since September 2003 (Table 1). Due the unbalanced design effects of plant species richness and functional group richness are partially confounded ($r = 0.438$, $p < 0.001$; Table 1). Two subplots (size 1×1 m) were located in close vicinity to each other (50 cm distance). Two treatments were established: ambient earthworm density (+ew) and earthworm density reduction (−ew). Subplots were enclosed with PVC shields (20 cm aboveground and 15 cm belowground) to decrease the re-colonization in earthworm reduction subplots [34]. Aboveground shields were removed two times a year during the mowing period. Earthworms were extracted from reduction subplots twice a year in spring (beginning of April) and autumn (end of September) by electro-shocking. A voltage was applied to the soil for 35 min *via* four octet devices [46] (DEKA 4000, Deka Gerätebau, Marsberg, Germany) powered by two 12 V batteries. During the application time the voltage was increased sequentially from 250 V (10 min) to 300 V (5 min), 400 V (5 min), 500 V (5 min) and 600 V (10 min). For more details on the arrangement of the steel rods of the octet devices and management of the earthworm subplots see Eisenhauer et al. [34]. Notably, steel rods were installed in both earthworm subplots controlling for potential side effects on infiltration. Two additional extraction campaigns on the −ew and on control subplots in 2006 (Eisenhauer, unpupl. data) confirmed that earthworm data from −ew subplots is an adequate measure of earthworm data in the +ew subplots. Extracted earthworms were identified, counted and weighted (with gut content) in the laboratory. Earthworms at the field site of the Jena Experiment mainly belong to two ecological groups [22]: anecic (*Lumbricus terrestris*) and endogeic (*Aporrectodea caliginosa, Octolasion tyrtaeum, Allolobophora chlorotica, and Aporrectodea rosea*) species. Only a small number of epigeic earthworms (*Lumbricus castaneus*) was extracted and therefore contributed to the total number and biomass of earthworms.

Infiltration Measurement

For *in situ* infiltration measurements we used a hood infiltrometer [47] (UGT, Müncheberg, Germany). These measurements do not require preparation of the soil and therefore can be applied on an undisturbed, vegetated soil surface. In 2011, we conducted three infiltration measurement campaigns (June, September and October) on plots containing 1, 4 and 16 plant species (Table 1). The first measurement campaign was conducted at end of June, the second at the beginning of September, and the third at the end of October. We carried out paired measurements in each plot: one on the reduced and one on the ambient earthworm density subplot. The first and second measurement campaigns were conducted about 65 and 160 days after the first earthworm extraction and the third measurement campaign 30 days after the second earthworm extraction. The extracted earthworm biomass in spring was related to the infiltration capacity in June and September. The earthworm biomass from the second extraction campaign, which was conducted after the second infiltration campaign, was related to the infiltration capacity in October.

A hood with a diameter of 16 cm was placed with the open side on the undisturbed soil surface. The contact between the soil and hood was sealed with wet sand. We conducted measurements at increasingly negative matric potentials (ψ_M) beginning at $\psi_M = 0$ m and reducing it stepwise by -0.02 m until the bubble point of the soil was reached. The bubble point refers to the matric potentials, upon which a pore channel allows for penetration of air into the hood and therefore the maximum applicable matric potential at this location. For a specific matric potential (ψ_M) the equivalent diameter (d_e) of the largest soil pore conducting water can be estimated after Jarvis et al. [48]. At $\psi_M = 0$ m the soil is saturated and the entire pore spectrum is potentially active. At smaller matrix potential larger pores are no longer active and infiltration capacity decreases. This allows evaluating infiltration capacity through different parts of the pore spectrum. At $\psi_M = -0.02$ m the largest active pores correspond to $d_e = 1.5$ mm. At each pressure level we recorded infiltration capacity until it was constant in time. This steady infiltration capacity was used for further analysis. Depending on the month most plots had a bubble point at $\psi_M < -0.04$ m. For data analysis of individual months, we only considered infiltration capacity for ψ_M up to -0.02 m. Infiltration rates at a given matric potential are directly linked to hydraulic conductivity [49]. The flow conditions in natural soils, however, are far from ideal with anisotropic behaviour, heterogeneous initial soil water contents and flow dynamics that do not correspond to the Richards equation near soil saturation [14]. Therefore, we refrained from deriving hydraulic conductivity from our infiltration rates, for example via Wooding's formula [49]. Instead, we worked with the observed infiltration rates, considering those as a surrogate for the capacity of the soil to conduct water at the applied matric potential.

Soil Texture and Moisture

Soil texture was determined from soil cores at 38 locations (average of 0–100 cm depth) distributed throughout the experimental site before plot establishment (G. Büchel, pers. comm., [50]) and values for each plot were interpolated by ordinary kriging. The fraction of sand and silt are negatively correlated (clay showing no spatial trend, Table S1). Thus, in the following statistical analysis for simplicity we used the sand fraction as factor representing soil texture.

The volumetric soil water content ($m^3\ m^{-3}$) was determined with an FDR probe (ML2x Theta Probe, Delta-T Devices, Cambridge, United Kingdom). The device was inserted from the top 6 cm deep (length of the prongs) into the soil surrounding the hood before the infiltration experiment. The average of three measurements was used for further analysis.

Table 1. The design of the present study in the frame of the Jena Experiment.

	Plant species richness			Plots
Plant functional group richness	**1**	**4**	**16**	
1	12	4	2	18
2	–	4	4	8
3	–	4	4	8
4	–	4	4	8
Plots	**12**	**16**	**14**	**42**

Combinations of plant species richness and plant functional group richness levels and the number of plots per diversity level for the earthworm subplots with 1, 4 and 16 plant species (n = 42, 84 subplots). Please note that this design is a selection of plots from the full design of the Jena Experiment [43]. Due to the non-orthogonal design, effects of plant species richness and functional groups richness are partially confounded (r = 0.438, p<0.001).

Statistical Analyses

Statistical analyses were performed using the statistical software R 2.13.1 (R Development Core Team, http.//www.R-project.org). Measures of infiltration capacity were log-transformed to account for heteroscedasticity and non-normality of errors. Analyses were performed with linear mixed-effect models in order to account for the nested design of our experiment (ambient and reduced earthworm density subplots nested within plots). For this we used the *lme* function implemented in the *nmle* package [51]. Analyses of variation in the infiltration capacity were performed for each month separately. Starting from a constant null model with plot identity as a random factor, we added the design variables of the Jena Experiment: block (as a factor, BL; 1, 2, 3, 4), plant species richness (log-linear term; SR; 1, 4, 16), plant functional group richness (linear-term; FG; 1, 2, 3, 4) and the earthworm treatment (E) as well as the interaction terms E×SR and E×FG as fixed effects and assessed their significance using likelihood ratio tests (L-ratio). 95% confidence intervals for fixed effects were calculated based on the full model. Contrasts for the presence/absence of grasses (GR; 0, 1), legumes (LEG; 0, 1), small herbs (SH; 0, 1) and tall herbs (TH; 0, 1) were fitted in series of alternative models after accounting for block, plant species and functional group richness effects. Confidence intervals for these contrasts are based on the separate models. As the effects of the presence (or absence) of certain plant functional groups on the response variable were considered as post-hoc tests, the respective p-values were adjusted according to the Holm procedure to avoid inflation of Type I error rates [52]. We used simple linear regressions to analyse the influence of texture (sand fraction in 0–10 cm depth), and soil water content before the measurement on the infiltration capacity.

Path analysis was used to investigate how the total earthworm biomass extracted from the earthworm reduction subplot in September, soil texture and the presence/absence of legumes and grasses directly and indirectly affected the infiltration capacity on subplots with reduced (−ew) and ambient (+ew) earthworm densities in October. The impact on infiltration capacity in reduced and ambient earthworm density subplots was calculated in a separate analysis. Path analysis allows testing direct and indirect relationships between variables in a multivariate approach [53]. Hence, by using path analysis we were able to test if certain plant functional groups, such as legumes and grasses, directly influence infiltration capacity or if infiltration is indirectly influenced by other variables such as changes in earthworm biomass. In the path analysis, arrows represent causal relationships, while rectangles represent manipulated (grasses and legumes) or measured variables (sand content, earthworm biomass and infiltration capacity). Non-signifcant Chi^2-test ($p>0.05$), low AIC and low RMSEA indicating an adequate model fit [53]. Beginning with the full model (including all possible pathways) the models were improved by stepwise removing of unimportant relationships based on AIC values [54]. Standardized path coefficients were derived based on the correlation matrix of standardized variables. Path analysis was performed using AMOS 5 (http://amosdeveleopment.com).

Results

Effect of Soil Texture and Moisture on Infiltration

The infiltration capacity in June and October did not differ between the four blocks (Table 2) encompassing the texture gradient. In September, infiltration capacity increased systematically for $\psi_M = 0$ and −0.02 m from block 1 (71.34±11.50 and 49.23±8.55 $*10^{-6}$ m/s) to block 4 (143.65±21.73 and 100.10±14.50 $*10^{-6}$ m/s). When replacing the variable 'block' with 'sand fraction in 0–10 cm depth' as a covariate in our analysis, texture became significant in September (95% CI = [−1.68 to −0.34], Table S2), but not in June and October. Surprisingly, the infiltration capacity at saturated conditions decreased with increasing sand content (Figure 1, centre). However, when testing only plots without legumes, we did not find a significant correlation between sand content and infiltration capacity (data not shown). Additionally, there was a significant interaction between sand content and legumes (95% CI = [−2.92 to −0.56], Table S2), highlighting the differential effect of legumes with varying soil texture. In our experiment the initial soil moisture measurements did not correlate significantly with the infiltration capacity (log $*10^{-6}$ m/s) at matric potential zero in June ($r = -0.207$, $p = 0.806$), September ($r = 0.439$, $p = 0.078$) and October ($r = 0.003$, $p = 0.993$). In order to assure that the latter result was not artefacts stemming from the non-orthogonal design of the observed plots, we also tested for confounding correlations between the texture (sand content) and the presence and absence of legumes. There was no correlation ($r = -0.117$, $p = 0.472$).

Effect of Plant Species Richness on Infiltration

Plant species richness (1, 4 and 16 species) did not significantly affect the infiltration capacity at any sampling date (Table 2). However, we observed a significant interaction effect between earthworms and plant species richness in October (95% CI = [0.072 to 0.280], Table 2), which was mainly caused by the mixture of 16 species (Figure 2). Additionally, the infiltration capacity was negatively correlated with plant species richness on plots with reduced earthworm densities ($r = 0.332$, $p = 0.036$). For infiltration on ambient earthworm density subplots a weak positive trend with plant species richness ($r = 0.217$, $p = 0.166$) was observed.

Effects of Plant Functional Groups on Infiltration

Infiltration capacity was significantly affected by the presence of certain plant functional groups, such as grasses and legumes when fitted after BL and SR, however functional group richness had no effect on infiltration capacity. In September, infiltration capacity increased at $\psi_M = 0$ m by 39% (95% CI = [0.07 to 0.423]) and $\psi_M = -0.02$ m by 40% (95% CI = [0.14 to 0.53]) in the presence of legumes (adjusted p-value = 0.023 and 0.012, respectively), while the presence of grasses caused only a small, non-significant decrease in infiltration capacity (adjusted p-value = 0.116 and 0.096, respectively; Figure 3, Tables 2 and S3). In October, both legumes and grasses significantly affected the infiltration capacity: in presence of legumes infiltration capacity increased at $\psi_M = 0$ m by 36% (95% CI = [0.14 to 0.53]) and $\psi_M = -0.02$ m by 38% (95% CI = [0.13 to 0.56]), while it decreased in the presence of grasses at $\psi_M = 0$ m by 23% (95% CI = [−0.46 to −0.06]) (adjusted p-value = <0.001 and 0.004, respectively) and at $\psi_M = -0.02$ m by 27% (95% CI = [−0.41 to −0.09]) (adjusted p-value = 0.045 and 0.018, respectively; Figure 3, Tables 2 and S3). No significant effects of plant functional groups were detected in June (Tables 2 and S3).

Because we worked on a short diversity gradient, the design is not completely orthogonal with respect to plant species richness and plant functional group number ($r = 0.439$, $p<0.001$), However, plant species richness and functional groups did not explain a signficant proportion of variation in infiltration capacity (Table 2) and therefore this correlation did not affect our results. Furthermore, there exist a positive relationship between plant species richness and presence of grasses ($r = 0.290$, $p = 0.062$) and legumes ($r = 0.223$, $p = 0.155$) respectively. In order to test whether our

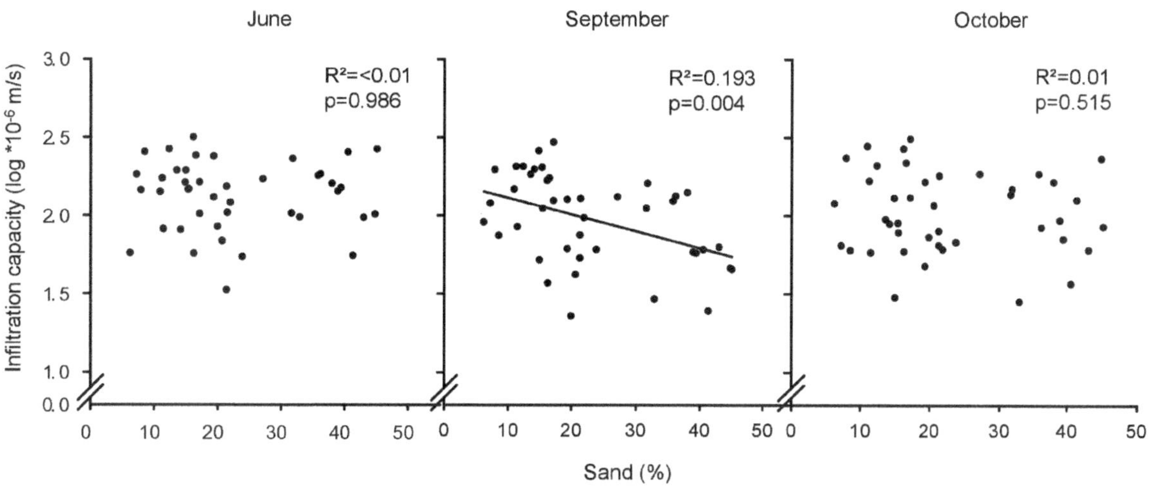

Figure 1. Relationship between infiltration capacity and soil texture for June, September, and October. Plot-level infiltration capacity at saturation plotted against soil texture (sand fraction in 0–10 cm depth in %) for the month June, September and October (as indicated). The regression line indicates significant relationships.

results were confounded by this correlation, we alternatively fitted legumes and grasses prior and after plant species richness, and found that fitting order did not affect the results (analyses not shown).

Effect of Earthworms on Infiltration

The biomass of endogeic earthworms accounted for 59% and 49% of total earthworm biomass in spring and fall, whereas the biomass of anecic species (*L. terrestris*) accounted for 36% and 49% of total biomass in spring and fall, respectively. The total number and biomass of earthworms increased significantly from spring (26.88±2.29 individuals m^{-2} and 13.86±2.03 g m^{-2}, respectively) to fall (40.55±4.02 individuals m^{-2} and 25.78±2.35 g m^{-2}, respectively; Table 3). Total earthworm biomass increased slightly, but not significantly in the presence of legumes in spring (t-test, p = 0.338), but significantly in fall (t-test, p = 0.001). In presence of grasses, total earthworm biomass decreased slightly but not significantly in spring and in fall (analyses not shown). For both extraction dates earthworm biomass did not correlate with plant species richness (analyses not shown).

In June 2011, the infiltration capacity was only marginally increased (linear mixed-effect models: n = 84, L-ratio = 4.14, p = 0.042, 95% CI = [−0.14 to −0.31]) in ambient earthworm density subplots (171.24±16.20 $*10^{-6}$ m/s) compared to the reduced earthworm density subplots (128.75±12.72 $*10^{-6}$ m/s). With decreasing matric potential ($\psi_M = -0.02$ m, exclusion of pores >1.5 mm) the effect of earthworms disappeared (Figure 4, Table 2, S3). For all other sampling dates and matric potentials, the earthworm treatment did not significantly affect infiltration capacity (Figure 4, Tables 2 and S3), while the earthworm biomass affected the infiltration capacity (see path analysis).

Path Analysis

The path analysis supported the results of the linear mixed-effect model approaches showing strong grass and legumes effects on the infiltration capacity. In addition, we could identify possible mechanisms shaping the infiltration capacity in ambient and reduced earthworm density subplots. The initial model for October (AIC = 49.10, Figure S1) was improved as described in the method section. In October the final model explained 13% of the infiltration capacity on subplots with ambient earthworm density (Figure 5A; $\chi^2 = 1.80$, p = 0.937, AIC = 29.80) and 29% of the infiltration capacity on subplots with reduced earthworm density (Figure. 5B; $\chi^2 = 1.10$ p = 0.954, AIC = 31.10). The total earthworm biomass increased in the presence of legumes and decreased in the presence of grasses. Earthworm biomass decreased with increasing sand content. Increasing total earthworm biomass increased the infiltration capacity on ambient earthworm density subplots directly (Figure 5A). Grasses decreased and legumes increased infiltration capacity on subplots with reduced earthworm biomass (Figure 5B). In summary, grasses had a stronger direct effect on the infiltration capacity on subplots with reduced earthworm density, whereas legumes had a stronger indirect effect on infiltration capacity on subplots with ambient earthworm density by increasing earthworm biomass (Figure 5B). For the infiltration on ambient earthworm density subplots we additionally included different ecological earthworm groups (anecic and endogeic) instead of the total earthworm biomass in separate models to disentangle the effects of different ecological earthworms groups on the infiltration capacity and to test the different effects of sand content and the presence/absence of legumes and grasses on them. Including only the biomass of anecic earthworms instead of total earthworm biomass in additional models, the effect of grasses and sand content disappeared, but the effect of legumes on the biomass of anecic earthworms increased (Figure S2A). In contrast, including the biomass of endogeic species instead of the total biomass, we observed similar results as for the total earthworm biomass model (Figure S2B).

Discussion

Effect of Earthworms on Infiltration

Our results indicate that biotic factors play a decisive role for soil hydraulic soil properties near saturation. In June infiltration capacity slightly increased on subplots with ambient earthworm densities compared to subplots with reduced earthworm densities. Remarkably, this effect was mainly caused by plots in block 1 and block 4 (data not shown) suggesting local changes of earthworm activity and considerable spatial variation due to texture and plant type that interact with infiltration [55]. In general, burrows of anecic species such as *L. terrestris* but also those of adult endogeic

Table 2. Results of linear mixed-effects models of the infiltration capacity for June, September and October.

	June					September				October			
Source	**df**	**0 m**		**−0.02 m**		**0 m**		**−0.02 m**		**0 m**		**−0.02 m**	
		L-ratio	**p**	**L-ratio**	**p**	**L-ratio**	**p**	**L-ratio**	**p**	**L-ratio**	**p**	**L-ratio**	**p**
BL	3	4.05	0.256	4.02	0.259	**10.77**	**0.013**	**11.76**	**0.008**	2.89	0.409	2.92	0.403
SR (log-linear)	1	0.28	0.596	0.12	0.724	0.04	0.834	0.06	0.800	0.76	0.385	0.50	0.480
FG	1	0.70	0.403	2.43	0.119	0.03	0.858	<0.01	0.991	0.05	0.825	0.03	0.870
GR	1	0.31	0.994	0.65	1.00	**4.27**	**0.116 ↓**	**4.62**	**0.096 ↓**	**6.39**	**0.045 ↓**	**7.59**	**0.018 ↓**
LEG	1	0.72	0.639	0.76	1.00	**7.60**	**0.023 ↑**	**8.96**	**0.012 ↑**	**10.95**	**<0.001 ↑**	**10.53**	**0.004 ↑**
SH	1	1.98	0.934	0.70	1.00	1.46	0.454	1.36	0.488	2.01	0.314	1.70	0.386
TH	1	1.02	0.934	0.72	1.00	0.24	0.627	0.10	0.758	0.33	0.563	0.58	0.446
E	1	**4.14**	**0.042 ↑**	1.89	0.169	0.23	0.630	0.46	0.498	0.88	0.349	0.34	0.557
E×SR(log-linear)	1	1.33	0.248	0.19	0.664	0.21	0.645	0.14	0.706	**4.69**	**0.030**	**5.50**	**0.019**
E×FG	1	0.60	0.437	1.80	0.406	0.26	0.614	0.49	0.484	1.24	0.264	**12.95**	**0.002**

Infiltration capacity as affected by block (BL), plant species richness (SR), plant functional group richness (FG), grasses (GR), legumes (LEG), small herbs (SH), tall herbs (TH), and earthworm treatment (E) in June, September and October separately for the matric potentials $\psi_M = 0$ m and $\psi_M = -0.02$ m.

Models were fitted by stepwise inclusion of fixed effects. Likelihood ratio tests were applied to assess model improvement (L-ratio) and the statistical significance of the explanatory terms (p values). Models always included 84 observations (n) on 42 plots, df = degrees of freedom (df) for each of the predictor variables. For GR, LEG, SH and TH the adjusted p-values according to the Holm procedure are given in the table. Significant effects are marked in bold. Arrows indicate increase (↑) or decrease (↓).

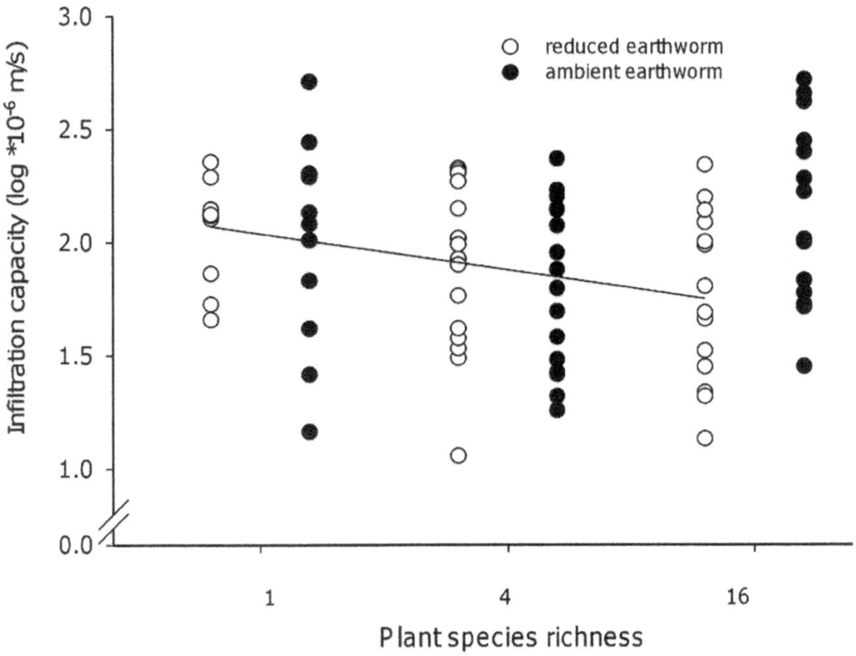

Figure 2. Interacting effects of plant diversity and earthworms on the infiltration capacity in October. Effect of plant species richness (1, 4, 16 species) and earthworm treatment (ambient earthworm density subplots and reduced earthworm density subplots) on the infiltration capacity at saturation. The regression line indicates a negative correlation between plant species richness and infiltration capacity on subplots with reduced earthworm densities (r = 0.33, p = 0.036).

species such as *A. caliginosa* are larger than 2 mm in diameter and this may explain why only water flow through larger pores (> 1.5 mm, at matric potential $\psi_M = 0$ m close to saturation), but not through smaller pores was directly affected by the presence of earthworms. This is also in line with the non-capillary nature of earthworm casts. Earthworm presence, particularly that of anecic species forming vertical burrows, facilitates larger pores conducting water (macropore flow) and contributes to infiltration when water is supplied in large quantities (tension free conditions) [56].

We measured infiltration capacity at elevated soil saturation, and it could be argued that saturation will never be reached in most natural soil environments. However, irrigation experiments with dye indicate that macropores, such as those formed by anecic earthworms, are activated at realistic rain intensities [19], [57], [58]. Beven and Germann [15] reported that rainfall intensities of 1–10 mm h^{-1} can initiate macropore flow. Applying an arbitrary threshold of 5 mm h^{-1} [59] to the high resolution (10 min) rainfall record of our field site (2003 to 2011) indicates that about 30% of the total rainfall could deliver macropore flow. One of the pressing questions for future work will be on how this macropore flow potential relates to actual water flow at greater depth. Furthermore, it is expected that precipitation patterns change due to climate change with decreasing precipitation in summer and increasing precipitation in fall/winter. It is therefore likely that higher frequency of extreme precipitation events increases the proportion of heavy rainfalls [60]. Macropores formed by earthworms and roots, and their interactions, thus likely to become more important for buffering strong precipitation events in the future.

Effects of Plant Functional Groups on Infiltration

At the end of the growing season infiltration capacity was strongly affected by the presence of certain plant functional groups, such as grasses and legumes. Legumes increased and grasses decreased infiltration capacity (Figure 3, Table 2). This is consistent with results of Archer et al. [61], who reported that legumes increased and grasses decreased hydraulic conductivity. Further, the significant effect of legumes was more pronounced at $\psi_M = 0$ m, whereas the effect of grasses was more pronounced effect at smaller matric potential ($\psi_M = 0.02$ m). It has been shown that decaying tap-roots of legumes form stable macropores and hence increase infiltration [27], [62]. This observation was supported by Mytton et al. [63], who showed that water infiltration was higher under clover compared to grass due to a higher fraction of soil pores greater than 60 µm with porosity being equal. Additionally, several studies showed that water flow through soil is enhanced by legumes or legumes-grass mixtures compared to pure grass stands due to root proliferation, which increased soil organic matter content and favored soil fauna such as earthworms [64]. The fibrous and rhizomatous roots of grass species tend to reduce infiltration by clogging soil pore space and blocking water flow [61].

Path Analysis

Path analysis helped to address underlying mechanism: the effect of legumes on infiltration capacity in ambient (control) earthworm density plots may be indirect by enhancing earthworm biomass (Figure 5A), suggesting that earthworm performance benefits from the presence of legumes [34], [35]. Additionally, the path analysis indicated that grasses directly influenced the infiltration capacity on subplots with reduced earthworm densities, while indirect effects of grasses via decreasing earthworm biomass were less important (Figure 5). For the Jena Experiment it was shown by Pérès et al. [29] that root biomass is strongly increased in presence of grasses while, as mentioned above, legumes signifcantly increased earthworm biomass (Figure 5A). This

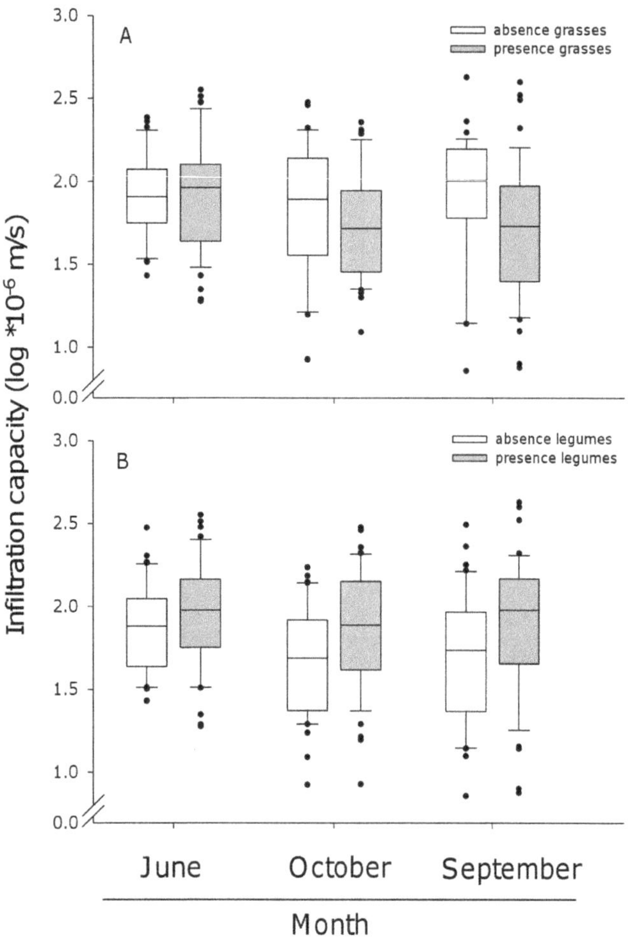

Figure 3. Effects of certain plant functional groups (presences of legumes and grasses) on the infiltration capacity. Variations in infiltration rates in June, September, and October as affected by (A) the presence of grasses, and (B) the presence of legumes at saturation. The box plots represent the median, the ends of the boxes defined the 25 and 75 quartiles. The error bars show the 10th and 90th percentiles of the data. Black dots outside of the 10th and 90th percentiles represent outliers.

suggests that the observed enhanced infiltration in plots with legumes was probably associated with a larger number of macropores caused by earthworms [65]. These results are in agreement with the findings of Abbott and Parker [66], who reported an increased infiltration of water due the activity of the geophagous earthworm species *Microscolex dubuis* in the presence of clover mulch. By contrast, decreased infiltration in grass plots was probably due to fine roots clogging soil pores. Thus, our results suggest that the presence of grasses and legumes affected hydraulic conductivity via different mechanisms, directly presumably via root activity and indirectly via altering earthworm biomass. The observed effects increased during the vegetation period (Table 2), probably due to the progressive increase of earthworm (Table 3) and root biomass [67].

The different ecological earthworm groups differ in forming of macro- and microaggregates by changing soil structure features (aggregate size, stability and soil organic content) and porosity (pore size distribution) [28], [68]. Interestingly, in October anecic earthworms were positively affected by the presence of legumes, whereas endogeic earthworms were negatively affected by sand content and by the presence of grasses. Additionally, both endogeic and anecic earthworm had a significant effect on the infiltration capacity, with a more pronounced effect of anecic earthworms (Figure S2). This is in line with other studies, which showed that the deep dwelling anecic earthworm enhance water infiltration rates [24], [69], while horizontal pores formed by endogeic earthworms limits the effectiveness in water flow through soil [70]. Generally, endogeic earthworms are considered as the major group improving soil aggregation, while anecic or compacting species destabilizes the soil due to their casting activity [28]. As shown by Lee and Foster [71] a mix of endogeic and anecic earthworms supports soil structural health.

In part our results are in contrast to other studies [72–74], which found a direct impact of earthworm treatment on infiltration. Studies on the influence of earthworms on infiltration capacity usually ignore gradients of texture and plant functional groups. This gradient increased the variance in our data and affected the earthworm populations in the subplots. However, our observation area was also rather small (240 cm^2) and the setup prevented from conducting additional measurements. Also, we compared subplots with ambient to subplots with reduced earthworm densities contrasting earlier studies which compared ambient conditions (by earthworm addition) with a control (no earthworms) [75], [76]. While our setup has the advantage to better reflect 'natural' conditions, it may have caused a comparatively smaller contrast between treatments. Since the used octet extraction method cannot remove earthworms completely (extraction reduced the surface activity of earthworms by about 38% five weeks after the last manipulation [77]), the differences between the treatments might not have been as strong as in other experiments, and re-colonization of earthworm reduction subplots may have weakened the contrast further. Unfortunately, we cannot exclude

Table 3. Number (individuals m^{-2}) and biomass (g m^{-2}) of earthworms of the Jena Experiment.

Earthworms	March	September
Number of anecics	3.14±0.53	5.47±0.63
Biomass of anecics	4.98±1.03	12.74±1.37
Number of endogeics	21.51±2.25	34.05±3.67
Biomass of endogeics	8.13±1.01	12.62±1.51
Number total earthworms	26.88±2.92	40.55±4.02
Biomass total earthworms	13.86±2.03	25.78±2.35

Total number and biomass of earthworms included also unidentifiable earthworms and the invasive earthworm *Lumbricus castaneus*.
Shown are the means (±1 SE) across plots in 2011 for the extraction dates in spring (March) and autumn (September).

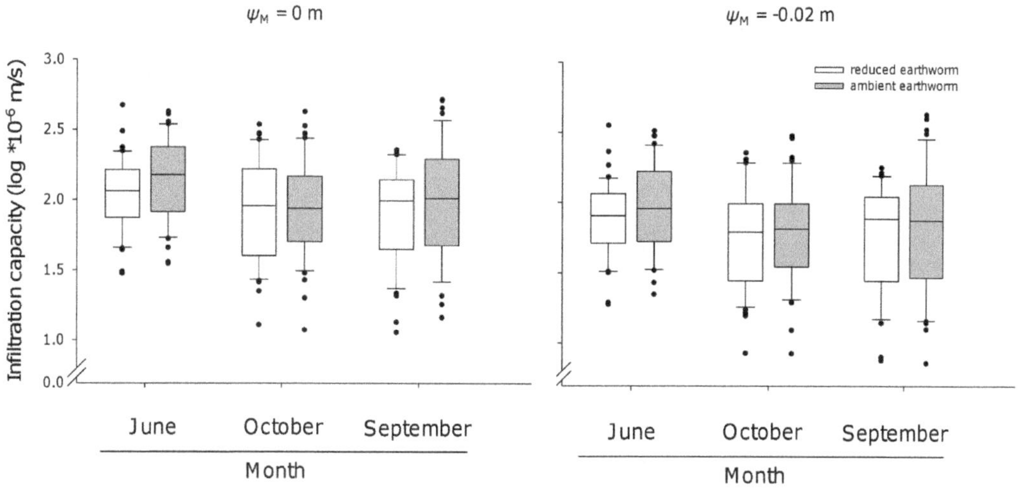

Figure 4. Effects of earthworms on the infiltration capacity. Variations in infiltration rates in June, September, and October as affected by earthworms (ambient and reduced earthworm density) for applied matric potential $\psi_M = 0$ m (left panel) and $\psi_M = -0.02$ m (right panel). The box plots represent the median, the ends of the boxes defined the 25 and 75 quartiles. The error bars show the 10th and 90th percentiles of the data. Black dots outside of the 10th and 90th percentiles represent outliers.

completely that extraction efficiency varies depending on plant community properties, because there exist no detailed studies that evaluated the extraction efficiency of the octet method as a function of plant community structure. Effects of earthworms on plant performance and the resulting impact on soil hydraulic properties depend also on other soil factors, such as texture and bulk density, leading to substantial spatial variation and interaction which probably masked earthworm effects. More observations on the pore structure in the different treatments are necessary to validate the proposed complex interactions in soil processes leading to the observed infiltration patterns.

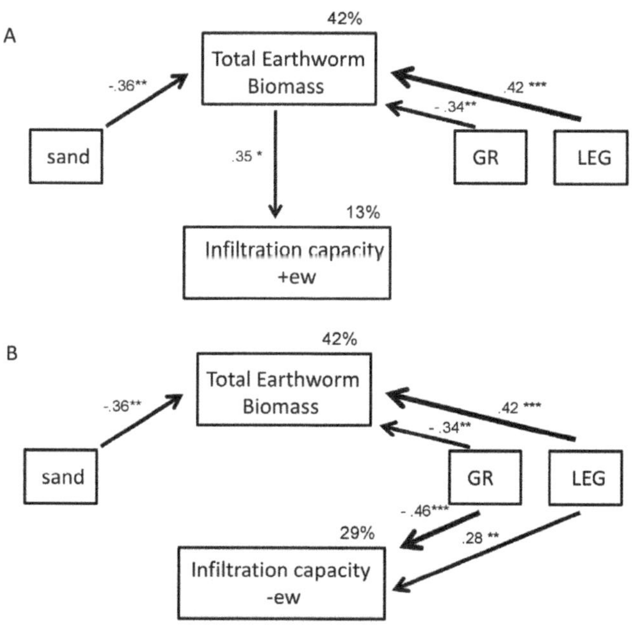

Figure 5. Path analysis of factors explaining infiltration capacity on plots with ambient or reduced earthworm density. Relationships between total biomass of earthworms (earthworm biomass), texture (sand in 10 cm depth) and plant functional groups (GR, grasses; LEG, legumes) for (A) infiltration rate on subplots with ambient (infiltration capacity +ew) and (B) reduced earthworm densities (infiltration capacity −ew) in October. The total earthworm biomass was extracted in September. Standardized path coefficients are given next to path arrows. Unexplained variation is denoted with e1–e3; *$p \leq 0.05$, **<0.01, ***$p = 0.001$. For details see text.

Effect of Plant Species Richness on Infiltration

In contrast to the pronounced effects of certain plant functional groups, such as legumes and grasses, plant diversity measures (plant species and functional group richness) had only small effects on the infiltration capacity and earthworm performance. In October, we found that plant diversity had only a marginally significant effect on the infiltration capacity on subplots with ambient earthworm density, but plant species diversity affected infiltration capacity significantly on subplots with reduced earthworm density, presumably due to plant roots clogging macropores [78], [79]. This is supported by data on the relationship between plant diversity and standing root mass in the Jena Experiment, showing that root biomass also increased with diversity level [38,80]. Additionally, infiltration capacity decreased over the growing season (data not shown) presumably by the growing roots clogging macropores [79]. Angulo-Jaramillo et al. [81] observed a decrease of hydraulic conductivity caused by sealing of interconnected pores at the soil surface. However, on subplots with ambient earthworm densities this clogging may have been counteracted by the activity of earthworms. These results suggest that water flow through soil is more strongly affected by earthworm biomass which is regulated by the presence of certain functional groups (particularly by legumes) then by loss of single plant species along the observed gradient.

Effect of Soil Texture and Moisture on Infiltration

Grain size distribution strongly influences hydraulic properties of porous media, and therefore texture has often been related to hydraulic conductivity [9], [82]. However, in our experiment saturated and near-saturated infiltration capacity in June and October were not affected by soil texture (Figure 1). Surprisingly, in September, infiltration capacity varied with soil texture, but unexpectedly it was lowest in coarse textured soils. However, the results are in agreement with Jarvis and Messing [12] and Lin et al. [3], who found higher hydraulic conductivity in finer as compared to coarser textured soils due to well-developed soil structure (earthworm burrows, root channels) and a high degree of macroporosity. As detailed above, legumes increased infiltration capacity, and interestingly the negative correlation between sand content and infiltration capacity disappeared when removing plots containing legumes. This is probably a result of the positive relationship between legumes and earthworms and resulting effects on infiltration. An alternative explanation for increased infiltration in the presence of legumes could be that roots affected the formation of aggregates and structure, particularly so in finer textured soils [5], [28]. Kördel et al. [83] suggested that the formed macropores are less stable over time in sandy soil. However, investigations on soil stability in the Jena Experiment itself [29] suggest the opposite effect, with legumes decreasing soil aggregate stability (indirectly, by increasing earthworm biomass and decreasing plant root biomass). At the same time, we found a lower number of earthworms in more sandy soils in 2011 (data not shown). This is in line with previous studies [84], [85] showing that sandy soils support smaller earthworm populations than clayey soils resulting in lower hydraulic conductivity in sandy soils. Thus, our results may be explained by the promotion of earthworm abundance by legumes and finer soil texture, but further research is necessary to support these findings. Our results underline the importance of soil structure as influenced by biotic processes and corroborate the findings of Bonsu [86] who suggested that the texture based calculation underestimates the hydraulic conductivity, in particular in fine textured soils.

One important source of error that can dominate or suppress other factors is the initial soil moisture content. Several studies have shown that the infiltration through soil is correlated with the initial soil moisture content [87], [88]. This relation was not expected in our setup, since the soil moisture is controlled by fixing the infiltration pressure head. At the beginning of an infiltration experiment, low initial soil moisture content enhanced the water flow through the soil because of larger gradients in matric potential and filling up of the soil storage. The influence of initial moisture content decreased over the duration of the experiment, when the pores probably were filled and infiltration reached a steady state [89]. A potential error source is finishing the experiment before steady infiltration rates are reached in dry soils and hence overestimating infiltration capacity. Thus, we checked whether infiltration rates were biased by initial soil moisture and this was not the case. Thus our infiltration measurements were performed correctly. Another factor potentially affecting infiltration rates is soil hydrophobicity. Hydrophobic exudates are produced by plant roots and soil microbes. Soto et al. [90] observed that the soil showed a tendency to be water repellent if the volumetric water content fell below $\theta_c = 22\%$ for medium textured soils, the so called 'critical soil water content' [91]. The majority of the measured volumetric water contents exceeded this threshold (data not shown). Since most of our observed effects enhanced during the growing season, while the chance for hydrophobicity decreased, we conclude that water repellency did not affect our results.

Conclusions

Despite large spatial variability of soil hydraulic properties, biotic factors emerged as significant agents for infiltration. The presence of legumes increased and the presence of grasses decreased infiltration capacity over the course of the growing season. The path analysis suggests that modifications in hydraulic conductivity are probably due to (i) roots directly, modifying the pore spectrum and (ii) indirectly, suppressing (grasses) or enhancing (legumes) earthworm activity. The results suggest that earthworm biomass is synchronized with other plant related processes such as root growth. Thus the observed effects of plant functional groups are attributable to earthworm and root activity.

Most predictions of near surface soil hydraulic properties are based on easily accessible soil properties such as soil texture. Our results suggest that biotic effects, especially the presence of certain plant functional groups affecting earthworm biomass, shape hydraulic conductivity and may even reverse effects of texture. Therefore, for explaining variations in hydrological processes, such as infiltration capacity, the structure of soil fauna and plant communities need to be considered.

Supporting Information

Figure S1 Initial path analysis of factors explaining infiltration capacity on subplots with ambient or reduced earthworm density. Relationships between earthworm biomass (total, anecic or endogeic as indicated), texture (sand in 10 cm depth) and functional groups (GR, grasses; LEG, legumes) and infiltration capacity at saturation for ambient (+ew) or reduced (−ew) earthworm density plots (as indicated).

Figure S2 Path analysis of the effects of the ecological earthworm groups (anecic and endogeic) on infiltration capacity. Path analysis showing the relationships between (A) endogeic and (B) anecic earthworm with texture (sand in 10 cm depth) and plant functional groups (GR, grasses; LEG, legumes) for infiltration capacity on subplots with ambient (+ew) earthworm densities in October. Standardized path coefficients are given next to path arrows. Unexplained variation is denoted with e1–e3; *$p \leq 0.05$, **<0.01, ***$p = 0.001$. For details see text.

Table S1 Block-wise variations in soil parameters (clay, silt and sand content) of the upper 10 cm. Plots were assembled into four blocks with block 1 nearest the river Saale and block 4 furthest from the river.

Table S2 Summary of mixed effects models for infiltration capacity at saturation in September. Results for the infiltration rate as affected by sand content in 0–10 cm depth (Sand), plant species richness (SR), plant functional group richness (FG), grasses (GR), legumes (LEG), small herbs (SH), tall herbs (TH), interaction Sand×LEG and earthworm treatment (E) as well as the interaction terms E×SR and E×FGfor measurements in September.

Table S3 Summary of the analysis of infiltration capacity at $\psi_M = 0$ m and 0.02 m. Estimates (est.) with 95% confidence intervals (lower and upper) for the effects of block (Block), plant species richness (SR), plant functional group richness (FG), grasses (GR), legumes (LEG), small herbs (SH), tall herbs (TH), and earthworm treatment (E) in June, September and October for the matric potentials $\psi_M = 0$ m and 0.02 m.

Acknowledgments

We thank all the people involved in the maintenance of the experiment, in particular the gardeners S. Eismann, S. Ferber, S. Hengelhaupt, U. Köber, K. Kunze, H. Scheffler as well as G. Kratzsch. Further, we thank Franziska

Guderle and many student helpers for the help during the infiltration measurements and data preparation.

Author Contributions

Conceived and designed the experiments: CF CR BJ NE SS WWW JB AH. Performed the experiments: CF BJ. Analyzed the data: CF CR NE AH JS. Contributed reagents/materials/analysis tools: CR NE SS WWW SA AH JS. Wrote the paper: CF CR NE SS WWW JB AH JS.

References

1. Kutílek M (2004) Soil hydraulic properties as related to soil structure. Soil Till Res 79: 175–184.
2. Dexter AR (1988) Advances in characterization of soil structure. Soil Till Res 11: 199–238.
3. Lin HS, McInnes KJ, Wilding LP, Hallmark CT (1999) Effects of soil morphology on hydraulic properties I. Quantification of soil morphology. Soil Sci Soc Am J 63: 948–954.
4. Ward AD, Trimble SW (2004) Environmental hydrology. Boca Ratan: CRC Press.
5. Angers DA, Caron J (1998) Plant-induced Changes in Soil Structure: Processes and Feedbacks. Biogeochemistry 42: 55–72.
6. Lipiec J, Kuś J, Słowińska-Jurkiewicz A, Nosalewicz A (2006) Soil porosity and water infiltration as influenced by tillage methods. Soil Till Res 89: 210–220.
7. Alaoui A, Lipiec J, Gerke HH (2011) A review of the changes in the soil pore system due to soil deformation: A hydrodynamic perspective. Soil Till Res 115–116: 1–15.
8. Hillel D (1998) Environmental Soil Physics. New York: CA Academic Press. 383–470 p.
9. Rawls WJ, Brakensiek DL, Saxto KE (1982) Estimation of soil water properties. Trans Am Soc Agric Eng 25: 1316–1320.
10. Wösten JHM, van Genuchten MT (1988) Using texture and other soil properties to predict the unsaturated soil hydraulic functions. Soil Sci Soc Am J 52: 1762–1770.
11. Vereecken H (1995) Estimating the unsaturated hydraulic conductivity from theoretical models using simple soil properties. Geoderma 65: 81–92.
12. Jarvis NJ, Messing I (1995) Near-saturated hydraulic conductivity in soils of contrasting texture measured by tension infiltrometers. Soil Sci Soc Am J 59: 27–34.
13. Watson KW, Luxmoore RJ (1986) Estimating macroporosity in a forest watershed by use of a tension infiltrometer. Soil Sci Soc Am J 50: 578–582.
14. Beven K, Germann P (2013) Macropores and water flow in soils revisited. Water Resources Research 49: 3071–3092.
15. Beven K, Germann P (1982) Macropores and water flow in soils. Water Resour Res 18: 1311–1325.
16. Messing I, Jarvis NJ (1990) Seasonal variation in field-saturated hydraulic conductivity in two swelling clay soils in Sweden. J Soil Sci 41: 229–237.
17. Zehe E, Flühler H (2001) Slope scale variation of flow patterns in soil profiles. Journal of Hydrol 247: 116–132.
18. Edwards WM, Shipitalo MJ, Owens LB, Norton LD (1990) Effect of *Lumbricus terrestris* L. burrows on hydrology of continuous no-till corn fields. Geoderma 46: 73–84.
19. Weiler M, Naef F (2003) An experimental tracer study of the role of macropores in infiltration in grassland soils. Hydrol Process 17: 477–493.
20. Bodner G, Loiskandl W, Buchan G, Kaul HP (2008) Natural and management-induced dynamics of hydraulic conductivity along a cover-cropped field slope. Geoderma 146: 317–325.
21. Ehlers W (1975) Observations on earthworm channels and infiltration on tilled and untilled loess soil. Soil Sci 119: 242–249.
22. Bouché MB (1977) Strategies lombriciennes. Ecological Bulletins 25: 122–132.
23. Pérès G, Cluzeau D, Curmi P, Hallaire V (1998) Earthworm activity and soil structure changes due to organic enrichments in vineyard systems. Biol Fert Soils 27: 417–424.
24. Edwards CA, Bohlen PJ (1996) The biology and ecology of earthworms. London: Chapman and Hall.
25. Shuster WD, Subler S, McCoy EL (2002) The influence of earthworm community structure on the distribution and movement of solutes in a chisel-tilled soil. Appl Soil Ecol 21: 159–167.
26. Tippkötter R (1983) Morphology, spatial arrangement and origin of macropores in some hapludalfs, West Germany. Geoderma 29: 355–371.
27. Meek BD, Rechel EA, Carter LM, DeTar WR (1989) Changes in infiltration under alfalfa as influenced by time and wheel traffic. Soil Sci Soc Am J 53: 238–241.
28. Six J, Bossuyt H, Degryze S, Denef K (2004) A history of research on the link between (micro)aggregates, soil biota, and soil organic matter dynamics. Soil Till Res 79: 7–31.
29. Pérès G, Cluzeau D, Menasseri S, Soussana JF, Bessler H, et al. (2013) Mechanisms linking plant community properties to soil aggregate stability in an experimental grassland plant diversity gradient. Plant Soil 373: 285–299.
30. Zacharias S, Wessolek G (2007) Excluding organic matter content from pedotransfer predictors of soil water retention. Soil Sci Soc Am J 71: 43–50.
31. Visa N (1992) Earthworm community response to tillage and residue management on different soil types in southern Finland. Soil Till Res 23: 221–239.
32. Lachnicht SL, Parmelee RW, McCartney D, Allen M (1997) Characteristics of macroporosity in a reduced tillage agroecosystem with manipulated earthworm populations: Implications for infiltration and nutrient transport. Soil Biol Biochem 29: 493–498.
33. Zartman RE, Moffet CA, Wester DB, Sosebee RE, Fish EB, et al. (2012) Influence of surface biosolids application on infiltration. Appl Environ Soil Sci 2012.
34. Eisenhauer N, Milcu A, Nitschke N, Sabais A, Scherber C, et al. (2009) Earthworm and belowground competition effects on plant productivity in a plant diversity gradient. Oecologia 161: 291–301.
35. Eisenhauer N, Milcu A, Sabais ACW, Bessler H, Weigelt A, et al. (2009) Plant community impacts on the structure of earthworm communities depend on season and change with time. Soil Biol Biochem 41: 2430–2443.
36. Tilman D, Reich PB, Knops J, Wedin D, Mielke T, et al. (2001) Diversity and productivity in a long-term grassland experiment. Science 294: 843–845.
37. Spehn EM, Hector A, Joshi J, Scherer-Lorenzen M, Schmid B, et al. (2005) Ecosystem effects of biodiversity manipulations in European grasslands. Ecol Monogr 75: 37–63.
38. Bessler H, Temperton VM, Roscher C, Buchmann N, Schmid B, et al. (2009) Aboveground overyielding in grassland mixtures is associated with reduced biomass partitioning to belowground organs. Ecology 90: 1520–1530.
39. Grevers MCJ, Jong ED (1990) The characterization of soil macroporosity of a clay soil under ten grasses using image analysis. Can J Soil Sci 70: 93–103.
40. Thompson SE, Harman CJ, Heine P, Katul GG (2010) Vegetation-infiltration relationships across climatic and soil type gradients. J Geophys Res 115: G02023.
41. Milcu A, Partsch S, Scherber C, Weisser WW, Scheu S (2008) Earthworms and legumes control litter decompostion in a plant diversity gradient. Ecology 89: 1872–1882.
42. Eisenhauer N, Milcu A, Sabais ACW, Bessler H, Brenner J, et al. (2011) Plant diversity surpasses plant functional groups and plant productivity as driver of soil biota in the long term. PLoS ONE 6: e16055.
43. Roscher C, Schumacher J, Baade J, Wilcke W, Gleixner G, et al. (2004) The role of biodiversity for element cycling and trophic interactions: An experimental approach in a grassland community. Basic Appl Ecol 5: 107–121.
44. Kluge G, Müller-Westermeier G (2000) Das Klima ausgewählter Orte der Bundesrepublik Deutschland: Jena. Berichte des Deutschen Wetterdienstes 213. Offenbach/Main, Germany.
45. FAO-Unesco (1997) Soil map of the world, Revised legend with corrections and update. ISRIC, Wageningen, Netherlands.
46. Thielemann U (1986) The octet-method for sampling earthworm populations. Pedobiologia 29: 296–302.
47. Schwärzel K, Punzel J (2007) Hood infiltrometer - A new type of tension infiltrometer. Soil Sci Soc Am J 71: 1438–1447.
48. Jarvis NJ, Leeds-Harrison PB, Dosser JM (1987) The use of tension infiltrometers to assess routes and rates of infiltration in a clay soil. J Soil Sci 38: 633–640.
49. Wooding RA (1968) Steady infiltration from a shallow circular pond. Water Resour Res 4: 1259–1273.
50. Ortlepp S (2004) Holozäne fluviale Dynamik im mittleren Saaletal. Sedimentologische Untersuchungen in der Unteraue, Jena: Friedrich-Schiller-Universität FSU Jena.
51. Pinheiro JC, Bates DM (2000) Mixed-effects models in S and S-Plus. New York: Springer-Verlag.
52. Bretz F, Hothorn T, Westfall P (2010) Multiple comparisons using R. Boca Baton: Chapman and Hall/CRC Press.
53. Grace JB (2006) Structural equation modeling and natural systems. Cambridge: Cambridge University Press.
54. Shipley B (2000) Cause and correlation in biology: A user's guide to path analysis, structural equations and causal Inference. Cambridge: Cambridge University Press. 317 p.
55. Kohler-Milleret R, Bayon R-C, Chenu C, Gobat J-M, Boivin P (2013) Impact of two root systems, earthworms and mycorrhizae on the physical properties of an unstable silt loam Luvisol and plant production. Plant Soil: 1–15.
56. Chan KY (2004) Impact of tillage practices and burrows of a native Australian anecic earthworm on soil hydrology. Appl Soil Ecol 27: 89–96.
57. Flury M, Flühler H, Jury WA, Leuenberger J (1994) Susceptibility of soils to preferential flow of water: A field study. Water Resour Res 30: 1945–1954.
58. van Schaik L, Palm J, Klaus J, Zehe E, Schröder B (2014) Linking spatial earthworm distribution to macropore numbers and hydrological effectiveness. Ecohydrology 7: 401–408.
59. Simmonds LP, Nortcliff S (1998) Small scale variability in the flow of water and solutes, and implications for lysimeter studies of solute leaching. Nut Cycl Agroecosyst 50: 65–75.

60. IPCC (2012) Summary for policymakers. In: Managing the risks of extrem events and disaster in advance climate change adaption. Cambridge, UK, and New York, NY, USA: Cambridge University Press. 1–19 p.
61. Archer NAL, Quinton JN, Hess TM (2002) Below-ground relationships of soil texture, roots and hydraulic conductivity in two-phase mosaic vegetation in South-east Spain. J Arid Environ 52: 535–553.
62. Mitchell AR, Ellsworth TR, Meek BD (1995) Effect of root systems on preferential flow in swelling soil. Commun Soil Sci Plant Anal 26: 2655–2666.
63. Mytton LR, Cresswell A, Colbourn P (1993) Improvement in soil structure associated with white clover. Grass Forage Sci 48: 84–90.
64. Obi ME (1999) The physical and chemical responses of a degraded sandy clay loam soil to cover crops in southern Nigeria. Plant Soil 211: 165–172.
65. Logsdon SD, Cambardella CA (2000) Temporal changes in small depth-incremental soil bulk density. Soil Sci Soc Am J 64: 710–714.
66. Abbott I, Parker CA (1981) Interactions between earthworms and their soil environment. Soil Biol Biochem 13: 191–197.
67. Pucheta E, Bonamici I, Cabido M, Díaz S (2004) Below-ground biomass and productivity of a grazed site and a neighbouring ungrazed exclosure in a grassland in central Argentina. Austral Ecol 29: 201–208.
68. Shipitalo MJ, Protz R (1988) Factors influencing the dispersibility of clay in worm casts. Soil Sci Soc Am J 52: 764–769.
69. Shipitalo MJ, Edwards WM, Redmond CE (1994) Comparison of water movement and quality in earthworm burrows and pan lysimeters. J Environ Qual 23: 1345–1351.
70. Ela SD, Gupta SC, Rawls WJ (1992) Macropore and surface seal interactions affecting water infiltration into soil. Soil Sci Soc Am J 56: 714–721.
71. Lee K, Foster R (1991) Soil fauna and soil structure. Soil Research 29: 745–775.
72. Stockdill SMJ (1966) The role of earthworm in pasture production and moisture conservation. Proceedings of the 28th Conference of the Zealand Grassland Association 13: 68–83.
73. Zachmann JE, Linden DR, Clapp CE (1987) Macroporous infiltration and redistribution as affected by earthworms, tillage and residue. Soil Sci Soc Am J 51: 1580–1586.
74. Wuest SB (2001) Earthworm, infiltration, and tillage relationships in a dryland pea–wheat rotation. Appl Soil Ecol 18: 187–192.
75. Zachmann JE, Linden DR (1989) Earthworm effects on corn residue breakdown and infiltration. Soil Sci Soc Am J 53: 1846–1849.
76. Joschko M, Söchtig W, Larink O (1992) Functional relationship between earthworm burrows and soil water movement in column experiments. Soil Biol Biochem 24: 1545–1547.
77. Eisenhauer N, Straube D, Scheu S (2008) Efficiency of two widespread non-destructive extraction methods under dry soil conditions for different ecological earthworm groups. Eur J Soil Biol 44: 141–145.
78. Barley KP (1954) Effects of root growth and decay on the permeability of a synthetic sandy loam. Soil Sci 78: 205–210.
79. Gish TJ, Jury WA (1983) Effect of plant root channels on solute transport. Trans Am Soc Agric Eng 26: 440–444.
80. Ravenek JM, Bessler H, Engels C, Scherer-Lorenzen M, Gessler A, et al. (2014) Long-term study of root biomass in a biodiversity experiment reveals shifts in diversity effects over time. Oikos: In press.
81. Angulo-Jaramillo R, Vandervaere J-P, Roulier S, Thony J-L, Gaudet J-P, et al. (2000) Field measurement of soil surface hydraulic properties by disc and ring infiltrometers: A review and recent developments. Soil Till Res 55: 1–29.
82. Saxton KE, Rawls WJ, Romberger JS, Papendick RI (1986) Estimating generalized soil-water characteristics from texture. Soil Sci Soc Am J 50: 1031–1036.
83. Kördel W, Egli H, Klein M (2008) Transport of pesticides via macropores (IUPAC Technical Report). Pure Appl Chem 80: 105–160.
84. Paoletti MG (1999) The role of earthworms for assessment of sustainability and as bioindicators. Agric, Ecosyst Environ 74: 137–155.
85. Bens O, Wahl N, Fischer H, Hüttl R (2007) Water infiltration and hydraulic conductivity in sandy cambisols: impacts of forest transformation on soil hydrological properties. Eur J For Res 126: 101–109.
86. Bonsu M (1992) A study of a texture-based equation for estimating the saturated hydraulic conductivity of an alfisol in the sudan savannah ecological zone, Ghana. Hydrol Sci J 37: 599–606.
87. Slater CS (1957) Cylinder infiltrometers for determining rates of irrigation. Soil Sci Soc Am J 21: 457–460.
88. Azooz RH, Arshad MA (1996) Soil infiltration and hydraulic conductivity under long-term no-tillage and conventional tillage systems. Can J Soil Sci 76: 143–152.
89. Blackburn WH (1975) Factors influencing infiltration and sediment production of semiarid rangelands in Nevada. Water Resour Res 11: 929–937.
90. Soto B, Basanta R, Benito E, Perez R, Diaz-Fierros F (1994) Runoff and erosion from burnt soils in northwest Spain. In: Sala M, Rubio JL., editors. Soil erosion and degradation as a consequence of forest fires. Logroño, Spain: Geoforma.
91. Dekker LW, Ritsema CJ (1994) How water moves in a water repellent sandy soil: 1. Potential and actual water repellency. Water Res Res 30: 2507–2517.

Effects of Nitrogen Addition and Fire on Plant Nitrogen Use in a Temperate Steppe

Hai-Wei Wei[1,2], Xiao-Tao Lü[1]*, Fu-Mei Lü[3], Xing-Guo Han[1,3]

1 State Key Laboratory of Forest and Soil Ecology, Institute of Applied Ecology, Chinese Academy of Sciences, Shenyang, China, **2** Graduate University of Chinese Academy of Sciences, Beijing, China, **3** State Key Laboratory of Vegetation and Environmental Change, Institute of Botany, Chinese Academy of Sciences, Beijing, China

Abstract

Plant nitrogen (N) use strategies have great implications for primary production and ecosystem nutrient cycling. Given the increasing atmospheric N deposition received by most of the terrestrial ecosystems, understanding the responses of plant N use would facilitate the projection of plant-mediated N cycling under global change scenarios. The effects of N deposition on plant N use would be affected by both natural and anthropogenic disturbances, such as prescribed fire in the grassland. We examined the effects of N addition (5.25 g N m^{-2} yr^{-1}) and prescribed fire (annual burning) on plant N concentrations and N use characters at both species and community levels in a temperate steppe of northern China. We found that N addition and fire independently affected soil N availability and plant N use traits. Nitrogen addition increased aboveground net primary productivity (ANPP), inorganic N, and N uptake, decreased N response efficiency (NRE), but did not affect biomass-weighed N concentrations at community level. Prescribed fire did not change the community level N concentrations, but largely decreased N uptake efficiency and NRE. At the species level, the effects of N addition and fire on plant N use were species-specific. The divergent responses of plant N use at community and species levels to N addition and fire highlight the importance of the hierarchical responses of plant N use at diverse biological organization levels to the alteration of soil N availability. This study will improve our understanding of the responses of plant-mediated N cycling to global change factors and ecosystem management strategies in the semiarid grasslands.

Editor: Fei-Hai Yu, Beijing Forestry University, China

Funding: This research was supported by the National Natural Science Foundation of China (41273094, 41173086, and 31170433) and the State Key Laboratory of Forest and Soil Ecology (LFSE2013-13). The funders had no role in study design, data collection and analysis, decision to publish, or preparation of the manuscript.

Competing Interests: The authors have declared that no competing interests exist.

* E-mail: lvxiaotao@iae.ac.cn

Introduction

Nitrogen (N) availability is the dominant limiting factor for primary productivity in most terrestrial ecosystems [1–4]. Consequently, the responses of plants to the variation in soil N availability would have great implications for plant growth and species competition. It is well established that plant N economy plays an important role in affecting plant productivity and ecosystem N cycling [5,6]. Both plant N concentration and N response efficiency (NRE) are widely used to evaluate the responses of plant N economy to changes of soil N availability. Changes of plant N concentration have great implications for their N use efficiency (NUE), which is generally defined as the dry mass productivity per unit N taken up from the soil [7]. N response efficiency is the amount of biomass produced per unit of plant available N in the soil [8–10]. While plant N concentration is a indicator that integrates plant physiological and morphological responses to soil N availability, NRE is an index that integrates ecological and biogeochemical responses of plants. There is increasing empirical evidence that both plant N use indices should be simultaneously examined as they involve different processes linking the plant-soil system [11].

Both increasing atmospheric N deposition and global warming would enhance soil N availability in terrestrial ecosystems [12,13], which exerts positive effects on plant growth and primary productivity [1,14]. It should be noted that, however, the relationship between increased N availability and plant N use characters may not be straightforward in natural ecosystems. For instance, the changes of NUE in response to N availability would be the balance between the responses of its two components, NPP and the amount of N taken up by plants. Similarly, the responses of NRE would be determined by the relative changes of NPP and soil N availability. A decrease of NRE in response to N addition means that N addition shows relatively stronger impacts on soil N availability than NPP. Previous studies showed that N addition generally resulted in an increase of plant N concentration and a decrease of NRE at species level [8,11,15,16]. It has been suggested that NRE would be a more representative measurement of plant responses to increasing soil N availability [11].

Most of previous studies mainly focused on changes of foliar N characters along nutrient availability gradients [6], which may not well mirror what is happening at higher biological organization levels. For example, there is significant variation of N concentrations between leaf and stem and also between aboveground and belowground parts at the individual plant level [17–18]. Furthermore, natural communities are always composed of plant species with various nutrient use strategies [19] and different nutrient limitation status [20]. There is increasing evidence that the response of plant N use to changes of nutrient availability would vary significantly among diverse biological scales, from foliar to community level [11].

Environmental disturbances and ecosystem management strategies may also affect plant N use through their impacts on soil N availability [21–23] and community composition [24–26]. Moreover, they have the potential to alter the magnitude and direction of the impacts of N addition on plant N use [27]. Fire, as an anthropogenic or natural disturbance, exerts great control on community composition and ecosystem functioning in several terrestrial ecosystems, including grassland [25,28,29]. Reduction in fire frequency is considered as one of the major reasons for the increases of woody plant abundance in grasslands all over the world [26,30]. Prescribed burning is used as a strategy to prevent shrub encroachment in grasslands [31]. Results from a meta-analysis showed that short-term fire would enhance soil N availability while long-term repeated fire would lead to a decrease of soil N [21]. Similarly, the effects of fire on plant N status and use characters would also depend on its frequency. In the short-term, fire would increase plant NUE due to its stimulation on photosynthesis [32], whereas plant NUE would decrease under long-term repeated burning [22]. Given the interactive effects of fire and N addition on soil nutrient availability and plant production as observed in other studies [27], it is reasonable to expect that they may interact to affect plant N use. However, empirical evidence for their interactive effects on plant N use is still scarce.

The main objective of this study was to determine the effects of N addition, fire, and their interaction on plant N concentrations and N use characters at both species and community levels. We conducted a manipulative experiment in a temperate steppe in northern China, which is a typical vegetation type in the Eurasian continent. Fire was an important natural disturbance factor before the 1980s in northern China, with a frequency of once very 3–5 years [33]. In previous studies, we found that N addition and fire exert influence on plant nutrient resorption [34] and C:N:P stoichiometry [35,36]. We tested the following hypotheses: (1) N addition would enhance plant N concentrations, but reduce plant N uptake efficiency and NRE. (2) Fire would also increase plant N concentrations, but decrease plant N uptake efficiency and NRE, as the short-term fire (annual burning for two years) usually enhance soil N availability. Furthermore, we expected that the responses of plant N use at species level would be different from that at community level.

Materials and Methods

Site description

This study was conducted near the Inner Mongolia Grassland Ecosystem Research Station (IMGERS, 116°42′E, 43°38′N), which is located in the middle reach of the Xilin River, northern China. Long-term (1970–2007) meteorological data indicated that the mean annual precipitation was approximately 345 mm and mean annual temperature was 0.3°C. The length of growing season is about 150 days. The soil of this site is classified as Haplic Calcisols according to the Food and Agriculture Organization classification. Mean bulk density (top 10 cm) is 1.3 g cm^{-3} and pH is 7.4. The vegetation is dominated by *Leymus chinensis* (Trin.) Tzvel., *Stipa grandis* P. Smirn., *Cleistogenes squarrosa* (Trin.) Keng., *Caragana microphylia* Lam., *Carex korshinskyi* Kom., and *Potentilla bifurca* L.

Experimental design

This study was conducted as part of the GFE (Grassland Fire Experiment) experimental setup near IMGERS. All necessary permits were gained from Institute of Botany, Chinese Academy of Sciences and local government before the beginning of this study. The design consists of nine blocks distributed across a grassland field, with each block containing a set of 10 m×10 m plots representing fully crossed treatments of fire frequency, N addition, and mowing frequency. Plots were separated by 1 m buffers. The experiment setup was established in 2005, with all burning, N addition, and mowing treatments were started since 2006. Burning was carried out in early or late April each year before the start of growing season. Nitrogen in the form of NH_4NO_3 was added in the rainy days of late June each year. Only the treatments of two burning frequencies (never burned vs annual burning) and two N addition levels (0 vs 5.25 g N m^{-2} yr^{-1}) that without mowing in 9 blocks were used in this study, with 4 treatments ×9 replications = 36 plots in all.

Sample collection and chemical analysis

Aboveground net primary production (ANPP) was determined by clipping all plants in randomly located 1 m×1 m quadrats in each plot during the peak of growing season in August 2007. All plant samples were oven-dried at 70°C for 48 h, and then weighed. Samples of each species per plot were combined and ground to pass through a 40-mesh sieve using a mechanical mill (Retsch MM 400, Retsch GmbH & Co KG, Haan, Germany). Total N concentrations were determined colorimetrically by the Kjeldahl acid- digestion method with an Alpkem auto-analyzer (Kjektec System 1026 distilling unit, Sweden) after extraction with sulfuric acid.

Three soil cores (topsoil, 0–10 cm) were sampled in each of those quadrats following plant sampling with a metal tube (5 cm in diameter) and pooled to obtain one composite soil sample per quadrat. To measure soil water content, a soil subsample from each quadrat was weighed before and after being oven dried at 105°C for 48 h. Soil inorganic N availability was determined by extracting 10 g of fresh, root-free soil with 50 mL 2M KCl. The soil-extractant mixture was shaken for 1 h and then filtered (Whatman No. 1 filter paper). NH_4^+-N was measured with the salicylate method while NO_3^--N was measured using the cadmium reduction method on a FIAstar 5000 Analyzer (Foss Tecator, Denmark). Total soil inorganic N was the sum of NH_4^+-N and NO_3^-N.

Calculations and statistical analysis

The community level biomass-weighed N concentration was calculated as:

$$N\ concentration_{community} = \frac{\sum N\ concentration_i \times ANPP_i}{ANPP} \quad (1)$$

Where i is the ith species of n species in the community.

N uptake efficiency was calculated as the proportion of available N in the soil that was taken up by the plants [8]

$$N\ uptake\ efficiency = \frac{N\ uptake}{N\ available} \quad (2)$$

NRE was calculated ANPP relative to the pool of available soil N [8–10]:

$$NRE = \frac{ANPP}{N\ available} \quad (3)$$

The Kolmogorov-Smirnov test was used to test for data normality.

Two-way ANOVAs with fire and N addition as the main factors were performed to examine main and interactive effects on soil inorganic N, soil water content, ANPP, N concentrations, N uptake, N uptake efficiency, and NRE at community level. Two-way ANOVAs were used to examine the effects of N addition and fire on the plant N use characters of five dominant species at species level. All statistical analyses were performed using SPSS software (version 13.0 for windows, SPSS Inc., Chicago, IL, USA).

Results

Community level responses

Nitrogen addition and fire significantly enhanced soil inorganic N by 94.2% ($P<0.001$) and 25.8% ($P<0.05$), respectively (Fig. 1a, Table 1). Fire significantly reduced soil water content ($P<0.001$), whereas N addition did not affect soil water content (Fig. 1b, Table 1). Nitrogen addition significantly increased ANPP by 36.3% ($P<0.001$; Fig. 1c, Table 1), whereas fire reduced ANPP by 14.3% ($P<0.05$; Fig. 1c; Table 1).

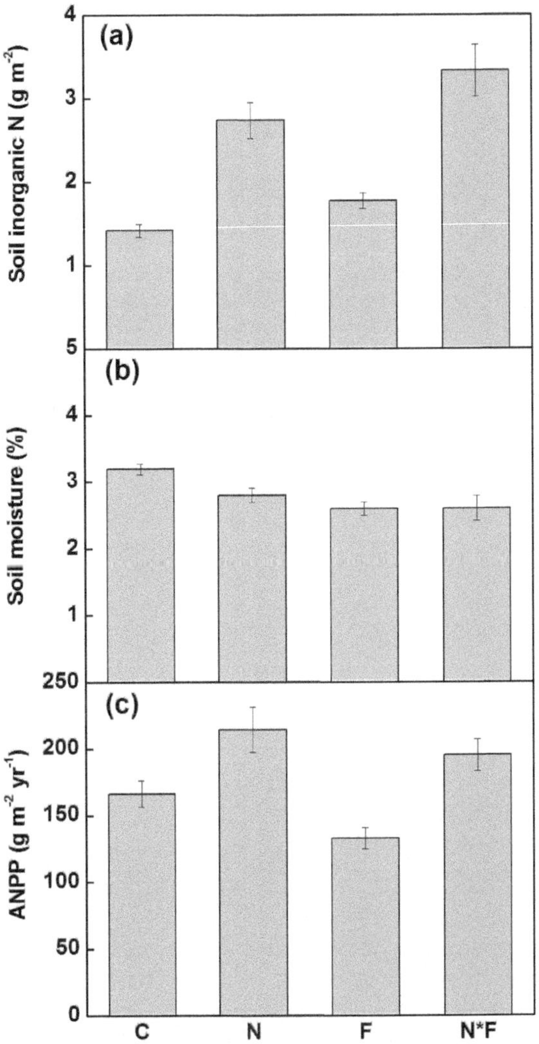

Figure 1. The effects of nitrogen addition and fire on (a) soil inorganic N concentrations, (b) soil water content, and (c) ANPP of plant community in a temperate steppe. C, control; N, nitrogen addition; F, fire; N*F, both N addition and fire. Data are adjusted means ± 1SE, where $n=9$ for all treatments.

Both N addition and fire had no impacts on biomass-weighed N concentrations at community level (Fig. 2a; Table 1). N addition enhanced plant N uptake by 41.6% ($P<0.01$; Fig. 2b; Table 1), while fire had no effects on plant N uptake (Fig. 2b; Table 1). Both N addition and fire decreased N uptake efficiency by 22.9% ($P<0.01$) and 18.2% ($P<0.05$; Fig. 2c; Table 1). NRE was significantly reduced by both N addition (31.5%, $P<0.001$, Fig. 2c; Table 1) and fire (36.6%, $P<0.001$, Table 1). There was no interaction between N addition and fire for all variables (Table 1).

Species level responses

Nitrogen addition significantly enhanced the ANPP of *C. korshinskyi* ($P<0.01$) and *L. chinensis* ($P<0.05$; Table 1; Fig. 3), but showed no impacts on that of *A. cristatum*, *C. squarrosa*, and *S. grandis*. When the species were analyzed individually, plant N use varied greatly among five dominant species (Fig. 3). N addition had no impacts on N concentration for all species, while fire significantly increased N concentration of all species except for *C. squarrosa* (Table 1; Fig. 3). N addition and fire interacted to affect N concentration of *A. cristatum* ($P<0.05$) and *S. grandis* ($P<0.01$; Table 1; Fig. 3). While N addition significantly enhanced N uptake in two out of all the five species (*A. cristatum* and *C. korshinskyi*), fire significantly reduced N uptake of *C. korshinskyi* but did not affect that of the other species (Table 1; Fig. 3).

Nitrogen addition significantly decreased N uptake efficiency of all species ($P<0.05$) except for *C. korshinskyi* (Table 1, Fig. 3), while fire only decreased the N uptake efficiency of *C. korshinskyi* ($P<0.001$). N addition significantly decreased NRE for all species ($P<0.05$), whereas fire only decreased that of *A. cristatum* ($P<0.05$) and *C. korshinskyi* ($P<0.001$). For all the species, there was no interactive effect between N addition and fire on NRE (Table 1).

Discussion

Results from this study showed that N addition and fire directly affected soil N availability, ANPP, plant N uptake and use traits at both species and community levels. We found no interaction between N addition and fire in affecting plant N use. Generally, N addition increased ANPP and plant N uptake, decreased N uptake efficiency and NRE, with no significant effect on plant N concentrations. In contrast, fire showed positive effects on plant N concentrations (especially at species level), negative effects on NRE, and neutral effects on plant N uptake.

We found that both N addition and fire increased soil inorganic N concentrations. A meta-analysis showed that the soil inorganic N would increase immediately after fire and then gradually declined to the level before fire within one year [21]. All the burning plots had been burned for two years before soils were sampled in this study. Our results suggest that short-term fire would stimulate soil N turnover in the temperate steppe. While N is assumed to be the primary limiting nutrient in this ecosystem [14], we found negative effects of fire on ANPP even it stimulated soil N availability. We suspect that water availability would play an importantly in mediating the effects of fire on ANPP in this semiarid grassland. Fire significantly reduced soil water content in this study. In the same experimental setup, Zhou et al. [33] also found that the soil moisture in unburned plots was significantly higher than that in the burned plots during the growing season. Furthermore, it has been shown that water addition instead of N addition significantly increased ANPP, indicting water availability would be the dominant limiting factor for plant growth in this ecosystem [37]. Consequently, it is reasonable to speculate that reduction of soil water availability after burning would account for

Table 1. Results (F-values) of two-way ANOVAs on the effects of nitrogen addition (N), fire (F), and their interaction on soil inorganic N, soil water content (SWC), ANPP, N concentration, N uptake, N uptake efficiency, and NRE at both community and species level in a temperate steppe.

Factor	Inorganic N	SWC	ANPP	N concentration	N uptake	N uptake efficiency	NRE
Community-level							
N	53.5***	2.5	24.2***	0	12.4**	7.9**	13.6***
F	6.0*	10.3**	6.2*	3.5	0.2	4.7*	20.7***
N*F	0.4	2.6	0.3	1.5	0.3	0.1	3
Species-level							
Agropyron cristatum							
N	-	-	3.6	0	4.6*	11.9**	10.9**
F	-	-	0.7	6.3*	0.1	1.5	5.8*
N*F	-	-	0.2	6.5*	0.5	0.2	2.5
Carex korshinskyi							
N	-	-	13.1**	3.5	13.5***	2.4	5.1*
F	-	-	16.3***	7.9**	8.4**	15.4***	27.6***
N*F	-	-	2.2	0.1	1.5	0.1	0.3
Cleistogenes squarrosa							
N	-	-	0.6	0.5	1	11.2**	12.5**
F	-	-	0.3	2.9	0	0.7	2.1
N*F	-	-	5.2*	2	5.4*	1.3	0.4
Leymus chinensis							
N	-	-	5.2*	2.3	0.9	7.0*	6.4*
F	-	-	0	6.7*	2.3	0.2	2.3
N*F	-	-	0.4	0	0	0	0
Stipa grandis							
N	-	-	0.1	1.9	0	5.1*	5.0*
F	-	-	3.8	18.2***	1.4	2.3	4.5
N*F	-	-	2	12.3**	3.7	0.7	0.1

The F-values are presented, together with their level of significance.
*$P<0.05$;
**$P<0.01$;
***$P<0.001$.
For all variables error $df = 36$.

the negative effects of fire on community ANPP as observed in the present study.

Nitrogen addition would elevate soil inorganic N availability and thus increase ANPP, which had been generally tested and confirmed by previous studies [1,14,16]. Similarly, our results showed that N addition (5.25 g N m^{-2} yr^{-1}) significantly enhanced soil inorganic N by 94.2%, and increased ANPP at both species and community levels. In the same area, Bai et al. [14] reported that N addition, even as low as 1.75 g N m^{-2} yr^{-1}, would significantly increase ANPP. It is notable that, however, the effects of N addition on ANPP would be largely dependent on soil water availability [37]. The positive effects of N addition on the growth of dominant species and the ANPP of whole community as observed in this study indicated that primary production was limited by N availability in this area. Given the projected increasing atmospheric N deposition, the primary productivity would be facilitated following the increasing N availability. We found no interactive effect of N addition and fire on primary production at community level, implying that the positive effects of N deposition on primary production would not be affect by natural fire or prescribed burning in the temperate steppe of northern China.

Our results showed that N addition significantly increased plant N uptake by 41.6%, whereas fire had no effects on plant N uptake. The positive effects of N addition on plant N uptake were resulted from the enhancement of ANPP, as we found no impacts of N addition on plant N concentrations at both species and community levels. Previous studies showed that fire would increase N concentrations in plant tissues [27]. In this study, we found two-year annual fire significantly enhanced biomass-weighed plant N concentrations at community level, which is consistent with the results of our previous study which was carried out in the same ecosystem [35]. Due to the offset between the increases of plant N concentration and the decreases of ANPP following fire, annual burning showed no significant effects on plant N uptake. Fire showed negative influence on N uptake efficiency as it had no effects on plant N uptake but enhanced soil inorganic N. It has been reported that N addition would decrease N uptake efficiency, as the increases of N uptake cannot commensurate with the increases of soil inorganic N in a peatland and a managed

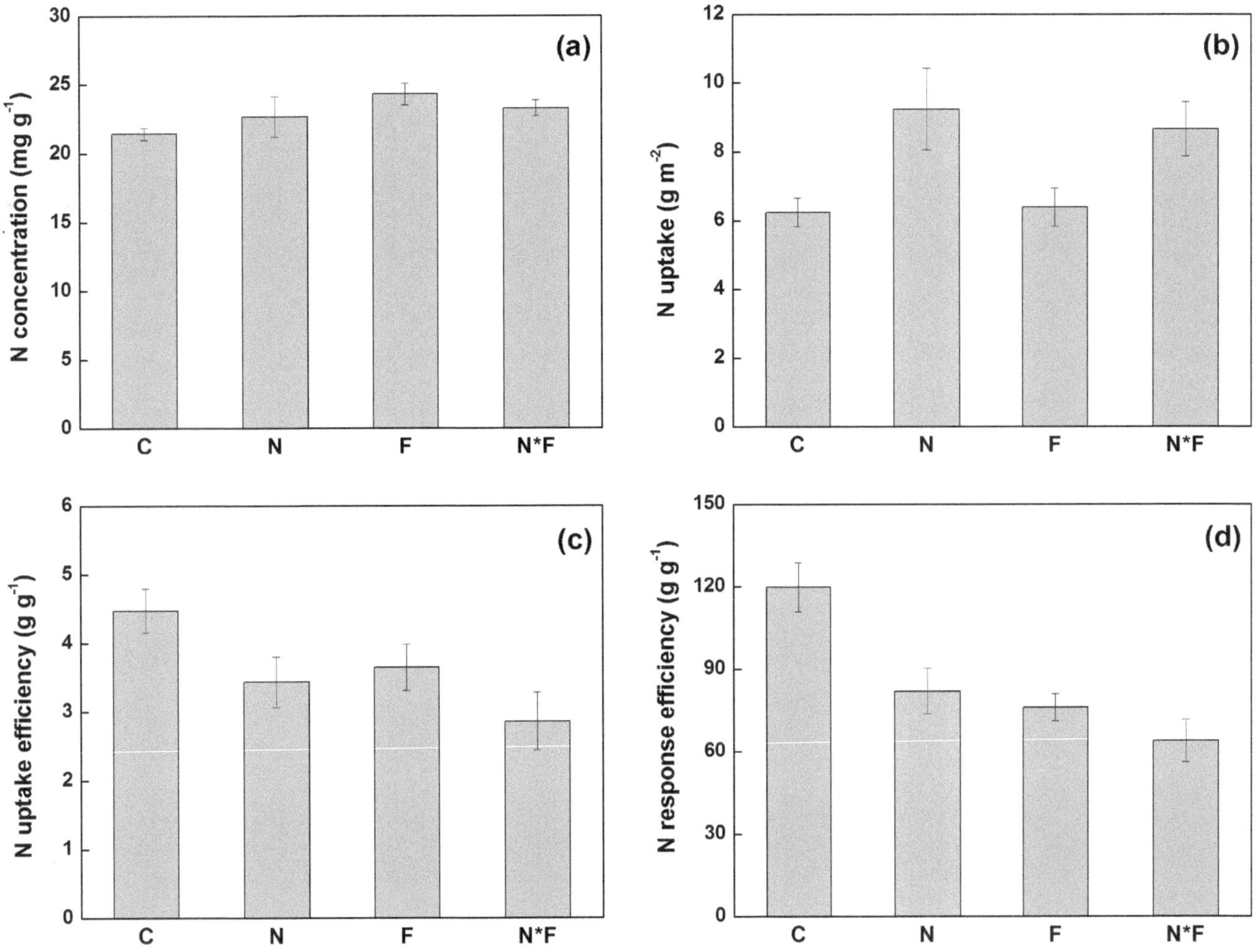

Figure 2. The effects of nitrogen addition and fire on (a) biomass-weighed N concentration, (b) N uptake, (c) N uptake efficiency, and (d) N response efficiency of plant community in a temperate steppe. C, control; N, nitrogen addition; F, fire; N*F, both N addition and fire. Data are adjusted means ± 1SE, where $n = 9$ for all treatments.

grassland [11,15]. Similarly, we found negative effects of N addition on plant N uptake efficiency at community level in this temperate steppe.

Previous studies reported that NUE often decreased with the increase of soil N availability [6,9,19], but not in other studies [38–40]. Contrary to our hypothesis, results from this study showed that N addition did not affect plant N concentrations at both species and community levels, implying that two-year N addition with the rate of 5.25 g N m^{-2} yr^{-1} did not affect NUE of plants in this temperate steppe. Our results were consistent with Reich et al. [41], who reported that increasing N availability did not change NUE in the forest ecosystem, but were different from Iversen et al. [11], who found that N addition (6 g N m^{-2} yr^{-1}) resulted in a decrease of NUE in peatlands. In temperate grassland, Yuan et al. [40] found that the responses of NUE to N addition were dependent on N addition rates. On the one hand, in our study, the N addition rate was 5.25 g N m^{-2} yr^{-1}, which might be not enough to induce a decrease of NUE. On the other hand, our results indicate that the efficiency of temperate steppe to use N would be consistent in the face of increasing N deposition, at least at the rate of 5.25 g N m^{-2} yr^{-1}. Our results that two years annual burning significantly enhanced biomass-weighed N concentrations at community level, indicating that fire would have negative effects of NUE in this community. The enhancement of plant N concentrations would have great means for litter quality and decomposition, as the litter decomposition rate and nutrient release are usually positive related to the N concentration in the litter [5,24,42].

Consistent with our hypothesis and findings from other studies [11,15], our results showed that N addition reduced NRE. Based on a natural soil N availability gradient in the semiarid grassland, Yuan et al. [43] found that NRE decreased monotonically with increasing N availability. The lower magnitude of the responses of ANPP to N addition than that of soil inorganic N availability resulted in a negative effect of N addition on NRE. Fire also led to a negative effect on plant NRE at community level, due to its negative effect on ANPP and positive effect on soil inorganic N. The negative effects of N addition and fire on community NRE suggests that part of soil inorganic N could not be taken up into plant biomass. Thus, more soil inorganic N would be lost by leaching and volatilization after N addition or burning. There is evidence that annually burning induce N loss through volatilization during fire [44], erosion, leaching, and denitrification [45]. It is notable that, the negative effects of N addition and fire on plant

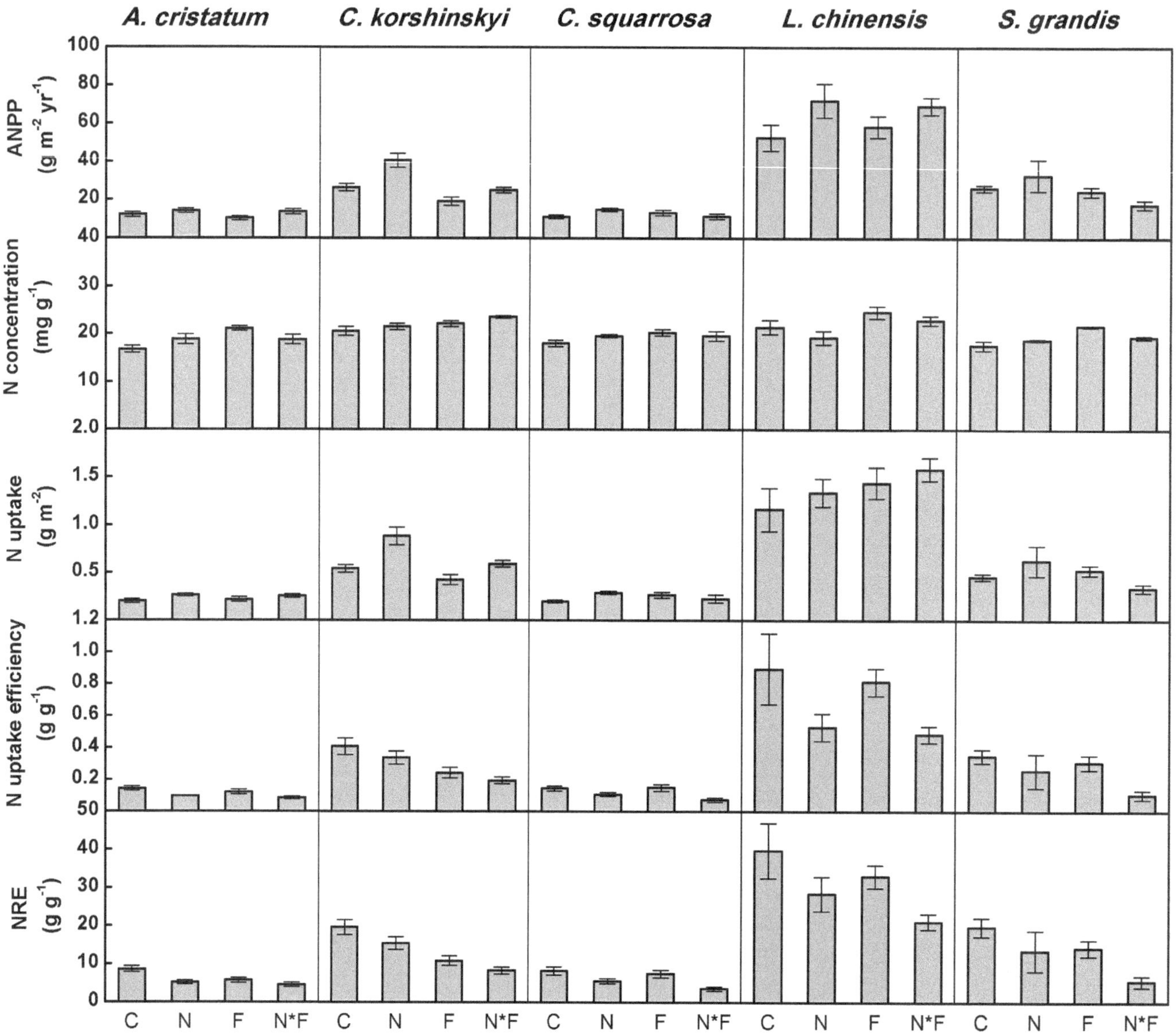

Figure 3. The effects of N addition and fire on plant N use of dominant species in a temperate stepp. C, control; N, nitrogen addition; F, fire; N*F, both N addition and fire. Data are adjusted means ±1SE, where $n = 9$ for all treatments.

community NRE were independent, rather than additive. Together, our results indicate that the temperate steppe would be less efficient in absorbing the increased soil inorganic N into plant biomass after N addition and burning.

The dominant species examined in this study greatly varied in their N use traits. In the control plots, the NRE of *L. chinensis* was almost five times higher than that of *A. cristatum* and *C. squarrosa*. The higher NRE of *L. chinensis* would help explain its dominant status in this N-limited grassland ecosystem. More soil N would be transferred to plant biomass if this ecosystem was mono-dominated by *L. chinensis*, as indicated by its higher NRE status. Consequently, any changes of community composition following N addition and fire would influence plant N use at community level. Moreover, the responses of plant N economy to N addition and fire were species-specific. While N addition significantly enhanced the N uptake of *A. cristatum* and *C. korshinskyi*, it showed no impacts on that of other species. Similarly, fire reduced NRE of *A. cristatum* and *C. korshinskyi*, but had no effects on other species. The idiosyncratic responses of different species to N addition and fire suggest that *A. cristatum* and *C. korshinskyi* were more sensitive to changes of soil N availability resulting from N addition and annual burning. Our results have great implications for the plant-mediated changes of biogeochemical cycling in response to global changes and ecosystem management strategies.

We hypothesized that the responses of plant N use traits to N addition and fire would be different at species and community levels. When all the dominant species were analyzed together, we found that the responses of plant N concentrations and NRE to N addition and fire did not different ranging from species to community level. Nitrogen addition did not alter plant N concentrations at both species and community levels, whereas it did reduce plant NRE at both levels. Fire decreased plant N concentrations and NRE at community level, while its effects on species-level plant N concentrations and NRE varied among different species. However, when those species were analyzed individually, we did found divergent responses of plant N use traits at species and community levels, which would be mainly resulted from the species-specific responses of plants to N addition and fire. In a recent study, Iversen et al. [11] reported that plant N use traits differed at different biological organizations ranging from individual plant tissue to the entire community in peatland ecosystems. Our study presented strong evidence for the hierarchical responses of plant N use to N addition and fire at different biological scales, especially for species and community levels in a grassland ecosystem.

Conclusions

Our results showed that both N addition and fire directly affected soil N availability and plant N use traits, with modest interaction between them. N addition increased ANPP, inorganic N, and N uptake, decreased NRE, and did not affect plant N concentrations. Prescribed fire, as an important management strategy in grasslands, enhanced plant N concentrations and decreased N uptake efficiency and NRE. The responses of plant N use to both N addition and fire would be different at species and community levels, mainly due to the species-specific responses. The divergence of plant N use traits among different plant species indicates that any factors leading to changes of community composition would indirectly alter plant N use at community level.

Acknowledgments

We thank Wu Wei-Jun, Li Qiang, and Li Li for assistance with field and laboratory work, two anonymous reviewers for their helpful comments on the manuscript, and Matthew Simmons for language improvement.

Author Contributions

Conceived and designed the experiments: XH XL FL. Performed the experiments: FL XL HW. Analyzed the data: XL HW. Contributed reagents/materials/analysis tools: FL XH XL. Wrote the paper: HW XL.

References

1. LeBauer DS, Treseder KK (2008) Nitrogen limitation of net primary productivity in terrestrial ecosystems is globally distributed. Ecology 89: 371–379.
2. Perakis SS, Kellogg CH (2007) Imprint of oaks on nitrogen availability and delta N-15 in California grassland-savanna: a case of enhanced N inputs? Plant Ecol 191: 209–220.
3. Cole L, Buckland SM, Bardgett RD (2008) Influence of disturbance and nitrogen addition on plant and soil animal diversity in grassland. Soil Biol Biochem 40: 505–514.
4. Vitousek PM, Howarth RW (1991) Nitrogen limitation on land and in the sea: How can it occur? Biogeochemistry 13: 87–115.
5. Vitousek P (1982) Nutrient cycling and nutrient use efficiency. Am Nat 119: 553–572.
6. Aerts R, Chapin FS (2000) The mineral nutrition of wild plants revisited: A re-evaluation of processes and patterns. Adv Ecol Res 30: 1–47.
7. Hirose T (2011) Nitrogen use efficiency revisited. Oecologia 166: 863–867.
8. Pastor J, Bridgham SD (1999) Nutrient efficiency along nutrient availability gradients. Oecologia 118: 50–58.
9. Shaver GR, Melillo JM (1984) Nutrient budgets of marsh plants: efficiency concepts and relation to availability. Ecology 65: 1491–1510.
10. Bridgham SD, Pastor J, McClaugherty CA, Richardson CJ (1995) Nutrient-Use-Efficiency: A litterfall index, a model, and a test along a nutrient availability gradient in North-Carolina peatland. Am Nat 145: 1–21.
11. Iversen CM, Bridgham SD, Kellogg LE (2010) Scaling plant nitrogen use and uptake efficiencies in response to nutrient addition in peatlands. Ecology 91: 693–707.
12. Melillo JM, Steudler PA, Aber JD, Newkirk K, Lux H, et al. (2002) Soil warming and carbon-cycle feedbacks to the climate system. Science 298: 2173–2176.
13. Wallenda T, Kottke I (1998) Nitrogen deposition and ectomycorrhizas. New Phytol 139: 169–187.
14. Bai Y, Wu J, Clark CM, Naeem S, Pan Q, et al. (2010) Tradeoffs and thresholds in the effects of nitrogen addition on biodiversity and ecosystem functioning: evidence from Inner Mongolia grasslands. Glob Change Biol 16: 358–372.
15. Keuter A, Hoeft I, Veldkamp E, Corre MD (2012) Nitrogen response efficiency of a managed and phytodiverse temperate grassland. Plant Soil 364: 193–206.
16. Chapin FS, Vitousek PM, Vancleve K (1986) The nature of nutrient limitation in plant-communities. Am Nat 127: 48–58.
17. Lu XT, Freschet GT, Flynn DFB, Han XG (2012) Plasticity in leaf and stem nutrient resorption proficiency potentially reinforces plant-soil feedbacks and microscale heterogeneity in a semi-arid grassland. J Ecol 100: 144–150.
18. Freschet GT, Cornelissen JHC, van Logtestijn RSP, Aerts R (2010) Substantial nutrient resorption from leaves, stems and roots in a subarctic flora: what is the link with other resource economics traits? New Phytol 186: 879–889.
19. Chapin FS (1980) The mineral nutrition of wild plants. Ann Rev Ecol Syst 11: 233–260.
20. Koerselman W, Meuleman AFM (1996) The vegetation N:P ratio: A new tool to detect the nature of nutrient limitation. J Appl Ecol 33: 1441–1450.
21. Wan SQ, Hui DF, Luo YQ (2001) Fire effects on nitrogen pools and dynamics in terrestrial ecosystems: A meta-analysis. Ecol Appl 11: 1349–1365.
22. Reich PB, Peterson DW, Wedin DA, Wrage K (2001) Fire and vegetation effects on productivity and nitrogen cycling across a forest-grassland continuum. Ecology 82: 1703–1719.
23. Pilkington MG, Caporn SJM, Carroll JA, Cresswell N, Phoenix GK, et al. (2007) Impacts of burning and increased nitrogen deposition on nitrogen pools and leaching in an upland moor. J Ecol 95: 1195–1207.
24. Watson PJ, Bradstock RA, Morris EC (2009) Fire frequency influences composition and structure of the shrub layer in an Australian subcoastal temperate grassy woodland. Aust Ecol 34: 218–232.
25. Bowman DM, Balch JK, Artaxo P, Bond WJ, Carlson JM, et al. (2009) Fire in the Earth system. Science 324: 481–484.
26. Bond WJ, Keeley JE (2005) Fire as a global 'herbivore': the ecology and evolution of flammable ecosystems. Trends Ecol Evol 20: 387–394.
27. Britton AJ, Helliwell RC, Fisher JM, Gibbs S (2008) Interactive effects of nitrogen deposition and fire on plant and soil chemistry in an alpine heathland. Environ Poll 156: 409–416.
28. Keeley JE, Pausas JG, Rundel PW, Bond WJ, Bradstock RA (2011) Fire as an evolutionary pressure shaping plant traits. Trends Plant Sci 16: 406–411.
29. Bradshaw SD, Dixon KW, Hopper SD, Lambers H, Turner SR (2011) Response to Keeley et al.: Fire as an evolutionary pressure shaping plant traits. Trends Plant Sci 16: 405–405.
30. Heisler JL, Briggs JM, Knapp AK, Blair JM, Seery A (2004) Direct and indirect effects of fire on shrub density and aboveground productivity in a mesic grassland. Ecology 85: 2245–2257.
31. Bowles ML, Jones MD (2013) Repeated burning of eastern tallgrass prairie increases richness and diversity, stabilizing late successional vegetation. Ecol Appl 23: 464–478.
32. Ojima DS, Schimel DS, Parton WJ, Owensby CE (1994) Long-term and short-term effects of fire on nitrogen cycling in tallgrass prairie. Biogeochemistry 24: 67–84.
33. Zhou L, Huang J, Lü F, Han X (2009) Effects of prescribed burning and seasonal and interannual climate variation on nitrogen mineralization in a typical steppe in Inner Mongolia. Soil Biol Biochem 41: 796–803.
34. Lü XT, Cui Q, Wang QB, Han XG (2011) Nutrient resorption response to fire and nitrogen addition in a semi-arid grassland. Ecol Engin 37: 534–538.
35. Cui Q, Lü XT, Wang QB, Han XG (2010) Nitrogen fertilization and fire act independently on foliar stoichiometry in a temperate steppe. Plant Soil 334: 209–219.
36. Lü XT, Lu FM, Zhou LS, Han X, Han XG (2012) Stoichiometric response of dominant grasses to fire and mowing in a semi-arid grassland. J Arid Environ 78: 154–160.
37. Lü XT, Kong DL, Pan QM, Simmons ME, Han XG (2012) Nitrogen and water availability interact to affect leaf stoichiometry in a semi-arid grassland. Oecologia 168: 301–310.
38. Aerts R, Decaluwe H (1994) Nitrogen use efficiency of *Carex* species in relation to nitrogen suplly. Ecology 75: 2362–2372.
39. Yasumura Y, Hikosaka K, Matsui K, Hirose T (2002) Leaf-level nitrogen-use efficiency of canopy and understorey species in a beech forest. Fun Ecol 16: 826–834.
40. Yuan ZY, Li LH, Huang JH, Han XG, Wan SQ (2005) Effect of nitrogen supply on the nitrogen use efficiency of an annual herb, Helianthus annuus L. J Integrat Plant Biol 47: 539–548.
41. Reich PB, Grigal DF, Aber JD, Gower ST (1997) Nitrogen mineralization and productivity in 50 hardwood and conifer stands on diverse soils. Ecology 78: 335–347.
42. Blair JM, Parmelee RW, Beare MH (1990) Decay-rates, nitrogen fluxes, and decomposer communities of single-species and mixed-species foliar litter. Ecology 71: 1976–1985.
43. Yuan ZY, Li LH, Han XG, Chen SP, Wang ZW, et al. (2006) Nitrogen response efficiency increased monotonically with decreasing soil resource availability: a case study from a semiarid grassland in northern China. Oecologia 148: 564–572.

44. Olear HA, Seastedt TR, Briggs JM, Blair JM, Ramundo RA (1996) Fire and topographic effects on decomposition rates and N dynamics of buried wood in tallgrass prairie. Soil Biol Biochem 28: 323–329.

45. Neary DG, Klopatek CC, DeBano LF, Ffolliott PF (1999) Fire effects on belowground sustainability: a review and synthesis. For Ecol Manag 122: 51–71.

13

Application of Organic Amendments to a Coastal Saline Soil in North China: Effects on Soil Physical and Chemical Properties and Tree Growth

Linlin Wang[1,2], Xiangyang Sun[1]*, Suyan Li[1]*, Tao Zhang[1], Wei Zhang[1], Penghui Zhai[1]

1 College of Forestry, Beijing Forestry University, Beijing, China, 2 Forestry Research Institute of Hulun Buir, Hulun Buir, Inner Mongolia, China

Abstract

The ability of the following four organic amendments to ameliorate saline soil in coastal northern China was investigated from April 2010 to October 2012 in a field experiment: green waste compost (GWC), sedge peat (SP), furfural residue (FR), and a mixture of GWC, SP and FR (1:1:1 by volume) (GSF). Compared to a non-amended control (CK), the amendments, which were applied at 4.5 kg organic matter m^{-3}, dramatically promoted plant growth; improved soil structure; increased the cation exchange capacity (CEC), organic carbon, and available nutrients; and reduced the salt content, electrical conductivity (EC), and exchangeable sodium percentage (ESP). At the end of the experiment in soil amended with GSF, bulk density, EC, and ESP had decreased by 11, 87, and 71%, respectively, and total porosity and organic carbon had increased by 25 and 96% respectively, relative to the CK. The GSF treatment resulted in a significantly lower Na^{+}+K^{+} content than the other treatments. CEC and the contents of available N, P, and K were significantly higher in the GSF-treated soil than in the CK and were the highest in all treatments. The FR treatment resulted in the lowest pH value and Ca^{2+} concentration, which decreased by 8% and 39%, respectively, relative to the CK. Overall, the results indicate that a combination of green waste compost, sedge peat and furfural residue (GSF treatment) has substantial potential for ameliorating saline soils in the coastal areas of northern China, and it works better than each amendment alone. Utilization of GWC and FR can be an alternative organic amendment to substitute the nonrenewable SP in saline soil amelioration.

Editor: Ben Bond-Lamberty, DOE Pacific Northwest National Laboratory, United States of America

Funding: This research was supported by the Eleventh Five-Year Plan of the National Scientific and Technical Supporting Programs (No. 2009BADB2B0504), and the Fundamental Research Funds for the Central Universities. The funders had no role in study design, data collection and analysis, decision to publish, or preparation of the manuscript.

Competing Interests: The authors have declared that no competing interests exist.

* E-mail: sunxy1965@126.com (XS); lisuyan@bjfu.edu.cn (SL)

Introduction

Soil salinization is a major obstacle to the optimal utilization of land resources. Salt-affected soils are widely distributed throughout the world, and about 20% of the world's cultivated land is salt-affected [1]. In the coastal regions of China, approximately 10^6 hm^2 of land is salt-affected because of sea water intrusion and the dry climate. The salt-affected soils in the coastal areas of northern China have several unique characteristics including a high water table (saline groundwater is 0.5–1.0 m below the soil surface), a high groundwater mineralization degree (7–10 $g \cdot L^{-1}$), serious secondary salinization, and a low nutrient content [2]. The present extent of salt-affected soils substantially restricts plant growth in these areas.

Sodium (Na^+) is the predominant soluble cation in most saline soils, particularly in the coastal areas. Most plants are hypersensitive to saline environments [3]. The tissues of plants growing in saline media generally exhibit an accumulation of Na^+ and Cl^- and/or the reduced uptake of mineral nutrients, especially Ca^{2+}, K^+, N, and P [4]. Previous studies have reported on the adverse effects of excessive amounts of salts on the physical and chemical properties of soils and on plant growth and yield [5,6]. Clark et al. (2007) [7] demonstrated that salt-affected soils exhibit poor soil structure resulting from soil physical processes such as slaking, swelling, and dispersion of clay. Tejada and Gonzalez (2005) [8] showed that an increase in electrical conductivity (EC) adversely affects soil total porosity, bulk density, and structural stability. Salinity also affects soil chemical properties such as pH, cation exchange capacity (CEC), exchangeable sodium percentage (ESP), soil organic carbon, and available nutrients [1]. Salinity also alters the osmotic and matric potential of the soil solution, thereby adversely affecting soil microbial communities and their activity [9]. Moreover, the changes in the proportions of exchangeable ions in the soil solution have osmotic and ion-specific effects that can produce imbalances in plant nutrients, including deficiencies of several nutrients or excessive levels of Na^+ [4]. Reclaiming these salt-affected soils will require new amelioration methods and improved management practices [10].

Leaching with water, chemical amendment, and phytoremediation are among the methods used to ameliorate saline soils [10]. The use of gypsum, calcite, calcium chloride, and other chemical agents that provide Ca, which tends to replace exchangeable Na, is effective for saline soil amelioration [11]. Another important practice is the application of organic matter conditioners, which can both ameliorate and increase the fertility of saline soils [12]. Salt-affected soils generally exhibit poor structural stability due to low organic matter content. Many researchers have suggested that the structural stability of soil can be improved by the addition of organic materials (e.g. green manures, farmyard and poultry manures, compost, and food processing wastes) [13,14]. Barzegar et al. (1997) [15] found that the application of organic matter to saline soils can accelerate Na^+ leaching, increase the percentage of water-stable aggregates, and decrease the ESP, EC, and soil salinity. Wichern et al. (2006) [16] reported that the addition of maize straw to a saline soil reduced the negative effects of salinity on the microbial community and mineralization.

In China, green waste is mostly produced by urban landscape maintenance and represents a significant portion of municipal solid waste [17]. In view of both environmental protection and economic development, composting is the best method to manage and utilize green waste. Compost has a unique ability to improve the chemical, physical, and biological characteristics of soils. It improves water retention and soil structure by increasing the stability of soil aggregates [18]. Green waste compost (GWC) is a biologically stable, humus-like substance that can be used for saline soil amelioration. Peat is a partly-decomposed plant material that slowly accumulates in pond and lake bottoms and swamp areas. Sedge peat (SP) and sphagnum peat moss are the most common types of peat used for soil remediation, and the former is a fine-textured and more decomposed type of peat. Sedge peat can hold water and nutrients ten to fifteen times their own weight when fully saturated, yet still can hold 40% air. They have high soil CEC and pH buffering capacity owing to their great soil specific surface and chelation capacity [18]. The utilization of sedge peat as a soil amendment in agriculture represents a potentially large market for peat.

However, the use of peat must be reduced, because it is considered as a nonrenewable resource [19]. Thus, an alternative to peat must be developed to achieve the function of peat amendment and meet consumer's environmental concern. Furfural residue (FR) is the by-product of furfural production; furfural is mainly obtained from corn cobs by acid catalysis at high temperature. Among organic amendments, FR has the highest potential for low-cost, high-volume production [20]. Because the cellulose and lignin in corn cobs are relatively stable, FR is enriched in cellulose and lignin [21]. The advantage of organic amendments (GWC, SP, and FR) is that it is unlikely to contain microorganisms that are pathogenic to humans, which is an important concern of such soil amendment [18,22].

Even though there are many studies dealing with organic amendments as degraded soil conditioners, very little is known about their effects on coastal saline soil. In view of the above, three organic materials (GWC, SP, and FR) and five treatments: no amendment, GWC alone, SP alone, FR alone and a mixture of GWC, SP, and FR (abbreviated as GSF) were applied in this study. The aim of the work was to evaluate the effects of different organic amendments on soil salinity and fertility of saline soil as well as plant growth in the coastal areas of northern China. The information obtained from this study will help provide guidance on selection of organic matters in ameliorating coastal saline soil while considering their environmental concerns.

Materials and Methods

Ethics Statement

All necessary permits were obtained for the described field studies. The study was approved by the Dagang Agriculture and Forestry Administration Bureau, Tianjin, China.

Experimental Site and Properties of the Organic Amendments

The research was conducted from April 2010 to October 2012 at the Coastal Salt-tolerant Plant Science and Technology Park (38°46′ N, 117°13′ E, 1.3 m a.s.l.), Dagang, Tianjin, China. The study site has a semi-humid continental monsoon climate. The mean annual temperature is 12.3°C, the highest monthly temperature is in July (26°C), and the lowest monthly temperature is in January (−4°C). On average, there are 2618 h of sunshine per year. The frost-free period is 211 d, and the global radiation is 121.1 $MJ\cdot m^{-2}$. The annual precipitation and evaporation are 593.6 mm and 1979.4 mm, respectively, and the mean relative humidity is 65%. Rainfall from June to September accounts for 84% of the total annual precipitation. The general properties of soil samples (0–25 cm depth) taken from the study site are shown in Table 1.

The organic amendments applied in this study were GWC, SP, FR, and GSF. The general properties of the organic amendments are shown in Table 2. The green waste used to prepare GWC consisted of fallen leaves and branch cuttings collected during greening maintenance. The waste was reduced to about 1 cm

Table 1. Properties of the saline soil before organic amendments were applied.

Property	Value
pH	7.75 (0.22)
Electrical conductivity ($dS\cdot m^{-1}$)	3.69 (0.17)
Clay ($g\cdot kg^{-1}$)[a]	96 (9)
Silt ($g\cdot kg^{-1}$)[a]	680 (22)
Sand ($g\cdot kg^{-1}$)[a]	224 (15)
Bulk density ($mg\cdot m^{-3}$)	1.54 (0.01)
Total porosity (%)	43.46 (0.18)
Total N ($g\cdot kg^{-1}$)	0.69 (0.03)
Total C ($g\cdot kg^{-1}$)	3.69 (0.13)
C/N ratio	5.35 (2.00)
Available N ($mg\cdot kg^{-1}$)	44.33 (1.17)
Available P ($mg\cdot kg^{-1}$)	5.38 (0.59)
Available K ($mg\cdot kg^{-1}$)	143.33 (12.02)
Na^+ ($g\cdot kg^{-1}$)[b]	2.40 (0.03)
Ca^{2+} ($g\cdot kg^{-1}$)[b]	1.25 (0.04)
Mg^{2+} ($g\cdot kg^{-1}$)[b]	0.38 (0.01)
K^+ ($g\cdot kg^{-1}$)[b]	0.06 (0.01)
Cl^- ($g\cdot kg^{-1}$)[b]	2.81 (0.02)
SO_4^{2-} ($g\cdot kg^{-1}$)[b]	2.38 (0.26)
HCO_3^- ($g\cdot kg^{-1}$)[b]	0.33 (0.01)
Total salt ($g\cdot kg^{-1}$)[c]	8.74 (0.30)
Cation exchange capacity ($cmol\cdot kg^{-1}$)	15.70 (0.05)
Exchangeable Na ($cmol\cdot kg^{-1}$)	2.5 (0.2)
Exchangeable sodium percentage (%)	15.8 (0.2)

Values are means of four samples and standard errors are presented in parentheses.
[a]Determined using the methods described by Bao (2005) [25].
[b]Soluble salt ions.
[c]Total salt = $Na^+ + K^+ + Ca^{2+} + Mg^{2+} + Cl^- + SO_4^{2-} + HCO_3^-$.

particle size with a grinder. Before composting, the moisture content of the raw material was adjusted to 60%, and the C/N ratio was adjusted to between 25 and 30 to optimize microbial activity [23]. In this experiment, the GWC was used when the primary fermentation was completed [24]. In SP treatment, sedge peat was obtained from The Garden Waste Consumption Centre of Chaoyang District (Beijing, China). FR was supplied by the Kain Furfural Corporation (Beijing, China). In GSF treatment, the GWC, SP, and FR were mixed at a ratio of 1:1:1 by volume.

Experiment

The experiment was conducted with five treatments : (1) a non-amended control (CK), (2) addition of 14.2 $kg \cdot m^{-3}$ of GWC, (3) addition of 9.1 $kg \cdot m^{-3}$ of SP, (4) addition of 5.5 $kg \cdot m^{-3}$ of FR, and (5) addition of 8.1 $kg \cdot m^{-3}$ of GSF. The quantities of amendments applied per m^3 were selected so that the same mass of organic matter (4.5 $kg \cdot m^{-3}$) was applied for each amendment. All treatments were arranged in a randomized complete block design with four blocks (78 m×12 m). Each block was again divided into five plots (12 m×12 m) and hence there were 20 (4×5) unit plots. The distance between two adjacent blocks and plots were 6 m and 3 m, respectively. The five treatments were randomly assigned to each plot within the individual blocks with a separate randomization for each block. The amendments were added to nine planting sites in each plot. The nine sites were evenly distributed based on a planting spacing of 3 m×3 m. Each planting site was 1 m×1 m×1 m. On April 25, 2010, the amendments were manually mixed with the soil in each site, and one 2-year-old Pagoda tree (*Sophora japonica* Linn) was planted in each site. Seedlings with similar height, basal diameter, and growth potential were obtained from the Coastal Salt-tolerant Plant Science and Technology Park. Each planting site was irrigated with 5 liters of water every other week from April 26, 2010 to June 25, 2010. All plots and planting sites were managed in the same manner. We measured the height and basal diameter of each Pagoda tree on October 10, 2010, October 12, 2011, and October 14, 2012.

Soil Sampling and Analyses

An auger was used to collect one soil core (0–25 cm depth, 5 cm diameter) from the amended zone of each of three randomly selected trees in each plot on October 11, 2010, October 13, 2011, and October 15, 2012. After drying, the soil samples were ground to pass through a 0.25-mm sieve and stored in sealed polyethylene bags at 4°C until chemical analysis.

Soil bulk density and total porosity were determined using the methods described by Bao (2005) [25]. pH and EC were determined with an MP521 pH/EC meter (Shanghai, China) for a slurry consisting of 1:5 (W/V) soil/distilled water. Soil organic carbon content was determined by the potassium dichromate wet-combustion method. Total N was determined using the micro-Kjeldahl method. Total P was determined with a colorimetric method [25]. Total K was measured with flame photometric detectors. Available N was determined with a micro-diffusion technique after alkaline hydrolysis. Available P was extracted using 0.5 M $NaHCO_3$ and was measured spectrometrically. Available K was measured by flame photometry after NH_4OAc neutral extraction [26].

Soluble Na^+ and K^+ were determined with a flame photometer, and Ca^{2+} and Mg^{2+} were measured by titration with EDTA [27]. The CEC was determined with a 1 M ammonium chloride solution in ethanol/water (60:40, v/v) at pH 8.2 [28]. Extracted Na was determined by flame photometry.

The ESP was determined by the formula,

$$ESP = 100 \times \left(\frac{Na_x}{CEC}\right)$$

Table 2. Properties of the organic amendments applied to the saline soil.

	Organic amendment			
Property†	GWC	SP	FR	GSF
pH	8.88 (0.03)	4.93 (0.08)	2.54 (0.03)	7.20 (0.03)
Electrical conductivity ($dS \cdot m^{-1}$)	6.82 (0.04)	1.54 (0.02)	2.99 (0.03)	3.36 (0.06)
Organic carbon ($g \cdot kg^{-1}$)	184 (5)	286 (10)	477 (15)	312 (8)
Total N ($g \cdot kg^{-1}$)	17.5 (0.04)	10.5 (0.01)	8.8 (0.12)	12.6 (0.22)
C/N ratio	10.5 (2.3)	27.2 (12.0)	54.2 (1.8)	30.0 (0.4)
Total P ($g \cdot kg^{-1}$)	3.20 (0.01)	0.07(0.01)	0.47 (0.03)	1.38 (0.03)
Total K ($g \cdot kg^{-1}$)	6.20 (0.03)	1.33 (0.02)	1.71 (0.02)	3.28 (0.11)
Na^+ ($g \cdot kg^{-1}$)	7.38 (0.07)	0.34 (0.01)	0.29 (0.02)	2.96 (0.07)
K^+ ($g \cdot kg^{-1}$)	11.38 (0.07)	0.03 (0.01)	1.04 (0.01)	4.63 (0.09)
Ca^{2+} ($g \cdot kg^{-1}$)	1.10 (0.07)	1.50 (0.05)	0.65 (0.03)	1.06 (0.09)
Mg^{2+} ($g \cdot kg^{-1}$)	2.26 (0.05)	0.98 (0.04)	7.81 (0.07)	3.76 (1.90)
Cl^- ($g \cdot kg^{-1}$)[a]	20.2 (4.2)	1.1 (0.1)	2.5 (0.6)	8.7 (0.3)
SO_4^{2-} ($g \cdot kg^{-1}$)[a]	13.90 (4.00)	5.90 (0.05)	35.90 (7.10)	19.00 (0.09)

Values are means of six samples and standard errors are presented in parentheses.
GWC, green waste compost; SP, sedge peat; FR, furfural residue; GSF, green waste compost plus sedge peat plus furfural residue (1:1:1 by volume).
†Based on a dry weight except for the moisture content; [a] Determined by titration method [25].

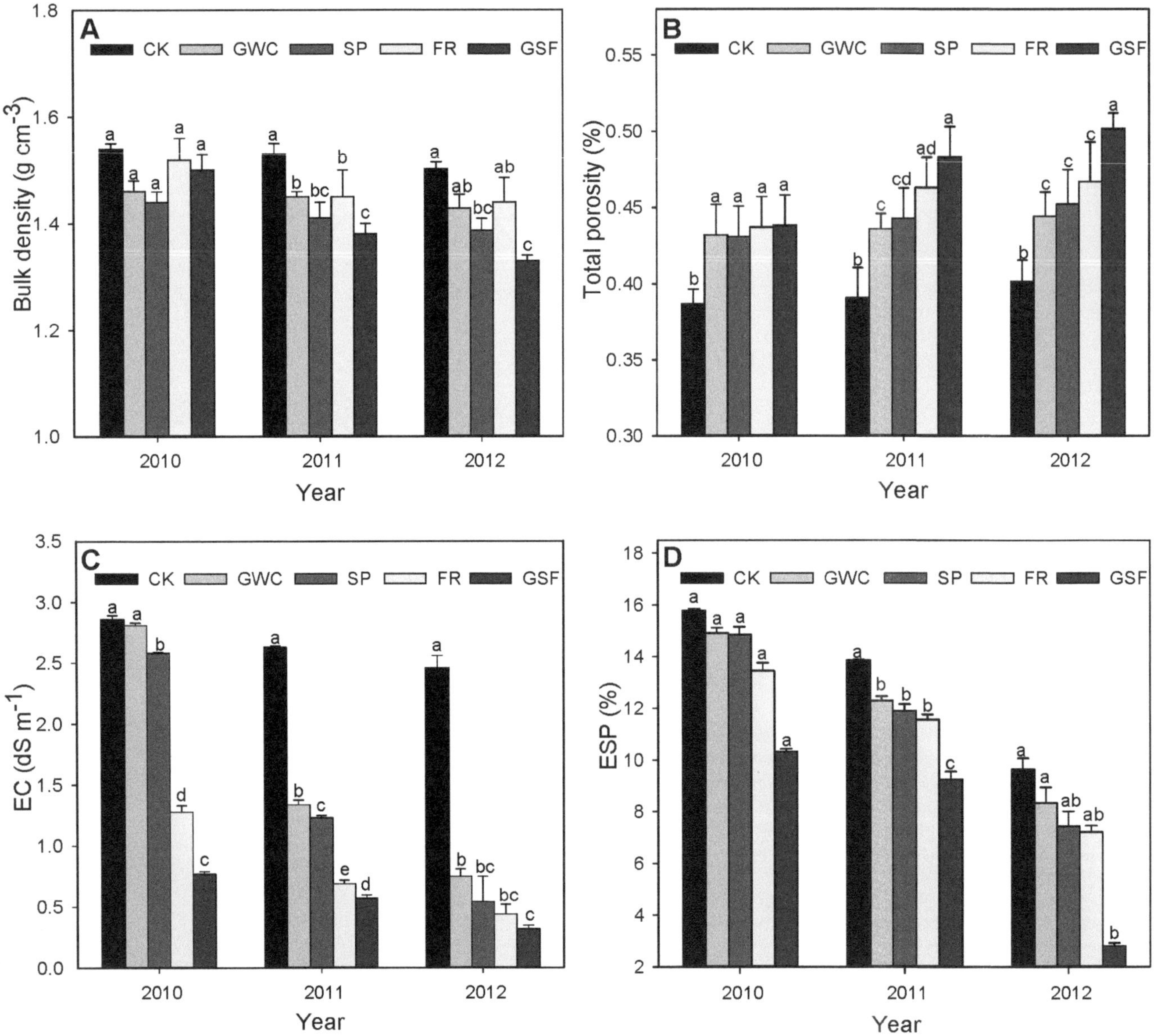

Figure 1. Effects of organic amendments on (A) bulk density, (B) total porosity, (C) electrical conductivity (EC), and (D) exchangeable sodium percentage (ESP) in saline soil. CK, non-amended control; GWC, green waste compost; SP, sedge peat; FR, furfural residue; GSF, green waste compost plus sedge peat plus furfural residue. Rates are indicated in the text but all amendments provided 4.5 kg of organic matter m^{-3}. Treatments were applied in April 2010, and samples were collected in October 2010, October 2011, and October 2012. Values are means+SEMs. Means in the same year followed by the same letter are not significantly different at $p \leq 0.05$ according to LSD.

where Na_x is the exchangeable sodium ($cmol \cdot kg^{-1}$) and CEC is the cation exchange capacity ($cmol \cdot kg^{-1}$). Exchangeable sodium (Na_x) was determined with 1 M ammonium acetate at pH 7 [29].

Data Analysis

To test the differences among the treatments, randomized block split-plot analysis of variance (ANOVA) was used with soil amendment treatments as the fixed factor and experimental blocks as the random factor. The probability level used to determine significance was $p<0.05$. If treatment effects were significant, means were compared with an LSD test at $p \leq 0.05$. All statistical analyses were performed using SPSS software package (version 20.0) (IBM Corporation, Armonk, New York).

Results

Bulk Density, Total Porosity, EC, and ESP as Affected by Organic Amendments

In October 2010, about 5.5 months after amendments were applied, there was no significant difference in soil bulk density among the treatments (Fig. 1A). In October 2011, bulk density was significantly lower in all amended plots than in the control (CK). In October 2012, bulk density was significantly lower in SP and GSF plots than in the control. In general, bulk densities were lowest in plots treated with GSF. All four amendments significantly increased the total soil porosity relative to the control (Fig. 1B). Total soil porosity was highest in GSF plots.

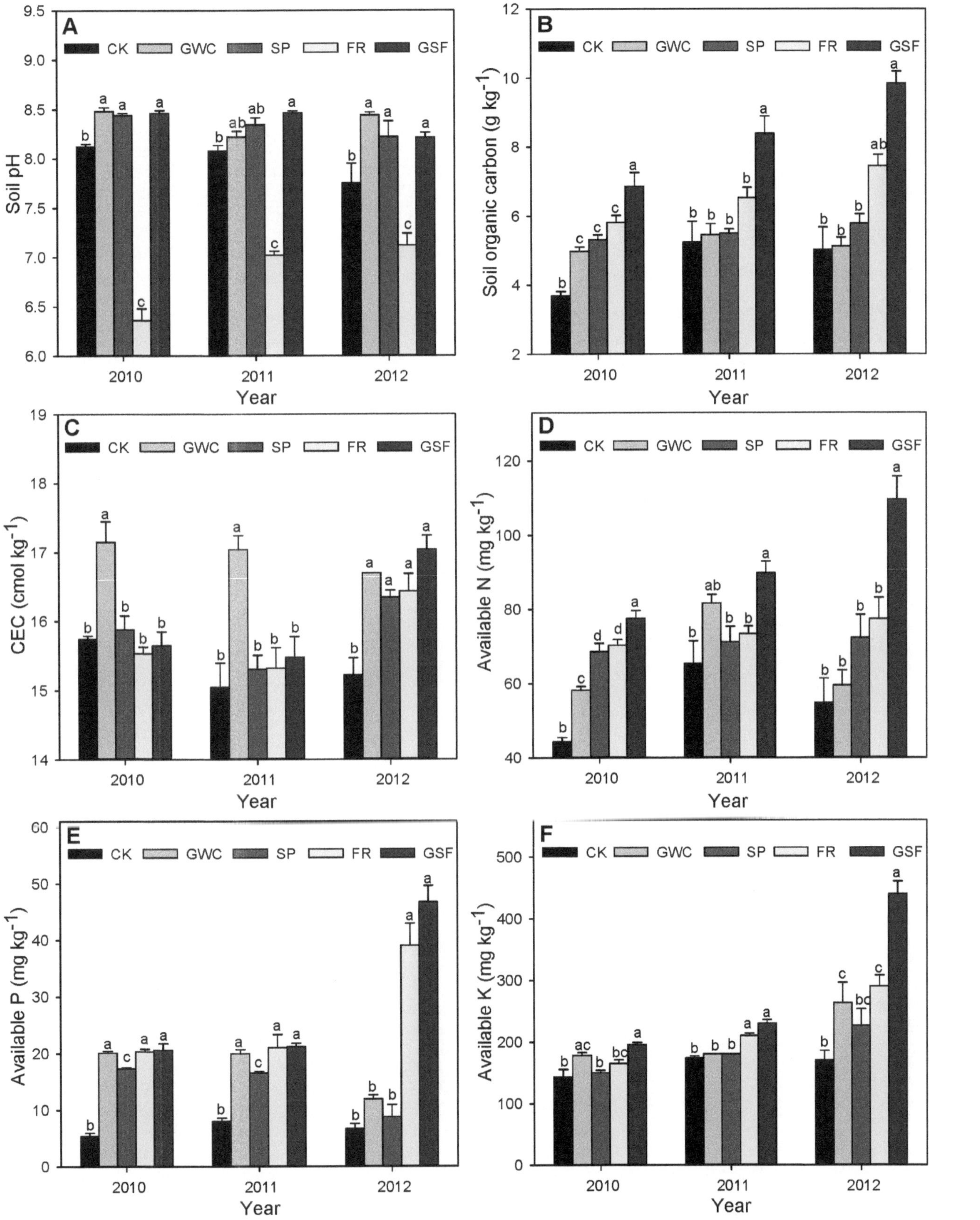

Figure 2. Effects of organic amendments on (A) soil pH, (B) soil organic carbon, (C) cation exchange capacity (CEC), (D) available N, (E) available P, and (F) available K in saline soil. CK, non-amended control; GWC, green waste compost; SP, sedge peat; FR, furfural residue; GSF, green waste compost plus sedge peat plus furfural residue. Rates are indicated in the text but all amendments provided 4.5 kg of organic matter

m^{-3}. Treatments were applied in April 2010, and samples were collected in October 2010, October 2011, and October 2012. Values are means+SEMs. Means in the same year followed by the same letter are not significantly different at $p \leq 0.05$ according to LSD.

By the second year of the study, all four amendments had significantly reduced soil EC values relative to the control (Fig. 1C). EC values dropped more slowly in GWC and SP plots than in FR and GSF plots, but by 2012, the EC values had dropped to a lower level in all plots treated with amendments.

The initial ESP value in the CK plots was 15.8%, which exceeded the critical sodicity value (≈15%) mentioned by Richards (1954) [29]. ESP tended to drop in all plots including the CK plots over time (Fig. 1D). In 2011, ESP values were lower in all amended plots than in the CK plots and were lowest in the GSF plots. In 2012, ESP values were lowest in the GSF plots and did not significantly differ among the other plots.

Soil pH, Organic Carbon, and CEC as Affected by Organic Amendments

In 2010 and 2012, FR decreased soil pH while the other three organic amendments increased pH relative to the control (Fig. 2A). The trends were similar in 2011 but were not always statistically significant. The effect of FR on pH tended to lessen from 2011 to 2012 and especially from 2010 to 2011. Soil pH did not significantly differ among GWC, SP, and GSF plots throughout the experiment.

In 2010, 5.5 months after the amendments were applied, soil organic carbon contents were significantly higher in all amended plots than in the CK plots and were highest in GSF plots (Fig. 2B). In 2011 and 2012, soil organic carbon was higher in GSF plots than in the other plots and did not significantly differ among the other plots, although soil organic carbon tended to also be high in FR plots.

In 2010 and 2011, soil CEC was higher in GWC plots than in the other plots and did not significantly differ among the other plots (Fig. 2C). In 2012, soil CEC was higher in all amended plots than in CK plots and did not significantly differ among the amended plots.

Available Soil Nutrients as Affected by Organic Amendments

The content of available N was significantly higher in the GSF plots than in the other plots, except in 2011, when the level of available N was higher in GSF plots than in all other plots except GWC plots (Fig. 2D). The content of available P was increased by all amendments in 2010 and 2011 (Fig. 2E); in 2012, available P increased to very high levels in FR and GSF plots but did not differ from that in CK plots for the other two amendments. In 2010, the content of available K was greater in GWC and GSF plots than in CK plots but did not differ among SP, FR, and CK plots (Fig. 2F). In 2011, available K was highest in GSF and FR plots and did not differ among the other plots. In 2012, available K increased much more in GSF plots than in the other plots. The level of available K was also significantly higher in GWC and FR plots than in CK plots.

Soluble Cations in the Soil as Affected by Organic Amendments

Table 3 shows the concentration of soluble cations in the soil on October 11, 2010 (about 5.5 months after amendments were applied) and at the end of the experiment on October 15, 2012 (about 29.5 months after treatments were applied). For all treatments including CK, the concentration of Na^{+}+K^{+} in October 2012 was lower than that in October 2010 ($p<0.05$). The Na^{+}+K^{+} concentration in October 2012 was highest in CK plots; intermediate in GWC, SP, and FR plots; and lowest in GSF plots. In addition, the proportion of soluble cations represented by Na^{+}+K^{+} (Na^{+}+K^{+}/total) tended to decline in plots treated with organic amendments but tended to increase in CK plots. The Na^{+}+K^{+}/total value in October 2012 was lowest in the GSF plots. Like Na^{+}+K^{+} concentrations, Ca^{2+} and Mg^{2+} concentrations tended to be lower in October 2012 than in October 2010; the drop in Ca^{2+} concentration was greatest in CK plots. The Ca^{2+} concentration in October 2012 was lowest in FR plots but did not statistically differ between the CK plots and those treated with organic amendments. Mg^{2+} concentrations did not significantly differ among treatments in October 2010 or in October 2012. Unlike the proportion of soluble cations represented by Na^{+}+K^{+} (Na^{+}+K^{+}/total), the proportion of soluble cations represented by Ca^{2+} (Ca^{2+}/total) tended to decrease in the CK plots and increase in the plots treated with organic amendments. In October 2012, the Ca^{2+}/total value was lowest in CK plots; intermediate in SP, FR, and GSF plots; and highest in GWC plots (Table 3).

Tree Growth as Affected by Organic Amendments

In our study, all amendments tended to enhance tree growth. In 2010, about 5.5 months after treatments were applied, tree height was greatest in GSF and FR plots but did not significantly differ among the other plots (Fig. 3A). The trends were similar in 2011 and 2012. Tree height did not statistically differ between GSF and FR treatments except in 2010. Basal diameter was higher in the GSF and FR plots than in CK plots, and did not differ between GWC, SP, and CK plots throughout the experiment (Fig. 3B). Like tree height, basal diameter did not statistically differ between GSF and FR treatments.

Discussion

The GWC, SP, FR, and GSF amendments in the current study decreased soil bulk density and EC value and increased soil total porosity. This agrees with Oo et al. (2013) [14] and Tejada et al. (2006) [13], who found that organic amendments improve soil structure because they promote the flocculation of clay minerals, which is essential for the aggregation of soil particles. The GSF-treated soil had the lowest soil bulk density, the lowest EC, and the highest soil total porosity. These results have been explained by Oo et al. (2013) [14], who reported that combinations of organic amendments resulted in substantial flocculation and in the formation of a large number of soil aggregates. Improvements in soil structure can explain decreases in bulk density and increases in soil porosity [22,30].

Soil EC values were negatively correlated with total porosity in the current study and in a previous study [14]. In addition, Kahlown et al. (2003) [6] and Barzegar et al. (1997) [15] found that the addition of organic matter to soil could increases leaching. This is probably related to the increased soil porosity resulting from the organic amendments [31]. Moreover, the four organic amendments used in our research, especially GSF and FR, contain calcium and magnesium (Table 2), which act as bases because they exist as oxides, hydroxides, and carbonates when applied to soil. As reported by Qadir et al. (2004) [32], an increase in the Ca^{2+} concentration in the soil solution results in the replacement of Na^{+} by Ca^{2+} at the cation exchange sites on the soil particles.

Table 3. Soluble cation contents of the soil in 2010 (about 5.5 months after amendments were applied) and in 2012 (about 29.5 months after amendments were applied).

	Na^++K^+ ($g\cdot kg^{-1}$)		Ca^{2+} ($g\cdot kg^{-1}$)		Mg^{2+} ($g\cdot kg^{-1}$)		Na^++K^+/Total[a] (%)		Ca^{2+}/Total[a] (%)	
Treatment	**2010**	**2012**	**2010**	**2012**	**2010**	**2012**	**2010**	**2012**	**2010**	**2012**
CK	2.46(0.03)a	2.17(0.11)a	1.25(0.13)a	0.28(0.02)ab	0.38(0.05)a	0.08(0.02)a	0.60(0.01)e	0.86(0.03)a	0.31(0.02)a	0.11(0.01)c
GWC	1.66(0.03)b	0.38(0.04)b	1.12(0.04)a	0.45(0.01)a	0.36(0.03)a	0.09(0.01)a	0.53(0.01)d	0.41(0.02)b	0.36(0.03)a	0.51(0.01)a
SP	1.61(0.02)b	0.36(0.03)b	0.98(0.02)ab	0.41(0.05)ab	0.29(0.02)a	0.06(0.01)a	0.56(0.02)c	0.43(0.02)b	0.34(0.02)a	0.49(0.03)a
FR	1.48(0.03)b	0.37(0.02)b	0.43(0.01)b	0.17(0.03)b	0.30(0.01)a	0.06(0.02)a	0.67(0.02)a	0.62(0.02)c	0.19(0.01)b	0.28(0.01)b
GSF	0.89(0.02)c	0.17(0.01)c	0.82(0.03)ab	0.20(0.04)ab	0.32(0.02)a	0.05(0.01)a	0.44(0.02)b	0.40(0.03)b	0.40(0.01)a	0.47(0.02)a

CK, non-amended control; GWC, green waste compost; SP, sedge peat; FR, furfural residue; GSF, green waste compost plus sedge peat plus furfural residue (1:1:1 by volume).
Values are means (SE). Means in a column followed by the same letter are not significantly different at $p \leq 0.05$ according to LSD.
[a]Total = Na^++ K^++ Ca^{2+}+ Mg^{2+}.

Meanwhile, the greater the total porosity, the greater the leaching of the exchanged Na^+ and the greater the subsequent reduction in soil salinity (as indicated by the EC value).

According to Mamedov et al. (2002) [33], ESP reflects soil aggregation. In the current study, the ESP value in the non-amended soil in 2010 was 15.8%, which indicates that swelling was the main mechanism determining its physical properties. In the amended soils, in contrast, ESP values were always <15%, suggesting that dispersion was probably the main mechanism determining their soil properties. Because ESP is inversely related to organic carbon content [13], it was not surprising that the plots with the lowest ESP values (the GSF plots) had the highest organic carbon contents.

Addition of GWC, SP, and GSF increased the pH of the saline soil, probably because these amendments had a high content of basic cations. Basic cations act in a similar manner as mineral lime, and they increase soil pH [34]. The FR plots had the lowest pH values probably because the FR was quite acidic (pH 2.54). The pH in the FR plots tended to increase over time, perhaps because the degradation of an organic amendment like FR that is rich in basic cations releases inorganic bases into the soil solution and thereby increases the soil pH [34].

Although the same mass of organic matter was added to the soils, the resulting increase in soil organic carbon differed depending on the amendment. This is consistent with Tejada et al. (2006) [30], who reported that the effect of organic amendments on soil organic carbon depended on the chemical nature of the amendments. On the one hand, the chemical nature of the organic amendment is likely to affect the rate at which it is decomposed by the microbial community [35]. On the other hand, the effects of the organic amendments on soil properties will affect plant growth and thus the input of plant residues into the soil [22,36]. The organic carbon content of soil amended with GSF and FR increased gradually over time, and we assume that this

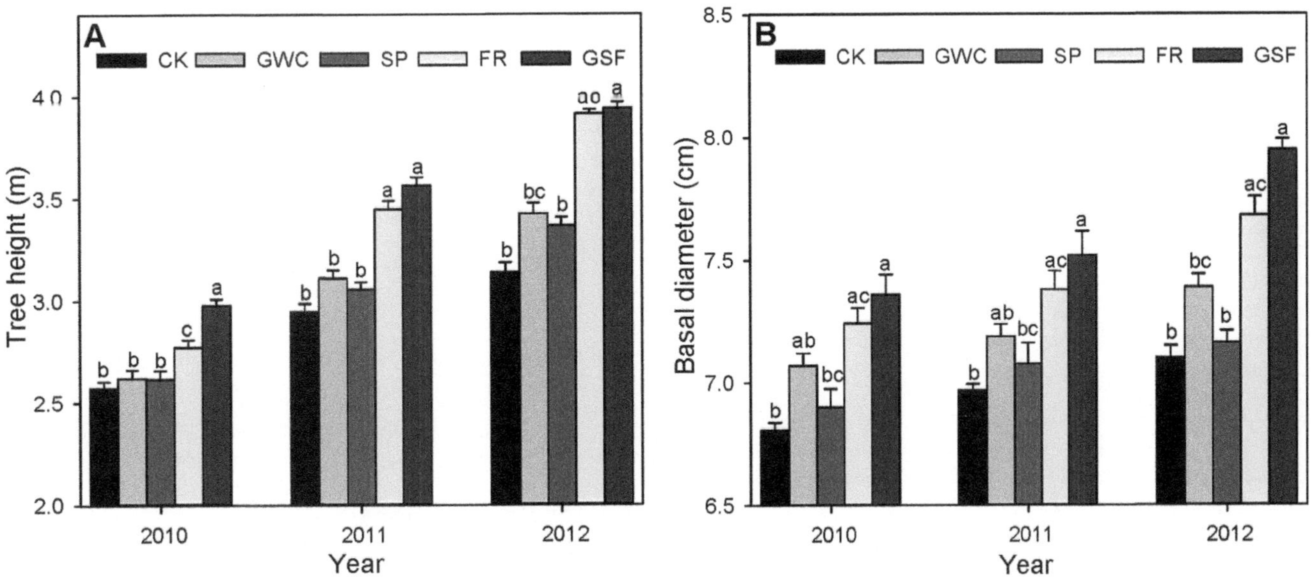

Figure 3. Effects of organic amendments on Pagoda tree (*Sophora japonica* Linn) (A) tree height, and (B) basal diameter in saline soil. CK, non-amended control; GWC, green waste compost; SP, sedge peat; FR, furfural residue; GSF, green waste compost plus sedge peat plus furfural residue. Rates are indicated in the text but all amendments provided 4.5 kg of organic matter m^{-3}. Treatments were applied in April 2010, and data was determined in October 2010, October 2011, and October 2012. Values are means+SEMs. Means in the same year followed by the same letter are not significantly different at $p \leq 0.05$ according to LSD.

increase reflected increased input from the trees (and perhaps reduced decomposition of that input).

CEC provides buffering against changes in pH, available nutrients, calcium levels, and soil structure [37]. The organic amendments used in our research led to an increase in soil CEC, especially in the GWC plots. This might have resulted from a high degree of organic matter oxidation in the GWC-amended soil, which could increase CEC values [38]. We also noted that, from 2010 to 2012, the CEC tended to increase in the SP, FR, and GSF plots. This may primarily be due to the long-lasting decomposition process of organic amendments [22].

Our findings showed that the organic amendments greatly increased the available N content. This agrees with previous reports that organic amendments increase available N [18,39]. However, the effects of organic amendments on nutrient availability depend on their chemical composition and decomposition rates [7]. In the current study, GSF resulted in the highest level of available N, because the mixed organic amendments contain higher percentage of macronutrients and considerable amounts of micronutrients [40]. The C/N ratio was lower in the GWC amendment than in the other organic amendments (Table 2), and a low C/N ratio (less than about 20:1) would result in intense microbial activity, and an increase in available N levels [41]. The available N content was relatively constant in the SP plots, this was probably because the highly-decomposed sedge peat decomposed slowly when added to soil [18]. The FR plots had the second highest level of available N at the end of the experiment, perhaps because the FR amendment had the highest organic matter content, and the organic matter in furfural is mineralized slowly over time [42].

The organic amendments significantly affected the levels of available P in the current study. Like available N content, available P content was highest in the GSF plots, perhaps because the mixed organic amendment has the ability to solubilize and mobilize P [43]. Moreover, P can be adsorbed by soil colloids, and the mixed organic amendment is able to bind large quantities of macronutrients and thereby to reduce their removal from soil by leaching [44]. P availability was also high in the FR plots, probably because the furfural residue was extremely acidic and thereby helped solubilize phosphate. In contrast, available P decreased in the GWC and SP plots probably because these amendments increased the pH.

The organic amendments also increased available K in the soil, which agrees with some researchers [45–47], who found relatively high levels of available K in soils treated with organic amendments. As was the case with available N and available P, available K was highest in the GSF plots and next highest in the FR plots. The difference in available K content among all amendments can be attributed to their chemical compositions. Thus, some K is released as a result of organic matter solubility and decomposition. Available K slightly increased in GWC and SP plots, perhaps because these amendments slowly release nutrients as they slowly decompose [18].

The organic amendments differed in their effects on the concentration of soluble cations. The application of organic amendments would probably increase the amount of Ca^{2+} derived from $CaCO_3$ because of the formation of organic acids [48]. Furthermore, potassium and sodium ions on the soil colloids would be replaced by calcium ions and leached from the soil. The results revealed that, by the end of the experiment, the dominant ions changed from Na^+ and K^+ to Ca^{2+} in GWC, SP, and GSF plots. Mg^{2+} content also tended to decrease in all plots over time. Previous studies have reported that Ca^{2+} could improve soil structure by formed cationic bridges between clay particles and soil organic matter [49]. In addition, Ca^{2+} can inhibit clay dispersion and the associated disruption of aggregates by replacing Na^+ and Mg^{2+} in clay and aggregates, thereby promoting aggregate stability [50]. These results indicate that GWC, SP, and GSF organic amendments could help reduce the salt content and otherwise improve saline soil. FR resulted in the lowest concentration of Ca^{2+} in 2010 and 2012, probably because the high SO_4^{2-} concentration in the furfural residue promoted the precipitation of calcium sulphate [51].

The increased tree height and basal diameter observed in the GSF and FR treatments might be due to better physiological growth of plants. GSF and FR treatments increased the availability of macronutrients as well as micronutrients and improved soil physical properties of saline soil. Increased net primary productivity would have returned higher amounts of plant residues to the soil, perhaps acting as a carbon and nutrient source to maintain a higher microbial biomass and activity [36]. This, in conjunction with the positive effects of organic amendments on soil biota (e.g. increased biodiversity and microbial activity), would lead to continued nutrients release from organic amendments and soil amelioration [22], and help meet the amelioration goal of establishment of self-sustaining ecosystem on saline soil.

Conclusions

The results of this study show that application of GWC, SP, FR, or GSF can improve the physical and chemical properties of saline soil and promote tree growth in the coastal areas of northern China. Application of GSF and FR substantially increased the organic carbon content of the soil and the availability of N, P, and K. The FR amendment effectively reduced the soil pH value. The GWC amendment increased the soil CEC and provided a long-term supply of nutrients. SP had a smaller effect on soil fertility than the other three amendments. Overall, the combination of GWC, SP and FR (the GSF treatment) provides better remediation results than each applied singly and has substantial potential for ameliorating coastal saline soils. The authors believe that SP could be fully or partially replaced by GWC and FR or their combinations in ameliorating coastal saline soil and that additional research should be conducted to optimize the mix proportion of GWC, SP, and FR. Considering the strong correlation between soil biological activities and soil quality [9,16], further studies are also needed to assess the effects of organic amendments on soil microbial community and enzymatic activities.

Acknowledgments

We thank Long Cheng and Donghua Zhang in the Coastal Salt-tolerant Plant Science and Technology Park of Dagang Agriculture and Forestry Administration for many sampling and analysis work.

Author Contributions

Conceived and designed the experiments: LW XS SL. Performed the experiments: LW. Analyzed the data: LW. Contributed reagents/materials/analysis tools: XS SL TZ WZ PZ. Wrote the paper: LW.

References

1. Sumner M (2000) Handbook of soil science. Boca Raton: CRC Press. 2148 p.
2. Wang Z, Zhu S, Yu R (1993) The saline soil in China. Beijing: Science Press. 573p.

3. Qadir M, Oster JD, Schubert S, Noble AD, Sahrawat KL (2007) Phytoremediation of sodic and saline-sodic soils. Advances in Agronomy 96: 197–247.
4. Kaya C, Kirnak H, Higgs D (2001) Enhancement of growth and normal growth parameters by foliar application of potassium and phosphorus in tomato cultivars grown at high (NaCl) salinity. Journal of Plant Nutrition 24: 357–367.
5. Liang Y, Yang Y, Yang C, Shen Q, Zhou J, et al. (2003) Soil enzymatic activity and growth of rice and barley as influenced by organic manure in an anthropogenic soil. Geoderma 115: 149–160.
6. Kahlown MA, Azam M (2003) Effect of saline drainage effluent on soil health and crop yield. Agricultural Water Management 62: 127–138.
7. Clark GJ, Dodgshun N, Sale PWG, Tang C (2007) Changes in chemical and biological properties of a sodic clay subsoil with addition of organic amendments. Soil Biology and Biochemistry 39: 2806–2817.
8. Tejada M, Gonzalez J (2005) Beet vinasse applied to wheat under dryland conditions affects soil properties and yield. European Journal of Agronomy 23: 336–347.
9. Rietz DN, Haynes RJ (2003) Effects of irrigation-induced salinity and sodicity on soil microbial activity. Soil Biology and Biochemistry 35: 845–854.
10. Qadir M, Ghafoor A, Murtaza G (2000) Amelioration strategies for saline soils: a review. Land Degradation and Development 11: 501–521.
11. Hanay A, Büyüksönmez F, Kiziloglu FM, Canbolat MY (2004) Reclamation of saline-sodic soils with gypsum and MSW compost. Compost Science and Utilization 12: 175–179.
12. Melero S, Madejón E, Ruiz JC, Herencia JF (2007) Chemical and biochemical properties of a clay soil under dryland agriculture system as affected by organic fertilization. European Journal of Agronomy 26: 327–334.
13. Tejada M, Garcia C, Gonzalez JL, Hernandez MT (2006) Use of organic amendment as a strategy for saline soil remediation: influence on the physical, chemical and biological properties of soil. Soil Biology and Biochemistry 38: 1413–1421.
14. Oo AN, Iwai CB, Saenjan P (2013) Soil properties and maize growth in saline and nonsaline soils using cassava-industrial waste compost and vermicompost with or without earthworms. Land Degradation and Development DOI: 101002/ldr2208.
15. Barzegar AR, Nelson PN, Oades JM, Rengasamy P (1997) Organic matter, sodicity, and clay type: influence on soil aggregation. Soil Science Society of America Journal 61: 1131–1137.
16. Wichern J, Wichern F, Joergensen RG (2006) Impact of salinity on soil microbial communities and the decomposition of maize in acidic soils. Geoderma 137: 100–108.
17. Liu G, Zhang R, El-Mashad HM, Dong R (2009) Effect of feed to inoculum ratios on biogas yields of food and green wastes. Bioresource Technology 100: 5103–5108.
18. Cooperband L (2002) Building soil organic matter with organic amendments. Center for Integrated Agricultural Systems. University of Wisconsin–Madison. USA.
19. Barkham JP (1993) For peat's sake: conservation or exploitation? Biodiversity and Conservation 2: 556–566.
20. Sun R, Song X, Sun R, Jiang J (2011) Effect of lignin content on enzymatic hydrolysis of furfural residues. Bioresources 6: 317–328.
21. Xing Y, Bu LX, Wang K, Jiang JX (2012) Pretreatment of furfural residues with alkaline peroxide to improve cellulose hydrolysis. Characterization of isolated lignin. Cellulose Chemistry and Technology 46: 249–260.
22. Larney FJ, Angers DA (2012) The role of organic amendments in soil reclamation: A review. Canadian Journal of Soil Science 92: 19–38.
23. Roca-Pérez L, Martinez C, Marcilla P, Boluda R (2009) Composting rice straw with sewage sludge and compost effects on the soil–plant system. Chemosphere 75: 781–787.
24. Zhang L, Sun X, Tian Y, Gong X (2013) Effects of brown sugar and calcium superphosphate on the secondary fermentation of green waste. Bioresource Technology 131: 68–75.
25. Bao SD (2005) Analysis of soil agrochemistry (3rd edition). Beijing, China: China Agriculture Press. 495 p.
26. Page AL, Miller RH, Keeney DR (1982) Methods of soil analysis. Part 2-Chemical and Microbiological Properties (2nd edition). Madison, Wisconsin: American Society of Agronomy and Soil Science. 1159 p.
27. Jackson ML (1962) Soil chemistry of analysis. Englewood Cliffe, New Jersey: Prentice Hall. 78 p.
28. Tucker BM (1954) The determination of exchangeable calcium and magnesium in carbonate soils. Australian Journal of Agricultural Research 5: 706–715.
29. Richards L (1954) Diagnosis and improvement of saline and alkali soils. In: Richards LA, editor. USA Department of Agriculture Handbook 60. Washington, DC: US Government Printing Office. 98–105.
30. Tejada M, Garcia C, Gonzalez JL, Hernandez MT (2006) Organic amendment based on fresh and composted beet vinasse: influence on soil properties and wheat yield. Soil Science Society of America Journal 70: 900–908.
31. Tejada M, Gonzalez JL, García-Martínez AM, Parrado J (2008) Effects of different green manures on soil biological properties and maize yield. Bioresource Technology 99: 1758–1767.
32. Qadir M, Oster JD (2004) Crop and irrigation management strategies for saline-sodic soils and waters aimed at environmentally sustainable agriculture. Science of the Total Environment 323: 1–19.
33. Mamedov AI, Shainberg I, Levy GJ (2002) Wetting rate and sodicity effects on interrill erosion from semi-arid Israeli soils. Soil and Tillage Research 68: 121–132.
34. Pocknee S, Sumner ME (1997) Cation and nitrogen contents of organic matter determine its soil liming potential. Soil Science Society of America Journal 61: 86–92.
35. Hahn AS, Quideau SA (2013) Long-term effects of organic amendments on the recovery of plant and soil microbial communities following disturbance in the Canadian boreal forest. Plant and Soil 363: 331–344.
36. Hyvönen R, Persson T, Andersson S, Olsson B, Ågren GI, et al. (2008) Impact of long-term nitrogen addition on carbon stocks in trees and soils in northern Europe. Biogeochemistry 89: 121–137.
37. Brady AC, Veil RR (2002) The nature and properties of soils, 13th edition. New Jersey: Prentice Hall. 960 p.
38. Liang B, Lehmann J, Solomon D, Kinyangi J, Grossman J, et al. (2006) Black carbon increases cation exchange capacity in soils. Soil Science Society of America Journal 70: 1719–1730.
39. Abbasi MK, Hina M, Khalique A, Khan SR (2007) Mineralization of three organic manures used as nitrogen source in a soil incubated under laboratory conditions. Communications in Soil Science and Plant Analysis 38: 1691–1711.
40. Ros M, Hernandez MT, García C (2003) Soil microbial activity after restoration of a semiarid soil by organic amendments. Soil Biology and Biochemistry 35: 463–469.
41. Yazdanpanah N, Pazira E, Neshat A, Mahmoodabadi M, Sinobas LR (2013) Reclamation of calcareous saline sodic soil with different amendments (II): Impact on nitrogen, phosphorous and potassium redistribution and on microbial respiration. Agricultural Water Management 120: 39–45.
42. Feng YJ, Li F, Wang XL, Liu XM, Zhang LN (2006) Principal chemical properties of artificial soil composed of fly ash and furfural residue. Pedosphere 16: 668–672.
43. Qadir M, Qureshi RH, Ahmad N (1998) Horizontal flushing: a promising ameliorative technology for hard saline-sodic and sodic soils. Soil and Tillage Research 45: 119–131.
44. Withers PJA, Bailey GA (2003) Sediment and phosphorus transfer in overland flow from a maize field receiving manure. Soil Use and Management 19: 28–35.
45. Warman PR, Termeer WC (2005) Evaluation of sewage sludge, septic waste and sludge compost applications to corn and forage: yields and N, P and K content of crops and soils. Bioresource Technology 96: 955–961.
46. Bhattacharyya P, Chakrabarti K, Chakraborty A, Nayak DC, Tripathy S, et al. (2007) Municipal waste compost as an alternative to cattle manure for supplying potassium to lowland rice. Chemosphere 66: 1789–1793.
47. Antil RS, Singh M (2007) Effects of organic manures and fertilizers on organic matter and nutrients status of the soil. Archives of Agronomy and Soil Science 53: 519–528.
48. Wong VNL, Dalal RC, Greene RSB (2009) Carbon dynamics of sodic and saline soils following gypsum and organic material additions: a laboratory incubation. Applied Soil Ecology 41: 29–40.
49. David R, Dimitrios P (2002) Diffusion and cation exchange during the reclamation of saline-structured soils. Geoderma 107: 271–279.
50. Zhang XC, Norton LD (2002) Effect of exchangeable Mg on saturated hydraulic conductivity, disaggregation and clay dispersion of disturbed soils. Journal of Hydrology 260: 194–205.
51. Amezketa E, Aragüés R, Gazol R (2005) Efficiency of sulfuric acid, mined gypsum, and two gypsum by-products in soil crusting prevention and sodic soil reclamation. Agronomy Journal 97: 983–989.

14

Spreading Topsoil Encourages Ecological Restoration on Embankments: Soil Fertility, Microbial Activity and Vegetation Cover

Desirée Rivera[1], Violeta Mejías[1], Berta M. Jáuregui[2], Marga Costa-Tenorio[3†], Ana Isabel López-Archilla[1], Begoña Peco[1*]

1 Departamento de Ecología, Universidad Autónoma de Madrid, Madrid, Spain, **2** Dirección Técnica, OHL Construcción, Madrid, Spain, **3** Departamento de Biología Vegetal I, Universidad Complutense, Madrid, Spain

Abstract

The construction of linear transport infrastructure has severe effects on ecosystem functions and properties, and the restoration of the associated roadslopes contributes to reduce its impact. This restoration is usually approached from the perspective of plant cover regeneration, ignoring plant-soil interactions and the consequences for plant growth. The addition of a 30 cm layer of topsoil is a common practice in roadslope restoration projects to increase vegetation recovery. However topsoil is a scarce resource. This study assesses the effects of topsoil spreading and its depth (10 to 30 cm) on two surrogates of microbial activity (β-glucosidase and phosphatase enzymes activity and soil respiration), and on plant cover, plant species richness and floristic composition of embankment vegetation. The study also evaluates the differences in selected physic-chemical properties related to soil fertility between topsoil and the original embankment substrate. Topsoil was found to have higher values of organic matter (11%), nitrogen (44%), assimilable phosphorous (50%) and silt content (54%) than the original embankment substrate. The topsoil spreading treatment increased microbial activity, and its application increased β-glucosidase activity (45%), phosphatase activity (57%) and soil respiration (60%). Depth seemed to affect soil respiration, β-glucosidase and phosphatase activity. Topsoil application also enhanced the species richness of restored embankments in relation to controls. Nevertheless, the depth of the spread topsoil did not significantly affect the resulting plant cover, species richness or floristic composition, suggesting that both depths could have similar effects on short-term recovery of the vegetation cover. A significant implication of these results is that it permits the application of thinner topsoil layers, with major savings in this scarce resource during the subsequent slope restoration work, but the quality of topsoil relative to the original substrate should be previously assessed on a site by site basis.

Editor: Andrew C. Singer, NERC Centre for Ecology & Hydrology, United Kingdom

Funding: Grants were received from the Ministry of Science and Innovation's Centre for Industrial and Technology Development (CDTI). The project is jointly funded by Madrid Regional Government grants for research groups, REMEDINAL 2 from the Spanish Ministry of Science and Innovation, and assistance from the Torres Quevedo Sub-program for the employment of Dr. BMJ by OHL, under the auspices of the Ministry of Science and Innovation. Thanks also go to the Science and Innovation Ministry for the Salvador de Madariaga Mobility grant awarded to BP. The funders had no role in the study design, data collection and analysis, decision to publish, or preparation of the manuscript.

Competing Interests: The authors have declared that no competing interests exist.

* Email: begonna.peco@uam.es

† Deceased.

Introduction

The construction of road, railway and other transport-related civil engineering infrastructure has greatly increased in recent decades. It has severe impacts on ecosystems, including habitat fragmentation, soil compaction, heavy erosion and sediment transport [1–3]. Topographic changes such as the creation of totally new artificial slopes are also associated with this engineering infrastructure. The angle, compaction and erosion potential of the new slopes hinder the rapid establishment of vegetation, which makes restoration enhancement necessary [4], especially in arid and semiarid conditions and in soils with low organic matter content [5,6]. Road slope colonization by plants is constrained by several factors such as seed dispersal capacity, slope type (roadcuts and embankments), aspect and soil physic-chemical properties [7,8]. Techniques such as hydroseeding, topsoil spreading, geotextile installation and planting [9–12] are used to encourage plant cover on roadslopes. Soil properties and vegetation cover can be improved by spreading topsoil [13] due to its content in seeds, plant propagules, nutrients and microorganisms which restore soil functionality [9,14]. The top 30 cm of soil is removed routinely at the start of road construction works. This substrate is stockpiled for the restoration project, and at the end of the works, a 30 cm layer of stockpiled topsoil is spread over the embankments. Few studies have analysed the effects of this restoration technique on soil microbial activity [15] and other soil fertility-related properties [13], despite the fact that topsoil is a scarce resource at many linear infrastructure construction sites, and issues such as the optimum topsoil depth to be spread during roadslope restoration have yet to be resolved.

Plant-soil interactions have a major influence on the structure and function of terrestrial ecosystems, increasing or decreasing the speed of plant succession and/or its trajectories [16–18]. To date, restoration management has mainly focused on manipulating succession in plant communities [18], although ecosystem restoration enhancement also requires consideration of the metabolic capacity of the soil (henceforth, soil functionality).

Several studies of primary succession in soils affected by heavy stress [18,19] have shown the importance of microorganisms in the restoration of soils and plant communities. Soil microbes are largely responsible for the cycling of elements within the soil [20] and are important drivers of plant diversity and productivity in terrestrial systems [21]. In disturbed areas, vegetation establishment depends on the effectiveness of the microorganisms to increase the availability of nutrients in the soil through the decomposition of organic matter, and also their ability to establish symbiotic relationships [22,23].

Soil functionality can be addressed through surrogate variables such as enzyme activity, related to the nutrient cycle, soil respiration and other physic-chemical properties related to soil fertility [24,25]. The most widely tested soil enzymes are those involved in the degradation of cellulose, the major components of plant litter, such as β-glucosidase [26,27], which makes it a good indicator of organic matter [28] and thus of the C, N, P and S available to plants [29]. β-glucosidase is also an effective predictor of organic C accumulation in soil in Mediterranean environments [30]. Extracellular phosphatases are also of interest due to their role in mineralizing P from nucleic acids, phospholipids and other ester phosphates [31]. Moreover, soil enzyme activity, particularly that of β-glucosidase and phosphatase, is a good indicator of the state of terrestrial ecosystems due to their sensitivity to changes in soil management [27,32]. On the other hand, soil respiration is the sum of root respiration and the decomposition of soil organic matter, plant litter, and root exudates by soil microorganisms. Soil surface CO_2 flux is an indirect measurement of the soil respiration rate, under the assumptions that the CO_2 flux is at a steady state and that all CO_2 flux is a result of respiration. This parameter provides a good measurement of the ecosystem's overall metabolism and indirectly, of microbial activity as well [33,34]. Additional soil properties such as pH, organic matter, N-P-K content and texture are also widely used parameters of soil fertility [35].

This study assesses the effects of topsoil spreading and depth on soil microbial activity, measured as enzymatic activity and soil respiration, plant cover, species diversity and species composition on a railway embankment in Central Spain. It also evaluates the difference between topsoil and the original embankment substrate in selected physic-chemical properties related to soil fertility.

We tested the following hypotheses: i) topsoil has higher fertility than the original embankment substrate, ii) topsoil spreading enhances enzymatic activity and soil respiration in disturbed soils, iii) added topsoil changes plant cover, species diversity and floristic composition on the embankments, and iv) the effects of topsoil spreading on microbial activity and vegetation can be modified by changing the depth of the topsoil added in restoration projects.

Material and Methods

Study area

The study area was a railway line under construction in the vicinity of Dehesa de Mari Martin, Navalcarnero, west of Madrid, Spain (40° 18 'N, 3° 58' W), located on tertiary sands resulting from the erosion of the granitic Sierra de Guadarrama Mountains. The area has a slightly undulating relief, with a 20 year old *Pinus pinaster* plantation as the predominant vegetation.

The local semiarid Mediterranean climate has cold winters, hot summers and summer drought. Average temperatures in the coldest and warmest months are 9 and 20°C respectively, with 425 mm average annual rainfall.

Experimental design

A recently built north-facing railway embankment with a 14° slope was selected for the experiment. A randomized 5 block design with three treatments was applied to 15 adjacent plots measuring 6×9 m. The treatments were: control (no added topsoil), T10 (addition of a 10 cm deep topsoil layer) and T30 (addition of a 30 cm deep topsoil layer).

Treatment T30 reflected current slope restoration practices, while T10 was selected for three reasons: a) to reduce erosion on the slopes, b) to reduce seed loss which might be occurring in the bank due to burial, given that few seeds can germinate and/or reach the surface from depths lower than 5 cm [36], and c) to optimize the topsoil, a scarce resource in great demand for linear infrastructure construction.

The topsoil used for the experiment was collected from the uppermost 30 cm of soil in the first week of December 2009 in a nearby pine forest through which the railway line will pass. Topsoil collection occurred after the summer drought and before significant autumn rains, and after the removal of woody vegetation (shrubs and trees). The topsoil was stockpiled for two days and spread on the experimental plots immediately afterwards.

Topsoil and soil fertility

To characterize the physic-chemical soil properties of the original embankment substrate and the stockpiled topsoil, 1 kg soil samples were randomly collected from the embankment ($n = 15$) and topsoil stockpiles ($n = 15$) prior to spreading, using eight 4 cm diameter by 10 cm deep soil cores for each composed sample. Samples were dried and sieved through a 2 mm pore diameter mesh, and analysed for their content in organic matter, total nitrogen, assimilable phosphorus and potassium, percentages of sand, silt and clay, and pH. The organic matter (OM) was analysed following the protocol described by Walkley and Black [37]. Total nitrogen was measured with the Kjeldahl method, and assimilable P and K were extracted with acetate-EDTA as per Läkanen and Erviö [38] and Cottenie et al. [39], respectively. The percentages of soil fractions (sand, silt and clay) were measured by the hydrometer method as per Day [40]. Finally, a 1:2.5 soil/water solution was used for the soil pH analysis.

Enzymatic activity

Six months after the experiment started (June 2010), at the time of expected maximum enzymatic activity due to the temperature and soil moisture conditions, soil samples were collected from the experimental plots (control, T10 and T30) with two replicate of soil cores (4 cm diameter by 10 cm depth), in three slope positions: the upper, middle and lower parts of each plot. We used 10 cm deep soil cores to study soil enzymatic activity, as other authors have done previously in semiarid Mediterranean grasslands [25,41], where annual vegetation roots are usually restricted to the first 10 cm below the soil surface. The collected material was air dried, sifted through a 2 mm pore diameter sieve and then analysed for phosphatase and β-glucosidase enzyme activity.

Phosphatase acid (EC 3.1.3.2) was determined colorimetrically as the amount of p-nitrophenol released per 0.5 g of soil after being incubated for 1 hour at 37°C with a solution of p-nitrophenyl phosphate substrate in a MUB pH 6.5 buffer [42]. The same method described for the phosphatase was used to determine the β-glucosidase (EC 3.2.1.21), with a p-nitrophenyl β-D-glucopyranoside substrate in a THAM pH 6 buffer [43,44].

For both enzymes, the spectrophotometric absorbance was measured in a Varian Cary 1C UV Spectrophotometer at 400 nm. The amount of p-nitrophenol produced in each sample (enzyme activity) was defined using calibration curves obtained from the patterns at different concentrations of p-nitrophenol, for phosphatase and β-glucosidase.

Soil respiration

Soil basal respiration was determined as the rate of CO_2 production, measured with an infrared gas analyser coupled to a flow chamber (IRGA, EGM-4 Environmental Gas Monitor for CO_2). The chamber was placed in three slope positions (upper, middle and lower) in each of the experimental plots, delimited by partially buried PVC pipes. Random measurements of soil respiration were taken in each plot once every month between from July 2010 and June 2011, between 10:00 and 16:00 h to avoid bias due to daily variations in temperature or soil moisture. Aboveground vegetation in the delimited areas was removed at least one week prior to the in situ measurements.

Plant cover, species diversity and composition of roadslope vegetation

The cover of each plant species was determined visually in spring 2010 and 2011, the optimum period for studying Mediterranean herbaceous communities. This involved setting up six 0.5 m quadrats in each plot (control, T10 and T30) in two slope positions, three quadrats set 1 m away from the edge of the upper plot and the other three 1 m from the edge of the lower plot. The total percentage cover of herbaceous plants was noted and two indicators of species diversity (species richness (S) and equitability (E = H'/ln S), [45]) were calculated.

Data analysis

The differences in the physic-chemical soil properties of topsoil vs. original embankment substrate was analysed with ANOVAs. β-glucosidase and phosphatase enzyme activity were analysed using a univariate general linear model (GLM). We produced a model for each parameter in which net enzyme activity (μmol of p-nitrophenol $g^{-1}·h^{-1}$) was included as the dependent variable, with the treatment (control, T10 and T30) and slope position (upper, middle, lower) and their interaction as fixed factors. The differences between factor levels were analysed with post-hoc Tukey tests. Mean soil respiration data (CO_2 production rate in g $CO_2·m^{-2}·h^{-1}$) per plot obtained during the twelve sample months was analysed with a repeated measures General Linear Model. The CO_2 production rate was the within-subject factor. Levels were months (12 measures) and the treatment (control, T10 and T30) and slope position (upper, middle, lower) were the between-subjects factors. Finally, total plant cover (square root transformed), species richness (log transformed) and equitability, were analysed by means of a repeated measures ANOVA, with the year as the within-subjects variable, and treatment (control, T10 and T30) and slope position (upper, lower) as the between-subjects factors. Where their interaction proved significant, a year by year ANOVA was performed and differences between factor levels were analysed by a Tukey test. For all statistical analysis we used SPSS 15.0 software (SPSS Inc., Chicago, IL, USA).

The response of floristic composition to the treatments was tested by Redundancy Analysis (RDA) using the CANOCO 4.5 program [46] and CANODRAW for Windows 4.12 [47] for the graphics. As environmental variables we used treatment, slope position and year. Forward automatic selection with Monte Carlo test (499 permutations) was used to find the environmental factors that best explained species cover values. We evaluate the final model with a Monte Carlo permutation test. The percentage cover values for species were log-transformed.

To compare the effects of treatments on vegetation over time, the Principal Response Curves method (PRC) was employed, using CANOCO (versión 4.5; [48]). Previously an RDA was carried out, introducing treatment interactions with years as environmental variables (excluding the interaction of period with the control which served as a reference series) and the year as covariable (in application terms). The statistical significance of this ordination was tested with a Monte Carlo permutations test (499 permutations). The significance of a second axis, when the first proved significant, was analysed with another RDA introducing the sample coordinates in the first axis as a covariable, in addition to the year [49].

Ethics Statement

Prior to the field studies, we obtained all the permits for the works. The Site Manager of Obrascón Huarte Lain S.A. and Madrid Regional Government were informed, since experimental plots were located on an embankment under construction. No protected species were sampled during the study.

Results

Topsoil and soil fertility

Most of the physic-chemical measured soil properties showed significant differences between topsoil and original embankment substrate (Table 1). Topsoil had significantly higher soil organic matter, total nitrogen content, assimilable P and lower soil pH than the original embankment substrate. Non-significant differences were observed for assimilable K. The percentage of silt was also higher in topsoil than in the original substrate, but there were no significant differences for percentage of clay or sand (Table 1).

Spread topsoil and microbial activity

Both of the tested enzymes, β-glucosidase and phosphatase, showed higher activity in the plots with spread topsoil than in the control plots ($F = 17.91$, $P<0.001$ for β-glucosidase and $F = 35.73$, $P<0.001$ for phosphatase) (Fig. 1.). Nevertheless, there were no significant differences in enzyme activity between topsoil depths (10 and 30 cm) (Post-hoc Tukey test, $P = 0.69$ for β-glucosidase and $P = 0.79$ for phosphatase). Differences in enzymatic activity according to soil location on the slopes (upper, middle and lower) were not significant for either β-glucosidase ($F = 2.36$, $P = 0.084$) or phosphatase ($F = 1.60$, $P = 0.214$).

Soil respiration varied through the year ($F = 98.47$; $P<0.001$). The lowest values of soil respiration were recorded in the driest or coldest months (summer and winter), while the highest values corresponded to the period with most moisture and mildest temperatures (Fig. 2). The repeated measures analysis showed a significant effect of the treatment ($F = 29.28$; $P<0.001$), with a generally higher CO_2 production in plots with the largest volume of fertile topsoil, T30 (0.23 ± 0.022 g $CO_2·m^{-2}·h^{-1}$), than in the control (0.12 ± 0.011 g·$CO_2·m^{-2}·h^{-1}$) and T10 treatments (0.17 ± 0.017 g·$CO_2·m^{-2}·h^{-1}$). In this case, plot position on the slope showed a significant effect on soil respiration ($F = 4.26$; $P< 0.05$), with higher soil respiration in the lower position (0.19 ± 0.018 g $CO_2·m^{-2}·h^{-1}$) than in the upper or middle plots (0.15 ± 0.014 and 0.18 ± 0.018 g $CO_2·m^{-2}·h^{-1}$, respectively). The

Table 1. Physic-chemical parameters (means ± SE) of original embankment substrate and topsoil used for the experiment.

Parameters	Original substrate	Topsoil	*F*	*P*
Available water (g/kg)	58.0±6.2[b]	41.4±6.8[a]	16.36	0.04
OM (g/kg)	1.2±0.8[a]	10.4±1.8[b]	107.47	<0.001
P (mg/100 g)	0.2±0.1[a]	0.4±0.1[b]	7.36	0.03
N (g/kg)	0.2±0.0[a]	0.4±0.1[b]	29.92	0.001
K (mg/100 g)	6.2±0.6	5.7±0.6	1.84	0.21
pH	6.4±0.1[b]	5.8±0.0[a]	152.81	<0.001
Clay (g/kg)	169.3±13.0	105.4±2.7	5.21	0.06
Sand (g/kg)	722.3±22.6	694.9±73.5	0.63	0.45
Silt (g/kg)	108.4±12.6[a]	199.7±12.9[b]	127.19	<0.001

OM: Organic matter; P: phosphorus content; K: potassium content; N: nitrogen content. Different letters (a, b) indicate significant differences ($P>0.05$) between treatments.

interaction (treatment x position) was not significant ($F = 6.35$; $P > 0.05$).

Topsoil spreading and total plant cover, species richness and equitability

Total plant cover was higher in the control than in the topsoil plots during the first year. Such cover increased in all plots during the second year until it was equal in all treatments. The 'period x treatment' interaction proved significant ($F = 9.36$; $P < 0.001$; Table 2) since the increase of herbaceous cover over time was greater in the topsoil plots than in the controls. The depth of spread did not affect total plant cover in either study year (Table 2).

In total, 121 species were identified over the two years. Of these, 79 species were found in the control plots and 94 in the topsoil plots (Table S1). *Avena barbata, Capsella bursa-pastoris, Carlina corymbosa, Diplotaxis erucoides, Plantago lagopus, Psilurus incurvus, Trifolium glomeratum* and *Trifolium scabrum* occurred exclusively on the control plots. Although species richness was similar under all treatments during the first year, it was higher in the topsoil plots than in the controls in the second year (period x treatment interaction $F = 10.46$; $P < 0.001$; Table 2). The depth of the spread topsoil did not affect species richness in either study year (Table 2).

Equitability was greater in the topsoil plots than in the controls during the first year (Table 2), but these differences disappeared during the second year (period x treatment interaction $F = 3.6$; $P = 0.03$). The depth of spread did not affect species equitability in either study year (Table 2).

Slope position was not significant in any of these models ($F = 0.18$; $P = 0.67$, for plant cover; $F = 0.02$; $P = 0.90$, for species richness and $F = 0.13$; $P = 0.72$, for species equitability).

Topsoil spreading and floristic composition

In the analyses of floristic composition, the environmental variables year and treatment were significant ($F = 51.05$; $P = 0.002$ and $F = 12.82$; $P = 0.02$, respectively) and were included in the model in the automatic selection procedure, while slope position was not significant ($F = 1.37$; $P = 0.09$) and was removed from the model. The first two RDA axes accounted for 83% of the species cover variation (Fig. 3), with 67% of the variation explained by the first axis. Floristic composition segregated during the first year (spring 2010) between the topsoil plots (T10–2010 and T30–2010) and the controls (C–2010). Three species, *Spergularia purpurea, Polygonum aviculare* and *Anthemis arvensis*, showed the highest cover

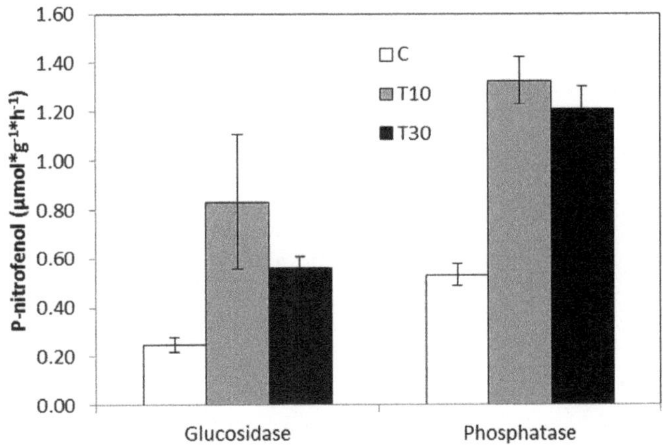

Figure 1. Enzymatic activity with topsoil. β-Glucosidase and Phosphatase activity in soil with added topsoil. C: control; T10: 10 cm deep layer of spread topsoil and T30: 30 cm deep layer of spread topsoil. Bars show standard error.

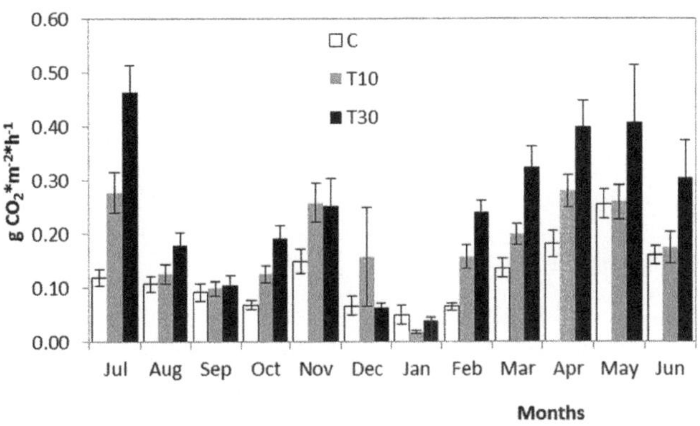

Figure 2. Annual CO_2 production rate. Monthly mean and standard errors of CO_2 production rate (g $CO_2 \cdot m^{-2} \cdot h^{-1}$) per treatment for the period July 2010-June 2011. C: control; T10: 10 cm deep layer of spread topsoil and T30: 30 cm deep layer of spread topsoil.

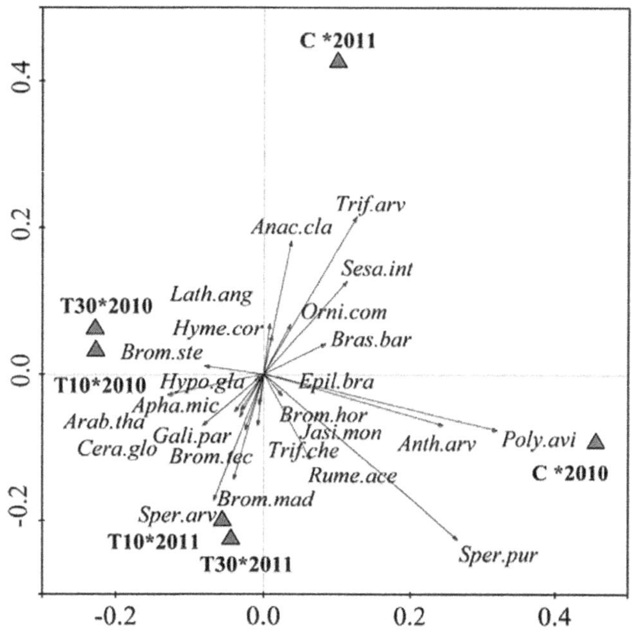

Figure 3. Floristic composition by year and treatment. RDA ordination diagram, first and second axes for species against treatments in each year (2010 and 2011). C, Control; T10, 10 cm topsoil depth; and T30, 30 cm topsoil depth. **Ana.cla.,** *Anacyclus clavatus*; **Anth.arv.,** *Anthemis arvensis*; **Apha.mic.,** *Aphanes microcarpa*; **Arab.tha.,** *Arabidopsis thaliana*; **Bras.bar.,** *Brassica barrelieri*; **Brom.hor.,** *Bromus hordeaceus*; **Brom.mad.,** *Bromus madritensis*; **Brom.ste.,** *Bromus sterilis*; **Brom.tec.,** *Bromus tectorum*; **Cera.glo.,** *Cerastium glomeratum*; **Epil.bra.,** *Epilobium brachycarpum*; **Gali.par.,** *Galium parisiense*; **Hyme.cor.,** *Hymenocarpus cornicina*; **Hypo.gla.,** *Hypochaeris glabra*; **Jasi.mon.,** *Jasione montana*; **Lath.ang.,** *Lathyrus angulatus*; **Orni.-com.,** *Ornithopus compressus*; **Poly.avi,** *Polygonum aviculare*; **Rume.ace.,** *Rumex acetosella*; **Sesa.int.,** *Sesamoides interrupta*; **Sper.-arv.,** *Spergula arvensis*; **Sper.pur.,** *Spergularia purpurea*; **Trif.arv.,** *Trifolium arvense*; and **Trif.che.,** *Trifolium cherleri*.

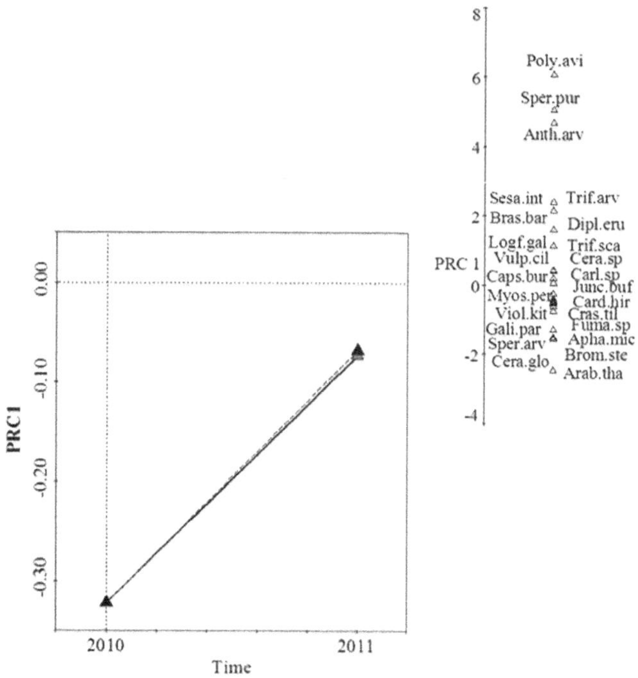

Figure 4. Species composition response to the depth of the spread topsoil. Diagram of the first (a) and second (b) PRC axes. The first two PRC axes proved significant ($F=13.09$; $P=0.002$ for axis 1 and $F=12.95$; $P=0.002$ for axis 2). The dotted line indicates the 10 cm topsoil treatment and the solid line the 30 cm treatment. In the ordination axis of the PRC diagram, the first component of the variation explained by the differences in treatment over time is represented, with time on the x axis. The weighting of each species on the principal axes, shown on the right of the PRC diagram, represents the affinity of each species to the treatments. Only species showing adjustments greater than 1 are included. **Anth.arv.,** *Anthemis arvensis*; **Apha.mic.,** *Aphanes microcarpa*; **Arab.tha.,** *Arabidopsis thaliana*; **Bras.bar.,** *Brassica barrelieri*; **Brom.ste.,** *Bromus sterilis*; **Caps.bur.,** *Capsella bursa-pastoris*; **Card.hir.,** *Cardamine hirsuta*; **Carl.sp.,** *Carlina sp.*; **Cera.sp.,** *Cerastium sp.*; **Cera.glo.,** *Cerastium glomeratum*; **Cras.til.,** *Crassula tillaea*; **Dipl.eru.,** *Diplotaxis erucoides* **Fuma.sp.,** *Fumaria* sp.; **Gali.par.,** *Galium parisiense*; **Junc.buf.,** *Juncus bufonius*; **Logf.gal.,** *Logfia gallica*; **Myos.per.,** *Myosotis personii*; **Poly.avi,** *Polygonum aviculare*; **Sesa.int.,** *Sesamoides interrupta*; **Sper.arv.,** *Spergula arvensis*; **Sper.pur.,** *Spergularia purpurea*; **Trif.arv.,** *Trifolium arvense*; **Trif.sca.,** *Trifolium scabrum*; **Viol.kit.,** *Viola kitaibeliana*; and **Vulp.cil.,** *Vulpia ciliata*.

values in the control plots during the first year, whereas those with highest cover values on the topsoil plots were *Bromus sterilis* and *Arabidopsis thaliana*. *Spergula arvensis* and *Bromus madritensis* had the highest cover values on the T10 and T30 plots in 2011, whereas *Trifolium arvense* and *Anacyclus clavatus* then provided the greatest cover on the control plots. The floristic composition on the T10 and T30 plots differed more from the control plots in the first year than in the second (Fig. 4). *Polygonum aviculare*, *Spergularia purpurea* and *Anthemis arvensis* showed the highest weightings on the first axis, in accordance with the RDA results (Fig. 3).

Species composition was not affected by the depth of the spread topsoil (Fig. 4). The ordination diagram of the second PRC axis did not provide any additional information to that of the first axis.

Discussion

Topsoil and soil fertility

According to our results, we can assume that the addition of topsoil to the original substrate in embankments will increase soil fertility and that this improvement will depend on the depth of the topsoil added to the original substrate. Organic matter content, total nitrogen and assimilable phosphorus were all higher in topsoil in comparison with the original embankment substrate. Organic matter was tenfold higher in the topsoil than in the original substrate, while N and P showed an increase of 50% (Table 1). This improvement was obviously linked to the quality of the spread topsoil, the depth from which it had been collected, the material used to build the embankments and the local climate. In more xeric conditions, Tormo et al. [13], for example, found more total N and assimilable P in plots with spread topsoil than without it, but found no differences in the organic matter content, which was very low in both cases (less than 1.5%). Similarly, the relatively low pH found in the analysis of the topsoil in the present experiment could be related to the origin of the material used in the experiment, a *Pinus pinaster* plantation, with many acidic compounds in the needles [50] which tend to acidify the soil.

Topsoil spreading and microbial activity

The results of this study are, to our knowledge, the first empirical evidence that topsoil spread on new slopes shaped during the construction of linear transport infrastructure improves microbial activity and soil functioning, measured through surro-

Table 2. Mean values (means ± SE) for total plant cover (%), species richness and equitability in spring 2010 and spring 2011 for each treatment.

Period	Treatment	Plant cover	Richness	Equitability
2010	Control	40.8±3.5 [b]	7.2±1	0.77±0.04[a]
	T10	21.7±2.0[a]	6.9±0.8	0.84±0.05[ab]
	T30	17.3±2.1[a]	6.2±1.3	0.85±0.09[b]
2011	Control	62.0±3.3	14.6±1.63[a]	0.83±0.03
	T10	52.2±3.8	17.8±1.95[b]	0.81±0.03
	T30	61.7±3.2	19.5±1.89[b]	0.82±0.03

a, b indicate significant differences ($P < 0.05$) between treatments within a year
T10 = topsoil applied to 10 cm depth, T30 = topsoil applied to 30 cm depth.

gate parameters, such as the activity of enzymes involved in vital nutrient cycles and soil respiration.

Plots where topsoil was spread showed more β-glucosidase and phosphatase activity than in the control plots, with an increase of 45% and 57% respectively, which supports our initial hypothesis. The differences found between the controls and the plots with topsoil were probably due to the higher organic matter content in the latter (Table 1). Several authors have found a positive relationship between the activity of these enzymes with respect to soil organic C content [32,51,52]. Inorganic nutrients required for plant and microorganism growth need to be released from the organic matter. When soil microorganisms detect a nutrient shortage, they start up their internal machinery and produce exo-enzymes, which release nutrients in an inorganic form from the dead organic matter. The addition of topsoil increases microbial activity and ultimately benefits plant growth, especially if it is rich in organic matter. However, the absolute values of the activity of these enzymes were lower than those found by other authors in similar climates [18,53]. This may be linked to the quality of the organic matter in the experimental topsoil, which was taken from a pine plantation and therefore may have contained a high proportion of extremely recalcitrant compounds [54], possibly limiting the microbial activity.

The observed decrease in soil pH after topsoil addition (Table 1) does not seem to have a negative effect on enzyme activity for β-glucosidase or phosphatase. Sinsabaugh et al. [24] found a negative relationship between soil pH and activity of these two enzymes. In the present case, the possible negative effect on enzyme activity due to the decrease in soil pH could be offset by the positive effect of increasing soil organic matter content in the treatments with topsoil addition.

Depth of spread topsoil and position on the slope seemed to have no effect on the activity of β-glucosidase or phosphatase. This result may be due to the fact that we measured enzymatic activity in the upper soil layer (10 cm depth), as recommended by other authors [25,41] who measured enzymatic activities to infer soil functioning in Mediterranean grasslands, as this is where most of the annual species' rizhosphere is located. The short time since the topsoil was spread could explain why no differences were found in enzyme activity on different parts of the slope (upper, middle and lower). Material from upper slope positions that builds up in the lower parts often generates a gradient of soil fertility and moisture, which affects enzyme activity [55] and vegetation. This process had not begun when the samples were taken, although one would expect the gradient to appear over time and be reflected in the form of more enzyme activity in the lower than the upper zone of the fill [56].

As in the case of the activity of β-glucosidase and phosphatase enzymes, soil respiration was greater in the presence of topsoil than in its absence (Fig. 2). However, contrary to the pattern found in enzyme activity, soil respiration increased with the spread topsoil depth, showing greater microbial activity in plots with 30 cm of spread soil than 10 cm. This result might be due to the way soil respiration was measured. The device used to quantify soil respiration measured the CO_2 produced in the top 10 cm, and also the CO_2 generated further below by microbial activity during measurement. The greater soil depth in T30 probably implies greater microbial density than the shallow soil layer (T10) and hence higher CO_2 production per surface unit, even if microbial density decreases with soil depth [57]. Seasonal fluctuations in microbial activity, measured via soil respiration, are mainly due to changes in the soil moisture and temperature conditions [58,59], as the rainiest months are the periods of greatest activity in disturbed soils [60,61]. Monthly soil respiration readings confirm this trend, with higher respiration, particularly in the treatments, during periods of mild temperatures and higher rainfall, normally autumn (November) and spring (February to July) in Mediterranean regions.

Monthly levels of soil respiration ranged between 0.038 and 0.464 g $CO_2 \cdot m^{-2} \cdot h^{-1}$ in soils with more organic matter, in contrast to 0.050 and 0.255 g $CO_2 \cdot m^{-2} \cdot h^{-1}$ recorded in the control plot soils. These levels are similar to those of temperate grasslands (0.050 g $CO_2 \cdot m^{-2} \cdot h^{-1}$) [62] and well-drained soils in temperate mixed forests (0.05 and 0.25 g $CO_2 \cdot m^{-2} \cdot h^{-1}$) [63], but slightly lower than those of forest soils (0.088 and 0.66 g $CO_2 \cdot m^{-2} \cdot h^{-1}$ minimum and maximum respectively) [64] and Mediterranean forests (0.268 and 0.887 g $CO_2 \cdot m^{-2} \cdot h^{-1}$) [59]. These differences may be due mainly to the heavy influence of forest tree roots on soil respiration [65,66], a factor that was absent from the present study.

Topsoil spreading and vegetation

Erosion control is one of the principal objectives of roadslope restoration. Rapid establishment of plant cover guarantees the stability of such surfaces [67,68] and is favoured by the application of topsoil [13]. However, this effect was not clear in our study. Contrary to expectations [8,13], in the spring following the topsoil application, plant cover was greater in the topsoil-free plots than in the others and there were no difference in species richness between treatments. This was basically due to the dominance of a few species: *Polygonum aviculare*, *Spergularia purpurea* and *Anthemis arvensis*, whose seeds have a persistent character and are present in deep soil levels [69], and thus may have been present on the embankment before topsoil was applied. These species attained higher cover values in plots with poorer soils since they are pioneer colonists, adapted to growth in nutrient-poor environments and less likely to thrive in nutrient-rich soils [70]. This result is congruent with the lower equitability of the control plots.

The embankment experienced a significant influx of seeds from neighbouring areas during summer–autumn 2010 [71–73]. This input can exceed the amount originating in the topsoil [8]. Thus, in the second spring, the topsoil plots showed a greater increase in plant cover than the controls and also a significant increase in species richness. Higher species richness may confer greater stability on roadslopes since it is presumed to increase the functionality of the ecosystem [74], broadening species phenology across critical periods. This increase in floristic diversity during the first stages of plant succession when restoring roadslopes has been described by other authors [73,75]. Generalist species greatly augment species richness through their ability to colonize bare and poor terrain. As these spaces become filled, interspecific competition leads the community to a more stable situation, with greater overall plant cover but lower species diversity, the outcome of competitive exclusion processes [12,73,76].

The depth of the spread topsoil does not seem to affect the restoration process significantly with respect to plant cover, species diversity (richness and equitability) or floristic composition. Other studies have described how seedling recruitment does not differ between applications of 10 cm or 30 cm of topsoil [9] and that the density of seedlings emerging from the seed bank is not affected by topsoil depth between 1 and 8 cm [77]. This is probably because seeds buried deeper than 5 cm experience difficulty in germination and seedling emergence [9,36,78]. Although other authors have noted an increase in plant cover [14] and lower species richness with increased topsoil depth [79], due to the increased availability of organic matter, nitrogen, phosphorus and available water, these findings are not confirmed by our study. This discrepancy is probably due to the fact that our communities were mainly composed of herbaceous annual plants, while the former studies described restoration in woody and perennial herbaceous communities that can take advantage of deeper fertile layers due to their deeper root systems.

Conclusions

The present study validates some of the initial hypotheses. Spread topsoil does seem to increase microbial activity and vegetation recovery on newly build road embankments. However, the 30 cm depth of the topsoil spread across the embankments to be restored does not seem to cause significant changes in species richness, equitability and floristic composition in comparison with the 10 cm depth. It is therefore recommended that topsoil should be spread to a depth of no more than 10 cm in order to optimize the use of this resource, enabling organic matter to be saved or used to treat a larger area. Nevertheless, the quality of the added soil relative to the original substrate must be checked for each particular case, especially for organic matter content, an essential soil component for the stimulation of microbial activity and hence plant growth, as highlighted by the findings of this and previous papers.

Supporting Information

Table S1 Mean and standard error (SD) of species cover (%) in the different plots and study years. C = Control, original substrate, T10 = topsoil applied to 10 cm depth, T30 = topsoil applied to 30 cm depth. Only species present in more than 1 quadrat have been listed.

Acknowledgments

This paper is a description of work by the Obrascón Huarte Lain S.A. Corporation (OHL) in collaboration with the Ecology Department of the Autónoma University of Madrid. It is part of a broader industrial research project, CENIT-OASIS (Safe, Intelligent and Sustainable Highway Operation), approved under CENIT tender 2008-2011.

Author Contributions

Conceived and designed the experiments: BP DR BMJ. Performed the experiments: DR VM BMJ MC AIL BP. Analyzed the data: DR VM BP. Contributed reagents/materials/analysis tools: DR VM BMJ AIL BP. Contributed to the writing of the manuscript: BP DR VM BMJ AIL. Important intellectual content: DR BMJ MC AIL BP.

References

1. Forman RTT, Alexander LE (1998) Roads and their major ecological effects. Annu Rev Ecol Syst 29: 207–C2.
2. Spellerberg IF, Morrison T (1998) The ecological effects of new roads: a literature review. Wellington NZ: Dept of Conservation.
3. Coffin AW (2007) From roadkill to road ecology: A review of the ecological effects of roads. J Transp Geogr 15: 396–406.
4. Hobbs RJ, Norton DA (1996) Towards a conceptual framework for restoration ecology. Restor Ecol 4: 93–110. DOI: 10.1111/j.1526-100X.1996.tb00112.x
5. Bochet E, García-Fayos P (2004) Factors controlling vegetation establishment and water erosion on motorway slopes in Valencia, Spain. Restor Ecol 15: 166–174.
6. Tejada M, Gonzalez JL (2007) Influence of organic amendments on soil structure and soil loss under simulated rail. Soil Till Res 93: 197–205.
7. García-Fayos P, García-Ventoso B, Cerdá A (2000). Limitations to plant establishment on eroded slopes in southeastern Spain. J Veg Sci 11: 77–86.
8. Mola I, Jiménez MD, López-Jiménez N, Casado MA, Balaguer L (2011) Roadside reclamation outside the revegetation season: management options under schedule pressure. Restor Ecol 19 (1): 83–92. DOI: 10.1111/j.1526-100X.2009.00547.x
9. Rokich DP, Dixon KW, Sivasithamparam K, Meney KA (2000) Topsoil handling and storage effects on woodland restoration in Western Australia. Restor Ecol 8: 196–208. DOI: 10.1046/j.1526-100x.2000.80027.x
10. Jochimsen ME (2001) Vegetation development and species assemblages in a long-term reclamation project on mine spoil. Ecol Eng 17: 187–198.
11. Mitchell DJ, Barton AP, Fullen MA, Hocking TJ, Zhi WB, et al. (2003) Field studies of the effects of jute geotextiles on runoff and erosion in Shropshire, UK. Soil Use Manage 19: 182–184. DOI: 10.1111/j.1475-2743.2003.tb00301.x
12. Matesanz S, Valladares F, Tena D, Costa-Tenorio M, Bote D (2006) Early dynamics of plant communities on revegetated motorway slopes from southern Spain: Is hydroseeding always needed? Restor Ecol 14: 297–307. DOI: 10.1111/j.1526-100X.2006.00132.x
13. Tormo J, Bochet E, García-Fayos P (2007) Roadfill revegetation in semiarid Mediterranean environments. Part II: topsoiling, species selection and hydroseeding. Restor Ecol 15: 97–102. DOI: 10.1111/j.1526-100X.2006.00194.x
14. Holmes PM (2001) Shrubland restoration following woody alien invasion and mining: Effects of topsoil depth, seed source, and fertilizer addition. Restor Ecol 9: 71–84. DOI: 10.1046/j.1526-100x.2001.009001071.x
15. Moynahan OS, Zabinski CA, Gannon JE (2002) Microbial community structure and carbon-utilization diversity in a mine tailings revegetation study. Restor Ecol. 10(1): 77–87. DOI: 10.1046/j.1526-100X.2002.10108.x
16. Bever JD, Westover KI, Antonovics J (1997) Incorporating the soil community into plant population dynamics: the utility of the feedback approach. J Ecol 85: 561–573.
17. Van der Putten WH, Bardgett RD, De Ruiter PC, Hol WHG, Meyer KM, et al. (2009). Empirical and theoretical challenges in aboveground-belowground ecology. Oecologia 161: 1–14. DOI: 10.1007/s00442-009.1351-8
18. García-Orenes F, Guerrero C, Roldán A, Mataix-Solera J, Cerdà A, et al. (2010) Soil microbial biomass and activity under different agricultural management systems in a semiarid Mediterranean agroecosystem. Soil Till Res 109: 110–115.
19. Bardgett RD, Richter A, Bol R, Garnett MH, Bäumler R, et al. (2007). Heterotrophic microbial communities use ancient carbon following glacial retreat. Biol Lett 3: 487–490. DOI: 10.1098/rsbl.2007.0242
20. Alvear M, Pino M, Castillo R, Trasar-Cepeda C, Gil-Sotres F (2006) Efecto de la cero labranza sobre algunas actividades biológicas en un alfisol del sur de Chile. J Soil Sci Plant Nutr 6: 38–53.
21. Van der Heijden MGA, Bardgett RA, Van Straalen NM (2008) The unseen majority: soil microbes as drivers of plant diversity and productivity in terrestrial ecosystems. Ecol Lett 11: 296–310. DOI: 10.1111/j.1461-0248.2007.01139.x

22. Abbott LK, Robson AD, Gazey C (1992) Selection of inoculant vesicular-arbuscular mycorrhizal fungi. In: Norris JR, Read DJ, Varma AK, editors. Methods in Microbiology. Vol. 24. Techniques for the study of mycorrhiza. London: Academic Press. pp. 1–21.
23. Jasper DA (2007) Beneficial soil microorganisms of the jarrah forest and their recovery in bauxite mine restoration in Southwestern Australia. Restor Ecol 15: 4 (supplement) 74–84. DOI: 10.1111/j.1526-100X.2007.00295.x
24. Sinsabaugh RL, Carreiro MM, Repert DA (2002) Allocation of extracellular enzymatic activity in relation to litter composition, N deposition, and mass loss. Biogeochemistry 60: 1–24.
25. Maestre FT, Puche MD (2009) Indices based on surface indicators predict soil functioning in Mediterranean semi-arid steppes. Appl Soil Ecol 41: 342–350. DOI:10.1016/j.apsoil.2008.12.007
26. Turner BL, Hopkins DW, Haygarth PM (2002a) β-glucosidase activity in pasture soils. App Soil Ecol 20: 157–162. DOI:10.1016/S0929-1393(02)00020-3
27. Zorzona R, Guerrero C, Mataix-Solera J, Arcenegui V, García-Orenes F, et al. (2006) Assessing air-drying and rewetting pre-treatment effect on some soil enzyme activities under Mediterranean conditions. Soil Biol Biochem 38: 2125–2134. DOI:10.1016/j.soilbio.2006.01.010
28. Ros M, Hernandez MT, García C (2003) Soil microbial activity after restoration of a semiarid soil by organic amendments. Soil Biol Biochem 35: 463–469.
29. Dick A, Tabatabai MA (1992) Significance and potential use of soil enzymes. In: Meeting FJB, editor. Soil microbial ecology: applications in agriculture and environmental management. NY, USA: Marcel Deckker. pp. 95–127.
30. Lagomarsino A, Moscatelli MC, Di Tizio A, Mancinelli R, Grego S, et al. (2009) Soil biochemical indicators as a tool to assess the short-term impact of agricultural management on changes in organic C in a Mediterranean environment. Ecol Indic 9: 518–527.
31. Turner BL, McKelvie ID, Haygarth PM (2002b) Characterisation of water extractable soil organic phosphorous by phosphatase hydrolysis. Soil Biol Biochem 34: 27–35. DOI:10.1016/S0038-0717(01)00144-4
32. Bandick AK, Dick RP (1999) Field management effects on soil enzyme activities. Soil Biol Biochem 31: 1471–1479.
33. Jensen LS, Mueller T, Tate KR, Ross DJ, Magid J, et al. (1996) Soil surface CO_2 flux as an index of soil respiration in situ: a comparison of two chamber methods. Soil Biol Biochem 28: 1297–1306.
34. Ryan MG, Law BE (2005) Interpreting, measuring, and modeling soil respiration. Biogeochemistry 73: 3–27. DOI: 10.1007/s10533-004-5167-7
35. Chapin III FS, Matson P, Money HA (2002) Principles of Terrestrial Ecosystem Ecology. New York: Springer.
36. Traba J, Azcárate FM, Peco B (2004) From what depth do seeds emerge? A soil seed bank function experiment with Mediterranean grassland species. Seed Sci Res 14: 297–303. DOI: http://dx.doi.org/10.1079/SSR2004179. Accessed 2014 Jun 10
37. Walkley A, Black IA (1934) An examination of the Degtjareff method for determining soil organic matter, and a proposed modification of the chromic acid titration method. Soil Sci 34: 29–38. DOI: 10.1097/00010694-193401000-00003
38. Läkanen E, Erviö R (1971) A comparison of eight extractants for the determination of plant available micronutrients in soils. Acta Agral Fenn 123: 223–232.
39. Cottenie A, Kiekens L, Verloo M (1975) Principles of soils and substrate analysis with regards to mobility of nutrient elements. Pedologie 25: 134–142.
40. Day PR (1965) Particle fractionation and particle-size analysis. In: Page AL, editor. Methods of soil analysis. Madison, WI: American Society of Agronomy. pp. 545–567.
41. Maestre FT, Puche MD, Guerrero C, Escudero A (2011) Shrub encroachment does not reduce the activity of some soil enzymes in Mediterranean semiarid grasslands. Soil Biol Biochem 43: 1746–1749. DOI:10.1016/j.soil-bio.2011.04.023
42. Tabatabai MA, Bremner JM (1969) Use of *p*-nitrophenol phosphate in assay of soil phosphatase activity. Soil Biol Biochem 1: 301–307.
43. Tabatabai MA (1982) Soil enzymes, in: Page AL, Miller EM, Keeney DR (Eds.), Methods of soil analysis. Part 2. Chemical and microbiological properties. Madison, WI: American Society of Agronomy. pp. 903–947.
44. Makoi JH Jr, Ndakidemi PA (2008) Review. Selected soil enzymes: examples of their potential roles in the ecosystem. Afr J Biotechnol 7 (3): 181–191. DOI: 10.5897/AJB07.590
45. Shannon CE, Weaver W (1963) The mathematical theory of communication. University of Illinois Press.
46. ter Braak CJF (1990) CANOCO – a FORTRAN program for canonical community ordination by [partial] [detrended] [canonical] correspondence analysis, principal components analysis and redundancy analysis (version 3.1). Ithaca, New York, US: Microcomputer Power
47. Smilauer P (1993) CanoDraw 3.0. Environmental Change Research Centre. London: University College.
48. ter Braak CJF, Smilaner P (2002) CANOCO reference manual and CanoDraw for Windows user's guide: software for canonical community ordination (version 4.5). Ithaca, New York, US: Microcomputer Power.
49. Van den Brink PJ, ter Braak CJF (1999) Principal response curve: analysis of time-dependent multivariate response of a biological community to stress. Environ Toxicol Veg Sci 19: 138–148. DOI: 10.1002/etc.5620180207
50. Arnold G (1992) Soil acidification as caused by the nitrogen uptake pattern of Scots pine (*Pinus sylvestris*). Plant Soil 142: 41–51. DOI: 10.1007/BF00010173
51. Jordan D, Kremer RJ, Bergfield WA, Kim KY, Cacnio VN (1995) Evaluation of microbial methods as potential indicators of soil quality in historical agricultural fields. Biol Fert Soils 19: 297–302.
52. Araújo ASF, Leite LFC, Santos VB, Carneiro RFV (2009) Soil microbial activity in conventional and organic agricultural systems. Sustainability 1: 268–276. DOI: 10.3390/su1020268
53. Melero S, López-Garrido R, Murillo JM, Moreno F (2009) Conservation tillage: Short- and long-term effects on soil carbon fractions and enzymatic activities under Mediterranean conditions. Soil Till Res 104: 292–298. DOI:10.1016/j.still.2009.04.001
54. Moro MJ, Domingo F (2000) Litter decomposition in four woody species in a Mediterranean climate: weight loss, N and P dynamics. Ann Bot-London 86: 1065–1071. DOI: 10.1006/anbo.2000.1269
55. Bergstrom DW, Monreal CM, Millette JA, King DJ (1998) Spatial dependence of soil enzyme activities along a slope. Soil Sci Soc Am J 62: 1302–1308. DOI:10.2136/sssaj1998.03615995006200050022x
56. Collins HP, Rasmussen PE, Douglas Jr CL (1992) Crop rotation and residue management effects on soil carbon and microbial dynamics. Soil Sci Soc Am J 56: 783–788. DOI:10.2136/sssaj1992.03615995005600030018x
57. Fierer N, Schimel JP, Holden PA (2003) Variations in microbial community composition through two soil depth profiles. Soil Biol Biochem 35: 167–176.
58. Davidson EA, Belk E, Boone RD (1998) Soil water content and temperature as independent or confounded factors controlling soil respiration in a temperate mixed hardwood forest. Glob Change Biol 4: 217–227. DOI: 10.1046/j.1365-2486.1998.00128.x
59. King JA, Harrison R (2002) Measuring soil respiration in the field: an automated closed chamber system compared with portable IRGA and alkali absorption methods. Commun Soil Sci Plan 33: 3–4. DOI:10.1081/CSS-120002753
60. Litton CM, Ryan MG, Knight DH, Stahl PD (2003) Soil surface carbon dioxide efflux and microbial biomass in relation to tree density 13 years after a stand replacing fire in lodgepole pine ecosystem. Glob Change Biol 9: 680–696. DOI: 10.1046/j.1365-2486.2003.00626.x
61. Castillo-Monroy AP, Maestre FT, Rey A, Soliveres S, García-Palacios P (2010) Biological soil crusts are a key contributor to soil CO_2 efflux, and modulate its spatio-temporal variability in a semi-arid Mediterranean ecosystem. Ph.D. Thesis: Ecología funcional de la costra biológica en ambientes semiáridos gipsícolas de la Península Ibérica. pp. 67–86.
62. Raich JW, Schlesinger WH (1992) The global carbon dioxide flux in soil respiration and its relationship to vegetation and climate. Tellus 44b: 81–99. DOI: 10.1034/j.1600-0889.1992.t01-1-00001.x
63. Davidson EA, Savage K, Bolstad P, Clark DA, Curtis PS, et al. (2002) Belowground carbon allocation in forest estimated from litterfall and IRGA-based soil respiration measurements. Agr Forest Meteorol 113: 39–51.
64. Dörr H, Münnich KO (1987) Annual variation in soil respiration in selected areas of the temperate zone. Tellus 39b: 114–121.
65. Hanson PJ, Edwards NT, Garten CT, Andrews JA (2000) Separating root and soil microbial contributions to soil respiration: a review of methods and observations. Biogeochemistry 48: 115–146.
66. Tang J, Baldocchi DD (2005) Spatial–temporal variation in soil respiration in an oak–grass savanna ecosystem in California and its partitioning into autotrophic and heterotrophic components. Biogeochemistry 73: 183–207. DOI: 10.1007/s10533-004-5889-6
67. Cerdà A (2007) Soil water erosion on road embankments in Eastern Spain. Sci Total Environ 378: 151–155.
68. Bochet E, García-Fayos P, Tormo J (2010) How can we control erosion of roadslopes in semiarid Mediterranean areas? Soil improvement and native plant establishment. Land Degrad Dev 21: 110–121. DOI: 10.1002/ldr.911
69. Thompson K, Bakker J, Bekker R (1997) The soil seed banks of North West Europe: methodology, density and longevity. United Kingdom: University Press of Cambridge.
70. Ortega M, Levassor C, Peco B (1997) Seasonal dynamics of Mediterranean pasture seed banks along environmental gradients. J Biogeogr 24: 177–195. DOI: 10.1046/j.1365-2699.1997.00080.x
71. Bochet E, García-Fayos P, Tormo J (2007) How can we control erosion of roadslopes in semiarid Mediterranean environments: Part I: seed dispersal and spontaneous colonization. Restor Ecol 15: 88–96. DOI: 10.1002/ldr.911
72. González-Alday J, Marrs RH, Martínez-Ruiz C (2009) Soil seed bank formation during early revegetation after hydroseeding in reclaimed coal wastes. Ecol Eng 35: 1062–1069. DOI:10.1016/j.ecoleng.2009.03.007
73. de la Riva EG, Casado MA, Jiménez MD, Mola I, Costa-Tenorio M, et al. (2011) Rates of local colonization and extinction reveal different plant community assembly mechanisms on road verges in central Spain. J Veg Sci 22: 292–302. DOI: 10.1111/j.1654-1103.2010.01248.x
74. Maestre FT, Quero JL, Gotelli NJ, Escudero A, Ochoa V, et al. (2012) Plant species richness and ecosystem multifunctionality in global drylands. Science 335: 214–218. DOI: 10.1126/science.1215442
75. Hall SH, Barton CD, Baskin CC (2010) Topsoil seed bank of an Oak-Hickory forest in eastern Kentucky as a restoration tool on surface mines. Restor Ecol 18: 834–842. DOI: 10.1111/j.1526-100X.2008.00509.x
76. Valladares F, Tena D, Matesanz S, Bochet E, Balaguer L, et al. (2008) Functional traits and phylogeny: What is the main ecological process determining species assemblage in roadside plant communities? J Veg Sci 19: 381–392. DOI: 10.3170/2008-8-18378

77. Zhang ZQ, Shu WS, Lan CY, Wong MH (2001) Soil seed bank as an input of seed source in revegetation of lead/zinc mine tailings. Restor Ecol 9: 378–385. DOI: 10.1046/j.1526-100X.2001.94007.x
78. Traba J, Azcárate FM, Peco B (2006) The fate of seeds in Mediterranean soil seed banks in relation to their traits. J Veg Sci 17: 5–10. DOI: http://dx.doi.org/10.1658/1100-9233(2006)017[0005:TFOSIM]2.0.CO;2. Accessed 2014 Jun 10
79. Bowen CK, Schuman GE, Olson RA, Ingram LJ (2005) Influence of topsoil depth on plant and soil attributes of 24-year old reclaimed mined lands. Arid Land Res Manag 19: 267–284. DOI:10.1080/15324980590951441

15

Soil Environmental Conditions and Microbial Build-Up Mediate the Effect of Plant Diversity on Soil Nitrifying and Denitrifying Enzyme Activities in Temperate Grasslands

Xavier Le Roux[1]*, Bernhard Schmid[2], Franck Poly[1], Romain L. Barnard[3], Pascal A. Niklaus[3], Nadine Guillaumaud[1], Maike Habekost[4], Yvonne Oelmann[5], Laurent Philippot[6], Joana Falcao Salles[1], Michael Schloter[7], Sibylle Steinbeiss[4], Alexandra Weigelt[8]

1 Université de Lyon, INRA, CNRS, Université Lyon 1, Microbial Ecology Centre (UMR 5557 CNRS, USC 1364 INRA), Villeurbanne, France, **2** Institute of Evolutionary Biology and Environmental Studies, University of Zurich, Zurich, Switzerland, **3** Institute of Plant Sciences, ETH Zurich, Zurich, Switzerland, **4** Max Planck Institute for Biogeochemistry, Jena, Germany, **5** Geoecology, University of Tübingen, Tübingen, Germany, **6** Laboratoire de Microbiologie des Sols, INRA, Dijon, France, **7** Research Unit for Environmental Genomics, Helmholtz Zentrum München, Neuherberg, Germany, **8** Institute of Ecology, Friedrich Schiller University, Jena, Germany

Abstract

Random reductions in plant diversity can affect ecosystem functioning, but it is still unclear which components of plant diversity (species number – namely richness, presence of particular plant functional groups, or particular combinations of these) and associated biotic and abiotic drivers explain the observed relationships, particularly for soil processes. We assembled grassland communities including 1 to 16 plant species with a factorial separation of the effects of richness and functional group composition to analyze how plant diversity components influence soil nitrifying and denitrifying enzyme activities (NEA and DEA, respectively), the abundance of nitrifiers (bacterial and archaeal *amoA* gene number) and denitrifiers (*nirK, nirS* and *nosZ* gene number), and key soil environmental conditions. Plant diversity effects were largely due to differences in functional group composition between communities of identical richness (number of sown species), though richness also had an effect per se. NEA was positively related to the percentage of legumes in terms of sown species number, the additional effect of richness at any given legume percentage being negative. DEA was higher in plots with legumes, decreased with increasing percentage of grasses, and increased with richness. No correlation was observed between DEA and denitrifier abundance. NEA increased with the abundance of ammonia oxidizing bacteria. The effect of richness on NEA was entirely due to the build-up of nitrifying organisms, while legume effect was partly linked to modified ammonium availability and nitrifier abundance. Richness effect on DEA was entirely due to changes in soil moisture, while the effects of legumes and grasses were partly due to modified nitrate availability, which influenced the specific activity of denitrifiers. These results suggest that plant diversity-induced changes in microbial specific activity are important for facultative activities such as denitrification, whereas changes in microbial abundance play a major role for non-facultative activities such as nitrification.

Editor: Vishal Shah, Dowling College, United States of America

Funding: The Jena Experiment is funded by the Deutsche Forschungsgemeinschaft (DFG, FOR 456) with additional support from the Friedrich Schiller University of Jena and the Max Planck Society. This work is a part of the DISCOVER project funded by the French ANR "Biodiversité" programme. B Schmid was additionally funded by the Swiss Science Foundation (grant number 31-65224-01). These funders had no role in study design, data collection and analysis, decision to publish, or preparation of the manuscript.

Competing Interests: The authors have declared that no competing interests exist.

* E-mail: Xavier.le-roux@univ-lyon1.fr

Introduction

In a context of unprecedented rates of species extinction and human-induced shifts in the composition of biological communities, understanding how changes in plant diversity impact ecosystem functioning is of paramount importance [1], [2]. During the last decades, several studies have provided evidence of the functional importance of biodiversity for ecosystem processes [3], [4], [5], but the specific mechanisms underlying the observed patterns have remained unclear. The question for plant diversity–ecosystem functioning research is thus no longer whether plant diversity matters, but how it matters [6], [7]. More detailed analyses have shown that among the different components of biodiversity, the number of species (species richness) and functional groups (functional group richness), the presence and abundance of particular species or functional groups, and the diversity of species functional traits are the main drivers of plant diversity effects on ecosystem functioning [7], [8], [9].

However, many biodiversity studies to date have focused on the effect of biodiversity on primary productivity, while key soil processes and associated microbial activities have less been studied, despite their importance for ecosystem functioning and plant growth, but see [10], [11], [12], [13], [14], [15]. In the present study, we focus on enzyme activities involved in nitrification and denitrification. These two key steps of the soil nitrogen (N) cycle exert key controls on ecosystems from local to global scale. In particular, nitrification and denitrification control (i) soil inorganic N availability and subsequent primary production, and (ii) N losses from ecosystems. The latter includes nitrate leaching, which plays a crucial role in groundwater quality, and emissions of nitrogen monoxide and nitrous oxide that impact atmospheric chemistry and global warming [16], [17]. Biodiversity effects on these processes are therefore not only important for the ecosystem's N budget and biomass production, but also for the way in which the ecosystem interacts with adjacent systems and the global climate [18]. Soil microorganisms are subject to specific ecological limitations and therefore may not respond like plant production to different components of biodiversity.

Soil nitrate is primarily produced by microbial oxidation of ammonium under aerobic conditions by autotrophic microorganisms during nitrification [19]. During the anaerobic process of denitrification [20], heterotrophic microorganisms utilize nitrate to oxidize available organic carbon compounds, releasing gaseous nitrous oxides and dinitrogen. Both processes respond sensitively to substrate availability and redox conditions. By altering the main driving variables of nitrification and denitrification (i.e. soil inorganic N availability, organic C availability, water content and oxygen availability, and pH), plant biodiversity can affect nitrification and denitrification. Several lines of evidence suggest that this is likely to take place:

(1) Soil nitrate concentration has been shown to decrease with increasing plant diversity [21], [22]. Moreover, nitrifying activity in grassland was reduced with increasing plant diversity, likely due to enhanced capture of ammonium by plants [21], though other studies found no effect of plant diversity on nitrification [23], [24]. It is also well established that the presence of legumes generally increases N availability to plant communities [25], thereby largely governing N cycling and ecosystem functioning in many grassland biodiversity studies [1], [22].
(2) The level or vertical distribution of soil water can be affected by plant diversity [26] which may subsequently affect nitrification or denitrification, though soil water content can be unaffected by plant diversity in temperate grasslands [27].
(3) Dissolved organic carbon (DOC) provides a readily available C source to denitrifying soil microorganisms. While the factors regulating DOC concentrations in soils are not yet well understood, the documented plant diversity effects on primary production and rooting patterns could possibly also affect DOC levels, and thereby the activity of denitrifiers.
(4) Plants produce compounds such as root exudates that affect microorganisms. In addition, nitrification and denitrification are sensitive to soil pH, and plants can acidify the rhizosphere and the surrounding soil. All these processes could vary with plant identity, which might affect nitrification and denitrification [28], [29].

Only a few studies have analysed the effect of plant diversity on potential nitrification and denitrification or the characteristics of nitrifying and denitrifying communities to date [11], [14], [15], [24], generally in mesocosms and on a short term (a few months). Moreover, these studies did not unravel the effects of plant species richness per se, species identity, and functional group presence/abundance, which requires to use a large pool of plant species, and did not hierarchize the importance of environmental factors and microbial abundances to explain the observed changes in potential microbial activities through causal relationships. Here we present a detailed analysis of plant diversity effects on nitrification and denitrification in a large-scale field trial (Jena Experiment) in which plant species richness and functional group presence/abundance were manipulated independently. All measurements were made in the 5th and 6th years after establishment of experimental plant communities, well clear of initial soil disturbance. The objectives of the study were (1) to identify the main components of plant diversity (richness vs. abundance or presence of key functional groups) that influence soil nitrifying and denitrifying enzyme activities, which are key characteristics influencing soil N dynamics; and (2) to assess the effects of plant diversity on nitrifying and denitrifying enzyme activities in relation to effects on putative environmental drivers of enzyme activities (soil moisture, C availability, nitrate and ammonium availability) and on abundances of nitrifiers and denitrifiers.

Materials and Methods

Ethics Statement

No specific permits were required for the described field studies. The Jena field site is a former arable land owned by an agricultural collective. The ground is leased by the Research consortium of the Jena experiment for scientific purpose including soil sampling and other experimental manipulations. The site is not protected in any way. The areas studied do not involve any species endangered or protected in Germany.

Study System and Experimental Design

The present study was carried out as part of a large biodiversity experiment located in Jena, Germany (50°57′N, 11°37′W, elevation 130 m a.s.l.), [30]. The experimental study system was derived from a typical Central European mesophilic grassland community as it was traditionally used for haymaking. Sixty common plant species of this community were used and divided by ordination using 17 species traits into four functional groups that could be identified *a posteriori* as grasses, legumes, "tall" non-leguminous herbs, and "small" non-leguminous herbs, [30].

The Jena Experiment was sown in 2002 and used an almost fully orthogonal species richness x functional group composition design where species of all functional groups occurred in monoculture and single functional groups could make up communities of up to 16 species (Figure S1), [30]. The experimental area was partitioned into four blocks, containing a total number of 88 large plots (20×20 m). Experimental communities of 1, 2, 4, 8 and 16 species were established in the plots. Plant species were chosen from a pool of 60 species with the constraint that all plant functional groups were evenly represented at each level of plant species richness. Functional group composition is defined as the combination of functional groups represented in a community. Functional group composition could be decomposed into the following contrasts: factors for the presence/absence of each functional group, covariables for the proportions of each functional group if present, and interactions. An alternative decomposition of functional group composition into contrasts was a full factorial with the four factors of presence/absence of each functional group and all interactions of these factors. Because only 16 of the 60 species were established as monocultures on the large plots, the species that were not established in the large plots were grown on 44 additional 3.5×3.5 m plots. Previous work using the same plots has shown that plot size is unimportant for biodiversity–ecosystem functioning relationships [31]. In all mixtures, plant species were sown at maximum evenness. All plots were manually weeded regularly, thus maintaining plant species richness at the planned levels or slightly below in cases where a plant species did not establish or died. The strong correlation between the number of sown species and the number of observed target species ($R^2>0.99$) verified the successful establishment of the species richness gradient. The experimental plant communities were mown twice per year and were not fertilized.

Soil Sampling

Soil samples (0–8 cm depth) were collected using corers (8 cm diameter) in October 2006 at the end of the growing season. Ten samples were randomly taken for each large plot (total of 880 soil cores), and five samples were randomly taken for each small plot (additional 220 soil cores). For each plot, a composite soil core was obtained by pooling one half of each core taken in the plot. The remaining half-cores were kept as individual soil samples. For each composite or individual sample, fresh soil was sieved using 2-mm mesh size, homogenized and stored at +4°C a few days before measurement. A sub-sample was immediately frozen at −20°C before DNA extraction. Soils were also similarly sampled from all the large plots during another field campaign in October 2007.

Denitrifying and Nitrifying Enzyme Activity Assays

Denitrifying enzyme activity, DEA, was measured in fresh soils according to Patra et al. [32] on all the composite soil samples in 2006 and 2007 and on individual soil samples (10 per plot) for 28 plots in 2006. DEA was determined as the linear rate of production of N_2O during a 8-hour incubation (28°C, 200 μg NO_3^--N g^{-1} and 1 mg C g^{-1} added) using a gas chromatograph (Agilent P200, Santa Clara, CA, USA). A 90:10 He-C_2H_2 atmosphere provided anaerobic conditions and inhibited N_2O-reductase activity. Nitrifying enzyme activity, NEA, was measured on fresh soil samples as detailed by Patra et al. [32]. Soil nitrate content was measured before and after a 7-hour aerobic incubation (28°C, 200 μgN-$(NH_4)_2SO_4$ g^{-1} added).

Quantification of nitrifier and denitrifier abundances. In early 2008, DNA was extracted for each frozen soil sub-sample of the October 2006 field campaign using the PowerSoil™ DNA Isolation Kit (MO BIO Laboratories, Carlsbad, CA, USA), and the abundance of denitrifiers was estimated by quantitative PCR (qPCR) targeting the genes encoding the catalytic subunit of the key enzymes of the denitrification pathway. Fragments of the *nirK*, *nirS* and *nosZ* genes encoding the copper and cd_1 nitrite reductases and the nitrous oxide reductase, respectively, were amplified as described by Baudoin et al. [33]. All assays were run using known copy number of linear plasmids containing targeted genes from either *Bradyrhizobium japonicum* USDA110, *Pseudomonas aeruginosa* PAO1, *Agrobacterium tumefaciens* C58 or *Sinorhizobium meliloti* 1021, containing known copy numbers of targeted genes as external standards [34].

Though nitrification is a two step-process, we targeted the first one, ammonia oxidation, as it is often assumed to limit nitrification rate [19]. The abundance of *amoA* gene copies from ammonia oxidizing bacteria (AOB) was quantified using a SybrGreenI-based real-time PCR technique as described in Leininger et al. [35] with the exception that the cloned *amoA* gene of *Nitrosomonas multiformis* ATCC25196 was used as standard. The abundance of *amoA* gene copies from ammonia oxidizing archeae (AOA) was measured by quantitative PCR targeting the archaeal *amoA* gene, according to the methodology described in Le Roux et al. [36]. Melting curve analysis confirmed the specificity of the amplification, and amplification efficiencies higher than 80% were obtained for all PCR reactions. All measurements were performed in triplicate.

For all quantitative PCR approaches, possible inhibitory effects of co-extracted humic compounds in soil extracts were checked by dilution series, but no inhibition was observed.

Soil Environmental Variables

Soil moisture was determined by the gravimetric technique for the samples collected in 2006. Soil ammonium and nitrate concentrations were determined by extracting a fresh soil sub-sample equivalent of 5 g dry soil with 50 mL of 2 M KCl for 30 min. After filtration, the extracts were analysed for ammonium and nitrate by continuous flow analysis (Skalar, Breda, Germany) according to Oelmann et al. [37]. Dissolved organic C (DOC) in soil solution was determined using an elemental analyzer (high-TOC, Elementar Analysensysteme GmbH, Hanau, Germany) in soil solution collected during a 2-week period around sampling dates (glass suction plates with 1 μm pore size installed at 10 cm depth; UMS GmbH, Munich, Germany). Soil pH was not affected by treatments so that pH was not included as a possible driver of observed changes in soil enzyme activities.

Statistical Analyses

The enzyme activity data were analyzed by multiple regression [38]. First, spatial variation was eliminated as a block effect and with covariables X and Y for north–south and west–east trends. We fitted a two-dimensional response surface of the type $X+Y+X*Y+X^2+Y^2+X^2*Y^2$ after the block term and then dropped all terms that were not significant from any particular model. After accounting for spatial variation by fitting these terms (which are summarized in the ANOVA tables 1, 2 and S2 in a single line), the treatment factors species richness, decomposed into a log-linear contrast (logSR) and deviation (SR), and the presence and initial sowing proportion of each of the four functional groups were fitted. Terms that were not significant were dropped from the model, resulting in the simplified models shown in tables 1 and S1 (the SR term was kept in the model for DEA in 2007, even though it was not significant, to help comparison with results obtained for 2006 activity data). Replacing initial sowing proportions of functional groups with actual aboveground biomass proportions yielded very similar results. To investigate the mechanisms explaining the observed plant diversity effects, we (i) added environmental soil covariables (soil moisture, soil nitrate, soil ammonium, and DOC) and microbial abundance data to the regression models, fitting them after the terms for spatial variation but before the plant diversity terms, and (ii) used path analysis to test potential causal relationships between plant diversity terms, environmental covariables, microbial abundance, and soil enzyme activities. The path analyses were calculated with variance-covariance matrices to assess significances of path coefficients; the standardized solutions (corresponding to the analysis of correlation matrices) are presented here.

Although technical replicates that went into each single data point were sometimes used (in particular for microbial abundance data), we never used technical replicates to compute error terms in the analysis but always averaged them for each biological replicate. Thus, the residual means squares in Tables 1 and 2 give the among-plot residual variation which is the appropriate error for our statistical tests, and in path analyses errors can be seen by the arrows on the dependent variables with no causal variable at the beginning of the arrow.

Table 1. ANOVA results of the multiple regression models fitted for the two dependent variables nitrifying and denitrifying enzyme activities for October 2006, using plant diversity components only as explanatory variables.

Nitrifying enzyme activity				
Source of variation	**d.f.**	**s.s.**	**m.s.**	**F pr.**
Spatial variation	4	1.237	0.309	<.001
LogSR	1	0.183	0.183	0.009
Legume abundance (% sown)	1	2.446	2.446	<.001
Residual	69	1.755	0.025	
Total	75	5.62	0.0749	
Denitrifying enzyme activity				
Source of variation	**d.f.**	**s.s.**	**m.s.**	**F pr.**
Spatial variation	6	14.29	2.38	<.001
logSR	1	0.96	0.96	0.004
SR	3	1.05	0.35	0.025
Presence of legumes	1	2.57	2.57	<.001
Grass abundance (% sown)	1	0.54	0.54	0.027
Residual	63	6.62	0.11	
Total	75	26.03	0.35	

LogSR and SR respectively refer to the decomposition of the species richness factor into a contrast for log-linear richness and remainder (deviation from log-linearity). For each table, values are presented only for variables that had a significant effect on activity. ANOVA results for 2007 are very similar and are presented in Table S1. d.f.: degree of freedom; s.s.: sum of squares; m.s.: mean sum of squares; F pr.: significance level (F-test).

Results

Effects of Components of Plant Diversity on NEA and DEA

NEA increased with the percentage of legumes in a community, but decreased with species richness, the latter effect size being smaller (Fig. 1 and 2, Table 1). On average, NEA increased by about 60% from communities without legumes to communities containing only legumes. DEA increased in the presence of legumes and with plant species richness, but decreased with the percentage of grasses (Fig. 1 and 2, Table 1). A small proportion of legumes was sufficient to notably increase DEA, while increasing the proportion of legumes beyond 25% had no further effect (Fig. 2). DEA was on average 20% higher at highest plant species richness than in monocultures. Multiple regression analysis revealed that the different components of plant diversity explained 60% and 44% of the overall variation in NEA and DEA, respectively (after having accounted for spatial variation). Functional group composition explained 13.4 and 1.6 times more variation of NEA and DEA, respectively, than plant species richness alone (Fig. 1).

We verified that these results remained valid when including all the monocultures (i.e. the 16 monocultures on large plots plus the 44 additional ones on small plots) in the multiple regression analysis (not shown). Furthermore, NEA and DEA were very similar between the two study years ($NEA_{2007} = 0.988 * NEA_{2006}$, $r^2 = 0.70$, $n = 88$; $DEA_{2007} = 0.953 * DEA_{2006}$, $r^2 = 0.69$, $n = 88$), and the relative effects of the different plant diversity components were also very similar (Table S1).

Table 2. ANOVA results of the multiple regression models fitted for the two dependent variables NEA and DEA for October 2006, using soil variables, microbial abundances and plant diversity components as explanatory variables.

Nitrifying enzyme activity				
Source of variation	**d.f.**	**s.s.**	**m.s.**	**F pr.**
Spatial variation	4	1.24	0.31	<.001
Ammonium	1	0.16	0.16	0.009
AOB abundance	1	0.65	0.65	<.001
logSR	1	0.02	0.02	NS
Legume abundance (% sown)	1	2.03	2.03	<.001
Residual	66	1.51	0.02	
Total	74	5.61	0.08	
Denitrifying enzyme activity				
Source of variation	**d.f.**	**s.s.**	**m.s.**	**F pr.**
Spatial variation	6	14.29	2.38	<.001
Moisture	1	1.51	1.51	<.001
Nitrate	1	2.27	2.27	<.001
nirK number*	1	0.25	0.25	NS
logSR	1	0.22	0.22	NS
Presence of legumes	1	0.97	0.97	0.002
Grass abundance (% sown)	1	0.47	0.47	0.03
Residual	62	5.98	0.10	
Total	74	25.95	0.35	

*results remain unchanged when considering nirS- or nosZ-harbouring bacteria. LogSR and SR respectively refer to the decomposition of the species richness factor into a contrast for log-linear richness and remainder (deviation from log-linearity). For each table, values are presented only for variables that had a significant effect on activity. In particular, dissolved organic carbon explained only little variation and was therefore not included. Abbreviations are as in Table 1.

Relationships between Soil Enzyme Activities and Abundance of (de)nitrifiers

The abundance of ammonia oxidizing bacteria, AOB, was five-fold higher than that of ammonia oxidizing archaea, AOA, in Jena soils (Fig. 3). A significant, although weak, correlation was observed between NEA and the AOB abundance: on average, soils with high nitrifying potential had significantly higher copy number of AOB *amoA* sequences (Fig. 3 top). Although the copy number of *amoA* sequences from AOA was significantly correlated to that of AOB ($AOAnumber = 0.21*AOBnumber$; $r^2 = 0.25$), no significant relationship was observed between NEA and AOA abundance (Fig. 3 bottom).

No significant correlation was observed between DEA and the abundance of denitrifiers (Fig. 4). This result was obtained for whatever targeted sequences we used to assess denitrifier abundance, i.e. for *nirS*-, *nirK*- or *nosZ*-harbouring denitrifiers. The abundance of the three sequence types were correlated with each other (*nirK* sequence number = 0.92**nirS* sequence number, $r^2 = 0.30$; and *nirS* sequence number = 5.66**nosZ* sequence number, $r^2 = 0.26$).

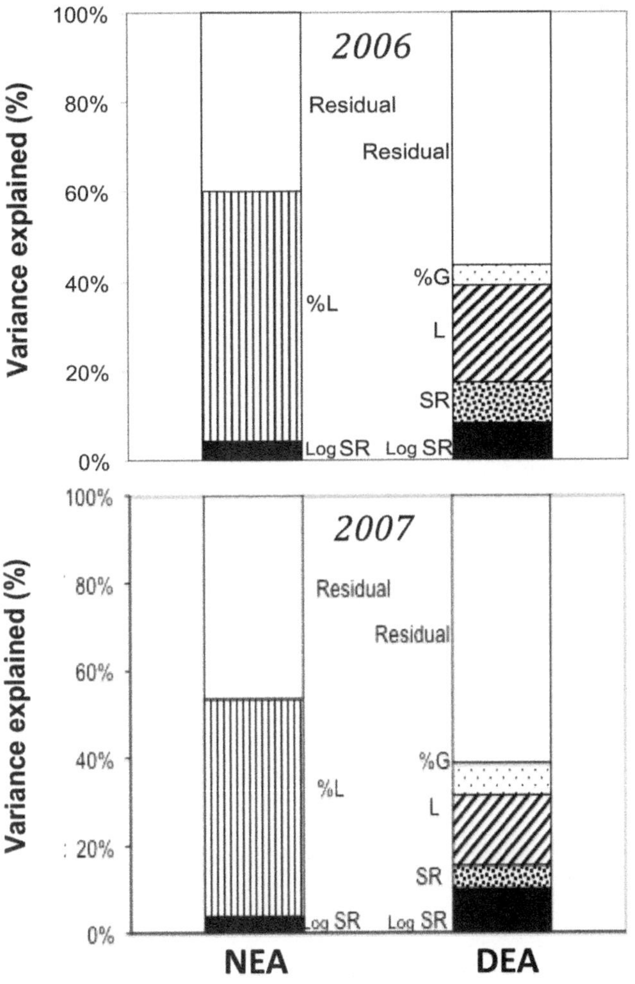

Figure 1. Plant Diversity Influence On Soil Enzyme Activities. Percentages of variation in nitrifying and denitrifying enzyme activities, respectively NEA and DEA, significantly explained by components of plant diversity. Data are presented (top) for 2006, and (bottom) for 2007. LogSR and SR refer to the decomposition of the species richness factor into a contrast for log-linear richness and deviation from log-linearity, respectively. L refers to presence of legumes, and %L and %G refer to the percentage of legumes and grasses, respectively. Significance levels of the effects of plant diversity components observed in 2006 and 2007 are presented in Table 1 and Table S1, respectively.

Role of Environmental Variables and Microbial Abundances for the Relationships between Components of Plant Diversity and Soil Enzyme Activities

Significant correlations were observed between each soil enzyme activity and some environmental variables. In particular, DEA was significantly correlated to both soil nitrate concentration ($p<0.001$, Figure S2) and soil moisture ($p<0.001$, not shown), whereas no significant relationship was observed with DOC. Similarly, NEA was significantly correlated ($p=0.009$) to soil ammonium concentration (not shown).

Including plant diversity components, soil environmental variables, and microbial abundances in our multiple regression analyses, we found that a substantial part of the plant diversity effect on NEA and DEA was attributable to plant-induced changes in key environmental variables and microbial abundances (Table 2, Fig. 5). In particular, all the effect of plant species richness on NEA was explained by an effect on the abundance of ammonia oxidizers: the richness effect was not significant anymore when AOB abundance was first accounted for. Furthermore, one fifth of the legume effect on NEA was explained by plant-induced variations in soil ammonium concentration. However, nitrifier abundance and soil ammonium explained only 19% of the variation in NEA, whereas an additional 46% of the variation was related to the percentage of legumes without being explained by any measured covariable.

All the effect of plant species richness on DEA was explained by an effect on soil moisture: the species richness effect was not significant anymore when moisture was first accounted for (Table 2, Fig. 5). Furthermore, plant diversity-induced changes in soil moisture and nitrate concentration explained 33.1% of the variation in DEA values, i.e. 76% of the observed effects of the different plant diversity components on DEA, whereas dissolved organic carbon was not identified as a key driver for DEA. Only 12.6% of the variation in DEA was related to the presence of legumes and the percentage of grasses without being explained by a measured soil environmental variable (Fig. 5).

Path analyses were used to provide a comprehensive analysis of predicted cause-effects relationships between plant diversity components, microbial abundance and soil environmental variables, and NEA or DEA (Fig. 6). Our results confirmed that all the negative effect of plant species richness on NEA was mediated by a decrease in the abundance of ammonia oxidizers. The positive effect of the percentage of legumes on NEA was partly explained by a positive effect on the abundance of ammonia oxidizers, and to a lesser extent on ammonium concentration (Fig. 6).

Similarly, path analysis showed that the positive effect of plant species richness on DEA was entirely mediated by an increase in soil moisture (Fig. 6). Both the positive effect of the presence of legumes and the negative effect of the proportion of grasses on DEA were partly explained by respectively a positive and a negative effect on nitrate concentration. However, apparent direct effects of the presence of legumes and the proportion of grasses on DEA remained significant (Fig. 6).

Discussion

Plant Diversity Effect on Soil Enzyme Activities: a Matter of Microbial Abundance or Specific Activity?

In this study, a significant correlation was observed between NEA and the abundance of nitrifiers, in particular ammonia oxidizing bacteria. The observed relationship between NEA and AOB abundance is consistent with the correlations obtained between nitrification levels and the abundance of AOB in different ecosystems, e.g. [32], [36], [39]; but see [40]. The data published so far have shown a higher abundance of AOA than AOB in a range of soils [35], with indications that the AOA:AOB ratio increases with increased nutrient limitation. Lamb et al. [15] reported that the richness of non-legume plant species increased the prevalence of AOA relative to AOB. In contrast, we found here that AOB abundance was 5-fold higher than AOA abundance in the Jena fertile grassland soils, though the abundance of AOB and AOA were correlated. This is consistent with the report that AOB and AOA abundances were correlated in another fertile grassland system [36]. Given the higher number of AOB as compared to AOA in Jena soils, and given that transformation rates for ammonium into nitrite by AOA are currently assumed to be lower than for AOB [41], it can be assumed that the major part of nitrification in the grassland soils studied was related to AOB activity. Our results

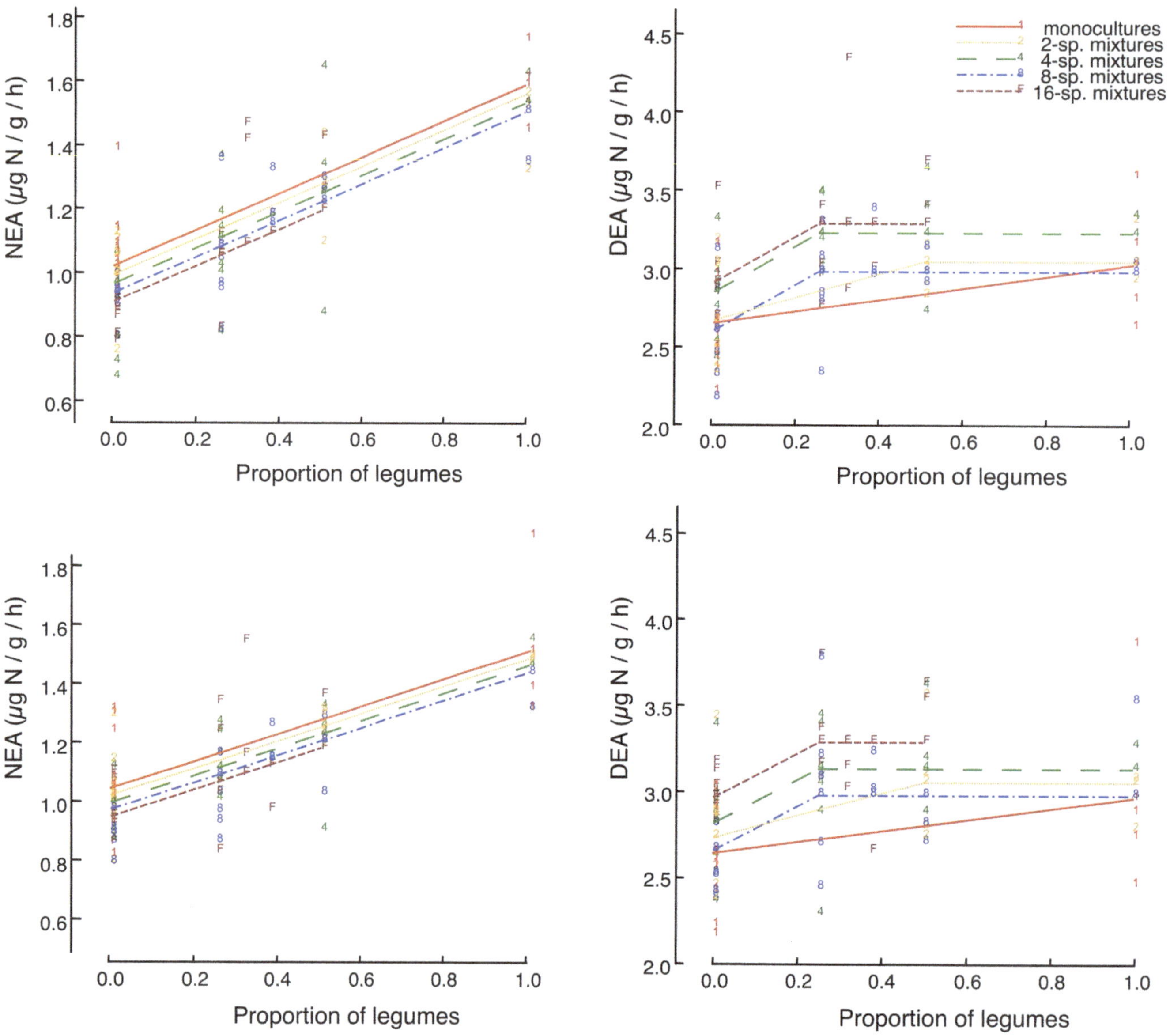

Figure 2. Unravelled effects of Legumes And Plant Species Richness On Soil Enzyme Activities. Changes in (left) NEA and (right) DEA as a function of the proportion of legumes within the plant community (x-axis) and plant species richness in the community (numbers associated to points). Data are presented for (top) 2006, and (bottom) 2007. The lines connect fitted values from corresponding multiple regression models with significance levels presented in Tables 1 and S1. For example, for the NEA-2006 plot (top, left), the increase of NEA with the proportion of legumes at a given richness level is significant at $p<0.001$ and the continuous increase of NEA with richness is significant at $p=0.009$. Note that for NEA the five regression lines are not all significantly different from each other, but reflect the significant change with log richness (for each doubling of species richness the lines move down by a constant distance), which is different for DEA.

actually show that plant diversity conditioned NEA at least partly through the build-up of nitrifying bacteria.

In contrast, no correlation was observed between DEA and the abundance of denitrifiers, although three different, relevant genes were targeted to assess denitrifier abundance. Lack of or weak correlation between DEA and denitrifier abundance has often been reported for a range soil systems, e.g. [42], though significant correlations were sometimes reported [43]. This shows that plant diversity conditioned DEA not through changes in denitrifier abundance, but by influencing the specific activity of denitrifiers, i.e. the mean activity per denitrifying cell.

The contrasting results obtained for the two activities can be explained by the roles of those activities for the two types of microoganisms. Indeed, nitrification is the key activity through which ammonia oxidizing organisms acquire energy and grow, so that a significant coupling between changes in the activity and abundance of ammonia oxidizers is expected. In contrast, denitrification is a facultative activity for denitrifying organisms that switch to another heterotrophic metabolism when oxygen availability is sufficient. Thus the enzyme pool per denitrifying cell can be strongly affected by plant diversity-induced changes in environmental conditions in soil. This shows that different representations are needed to model the relationships between plant diversity and soil functions: for nitrification, nitrifier abundance is a key variable affected by plant diversity-induced

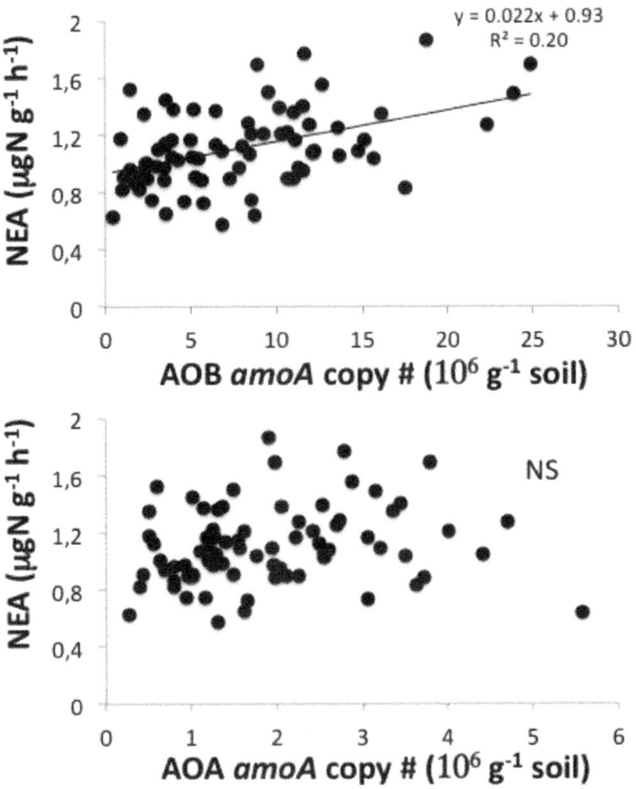

Figure 3. Relationship Between Plant Diversity-induced Changes In Nitrifier Abundance And Nitrifying Enzyme Activity. Correlations between NEA and (top) the abundance of ammonia oxidizing bacteria assessed by the number of their *amoA* sequences, and (bottom) the abundance of ammonia oxidizing archaea assessed by the number of their *amoA* sequences, in 2006. When significant ($p<0.05$), the linear regression is drawn.

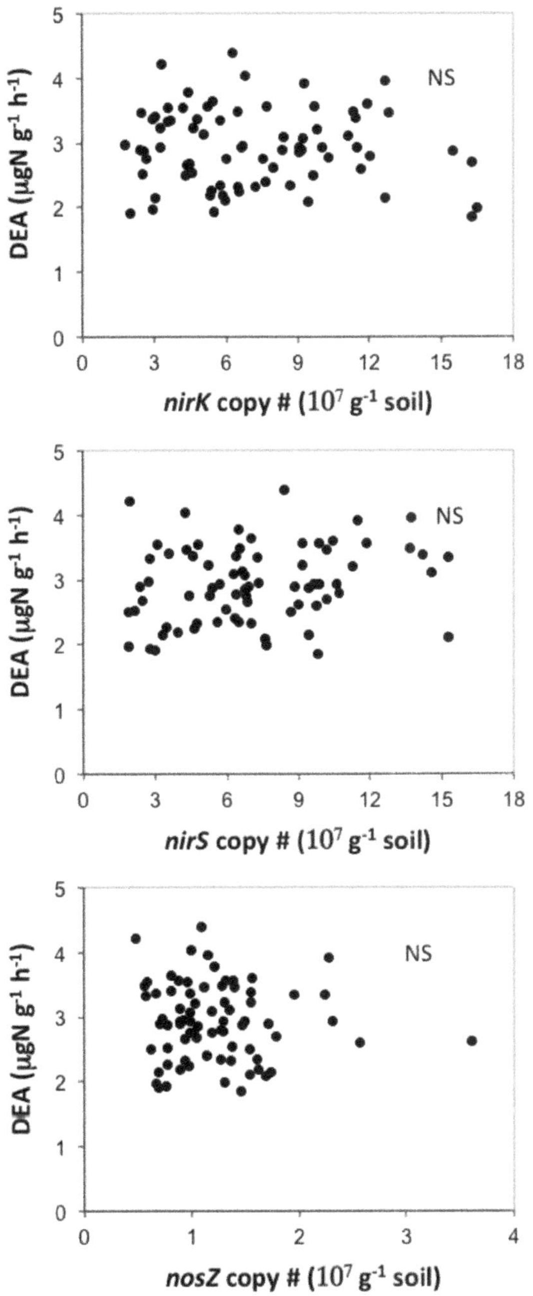

Figure 4. Relationship Between Plant Diversity-induced Changes In Denitrifier Abundance And Denitrifying Enzyme Activity. Correlations between DEA and the abundance of (top) *nirS*–harbouring denitrifiers, (middle) *nirK*-harbouring denitrifiers and (bottom) *nosZ*-harbouring denitrifiers in soil, in 2006. No significant correlation was observed (all $p>0.05$).

changes in soil conditions, whereas denitrifier specific activity is the key variable affected for denitrification.

Underlying Drivers of the Effects of Plant Diversity Components on Enzyme Activities

The results of the multiple regression and path analyses show that the presence or proportion of legumes, the proportion of grasses along with plant species richness were key determinants of NEA and DEA, legumes playing a major role for NEA. Several studies have already reported the key role of legumes on ecosystem processes, via an associated enhancement of soil N availability, in biodiversity experiments [22], [23], [37], [44], [45], though climatic and soil nutritional limitations can dampen legume effects [9], [21], [46]. In particular, an effect of legumes was reported on nitrification, denitrification and nitrous oxide fluxes [24] and nitrifiers [14]. In addition, DEA was negatively related to the proportion of grasses in a plant community. This is consistent with previous reports showing that a legume-induced increase in soil mineral N can be compensated by the ability of grasses to effectively take up mineral N even at very low concentrations in the soil due to their extensive rooting system [23], [37], [44], [47]. However, multiple regression and path analyses show that increased ammonium and nitrate concentrations in soil only partly explained the legumes effects on NEA and DEA, respectively. Similarly, decreased soil nitrate concentration only partly explained the grasses effects on DEA. Actually soil mineral N concentrations under field conditions result from various N processes depleting or increasing the mineral N pool; snapshot measurements of mineral N concentrations reflect quasi-steady state concentrations rather than turnover and thus are questionable proxies of actual N availability [48]. For example, Malchair et al. [14] reported that soil ammonium concentration does not reflect ammonium availability to nitrifiers, which could explain the lack of relationship between NEA or nitrifier abundance

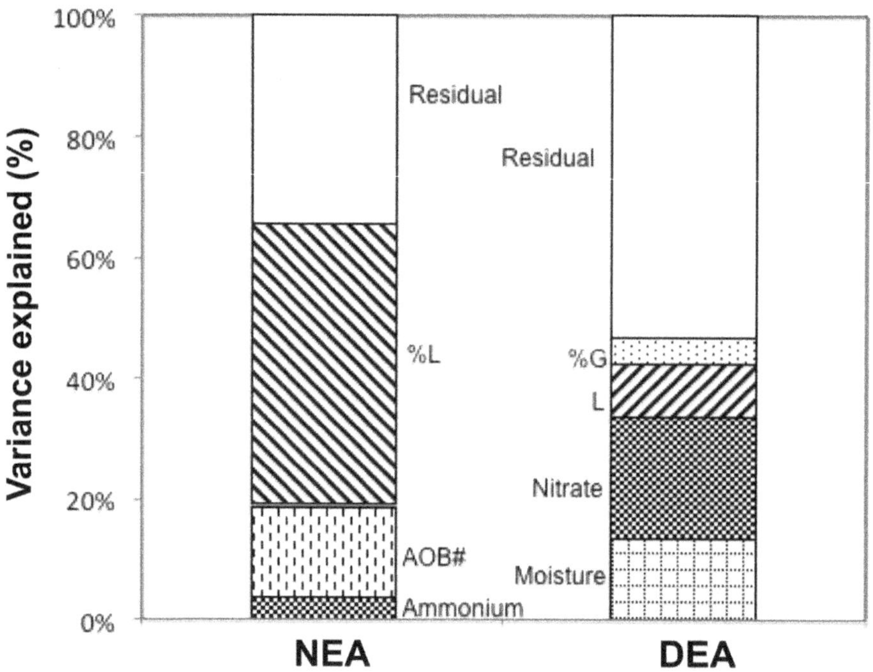

Figure 5. Soil Conditions And Microbial Abundances Partly Explaining Plant Diversity Effect On Soil Enzyme Activities. Percentages of variation in nitrifying and denitrifying enzyme activities, respectively NEA and DEA, significantly explained by environmental soil variables and microbial abundance, and remaining percentages of variation significantly explained by components of plant diversity (2006 data). See Fig. 2 for legend, and table 2 for full regression analyses and significance levels.

and ammonium concentration. This would explain why the residual direct effect of legumes or grasses on NEA or DEA is still significant even after accounting for an indirect effect through ammonium or nitrate concentration (see figure 5). Interestingly, our results show that legumes increased DEA regardless of their abundance, likely due to a change in the factor(s) limiting the specific activity of denitrifers (i.e. activity per denitrifying cell). The limiting factor would be N availability for non-legume plant mixtures, but 25% of legumes in a community (Figure 2) would be enough to obtain non limiting nitrate concentrations; C or anaerobic microsite availability would thus become limiting for denitrifiers in plant mixtures with at least 25% of legumes.

Our results also show a positive effect of plant species richness sensu stricto on DEA consistently with a few other studies [49]. We show that this effect was entirely mediated by

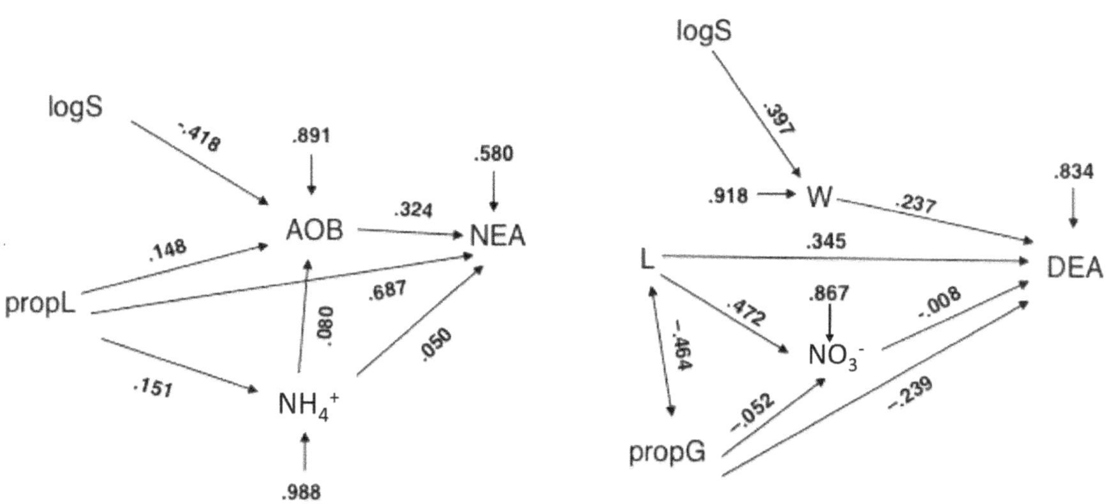

Figure 6. Path Diagram: Plant Diversity Components, Environmental Soil Variables, Microbial Abundances and Enzyme Activities. Path diagram for effects of plant diversity components on NEA and DEA, showing how observed effects can be explained entirely or partly by indirect effects via microbial abundance and key environmental drivers for (right) nitrifiers and (left) denitrifiers (2006 data). Lack of fit of path models was not significant for NEA ($\chi 2 = 1.38$, d.f. = 3, P = 0.71) or significant for DEA ($\chi 2 = 19.53$, d.f. = 7, P = 0074). Path coefficients indicate by how many standard deviations the effect variable would change if the causal variable were changed by one standard deviation.

increased soil moisture. This is consistent with a recent study showing that plant species richness effects on in-situ ammonification rates were mediated through soil moisture [50]. For a Mediterranean grassland it has been reported that soil moisture explained 60% of the variation in DEA across a full factorial combination of simulated components of global change [51]. Here, we show that this also applies across a gradient of plant species richness, though soil nitrate had also an important role. In contrast, NEA was negatively related to plant species richness: this was entirely explained by a negative effect of richness on the abundance of ammonia oxidizers, which could be partly due to the richness-induced increase in soil moisture and/or decrease in soil ammonium availability, but this could not be captured in our path analysis.

Though plant diversity effects cannot be generalised without considering the level of soil resources [52], the experimental setup of our study allowed us to efficiently disentangle the respective effects of plant species richness and plant functional group composition (the presence and relative abundance of each functional group in the plant community) on NEA and DEA. The interaction between legume effect and effects of other plant diversity components was not significant: a positive effect of legume relative abundance occurred independently of species richness, and the effects of species richness occurred independently of legume abundance. Moreover, our results show that these plant diversity components conditioned NEA at least partly through the build-up of nitrifying bacteria, likely driven by changes in soil ammonium availability. In contrast, plant diversity components conditioned DEA mainly through changes in soil moisture and nitrate that influenced the specific activity of denitrifiers rather than their abundance.

Accounting for Temporal and Spatial Variability of NEA and DEA

Nitrogen dynamics and availability in the soil are governed by many interacting processes such as ammonification, nitrification, uptake by plants and microorganisms, denitrification, leaching and volatilization [53]. These processes can be highly variable in time and space [54], [55]. To evaluate our accounting for spatial variability of soil enzyme activities in the present study, we compared - for 28 plots randomly selected - the mean of activity values obtained in the 10 individual soil cores sampled per plot to the value obtained in the composite soil sample derived from the same plot during the 2006 field campaign. Strong correlations were observed between the two for both DEA ($DEA_{composite} = 1.06 * DEA_{mean}$, $r^2 = 0.92$, n = 28 plots) and NEA ($NEA_{composite} = 1.00 * NEA_{mean}$, $r^2 = 0.79$, n = 28 plots), thus demonstrating that our composite soil samples adequately integrated within-plot spatial variability in soil activities.

Temporal variability in nitrification and denitrification can be large: for instance, large emission bursts of nitrous oxide from soil in the field are triggered within hours of wetting events, making them difficult to integrate [55]. However, NEA and DEA are related to nitrification and denitrification by integrating the activity of living nitrifiers and denitrifiers over periods in the order of at least a few days in the field. Moreover, the observed patterns were consistent over time: (i) activities measured in both years were strongly correlated, and (ii) the drivers identified were the same in both years. Additional evidence for the constancy of these drivers is indicated by the consistency over seasons and years of both the soil moisture patterns and the effects of legumes and grasses on soil N concentrations in Jena plots [37]. Overall, these results support the robustness of our findings.

Conclusion

The main objective of biodiversity experiments with artificially established model ecosystems such as the Jena experiment is to shed light on the driving mechanisms of biodiversity–ecosystem functioning relationships [56]. Here, we identified and ranked the key components of plant diversity and associated mechanisms that drive enzyme activities associated to two belowground processes (nitrification and denitrification), which is rarely quantified in biodiversity–ecosystem functioning studies. Legume effect and effect of plant species richness per se represented 93% and 7% of total plant diversity effects on NEA, respectively. In contrast, the effects of legumes, grasses and plant species richness per se represented 50%, 11% and 39% of total plant diversity effects on DEA, respectively. Our results highlight the main biological processes explaining those patterns: plant diversity components conditioned NEA largely through the build-up of nitrifying bacteria likely responsive to soil ammonium availability. In contrast, plant diversity components conditioned DEA mainly through changes in the specific activity of denitrifiers that was influenced by changes in soil moisture and nitrate, the latter explaining 76% of the total plant diversity effects observed on DEA.

Our results demonstrate that the response of soil microbial enzyme activities to plant diversity varies qualitatively according to the type of activity, which should be well represented in ecosystem models. For non-facultative functions like nitrification, accounting for the changes in the abundance of microbial groups is a major step forward in understanding and predicting changes in soil functioning, which are expected to be relatively long lasting since significant changes in nitrifier abundance in such grasslands typically operate over period of months under moderate disturbance regime [36]. For facultative functions like denitrification, plant diversity-induced changes in the microbial activity depends mainly on changes in the specific activity of microorganisms modulated by changed soil conditions (here oxygen and to a lesser extent nitrate availability), which is likely more dynamic and, e.g., strongly weather-dependent. Ecosystem models should take this into account to adequately represent aboveground/belowground couplings induced by changes in plant diversity.

Supporting Information

Figure S1 Designs used to manipulate different components of plant diversity. The figure compares the design of the Jena Experiment used to manipulate different components of plant diversity to typical designs used in previous grassland plant species assemblage experiments aiming at studying biodiversity-ecosystem functioning relationships.

Figure S2 Relationship between denitrifying enzyme activity and nitrate concentration in soil for October 2006.

Table S1 ANOVA results of the multiple regression models fitted for the two dependent variables nitrifying and denitrifying enzyme activities for October 2007, using plant biodiversity components only as explanatory variables.

Acknowledgments

We thank E.D. Schulze and W.W. Weisser for allowing us to work in the framework of the Jena Experiment. We also thank E. Marquard for the comparison of sown and realized diversity in the Jena Experiment, and S. Rosenkranz for data provision. Nitrification and denitrification measurements were performed at the AME platform (UMR5557-USC1364, Lyon).

Author Contributions

Conceived and designed the experiments: XLR BS FP AW. Performed the experiments: XLR FP RB PN NG MH YO LP JS MS SS AW. Analyzed the data: XLR BS FP. Contributed reagents/materials/analysis tools: XLR FP RB PN NG MH YO LP JS MS SS AW. Wrote the paper: XLR BS FP RB PN NG MH YO LP JS MS SS AW.

References

1. Hooper DU, Chapin FS, Ewel JJ, Hector A, Inchausti P, et al. (2005) Effects of biodiversity on ecosystem functioning: A consensus of current knowledge. Ecological Monographs 75: 3–35.
2. Hector A, Bagchi R (2007) Biodiversity and ecosystem multifunctionality. Nature 448: 188–191.
3. Loreau M, Naeem S, Inchausti P, Bengtsson J, Grime JP, et al. (2001) Ecology - Biodiversity and ecosystem functioning: Current knowledge and future challenges. Science 294: 804–808.
4. Balvanera P, Pfisterer AB, Buchmann N, He JS, Nakashizuka T, et al. (2006) Quantifying the evidence for biodiversity effects on ecosystem functioning and services. Ecology Letters 9: 1146–1156.
5. Cardinale BJ, Srivastava DS, Duffy JE, Wright JP, Downing AL, et al. (2006) Effects of biodiversity on the functioning of trophic groups and ecosystems. Nature 443: 989–992.
6. Rosenfeld JS (2002) Logical fallacies in the assessment of functional redundancy. Conservation Biology 16: 837–839.
7. Naeem S, Wright JP (2003) Disentangling biodiversity effects on ecosystem functioning: deriving solutions to a seemingly insurmountable problem. Ecology Letters 6: 567–579.
8. Reich PB, Tilman D, Naeem S, Ellsworth DS, Knops J, et al. (2004) Species and functional group diversity independently influence biomass accumulation and its response to CO_2 and N. PNAS 101: 10101–10106.
9. Spehn EM, Hector A, Joshi J, Scherer-Lorenzen M, Schmid B, et al. (2005) Ecosystem effects of biodiversity manipulations in European grasslands. Ecological Monographs 75: 37–63.
10. Chung HG, Zak DR, Reich PB, Ellsworth DS (2007) Plant species richness, elevated CO_2, and atmospheric nitrogen deposition alter soil microbial community composition and function. Global Change Biology 13: 980–989.
11. Bremer C, Braker G, Matthies D, Beierkuhnlein C, Conrad R (2009) Plant presence and species combination, but not diversity, influence denitrifier activity and the composition of nirK-type denitrifier communities in grassland soil. FEMS Microbial Ecology 70: 377–387.
12. Salles JF, Poly F, Schmid B, Le Roux X (2009) Community niche predicts the functioning of denitrifying bacterial assemblages. Ecology 90: 3324–3332.
13. Salles JF, Le Roux X, Poly F (2012) Relating phylogenetic and functional diversity among denitrifiers and quantifying their capacity to predict community functioning. Frontiers in Microbiology 3: article 209, doi:10.3389/fmicb.2012.00209.
14. Malchair S, De Boeck HJ, Lemmens CMHM, Merckx R, Nijs I, et al. (2010) Do climate warming and plant species richness affect potential nitrification, basal respiration and ammonia-oxidizing bacteria in experimental grasslands? Soil Biology Biochemistry 42: 1944–1951.
15. Lamb EG, Kennedy N, Siciliano SD (2011) Effects of plant species richness and evenness on soil microbial community diversity and function. Plant and Soil 338: 483–495.
16. Firestone MK, Firestone RB, Tiedje JM (1980) Nitrous-oxide from soil denitrification - Factors controlling its biological production. Science 208: 749–751.
17. Le Roux X, Abbadie L, Lensi R, Serça D (1995) Emission of nitrogen monoxide from African tropical ecosystems: Control of emission by soil characteristics in humid and dry savannas of West Africa. Journal of Geophysical Research 100, D11: 23133–23142.
18. Hungate BA, Dukes JS, Shaw MR, Luo YQ, Field CB (2003) Nitrogen and climate. Science 302: 1512–1513.
19. Prosser JI (1989) Autotrophic nitrification in bacteria. Advances in Microbial Physiology 30: 125–181.
20. Tiedje JM (1988) Ecology of denitrification and dissimilatory nitrate reduction to ammonium. In: Zehnder AJB, editor, Environmental Microbiology of Anaerobes. John Wiley and Sons, N.Y., 179–244.
21. Niklaus PA, Kandeler E, Leadley PW, Schmid B, Tscherko D, et al. (2001) A link between plant diversity, elevated CO_2 and soil nitrate. Oecologia 127: 540–548.
22. Scherer-Lorenzen M, Palmborg C, Prinz A, Schulze ED (2003) The role of plant diversity and composition for nitrate leaching in grasslands. Ecology 84: 1539–1552.
23. Hooper DU, Vitousek PM (1997) The effects of plant composition and diversity on change. Science 302: 1512–1513.
24. Niklaus PA, Wardle DA, Tate KR (2006) Effects of plant species diversity and composition on nitrogen cycling and the trace gas balance of soils. Plant and Soil 282: 83–98.
25. Mallarino AP, Wedin WF (1990) Effect of species and proportion of legume on herbage yield and nitrogen concentration of legume-grass mixtures. Grass and Forage Science 45: 393–402.
26. Caldeira MC, Ryel RJ, Lawton JH, Pereira JS (2001) Mechanisms of positive biodiversity-production relationships: insights provided by $d^{13}C$ analysis in experimental Mediterranean grassland plots. Ecology Letters 4: 439–443.
27. Niklaus PA, Alphei J, Kampichler C, Kandeler E, Körner C, et al. (2007) Interactive effects of plant species diversity and elevated CO_2 on soil biota and nutrient cycling. Ecology 88: 3153–3163.
28. Le Roux X, Bardy M, Loiseau P, Louault F (2003) Stimulation of soil nitrification and denitrification by grazing in grasslands: do changes in plant species composition matter? Oecologia 137: 417–425.
29. Patra AK, Abbadie L, Clays A, Degrange V, Grayston S, et al. (2006) Effects of management regime and plant species on the enzyme activity and genetic structure of N-fixing, denitrifying and nitrifying bacterial communities in grassland soils. Environmental Microbiology 8: 1005–1016.
30. Roscher C, Schumacher J, Baade J, Wilcke W, Gleixner G, et al. (2004) The role of biodiversity for element cycling and trophic interactions: an experimental approach in a grassland community. Basic and Applied Ecology 5: 107–121.
31. Roscher C, Temperton VM, Scherer-Lorenzen M, Schmitz M, Schumacher J, et al. (2005) Overyielding in experimental grassland communities - irrespective of species pool or spatial scale. Ecology Letters 8: 419–429.
32. Patra AK, Abbadie L, Clays A, Degrange V, Grayston S, et al. (2005) Effects of grazing on microbial functional groups involved in soil N dynamics. Ecological Monographs 75: 65–80.
33. Baudoin E, Philippot L, Cheneby D, Brauman A (2009) Direct seeding mulch-based cropping increases both the activity and the abundance of denitrifier communities in a tropical soil. Soil Biology Biochemistry 41: 1703–1709.
34. Henry S, Bru D, Stres B, Hallet S, Philippot L (2006) Quantitative detection of the *nosZ* gene, encoding nitrous oxide reductase, and comparison of the abundances of *16S rRNA, narG,* Applied Environmental Microbiology 72: 5181–5189.
35. Leininger S, Urich T, Schloter M, Schwark L, Qi J, et al. (2006) Archaea predominate among ammonia-oxidizing prokaryotes in soils. Nature 442: 806–809.
36. Le Roux X, Poly F, Currey P, Commeaux C, Hai B, et al. (2008) Effects of aboveground grazing on coupling among nitrifier activity, abundance and community structure. ISME Journal 2: 221–232.
37. Oelmann Y, Wilcke W, Temperton VM, Buchmann N, Roscher C, et al. (2007) Soil and plant nitrogen pools as related to plant diversity in an experimental grassland. Soil Science Society of America Journal 71: 720–729.
38. Schmid B, Joshi J, Schlapfer F (2002) Empirical evidence for biodiversity-ecosystem functioning relationships. In: Kinzig A, Pacala SW, Tillman RW editors. Linking biodiversity and ecosystem functioning. Princeton University Press, Princeton, 120–150.
39. Dollhopf SL, Hyun JH, Smith AC, Adams HJ, O'Brien S, et al. (2005) Quantification of ammonia-oxidizing bacteria and factors controlling nitrification in salt marsh sediments. Applied Environmental Microbiology 71: 240–246.
40. Phillips CJ, Harris D, Dollhopf SL, Gross KL, Prosser JI, et al. (2000) Effects of agronomic treatments on structure and function of ammonia-oxidizing communities. Applied Environmental Microbiology 66: 5410–5418.
41. Könneke M, Bernhard AE, de la Torre JR, Walker CB, Waterbury JB, et al. (2005) Isolation of an autotrophic ammonia-oxidizing marine archaeon. Nature 437: 543–546.
42. Attard E, Recous S, Chabbi A, De Berranger C, Guillaumaud N, et al. (2011) Soil environmental conditions rather than denitrifier abundance and diversity drive potential denitrification after changes in land-uses. Global Change Biology 17: 1975–1989.
43. Petersen DG, Blazewicz SJ, Firestone M, Herman DJ, Turetsky M, et al. (2012) Abundance of microbial genes associated with nitrogen cycling as indices of biogeochemical process rates across a vegetation gradient in Alaska. Environmental Microbiology 14: 993–1008.
44. Craine JM, Tilman D, Wedin D, Reich P, Tjoelker M, et al. (2002) Functional traits, productivity and effects on nitrogen cycling of 33 grassland species. Functional Ecology 16: 563–574.
45. Mulder CPH, Jumpponen A, Hogberg P, Huss-Danell K (2002) How plant diversity and legumes affect nitrogen dynamics in experimental grassland communities. Oecologia 133: 412–421.
46. Tilman D, Knops J, Wedin D, Reich PB, Ritchie M, et al. (1997) The influence of functional diversity and composition on ecosystem processes. Science 277: 1300–1302.
47. Palmborg C, Scherer-Lorenzen M, Jumpponen A, Carlsson G, Huss-Danell K, et al. (2005) Inorganic soil nitrogen under grassland plant communities of different species composition and diversity. Oikos 110: 271–282.

48. Robson TM, Lavorel S, Clement JC, Le Roux X (2007) Neglect of mowing and fertilisation leads to slower nitrogen cycling in subalpine grasslands. Soil Biology & Biochemistry 39: 930–941.
49. Sutton-Grier AE, Wright JP, McGill BM (2011) Plant trait diversity buffers variability in denitrification potential over changes in season and soil conditions. PLoS One 5: e11618, doi:10.1371/journal.pone.0011618.
50. Rosenkranz S, Wilcke W, Eisenhauer N, Oelmann Y (2012) Net ammonification as influenced by plant diversity in experimental grasslands. Soil Biology & Biochemistry 48: 78–87.
51. Barnard R, Le Roux X, Hungate BA, Cleland EE, Blankinship JC, et al. (2006) Several components of global change alter nitrifying and denitrifying activities in an annual grassland. Functional Ecology 20: 557–564.
52. Sutton-Grier AE, Wright JP, McGill BM (2011) Environmental conditions influence the plant functional diversity effect on potential denitrification. PLoS One 6: e16584. doi:10.1371/journal.pone.0016584.
53. Schimel JP, Bennett J (2004) Nitrogen mineralization: Challenges of a changing paradigm. Ecology 85: 591–602.
54. Corre MD, Schnabel RR, Stout WL (2002) Spatial and seasonal variation of gross nitrogen transformations and microbial biomass in a Northeastern US grassland. Soil Biology and Biochemistry 34: 445–457.
55. McClain ME, Boyer EW, Dent L, Gergel SE, Grimm NB, et al. (2003) Biogeochemical hot spots and hot moments at the interface of terrestrial and aquatic ecosystems. Ecosystems 6: 301–312.
56. Allan E, Weisser WW, Fischer M, Schulze ED, Weigelt A, et al. (2013) A comparison of the strength of biodiversity effects across multiple functions. Oecologia (published online in February 2013) doi:10.1007/s00442-012-2589-0.

16

Biochar Decelerates Soil Organic Nitrogen Cycling but Stimulates Soil Nitrification in a Temperate Arable Field Trial

Judith Prommer[1], Wolfgang Wanek[1]*, Florian Hofhansl[1], Daniela Trojan[2], Pierre Offre[2], Tim Urich[2], Christa Schleper[2], Stefan Sassmann[3], Barbara Kitzler[4], Gerhard Soja[5], Rebecca Clare Hood-Nowotny[1]*

1 Department of Microbiology and Ecosystem Science, University of Vienna, Vienna, Austria, **2** Department of Ecogenomics and Systems Biology, University of Vienna, Vienna, Austria, **3** Core Facility of Cell Imaging and Ultrastructure Research, University of Vienna, Vienna, Austria, **4** Institute of Forest Ecology and Soil, Federal Research and Training Centre for Forests, Natural Hazards and Landscape, Vienna, Austria, **5** Department of Health and Environment, Austrian Institute of Technology, Tulln, Austria

Abstract

Biochar production and subsequent soil incorporation could provide carbon farming solutions to global climate change and escalating food demand. There is evidence that biochar amendment causes fundamental changes in soil nutrient cycles, often resulting in marked increases in crop production, particularly in acidic and in infertile soils with low soil organic matter contents, although comparable outcomes in temperate soils are variable. We offer insight into the mechanisms underlying these findings by focusing attention on the soil nitrogen (N) cycle, specifically on hitherto unmeasured processes of organic N cycling in arable soils. We here investigated the impacts of biochar addition on soil organic and inorganic N pools and on gross transformation rates of both pools in a biochar field trial on arable land (Chernozem) in Traismauer, Lower Austria. We found that biochar increased total soil organic carbon but decreased the extractable organic C pool and soil nitrate. While gross rates of organic N transformation processes were reduced by 50–80%, gross N mineralization of organic N was not affected. In contrast, biochar promoted soil ammonia-oxidizer populations (bacterial and archaeal nitrifiers) and accelerated gross nitrification rates more than two-fold. Our findings indicate a de-coupling of the soil organic and inorganic N cycles, with a build-up of organic N, and deceleration of inorganic N release from this pool. The results therefore suggest that addition of inorganic fertilizer-N in combination with biochar could compensate for the reduction in organic N mineralization, with plants and microbes drawing on fertilizer-N for growth, in turn fuelling the belowground build-up of organic N. We conclude that combined addition of biochar with fertilizer-N may increase soil organic N in turn enhancing soil carbon sequestration and thereby could play a fundamental role in future soil management strategies.

Editor: Shibu Jose, University of Missouri, United States of America

Funding: The authors highly appreciate the funding through FWF Project V138-B17 (to RCHN) as well as the FFG Project 825438 (to GS). The funders had no role in study design, data collection and analysis, decision to publish, or preparation of the manuscript.

Competing Interests: The authors have declared that no competing interests exist.

* E-mail: rebecca.clare.hood-nowotny@univie.ac.at (RCHN); wolfgang.wanek@univie.ac.at (WW)

Introduction

Carbon farming opportunities in temperate agricultural systems are limited to no-till farming and conservation agriculture strategies. In these systems careful carbon management is required to avoid problems of nutrient, particularly N, immobilization by the soil microbial community, which can lead to substantial crop yield depressions. Biochar application to temperate agricultural soils offers the prospect for substantial carbon sequestration without the concomitant yield losses or greenhouse gas emissions associated with other strategies [1–4]. The pyrolysis derived aromatic-macromolecular structure of biochar [5] makes it more recalcitrant to microbial decomposition, which means it does not immobilize significant quantities of valuable nutrients when added to soil, rendering biochar a viable long-term carbon sink option for temperate agriculture [6–7]. Promoting novel carbon farming options in a wider agronomic context requires facing risk-benefit assessments by the soil custodians, namely farmers and other environmentally concerned stakeholders. As a result, an in-depth understanding of the impact of biochar on the soil N cycle, the most economically relevant biogeochemical cycle to farmers in mid to high latitudes, will be required to facilitate decision-making and allow for improved biochar management and subsequent adoption as a carbon sequestering strategy.

There is increasing evidence suggesting that biochar addition to soil impacts a number of processes of the soil N cycle [8] and that the impacts of biochar addition are strongly dependent on the biochar feedstock and the pyrolysis conditions, in addition to the respective soil properties and local environmental and climatic conditions [9]. For instance, biochar was shown to decrease nitrate leaching [10–11], to increase soil N immobilization [12–13], to increase ammonification, the conversion of organic N into ammonium [14–15], to increase nitrification, the oxidation of ammonium to nitrate [16–17], to promote volatilization, the loss of ammonia to the atmosphere [18], and to increase plant N uptake [19–21]. The mechanisms posited to explain the differences in N cycling processes when biochar is added to soil are mostly direct ones, resulting from the unique properties of the biochar such as its highly porous structure, the large surface area and its ion-exchange capacity [22], with associated beneficial effects on nutrient retention and water holding capacity. Some N cycle processes like NH_3 volatilization can be stimulated by inherent biochar properties such as the potential of alkaline biochar to increase soil pH [23]. Moreover biochars can have high nutrient contents and increase soil nutrient availability directly or through the effects of priming which may increase the bio-availability of soil nutrients [24–25]. In addition, indirect mechanisms associated with changes in the soil microbial community can be triggered by the properties specific to biochar and thereby have strong implications for soil microbial N processing [16,26]. To date the majority of biochar studies were carried out as short-term pot experiments, many of these focusing on poor, acidic and nutrient depleted soils where the addition of biochar has the potential to cause largest effects including fundamental changes in soil nutrient cycles, often resulting in pronounced increases in crop production [27–28]. Long-term field trials examining the effects of biochar on intensively managed and fertile agricultural soils in temperate regions are therefore urgently needed to advance our predictive capabilities of biochar - soil N cycle interactions [8,20].

The impacts of biochar on soil N transformation processes are extremely important, given the dominant role N nutrition plays in regulating crop production in temperate regions, as it has to be replenished either as inorganic N fertilizer or it is cycled back to the soil in a complex organic form such as crop residues. In the past two decades there has been a major paradigm shift in our conceptualization of the soil N cycle [29]. It places greater emphasis on the organic turnover processes in the soil N cycle such as the depolymerization of organic matter, which is now regarded as the rate limiting step of dissolved organic N production and subsequently of soil N mineralization and nitrification. As a result of the biochar characteristics described above biochar addition to soil could have major impacts on the organic N turnover component of the soil N cycle. The multitude of simultaneously occurring soil N processes (Figure 1) make measuring and understanding soil N transformation rates difficult and thus requires isotopic tools [30–31]. Using an array of parallel established and novel ^{15}N isotope pool dilution techniques [32–33], it should be possible to study each key transformation process independently. These techniques in consortium allow determining the impact of biochar on multiple soil processes in a synchronized manner and with a high degree of resolution and reliability. In the context of biochar addition this issue of complexity is of particular relevance as biochar addition could have conflicting impacts on different control points of the soil N cycle.

Our major objective therefore was to investigate the effects of biochar on key inorganic and organic N transformation processes, on soil C and N pools and on the soil microbial community. A hardwood-based biochar (pyrolysis at 500°C for two hours) was added in March 2011 to an arable field on calcareous Chernozem in Traismauer, Austria. The field trial consisted of four treatments replicated four times in circular plots of 6.5 m, comprising NPK fertilized plots as controls combined with biochar addition rates of 0, 24 and 72 t dry mass ha^{-1}. As agricultural systems rely on fertilization to grant crop production, fertilized plots were taken as controls (NPK, no biochar) and compared to the two biochar addition rates i.e. 24 t BC ha^{-1} with NPK amendment (BC1N) and 72 t BC ha^{-1} with NPK amendment (BC3N). A treatment with 72 t BC ha^{-1} lacking NPK amendment was also setup (BC3) to check for nutrient immobilization effects of BC-alone treatments. In this study we focused on the control (NPK) and the highest biochar treatment (BC3N). Soil samples were taken

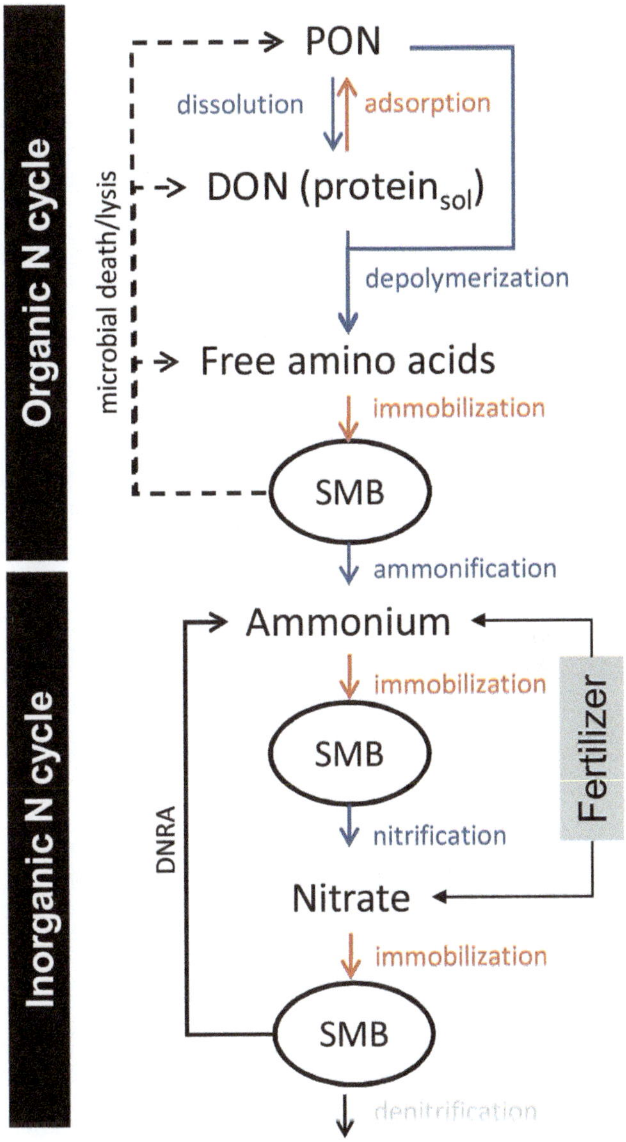

Figure 1. Overview of the agricultural soil N cycle. The organic N and inorganic N cycles are highlighted. DNRA...dissimilatory nitrate reduction to ammonium, DON...dissolved organic N, PON...particulate organic N such as detritus and soil organic matter, SMB...soil microbial biomass. Not shown - heterotrophic nitrification which results from microbial uptake of e.g. free amino acids, oxidation of reduced N compounds and release of nitrate into the soil.

four months after biochar application and sampling was repeated several times during a period of fourteen months. We applied a novel array of isotope pool dilution assays to measure gross rates of protein, amino acid, ammonium and nitrate dynamics (Figure 1) and molecular techniques to investigate nitrifier (ammonium oxidizer) population sizes. In addition we used scanning electron microscopy to examine the physical structure of the biochar in order to understand its possible role in shaping the soil microbial community.

Results

Biochar Porosity

The cellular structure of wood was well conserved in the biochar as seen in Figure S1. The mean pore coverage of xylem cross sections was 47.6±5.9% (mean±SD) for the biochar applied in this field trial. Overall pore size number distribution showed a maximum at approximately 0.7 µm diameter (related to simple xylem pits), and a second and third peak which are closely related to bordered xylem pits and parenchyma cells (around 3 µm) and xylem vessels (15–25 µm; Fig. 2, Fig. S2). Beech has a diffuse-porous wood structure [34] and early-wood and late-wood had distinctly different mean vessel diameters (24±12 and 15±9 µm, respectively), resulting in a decrease in mean pore coverage from 61.8±6.1% for early-wood to 40.2±4.1% for late-wood. Secondary cortex cells showed a wide range of diameters, with a peak ranging from 3 to 5 µm (data not shown). On an area basis the contribution of small pores decreased and the major contribution to total porosity shifted towards larger pores with diameters >10 µm i.e. xylem vessels (Fig. 2b).

Effect of Biochar on General Soil Properties

The effect of biochar on soil pH and cation exchange capacity was minimal. Soil pH decreased by 0.1 units from 7.5 to 7.4 in biochar-treated soils (Table 1). Soil cation-exchange capacity decreased slightly from an initial 22.5 to 20.8 cmolc kg^{-1} in biochar-treated soils but in both cases was dominated by Ca^{2+} (~90%) and Mg^{2+} (~7%), with a trend towards an increase of K^+ saturation from 1.6 to 4.8% (measured on one composite soil sample per treatment, Table 1).

Bulk Soil C and N Contents and Soil Water Relations

Biochar amendment resulted in increased soil organic C and there was a trend of enhanced total soil N concentration with biochar addition. Soil bulk densities and gravimetric soil water contents declined while water-filled pore space remained unaltered by biochar addition (two-way mixed ANOVA for effects of treatment and time, Table 2, Figure 3). Time affected most of these and the following soil parameters (Table 2), but emphasis here is given on treatment effects and treatment × time interactions; the latter might indicate a decreasing or increasing biochar-effect on soil properties or fluxes. However, all time × treatment interactions effects investigated in this study were either non-significant or not related to increasing or decreasing biochar-effects over time. Direct time effects were most likely caused by seasonal changes, agricultural management and subsequent plant growth effects. Soil organic C was significantly higher in biochar-treated soils (47.4 mg C g^{-1}) than in non-biochar controls (18.7 mg C g^{-1}; $p = 0.001$) (Table 2). Total soil N responded analogously, with a slight but non-significant increase due to biochar-treatment (1.47 to 1.64 mg N g^{-1}; $p = 0.062$). Concurrent with increases in soil organic C the soil bulk density decreased from 1.17 to 1.04 g cm^{-3} following biochar amendment ($p<0.001$). Along the same lines biochar amendment caused an increase in soil porosity from 56.0 to 60.6% ($p<0.001$) and in gravimetric soil water content from 0.140 to 0.163 g H_2O g^{-1} dry soil ($p<0.001$). As a result of increased soil water content and soil porosity with biochar addition, water-filled pore space (and air-filled pore space) did not change between controls (29.8%) and biochar-treated soils (28.9%, $p = 0.229$). Interestingly, by comparing all four treatments we found the biochar-effect to be amount-dependent i.e. the magnitude of the effects increased or decreased with the amount of biochar added in the field (Table S1). Soil organic C increased with biochar rate, bulk density decreased and soil porosity and soil water content increased with biochar rate (Table S1).

Extractable C and N Concentrations

Dissolved organic C in soil extracts was slightly but significantly lower in the biochar treatment compared to the controls (111 vs. 123 µg C g^{-1} respectively; $p = 0.010$) (Figure 3, Table 2). Concentrations of extractable ammonium were low compared to extractable nitrate and not significantly different between the control and biochar treatments, with 1.66 and 1.55 µg NH_4^+-N g^{-1} respectively ($p = 0.383$). In contrast soil nitrate concentrations in the biochar (BC3N) treated soils were significantly lower than those from the control soils ($p = 0.001$), with 8.40 and 14.4 µg NO_3^-N g^{-1} respectively. Dissolved organic N concentrations in the soil extracts decreased, though not significantly, from 9.62 in biochar-treated soils to 8.35 µg N g^{-1} in control treatments ($p = 0.187$). The ratio of dissolved organic C: dissolved organic N varied around 16:1 and was not affected by biochar addition ($p = 0.909$). Nitrate dominated the extractable N pool, accounting for between 67% (BC3N) and 76% (controls). The decrease in soil nitrate therefore translated into a significant decline of total dissolved N from 25.0 to 17.0 µg N g^{-1} in the biochar treatment ($p = 0.001$). Free amino acids were low in concentration (around 0.7 µg N g^{-1}) and were not affected by treatment ($p = 0.110$). Measured only once (July 2011, four months after the start of the field experiment) we found no significant differences in the concentrations of different size classes of dissolved organic N between the two treatments, with extractable high-molecular weight organic N concentrations around 3.3 µg N g^{-1} and low-molecular weight organic N concentrations of around 6 µg N g^{-1} (Table 2). Free amino acids accounted for approximately 12% of the low molecular weight organic N pool and comprised 8% of dissolved organic N in both treatments. Around 40% of the dissolved organic N and all the high-molecular weight organic N fraction (>3 kDa; 105–107%) was protein, although this measurement is based only on the July 2011 sampling. The negative effect of biochar on concentrations of dissolved organic C, total dissolved N and nitrate was consistent even when all four treatments were analyzed (Table S1), with decreases in dissolved organic C, nitrate and free amino acids with increasing biochar addition rate (Table S1).

Soil N Transformation Rates

The full set of gross N transformation processes was measured once, and one-way ANOVA results are presented in Table 3 and Figure 4; nitrate fluxes were measured on three occasions throughout the seasons (Table 3, Figure 4, see circles). Measured rates of protein influx into the extractable organic N pool were significantly ($p = 0.016$) higher in the NPK treatment (80.0 µg N g^{-1} d^{-1}) and almost double the rates in the biochar treatment (44.2 µg N g^{-1} d^{-1}). Biochar showed a trend ($p = 0.097$) of depressing protein efflux rates out of the extractable organic N pool, with rates in the biochar treatment (53 µg N g^{-1} d^{-1}) being half that of the control treatment (115 µg N g^{-1} d^{-1}). The decline in amino acid transformation rates was even more pronounced. Amino acid influx rates into the organic N pool were significantly ($p = 0.0004$) lower in biochar treated soils (5.4 µg N g^{-1} d^{-1}), more than fivefold lower than those in the control treatment (28.5 µg N g^{-1} d^{-1}) and so too were amino acid consumption rates. Amino acid consumption rates declined more than six-fold from 50.3 µg N g^{-1} d^{-1} in the control treatment to 7.9 µg N g^{-1} d^{-1} in the biochar treatment ($p = 0.039$). Gross mineralization rates were not significantly different between the two treatments ($p = 0.616$) nor were ammonium consumption rates ($p = 0.526$;

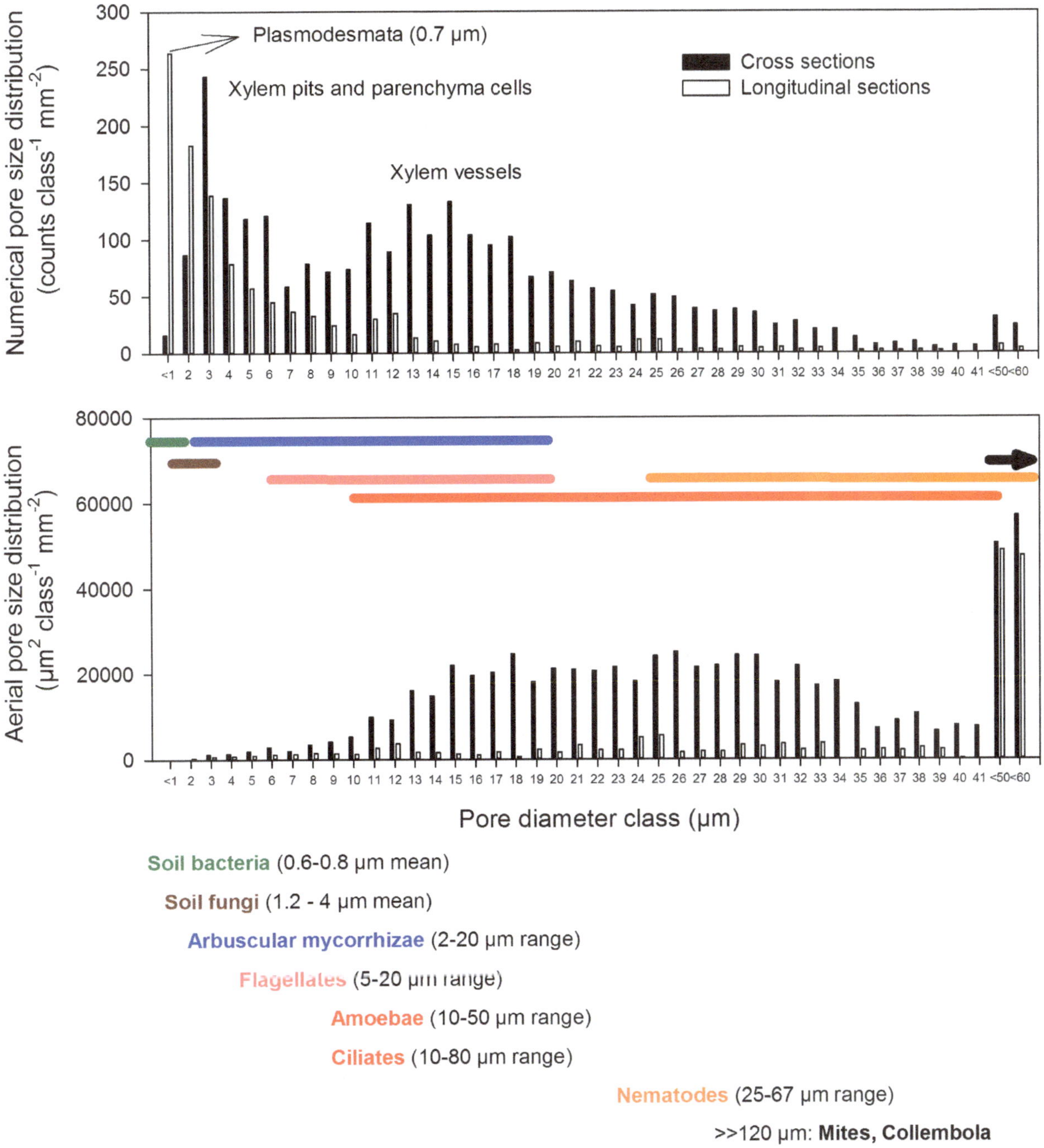

Figure 2. Pore size distribution of beech wood biochar as determined by scanning electron microscopy. In total, mean diameters of 7715 pores were measured in 6 longitudinal and 7 cross sections of charcoal pieces, each image covering 0.35 mm^2. Pore size distributions are presented (1, top panel) as counts mm^{-2} per 1 µm size class (top) and (2, middle panel) on an area basis, i.e. µm^2 mm^{-2} per 1 µm size class. Typical mean diameters (soil bacteria, soil fungi) and diameter ranges (arbuscular mycorrhizal fungi; protozoan grazers such as flagellates, amoebae, ciliates; and mesofaunal grazers and predators such as nematodes, mites and collembola) are presented color-coded (bottom panel) and with range lines in the middle panel.

Figure 3), and both processes ranged between 1.5 and 2.1 µg N g^{-1} d^{-1}. In contrast, gross nitrification and gross nitrate immobilization were strongly enhanced by biochar application. Gross nitrification rates in the biochar treatment (10.2 µg N g^{-1} d^{-1}) were fourfold higher than the rates in the control treatment (2.4 µg N g^{-1} d^{-1}), and significantly different ($p<0.001$). A similar trend ($p=0.0573$) was evident for nitrate consumption rates which almost doubled from 6.8 to 11.3 µg N g^{-1} d^{-1} in the biochar treatment.

Rates of nitrification and nitrate consumption were re-measured twice. Over the course of time nitrification rates were consistently higher (threefold) in the biochar treated compared to the control plots ($p<0.001$, Table 3). The same pattern was apparent for nitrate consumption rates which were consistently negatively affected by biochar amendment ($p=0.001$). On the sampling date

Table 1. Soil properties in control (NPK) and biochar-treated (BC3N) soils of the Traismauer biochar field trial.

Parameter	Unit	Control (NPK)	Biochar (BC3N)
pH ($CaCl_2$)		7.5	7.4
$CaCO_3$	%	15.8	15.2
Humus	%	2.4	18.1
Total N	%	0.148	0.203
P (CAL)	mg kg^{-1}	49	84
P_{tot} (acid digest)	g kg^{-1}	5.46	5.54
Sand	%	18.3	n.d.[1]
Silt	%	57.2	n.d.
Clay	%	24.5	n.d.
CEC	cmol kg^{-1}	22.5	20.8
Ca (CEC)	cmol kg^{-1}	20.7	18.2
Mg (CEC)	cmol kg^{-1}	1.46	1.53
K (CEC)	cmol kg^{-1}	0.36	0.99
Na (CEC)	cmol kg^{-1}	<0.04	<0.04
Al (CEC)	cmol kg^{-1}	<0.06	<0.06
Fe (CEC)	cmol kg^{-1}	<0.01	<0.01
Mn (CEC)	cmol kg^{-1}	<0.01	<0.01
H (CEC)	cmol kg^{-1}	0.002	0.002
Fe (EDTA)	mg kg^{-1}	40	67
Mn (EDTA)	mg kg^{-1}	107	128
Cu (EDTA)	mg kg^{-1}	7.2	7.1
Zn (EDTA)	mg kg^{-1}	2.3	7.5

All parameters were measured on composite samples collected in July 2011 from the two treatments (n = 1). n.d. - not determined.
[1]Soil texture is not shown for biochar treated soils. At high contents humus biases the particle size distribution and is therefore conventionally decomposed by prolonged H_2O_2 (or similar oxidant) treatment prior to soil texture analysis; biochar/charcoal is largely resistant against this oxidant treatment.

where we had gross nitrification rates measured for all four treatments nitrification rates were significantly higher in all the biochar treatments ($p<0.001$, Table S1) with rates increasing significantly with increasing rates of biochar application.

Microbial Community Size and Abundance of Archaeal and Bacterial Ammonia-oxidizers

Soil DNA concentration was significantly higher in the biochar treatment (18.0 μg DNA g^{-1} soil) compared to the controls (12.4 μg DNA g^{-1}; $p = 0.042$) (Figure 5, Table 2). Ammonia-oxidizing archaea (AOA) outnumbered bacterial ammonia-oxidizers (AOB) by a factor of 3.2 to 3.6, in both treatments ($p = 0.402$). At the same time archaeal amoA gene copy numbers (AOA) and bacterial amoA gene copy numbers (AOB) expressed on a dry soil mass basis increased through biochar application (AOA: $P = 0.045$, AOB: $p = 0.020$). AOA increased 1.5-fold from 1.68×10^7 to 2.47×10^7 amoA copy numbers g^{-1} soil and AOB 1.7-fold from 5.01×10^6 to 8.45×10^6 amoA copy numbers g^{-1} soil through biochar amendment. As a result of increasing soil DNA concentration and amoA copy numbers per gram soil we did not find a significant effect of biochar on the abundance of AOA ($p = 0.637$) and AOB ($p = 0.436$) expressed on a DNA basis, i.e. the relative share of ammonia oxidizers in the microbial community did not change according to treatment. Across all four treatments we did not find an effect of biochar rate on either soil DNA concentration or any measure of AOA or AOB abundance (Table S1). However, archaeal and bacterial amoA gene copy numbers were strongly positively related ($R^2 = 0.83$, $p = 0.0114$; data not shown). Moreover, we found a trend towards a positive relationship between both, AOA or AOB numbers on a dry soil mass basis and gross nitrification rates (AOA: $R^2 = 0.65$, $p = 0.051$, AOB: $R^2 = 0.60$, $p = 0.072$; Figure 6). The "weakness" of this relationship derived mainly from the control treatment where nitrification rates were relatively constant between the two measurement dates while amoA copy numbers varied more strongly. Nitrification rates were significantly positively correlated with the total number of amoA copy numbers on a soil dry mass basis (AOA plus AOB; $R^2 = 0.66$, $p = 0.049$, data not shown).

Discussion

Soil Organic N Cycling

Biochar addition had a strong negative effect on soil organic N cycling rates as it reduced protein production and protein consumption and concurrently decreased free amino acid production and consumption (Figure 4). This is important given that protein depolymerization plays the predominant role in organic N recycling with approximately 50% of soil organic matter and >90% of plant detritus being composed of protein [35–36], and with proteins representing around 40% of dissolved organic N in this study. Proteins are therefore considered the most important substrate for microbial N mobilization and N use [33], and thus any change in their dynamics strongly impacts whole-system N cycling. The influx into the K_2SO_4 extractable protein pool (representing a labile part of the total protein pool in soils) is mainly fuelled by microbial turnover and protein desorption processes while efflux from this pool relates to protein depolymerization or protein adsorption (Figure 1).The concomitant reductions in free amino acid turnover were most likely a direct consequence of the reduced protein-N cycling in biochar amended soils. Amino acid production is driven by depolymerization of accessible proteins through extracellular proteases and peptidases which release oligopeptides and free amino acids that are then taken up by microbes (free amino acid consumption) [33]. Though oligopeptides also represent a significant source of N for microbes and plants [37] we did not measure their dynamics, which would have been expected to change accordingly. We offer two possible explanations for the observed negative effects of biochar on organic N (protein) dynamics related to the surface properties of biochar and its unique pore structure.

(1) The observed decline in protein production rates might result from protein adsorption to biochar surfaces. The reactive surface of biochar has been shown to adsorb numerous low and high molecular weight organic molecules [38], including proteins and enzymes [39]. Hence, the reduced motility and accessibility of proteins in the biochar treated soils could explain the decrease in protein influx rates observed with resultant implications for amino acid dynamics. Very strong H-bonding between proteins and polyphenols are well known [40], and the abundant phenolic groups on low temperature biochars [41] provide ample ligands for strong sorption of proteins and peptides [9].

(2) Biochar could also provide structural protection, by the entrapment of organic matter within its inherent pore network. The "mesopore protection hypothesis" by Zimmerman et al. [42] suggests that small organic molecules can be occluded within mineral mesopores (2–50 nm diameter),

Table 2. Results of ANOVA of soil physicochemical data for the factors treatment and time in control (NPK) and biochar (BC3N) treatments.

Pools	Unit	NPK	BC3N	Treatment		Time		Interaction	
		mean	mean	F	P	F	P	F	P
C_{org}	mg C g^{-1} DW	18.66	47.43	(1, 6) 43.7	0.001	(1, 6) 0.0	0.951	2.5	0.165
N_{tot}	mg N g^{-1} DW	1.47	1.64	(1, 6) 5.3	0.062	(2, 12) 37.5	<0.001	0.3	0.778
BD	g cm^{-3}	1.17	1.04	(1, 6) 48.4	<0.001	(1, 6) 55.3	<0.001	0.3	0.595
Grav. WC	g H_2O g^{-1} DW	0.14	0.17	(1, 6) 46.7	<0.001	(6, 36) 218.7	<0.001	6.2	<0.001
Porosity	%	56.03	60.61	(1, 6) 48.6	<0.001	(1, 6) 55.3	<0.001	0.3	0.594
WFPS	%	29.79	28.92	(1, 6) 1.8	0.229	(6, 36) 215.4	<0.001	2.3	0.059
AFPS	%	70.21	71.08	(1, 6) 1.8	0.229	(6, 36) 215.4	<0.001	2.3	0.059
DOC	µg C g^{-1} DW	122.56	111.21	(1, 6) 13.6	0.010	(5, 30) 205	<0.001	2.9	0.028
DOC:DON		16.10	15.92	(1, 6) 0.0	0.909	(1.6, 9.9) 14.4	0.002	0.8	0.454
DON	µg N g^{-1} DW	9.62	8.35	(1, 6) 2.2	0.187	(2.3, 13.9) 28.5	<0.001	0.3	0.782
NO_3^-	µg N g^{-1} DW	14.43	8.40	(1, 6) 39.6	0.001	(6, 36) 95.9	<0.001	8.6	<0.001
NH_4^+	µg N g^{-1} DW	1.66	1.55	(1, 6) 0.9	0.383	(2.6, 15.5) 34.8	<0.001	1.0	0.417
FAA	µg N g^{-1} DW	0.77	0.70	(1, 6) 3.5	0.110	(1.6, 9.8) 3.2	0.091	0.9	0.429
TDN	µg N g^{-1} DW	24.95	17.03	(1, 6) 33.3	0.001	(5, 30) 97.2	<0.001	1.3	0.285
HMW-N_{org}	µg N g^{-1} DW	3.24	3.47	(1, 23) 1.1	0.306				
LMW-N_{org}	µg N g^{-1} DW	6.78	5.68	(1, 23) 0.7	0.401				
Protein	µg N g^{-1} DW	3.46	3.65	(1, 23) 1.4	0.253				
DNA	µg g^{-1} DW	12.40	18.04	(1, 6) 6.6	0.042	(1.1, 6.9) 0.1	0.774	2.6	0.154
AOA/DNA	amoA copy no µg^{-1} DNA	1.41E6	1.36E6	(1, 6) 0.2	0.637	(2, 12) 8.2	0.006	3.3	0.070
AOA/soil	amoA copy no g^{-1} soil DW	1.68E7	2.47E7	(1, 6) 6.3	0.045	(1.1, 6.6) 3.5	0.103	0.9	0.378
AOB/DNA	amoA copy no µg^{-1} DNA	419075	461583	(1, 6) 0.7	0.436	(1.2, 7) 7.2	0.028	1.6	0.246
AOB/soil	amoA copy no g^{-1} soil DW	5.01E6	8.45E6	(1, 6) 9.9	0.020	(2, 12) 4.5	0.035	1.3	0.319
AOA:AOB		3.64	3.19	(1, 6) 0.8	0.402	(2, 12) 0.4	0.681	0.0	0.970

Data were analyzed by two-way mixed ANOVA. Data for protein, HMW-N_{org} and LMW-N_{org} were available only for one sampling date and therefore were analyzed by one-way ANOVA. Abbreviations: C_{org}, soil organic C; N_{tot}, total soil N; BD, soil bulk density; Grav. WC, gravimetric soil water content; WFPS, water filled pore space; AFPS, aire filled pore space; DOC, dissolved organic C; DON, dissolved organic N; NO_3^-, nitrate; NH_4^+, ammonium; FAA, total free amino acids; TDN, total dissolved N; HMW-N_{org}, high molecular weight organic N; LMW-N_{org}, low molecular weight organic N; DNA, soil DNA content; AOA/DNA, archaeal amoA copy numbers on DNA basis; AOA/soil, archaeal amoA copy numbers on dry soil basis; AOB/DNA, bacterial amoA copy numbers on DNA basis; AOB/soil, bacterial amoA copy numbers on dry soil basis.

physically protecting organic molecules such as proteins (<10 nm) and amino acids (<2 nm) from breakdown by excluding decomposers and their extracellular enzymes. Mesopore surface area and volume are mainly determined by gas sorption analysis but are not targeted by scanning electron microscopy, and therefore have not been measured in this study. Nevertheless, low-temperature wood biochars produced at <650°C, just as the Romchar used in this study, have been shown to have a low mesopore volume and surface area as measured by gas sorption [39]. Physical protection of organic matter in mesopores by occlusion from decomposing microbes may thus have played a minor role in this study.

Just like the decrease in protein production, the decline in free amino acid production in biochar treated soils could also be triggered by protease and peptidase adsorption to or occlusion in biochar. Enzyme adsorption to solid support materials (e.g. soil matrix or biochar) has been reported to decrease soil enzymatic activity [43], and adsorbed substrates such as proteins and oligopeptides are also less accessible and bioavailable for further breakdown to free amino acids. Decreased aminopeptidase activity has been observed in experiments with fast pyrolysis biochar of switchgrass and was ascribed to sorption reactions between substrate and biochar [44]. Another study found accelerated N-mineralization of the recalcitrant soil organic N pool in arable soils treated with maize biochar [14]. The inconsistent effects of biochar on soil enzymes may be the result of several factors such as biochar feedstock, pyrolysis conditions, soil type as well as the respective enzyme and/or substrate [44]. For example, contrasting results obtained in our experiment and that of Nelissen et al. (2012) [14] could be explained by structural differences between the biochars used in these two studies, leading to variable enzyme diffusion impedance. Indeed, the Romchar employed here maintained its native wood-like structure according to the scanning electron microscopy images (Figure S1) while non-woody materials such as maize biochar have an amorphous structure with little evidence of retained pores [45].

Soil N Mineralization

Given that biochar application significantly decelerated the production and consumption of low molecular weight organic N in our field experiment, we would have expected a decrease in soil N mineralization rates. Somewhat surprisingly biochar had no significant effect on gross N mineralization and gross ammonium immobilization rates in our study, in line with a laboratory incubation of cultivated black Chernozem soil with wheat biochar

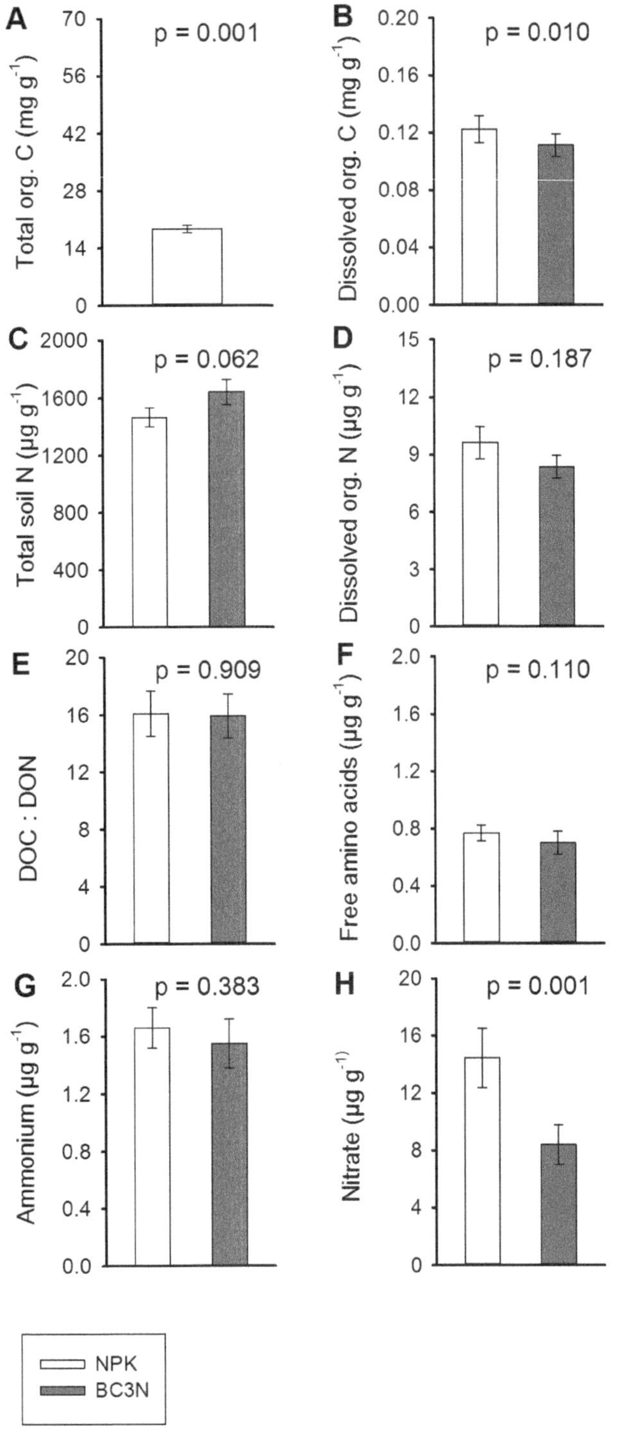

Figure 3. Soil C and N pool sizes for control (NPK) and biochar (BC3N) treatments. Bars represent means ±1 SE (n = 4) from seven measurements between July 2011 and September 2012 for extractable compounds, and from two measurements (September 2011 and 2012) for soil organic C and total soil N. Open bars, control treatment (NPK); grey bars, biochar treatment (BC3N). Units in mg C g^{-1} dry soil and µg N g^{-1} dry soil. P values are from two-way mixed ANOVA (see Table 2).

[46], but in contrast to observed accelerated rates of gross N mineralization of the recalcitrant soil organic N pool in an arable soil treated with maize biochar [14]. Interestingly mineralization (respiratory use) of amino acids and peptides, measured as the uptake of ^{14}C-labelled organic N and subsequent $^{14}CO_2$ release, was not affected by application of a wood-based biochar [15]. Divergent, positive, neutral or negative effects of biochar on soil N mineralization may be the upshot of different biochar types along with impacts on the C and N status of the soil microbial community. Carbon-rich biochar, poor in nitrogen, such as beech wood biochar (C:N ratio ~200:1) might promote N limitation, leading to the retention of catabolically produced ammonium in the N-limited microflora, which therefore results in a decrease of the amount of ammonium released to the soil [29]. Conversely, nitrogen-rich biochars with low C:N ratios such as maize-biochars [14] or manure-biochars [47] promote microbial C limitation, causing the excess of N to be mineralized and therefore N mineralization rates to increase. In our case the ammonium released (N mineralization) relative to organic N taken up by microbes (free amino acid consumption) was higher in the biochar treated soils than in controls (Figure 4), indicating that biochar addition did not lead to microbial N limitation as a consequence of the increased organic C content in soil. This suggests that the majority of organic C in the biochar was not metabolized or assimilated by the soil microbial community, and confirms that biochar is an efficient tool for long-term C sequestration in soil, with half-lifes >100 years [48]. The absence of microbial N limitation in the biochar treatment might be directly related to the field applications of fertilizer and might also be related to the slightly lower concentrations of dissolved organic C in the biochar treated plots.

Soil Nitrification

Apart from the negative effect that biochar had on soil organic N cycling, the most striking result was a strong and persistent (1.5 years) stimulation of gross nitrification. Several studies have examined the impact of biochar on soil net nitrification [8] but only few have reported gross nitrification rates, which allow the simultaneous analysis of both, productive and consumptive processes for nitrate. Biochar addition increased gross nitrification rates in two forest soils and one arable soil [14,26,49] but had no effect in other experiments with grassland and arable soils [46,49]. Several hypotheses have been raised to explain biochar-related (positive) effects on nitrification rates.

One of the dominant theories to explain the increase in nitrification rates following biochar amendment is the liming effect [14,26]. Biochar commonly has a neutral to alkaline pH [50] and therefore may increase the pH of acidic soils with a concurrent positive response of soil nitrifiers whose pH optimum is slightly acidic to neutral pH [51–52]. However the pH of the soil used in this study was slightly alkaline and did not increase but rather decreased in response to the addition of biochar.

Equally we can disregard the tenet that biochar addition leads to an increase in nitrification due to adsorption of nitrification inhibitors. In contrast to forest and grassland soils [49,53–54], intensively managed agricultural soils usually contain low amounts of naturally occurring nitrification inhibitors [55] and their effects might have been of minor importance in this experiment.

Nitrification rates have been shown to be strongly controlled by the ammonium supplied by soil N mineralization in a wide variety of ecosystem types including woodland, grassland and agricultural soils [56]. In this study, biochar amendment neither had an effect on gross ammonium production rates nor on the pool size of extractable ammonium, which obviates a simple substrate-driven explanation of increased nitrification rates in this field experiment. Whereas nitrification rates in the control soils were of the same order of magnitude as the mineralization rates, biochar treated

Table 3. Results of ANOVA of soil N transformation rates for the factor treatment in control (NPK) and biochar (BC3N) treatments.

Processes	Unit	NPK	BC3N	Treatment		Time		Interaction	
		mean	mean	F	P	F	P	F	P
One date									
Prot_GP	mg N kg^{-1} d^{-1}	79.99	44.15	(1, 11) 8.4	0.016				
Prot_GC	mg N kg^{-1} d^{-1}	115.06	52.37	(1, 11) 3.4	0.097				
AA_GP	mg N kg^{-1} d^{-1}	28.54	5.36	(1, 11) 26.7	0.000				
AA_GC	mg N kg^{-1} d^{-1}	50.25	7.91	(1, 11) 5.6	0.039				
N min	mg N kg^{-1} d^{-1}	1.88	2.10	(1, 11) 0.3	0.616				
NH_4^+_GC	mg N kg^{-1} d^{-1}	1.86	1.52	(1, 11) 0.4	0.526				
Nitrification	mg N kg^{-1} d^{-1}	2.42	10.16	(1, 11) 50.6	<0.001				
NO_3^-_GC	mg N kg^{-1} d^{-1}	6.83	11.27	(1, 11) 4.6	0.057				
Three dates									
Nitrification	mg N kg^{-1} d^{-1}	3.38	9.09	(1, 6) 82.2	<0.001	(2, 12) 0.4	0.698	3.6	0.061
NO_3^-_GC	mg N kg^{-1} d^{-1}	4.92	11.01	(1, 6) 35.2	0.001	(2, 12) 1.7	0.225	0.6	0.571

All transformation rates were measured and analyzed by one-way ANOVA for July 2011 data ("one date"). Nitrification and nitrate consumption was measured three times for NPK and BC3N treatments ("three dates") and analyzed by two-way mixed ANOVA. Abbreviations: GP, gross production (influx); GC, gross consumption (efflux); Prot, protein; AA, amino acid; N min, N mineralization; NH_4^+, ammonium; NO_3^-, nitrate.

soils exhibited nitrification rates four to five times higher than mineralization rates. This dichotomy between ammonium supply and nitrification rates could point to increased importance of heterotrophic nitrification fuelled by organic N compounds in biochar-treated soils. However heterotrophic nitrification mainly occurs in grassland and forest soils and especially in acidic soils [57–58], although this process has been recently detected in acidic arable and neutral soils [59–60]. This nitrification pathway will have to be investigated in follow-up incubation experiments, using acetylene as an inhibitor of autotrophic (archaeal and bacterial) nitrification [61] which does not affect heterotrophic nitrification [62–63].

Increased nitrification rates in soils treated with biochar could be attributed to the alteration of the soil water status, including its distribution and associated changes in soil oxygenation, since nitrification is dependent on soil O_2 availability. The soil air/water balance is essential for nitrification, which has been found to peak at 60% water-filled pore space [64]. Altered soil moisture conditions have been reported in various biochar studies [17,65–66]. However, in this study air-filled pore space was highly variable and on average greater than 70%, independent of treatments; therefore higher nitrification rates (Figure 4) could not be explained by improved oxygen availability.

While none of the proposed drivers of soil nitrification could explain the observed pattern of enhanced nitrification in biochar treated soils, the size of the ammonia-oxidizing community increased with biochar amendment (Fig. 5). Recently, the oxidation of ammonia to nitrite, the first and rate limiting step of nitrification, was shown to be performed by prokaryotes in both, the bacterial and archaeal domains [67]. Both ammonia oxidizing archaea (AOA) and ammonia oxidizing bacteria (AOB) abundances were significantly higher in the biochar treated soils, AOA outnumbering AOB in both treatments. There are numerous studies showing that AOA are often significantly more abundant than AOB in soils of different origins [68–69]. However the relative importance and contribution of these different groups to the nitrification process in natural environments is still unclear and to our knowledge there are no published studies available on the interaction of biochar and AOA, at least in temperate ecosystems. AOB abundances have been found to increase after natural forest fire events most likely as a consequence of the liming effect of charcoal and its sorption of nitrification inhibitors [26]. In a pot experiment the AOB community structure was unaffected by the sole addition of *Eucalyptus* wood biochar but changed when biochar was added together with an inorganic or organic N source [70], a possible result of substrate-limited nitrification as the ammonium concentrations declined in sole biochar treatments. The stimulation of both, AOA and AOB abundances through biochar addition in this field experiment was only significant when amoA gene copy numbers were expressed on a soil mass basis but not when expressed on a soil DNA basis. The microbial community size, indicated by total soil DNA content, increased with biochar addition revealing a general positive effect of biochar on the whole microbial community rather than a specific stimulation of the autotrophic nitrifiers. This unspecific stimulation of microbes and concomitant increase of nitrifiers is corroborated by the lack of biochar effects on more specific soil nitrification controls such as soil pH, ammonia and oxygen availability, and presence of nitrification inhibitors. The biochar effect was also unspecific in terms of stimulation of AOA and AOB; both AOA and AOB abundances increased with biochar and were strongly positively related across all soil samples independent of treatment. In addition ratios of AOA: AOB were unaffected by the biochar amendment.

AOA and AOB abundances were broadly ($P<0.1$) correlated with gross nitrification rates, suggesting that bacterial and archaeal abundance increased with increasing biochar addition and providing a functional link to the process data (Fig. 6). This and the total DNA data add weight to an underlying hypothesis, i.e. that the structural nature of the biochar provides a protective habitat for bacterial and archaeal growth. It is conceivable, that biochar promotes bacterial and archaeal community size by enhancing the colonizable surfaces in the soil and providing habitable protected pores. The unique pore-size distribution typically found in biochar with pores ranging from 1 nm to 100 μm [71] is believed to provide various beneficial functions to

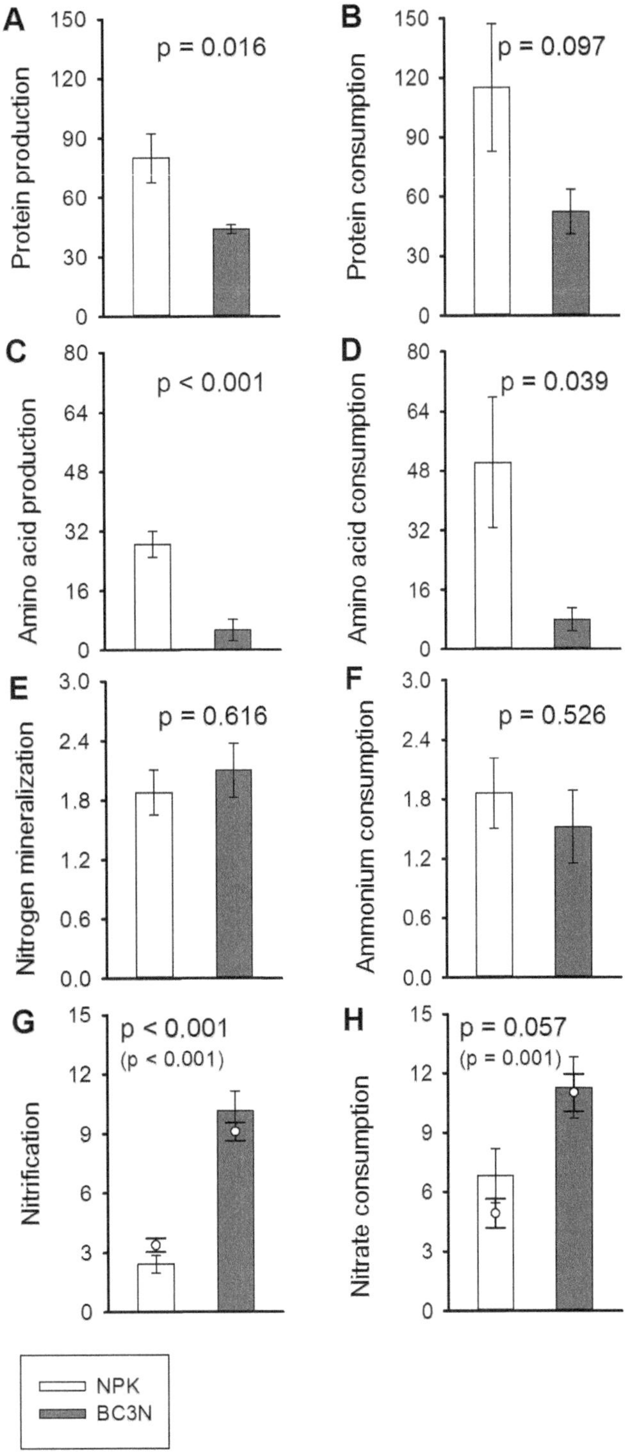

Figure 4. Soil N transformation rates for control (NPK) and biochar (BC3N) treatments. Bars represent means ±1 SE (n = 4) of measurements from July 2011. Open bars, control treatment (NPK); grey bars, biochar treatment (BC3N). Open circles represent means ±1 SE (n = 4) of nitrification rates from three measurement dates. Left panels represent gross influx rates into the target pools (production rates), right panels gross efflux rates (consumption rates). Units in µg N g^{-1} dry soil d^{-1}, equivalent to mg N kg^{-1} d^{-1}. P values are from one-way ANOVA (all rates) or two-way mixed ANOVA (nitrification for three time points, in brackets).

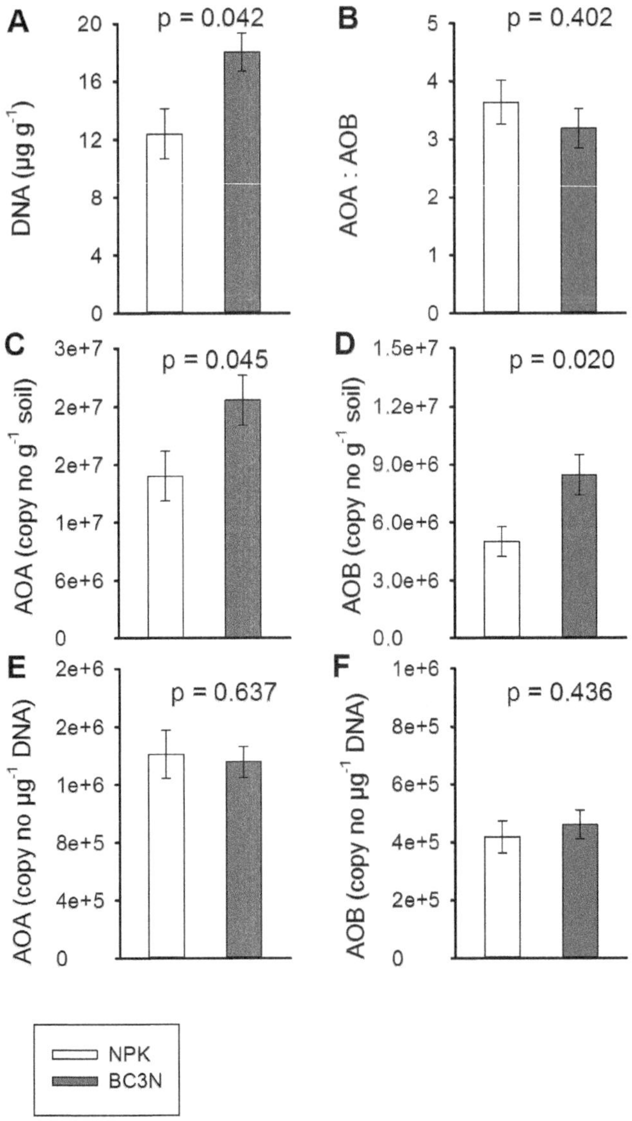

Figure 5. Soil microbial community size and ammonia-oxidizing community structure. Bars represents means ±1 SE (n = 4). Open bars, control treatment (NPK); grey bars, biochar treatment (BC3N). DNA, soil DNA content. AOA, archaeal amoA copy numbers; AOB, bacterial amoA copy numbers. AOA and AOB copy numbers are presented on soil dry mass and DNA basis. P values from two-way ANOVA (see Table 2).

soil bacteria and archaea [72]. Soil prokaryotes typically have mean cell sizes of 0.6 to 0.8 µm in diameter [73–74] with a length: diameter ratio of 1.5:1 [75] and were shown to inhabit pores greater than 0.6 µm [76]. Soil fungi have slightly larger mean diameters, of between 1.2 to 4 µm [74,77] but mycorrhizal hyphae vary enormously in diameter ranging between 2 to 20 µm [78]. As a consequence of the high proportion of colonizable pores, addition of *Eucalyptus* charcoal to a coarse textured soil increased habitable pore space surface area by 11% at a biochar rate of 25 t ha^{-1} [70] and habitable pore volume by 6.5% at 50 t ha^{-1} [79]. Indeed we found an increase of pore volume of 8% at biochar rates of 72 t ha^{-1}. More importantly it has been shown that interior surfaces and pores are rapidly colonized, microbiota showing no differences in densities between outer and internal biochar surfaces 3 years after application [79]. It has been

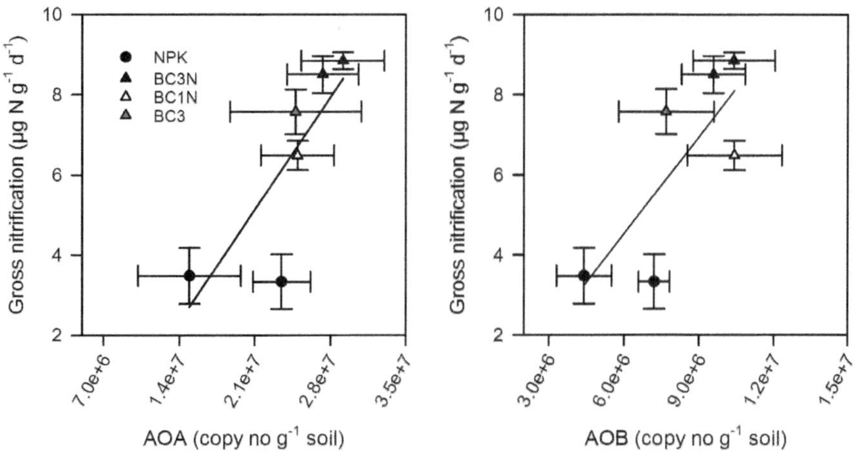

Figure 6. Relationship of archaeal (AOA) and bacterial amoA abundance (AOB) with gross soil nitrification rates. The relationships between archaeal amoA copy numbers (AOA) and bacterial amoA copy numbers (AOB) with gross nitrification rates were marginally significant for AOA ($p = 0.051$) and AOB ($p = 0.072$). Symbols represent the four treatments measured in September 2012. NPK and BC3N were also measured in June 2012. Values are means ±1 SE ($n = 4$).

suggested that soil microbes are protected from being grazed upon their natural predators by colonizing pore habitats in biochar that are inaccessible for soil microfauna (Fig. 2). By this mechanism biochar pores can act as a safe microhabitat for various soil microbes through the exclusion of larger predators [80–81] such as nematodes, mites and collembolans [73] but also microfaunal protozoan grazers and predators, i.e. flagellates, amoebae and ciliates [82–83]. Our scanning electron microscopy data suggest that there was an abundance of colonizable pores, particularly in the 0.3 to 40 µm range with maxima around 0.7, 3 and 15 µm. The abundant sack like structures of parenchyma cells and the long cave like structures of xylem vessels (Figure S1) could therefore offer physical protection for soil prokaryotes and fungi from grazers such as nematodes and amoeba, which are just physically too big to enter the cave or pore. This in turn could reduce microfaunal grazing, which leads to a decrease in mineralization of organic N via the soil "microbial loop" [84]. This scenario provides a possible explanation for the observed increases in nitrification rate, with bacteria, fungi, AOA and AOB taking refuge in the biochar. In the biochar pores AOA and AOB have ample access to carbon dioxide and available ammonia, given the high pH of this soil; this allows them to proliferate, but protected from grazing, thus not contributing significantly to N mineralization. Interestingly this scenario would suggest possible top down control of nitrification, with predators having been shown to consume up to 60% of bacterial production in some soils [85]. In addition to the provision of protected habitat provided by biochar co-location of nutrients and water on biochar surfaces and in pores may increase substrate availability, and thus may further stimulate microbial colonization [72] and nitrification.

Conclusions and Outlook

The main findings of this field study are that application of wood biochar led to marked decreases in soil organic N transformation rates while soil nitrification was significantly stimulated in this type of agricultural soil. This field study therefore indicates a de-coupling of the intrinsic organic and inorganic soil N cycles in the short- to medium-term (four to 18 months) as a consequence of biochar addition. Assuming these effects persist, the deceleration of soil organic N cycling ultimately leads to the dichotomy of two intertwined agronomic processes; the build-up of soil organic N and the slowing-down of inorganic N release from soil organic N essential for crop growth. Addition of fertilizer-N in combination with biochar effectively decouples these processes, with plants and microbes utilizing fertilizer-N for growth, in turn fuelling the belowground build-up of soil organic N. Since soil organic N and particularly refractory soil organic matter are peptide dominated materials [86], the accumulation of soil organic N is expected to enhance soil carbon sequestration due to increases in non-biochar derived soil organic matter. However, it remains an open question whether adsorptive capacity of biochar will saturate and pore-filling of biochar slows or curtails this process, though recent investigations with aged and fresh biochar suggested that the effects of biochar on organic compound sorption may be long lasting [87].

Stimulation of soil nitrification can have positive effects on plant N availability and crop growth but concurrently increases the risk of soil N losses through nitrate leaching and gaseous losses by denitrification and nitrifier-denitrification. The decrease in soil nitrate concentration and increase in microbial nitrate utilization however point to a relatively conservative N cycling in biochar-treated arable soils. It will therefore be of great importance to study the responses of soil organic N, nitrification, soil N losses and plant growth over longer time periods and in various agricultural settings, to better understand the long-term effects of biochar amendments on soil N cycling.

Given the central role of soil organic N cycling and nitrification in controlling global terrestrial N cycles, the implications of this work are of major importance. If it is more broadly proven that addition of biochar combined with fertilizer-N leads to a greater than sum increase in non-biochar derived soil organic matter, as a consequence of the mechanisms described here, it may prove to be a fundamental factor in future soil management: as building soil organic matter through carbon farming is the key step required to kick start the next black/green revolution.

Materials and Methods

Description of Field Trial

The field experiment was established in Traismauer, Lower Austria, Austria (48° 19′ 886″ N, 15° 44′ 308″ E; 268 m a. s. l) in

March 2011 by the Austrian Institute of Technology (AIT) Tulln, Austria. The temperate climate shows strong continental influence (Pannonian climate), with mean annual precipitation of 550 mm and mean annual temperature being 10°C. The arable soil was a deep sandy to loamy silt (18.3% sand, 57.2% silt and 24.5% clay), classified as a calcareous Chernozem on loess. Soil pH of non-biochar treated soil was 7.5 in 10 mM $CaCl_2$, the carbonate content was 15.8% and humus content 2.4% (Table 1). Cation exchange capacity of the soil was 22.5 cmol kg^{-1}, dominated by Ca^{2+} (~90%) and Mg^{2+} (~7%), with a small contribution by K^+ (2%). The site was used for agriculture for >100 years, with spring barley grown in 2011 and sunflower in 2012. Hardwood-derived biochar (80% beech, rest derived from other hardwoods) was produced by pyrolysis at 500°C for two hours and was purchased from S.C. Romchar S.R.L. (Harghita, Romania). This biochar had a pH of 8.6 in water and 7.5 in 10 mM $CaCl_2$ and comprised the following elemental concentrations: 80.3% C, 9.9% O, 0.4% N, and 1.6%H contents (Table 4). Volatiles made up 11.1% and ashes 15.2% of biochar dry mass (both determined at 550°C, Table 4). Biochar was applied once to the field at a dry mass rate of 0, 24 and 72 t ha^{-1} in March 2011 and incorporated to a depth of 15 cm. NPK fertilizer was applied annually in spring providing an annual N input of 120 (2011) and 75 kg ha^{-1} (2012), with corresponding annual rates of P input of 26 (2011) and 31 kg ha^{-1} yr^{-1} (2012) and K input of 50 (2011) and 100 kg ha^{-1} yr^{-1} (2012). The four treatments were: control i.e. NPK (no biochar application, NPK added), BC1N (24 t biochar ha^{-1}, NPK added), BC3N (72 t biochar ha^{-1}, NPK added) and BC3 (72 t biochar ha^{-1}, no NPK added). The four treatments were arranged according to a Latin square and replicated four times in circular plots of 6.5 m diameter.

The field is privately owned and not protected in any way. The local winegrower association Traisen valley (Weinbauverband Traisental) gave us the written permission to conduct the field experiment at the premises owned by Rudolf Hofmann who also agreed in his role as landowner to set up the experimental plots on his land. No governmental permits were required for the described field study, and the study did not involve endangered or protected species.

Table 4. Element contents of beech wood biochar pyrolyzed for two hours at 500°C.

Element	Content (% d.w.)
C	80.30
O	9.88
Ca	3.75
H	1.60
Si	1.58
Fe	0.79
K	0.73
N	0.40
Al	0.243
Mg	0.198
Mn	0.091
P	0.076
Si	0.053
Cl	0.010
Zn	0.009
Cr	0.005
Co	0.002
Ashes (550°C)	15.2
Volatiles (550°C)	11.1

Soil sampling was performed several times in 2011 and 2012 (see Table S2 for sampling dates and for the treatments sampled and the pools and processes measured at each date). Soil samples were taken from the topsoil layer (0–10 cm) using a soil corer with 8 cm diameter. From each treatment plot three soil samples were collected randomly and pooled to form one composite sample. The fresh samples were then weighed for bulk density calculation, sieved to pass 2 mm and stored at 4°C for a maximum of 2–3 days prior to analyses.

Soil Physicochemical Analyses

General soil properties. Soil water content (SWC, in g H_2O g^{-1} dry soil) was determined gravimetrically after drying fresh soil at 105°C for 24 hours in a forced-air drying oven. Bulk density (BD, in g cm^{-3}) was calculated from the fresh weight divided by soil core volume times dry weight: fresh weight ratio. The percentage of water-filled pore space (%WFPS) was calculated as follows:

$$\%WFPS = \frac{SWC \times BD \times 100}{1 - \frac{BD}{2.65}}$$

where 2.65 g cm^{-3} represents the mean density of rock.

Soil total C and N contents were measured on oven-dried (105°C) and homogenized soil samples. Samples were ground to a fine powder with a ball mill (MM2000, Retsch, Germany) and aliquots of 10 mg weighed into tin capsules. Total soil C and N contents were determined by an elemental analyzer (EA 1110; CE Instruments, Italy). For quantification of the soil organic carbon content samples were pretreated with excess acid (2 M HCl) for 60 min and then dried at 105°C to remove carbonate from the calcareous soil. Other basic soil properties, such as soil pH, soil organic matter content, carbonate content, soil texture, total P content, cation-exchange capacity, base saturation, and exchangeable cations (see Table 1) were only measured for control (NPK) and biochar-treated soils (BC3N) at the Institute of Sustainable Plant Production, Federal Agency for Food Security (AGES) in Vienna, Austria, according to standard ÖNORM procedures (www.austrian-standards.at/en).

Biochar characterization. The biochar (beech wood charcoal) was characterized for element, volatiles and ash contents by the German Biomass Research Center (DBFZ, Leipzig, Germany) and results are presented in Table 4. Biochar pH was measured in an aqueous slurry (5% (w/v)) using a pH electrode. For measurements of the porosity (pores >300 nm diameter) and the pore size distribution of the biochar, charcoal samples were mounted on 0.5″ aluminum specimen stubs equipped with SEM-Carbon foils (Agar scientific, UK). Since the specimens were conductive no additional coating was performed after air drying at 40°C for 24 hours. Micrographs were taken on a Philips XL 20 scanning electron microscope (SEM; Philips, Eindhoven, The Netherlands) at 350× and 2000× magnification with an acceleration voltage of 15 kV. For pore size and pore coverage measurements the software Scandium Version 5.0 (Soft Imaging Systems GmbH, Olympus, Vienna, Austria) and the image analyzing software GSA Image Analyzer (GSA, Gesellschaft für Softwareentwicklung und Analytik mbH, Germany) were used. To estimate pore size distribution in total, 7715 pore measurements in

6 longitudinal and in 7 cross-sectional images covering 0.35 mm^2 and in 25 longitudinal images covering 0.0112 mm^2 were performed. Pores down to a diameter of 300 nm could be safely measured. Pore size distributions were converted to counts mm^{-2} or circular area in μm^2 mm^{-2} and summed up for each 1 μm size class.

Pool sizes of extractable organic C, inorganic N and organic N forms. For measurements of extractable C and N pools fresh soils (aliquots of 4 g) were extracted for 30 min with 30 ml 0.5 M K_2SO_4, subsequently filtered through ash-free cellulose filters and extracts kept frozen at −20°C for later analyses. Total dissolved N and dissolved organic carbon (measured as non-purgeable organic carbon) were measured by high temperature catalytic oxidation on a Shimadzu TOC-VCPH with TNM-1 unit and ASI-V autosampler (Shimadzu Austria, Korneuburg, Austria). Ammonium and nitrate were determined with colorimetric methods, NH_4^+ with a modified indophenol reaction and NO_3^- with the VCl_3/Griess assay, as described in detail by Hood et al. [88]. Amino acid concentrations were quantified by a modified fluorometric OPAME procedure based on Jones et al. [89], optimized for free amino acid measurement in protein hydrolysates. The modified OPAME reagent was made up of 10 mg o-phthaldialdehyde dissolved in 1 ml of HPLC grade methanol to which 20 μl of 3-mercaptopropionic acid was added. This reagent was mixed with 40 ml potassium borate buffer (0.2 M; pH 9.5) and left to stand overnight. Sample aliquots (50 μl) were mixed with 200 μl reagent in the microtiter plate and measured after 10 min at an excitation wavelength of 340 nm and emission wavelength of 450 nm with a fluorescence microplate reader (Tecan Infinite M200, Tecan Austria GmbH, Grödig, Austria). As NH_4^+ in the soil extracts also gives a fluorescence signal, NH_4^+ concentration had to be determined [88] and its fluorescence contribution was determined via a NH_4^+ concentration series and subtracted from the total fluorescence of the samples. Total free amino acid-N concentration was then calculated using glycine as a standard and assuming that each mol of amino acid contains 1.37 mol N. Proteins were quantified in K_2SO_4 extracts by acid hydrolysis and subsequent amino acid analysis (modification of [89–90]; see amino acid determination above). Protein hydrolysis to free amino acids was performed in 6 M HCl and heating was performed for 15 hours at 100°C in a drying oven with Bovine serum albumin (BSA containing 15.8% N) as internal standard. Samples were then dried in an N_2 stream and redissolved in deionized water. Determination of the NH_4^+ concentration to correct the amino acid fluorescence was done as in the case of amino acid determination, with the exception that the sodium nitroprusside solution (the color reagent) contained 1.5 M NaOH instead of the original 0.3 M NaOH to neutralize residual acid in the hydrolysates.

Isotope Pool Dilution Assays

Gross rates of N transformation processes were determined using the ^{15}N isotope pool dilution (IPD) technique [31,91]. This method entails the addition of an isotopic tracer to a target pool and measuring the isotopic composition and concentration over time which allows estimating gross rates of influx to (production) and efflux from (consumption, immobilization) the target pool. In order to avoid the stimulation of consumption rates due to excessive addition of ^{15}N labeled substrate to the respective target pool, each pool size was determined prior to labeling the soil samples. Thereafter to each target pool 10% of the pool size was added in the form of ^{15}N tracer (98 at% ^{15}N). Before starting the IPD assays, all soils were adjusted to 40% water holding capacity and then the ^{15}N tracer solutions were added to bring the soils to 70% water holding capacity. After addition of the tracer to the soils the vials were shaken vigorously to homogenously distribute the label. The incubations were performed at room temperature (~20°C) and terminated by adding 0.5 M K_2SO_4 and subsequent extraction for 30 min on a rotary shaker followed by filtration through ash-free filter papers into new vials. Extracts were immediately frozen and kept for further analyses. An overview of the soil preparation pipeline for these isotope pool dilution assays is presented in Figure S2.

Gross N mineralization and NH_4^+ immobilization. Each 2 g soil sub sample was weighed into disposable HDPE vials (20 ml scintillation vials) and labeled with $^{15}NH_4Cl$ (98 at% ^{15}N purchased from Isotec-Sigma Aldrich, Vienna, Austria). The incubations were terminated after 4, 24 and 48 hours by extraction with 15 ml 0.5 M K_2SO_4. Prior to isotopic analyses ammonium concentrations in the labeled extracts were measured photometrically. Ammonium was isolated from the extracts by microdiffusion [92]. In brief, this method implies the conversion of ammonium into ammonia gas which is then trapped on an acidified cellulose filter disc enclosed in a semi-permeable Teflon membrane (acid trap), where the ammonia gas dissolves and redissociates to ammonium. The interconversion of ammonium and ammonia is a consequence of the pH-dependent equilibrium between these two N forms. To each 10 ml of soil extract one acid trap and 100 mg MgO were added to raise the pH to >9.5, the vials were closed immediately and left to stand at room temperature for three days. In cases where ammonium concentrations exceeded the optimal range of N (10–25 μg NH_4^+-N 10 ml^{-1}) for elemental analyzer-isotope ratio mass spectrometry (EA-IRMS) measurement, extracts were diluted with 0.5 M K_2SO_4 prior to microdiffusion. After microdiffusion acid traps were transferred to 2 ml HDPE screw-cap vials and dried in an evacuated desiccator over concentrated sulphuric acid for two days. The cellulose discs were retrieved and transferred into tin capsules and subjected to EA-IRMS. Conversely extracts with ammonium-N concentrations less than 2 μg NH_4^+-N 10 ml^{-1} were prepared to be measured with purge-and-trap isotope ratio mass spectrometry (PT-IRMS) [93]. For this purpose acid traps were made from glass fiber filters. Acid traps to be measured with PT-IRMS were opened and the filter discs transferred to 1.5 ml HDPE snap-cap vials containing 1 ml deionized water and shaken on a Vortex to redissolve the trapped ammonium. The alkaline persulfate oxidation with subsequent VCl_3/azide reactions were then applied to convert NH_4^+ via NO_3^- to N_2O for analysis by PT-IRMS (see below) [93].

Gross nitrification and NO_3^- immobilization. Soil labeling was done with $K^{15}NO_3$ (98 at% ^{15}N purchased from Isotec-Sigma Aldrich) and carried out as a mentioned above for gross N mineralization. As in the case of gross N mineralization a reduction-microdiffusion method was used to isolate nitrate-N after conversion by Devarda's alloy to ammonium-N from the extract into acid traps [92,94] and nitrate concentrations were determined before starting the microdiffusions. In a first step ammonium was eliminated from the extracts by adding one acid trap and MgO (100 mg) to 10 ml sample, then the vials were closed and microdiffusion proceeded for three days at room temperature. After retrieving this acid trap, a new one was added together with 50 mg Devarda's alloy (45% aluminum, 50% copper and 5% zinc), a reducing catalyst converting nitrate to ammonium, to the soil extracts. After five days of microdiffusion at room temperature to collect the ammonia produced from nitrate the second set of acid traps was treated as mentioned above for gross N mineralization and isotope ratios measured via EA-IRMS (see below) to estimate gross nitrification and gross nitrate immobilization.

The organic N-IPD assays described below are novel and were applied here for the first time to measure gross protein and free amino acid fluxes in soils (but see Wanek et al. [32] for the first amino acid IPD by gas chromatography-mass spectrometry).

Gross amino acid production and immobilization. The free amino acid pool was labeled by addition of a ^{15}N-labeled algal amino acid mixture (U-^{15}N-98%, Cambridge Isotope Laboratories Europe, Radeberg, Germany). For this, each 4 g fresh soil were weighed into 50 ml HDPE centrifuge tubes and the ^{15}N tracer was added. Samples were extracted for half an hour with 30 ml 0.5 M K_2SO_4 after 5 and 30 minutes incubation time. An ultrafiltration method was used to separate low-molecular weight organic N from high molecular weight organic N. Amicon Ultra-2 pre-launch centrifugal filter devices (Millipore Corporation) with a nominal molecular weight limit (NMWL) of 3 kDa were taken to collect the low molecular weight organic N fraction (mainly amino acids and oligopeptides) together with the inorganic N forms (NO_3^- and NH_4^+) in the permeate and the high molecular weight organic N forms (mainly proteins) in the retentate. Ultrafiltration was performed on a Beckman centrifuge (Beckman J2–21, Beckmann Coulter, Krefeld, Germany). From the labeled amino acid extract aliquots of 1.7 ml were centrifuged at 4000*g* for one hour at room temperature with a swinging bucket rotor (Beckman JS-13.1). In order to get rid of inorganic N from low molecular weight organic N, each 1 ml of permeate was filled into a 2 ml cryo-vial and 20 mg Devarda's and 20 mg MgO were added together with an acid trap. The closed cryo-vials were kept at 37°C for 3 days; thereafter the acid traps were discarded. Via alkaline persulfate oxidation and VCl_3/azide reaction the low molecular weight organic N was then converted to NO_3^- and finally to N_2O for analysis by PT-IRMS (see below). Moreover free amino acid concentrations were determined in each extract using the modified OPAME procedure as detailed above, to correct the isotope enrichment of free amino acids for the contributions of other low molecular weight organic N forms using a simple two-source isotope pool mixing model [95]. Free amino acid concentrations and the corrected isotope enrichments were used for later pool dilution calculations.

Gross protein production and consumption. Soil labeling was performed as in the case of the amino acid-IPD assays but with the ^{15}N tracer being a purified algal crude protein (98 at%^{15}N purchased from Isotec-Sigma Aldrich) and with incubation times of 5 min, 30 min, 4 and 24 hours. After ultrafiltration (similar to amino acid-IPD but centrifugation parameters altered) of each 1.7 ml of the labeled extract at 7500 *g* for one hour at room temperature with a fixed angle rotor (Beckman JA-20), the permeate was put aside. The high molecular weight organic N containing retentate was washed twice with each 1 ml 50 mM K_2SO_4 via 30 min centrifugation at room temperature and 7500 *g*, to remove inorganic N and low molecular weight organic N. Thereafter the retentate was recovered by 2 min centrifugation at 1000 *g* at room temperature and diluted to 0.5 ml with 50 mM K_2SO_4. The alkaline persulfate oxidation and subsequent VCl_3/azide reactions were then used to convert the high molecular weight organic N via NO_3^- to N_2O for analysis by PT-IRMS (see below).

Alkaline persulfate oxidation (APO) of N_{org} (NH_4^+) to NO_3^-. The APO reaction was used to convert ammonium and organic N forms into nitrate [93]. At elevated temperature persulfate decomposes forming persulfate radicals that efficiently oxidize organic and inorganic compounds, in the case of organic N and ammonium producing nitrate. The persulfate reagent was prepared by dissolving 44 g $Na_2S_2O_8$, 16.8 g NaOH and 30 g H_3BO_3 in deionized water with an end volume of 1 L [96]. For digestion aliquots of 750 µl of sample were pipetted into 1.5 ml HPLC glass vials and each 750 µl of persulfate solution were added. Vials were closed immediately (to avoid losses of ammonium due to the alkaline pH) and autoclaved for one hour at 120°C. Digested samples were stored at 4°C until further analyses.

VCl_3/azide conversion of NO_3^- to N_2O. The isotope ratios of nitrate after persulfate digestion were measured by conversion of nitrate to N_2O by a coupled VCl_3/azide reaction in which in a first step, VCl_3 is used to reduce nitrate to nitrite, which in a second step is further reduced to N_2O by sodium azide [93]. For this, aliquots of 1 ml sample were pipetted into 12-ml screw cap exetainers with butyl rubber septum (IVA-VC309; IVA Analysetechnik, Merbusch, Germany) which were closed tightly and purged with helium for 10 min to get rid of air-N_2O in the headspace. Thereafter 160 µl sodium azide and finally 1 ml VCl_3 were added using gas-tight syringes by injection through the septa. Exetainers were put on a rotary shaker and kept at 37°C for 18 hours. The reaction was stopped by neutralizing and deactivating the sodium azide through injecting each 160 µl 6 M NaOH into the exetainers and N_2O was analyzed by PT-IRMS.

PT-IRMS. The exetainer vials were then loaded into a 96-slot autosampler with a double-hole needle (GC-PAL, CTC Analytics, Zwingen, Switzerland) connected via a Gasbench II headspace analyzer (Thermo Fisher, Bremen, Germany) and a PreCon cryotrapping device (Thermo Fisher) to a gas isotope ratio mass spectrometer (IRMS, Delta V Advantage, Thermo Fisher) [93]. Through the double needle of the autosampler an auxiliary helium flow purged the contents of the exetainer vials through a water and CO_2 scrubber into a sequence of two cryo-traps held at liquid N_2 temperature, and the N_2O was then released and residual CO_2 and air (O_2, N_2) separated by gas chromatography. The purified sample N_2O was then transferred into the mass spectrometer via an open split with repeated pulses of standard N_2O gas for internal calibration (Air Liquide, Vienna, Austria). For calibration of at%^{15}N respective standards varying in ^{15}N:^{14}N ratios were run throughout the sample preparation methods. For further details on PT-IRMS analyses see Lachouani et al. [93].

EA-IRMS. Acid traps and homogenized dry soils were transferred into tin capsules and directly analyzed for N content and at%^{15}N by EA-IRMS. The system consisted of an elemental analyzer (EA 1110, CE Instruments, Milan, Italy) connected via a ConFlo III interface (Thermo Fisher) to the IRMS (Delta^{PLUS}, Thermo Fisher). External calibration was performed using laboratory standards for isotope composition and N concentration [93].

Isotope pool dilution calculations. Gross production (GP) and gross consumption (GC) rates, i.e. influx into and efflux from the target pools that was labeled with ^{15}N in the course of IPD experiments was then calculated based on the isotope pool dilution equations developed by others [97]:

$$GP = \frac{C_{t_2} - C_{t_1}}{t_2 - t_1} \times \frac{\ln\left(\frac{APE_{t_1}}{APE_{t_2}}\right)}{\ln\left(\frac{C_{t_2}}{C_{t_1}}\right)}$$

$$GC = \frac{C_{t_2} - C_{t_1}}{t_2 - t_1} \times \left(1 + \frac{\ln\left(\frac{APE_{t_2}}{APE_{t_1}}\right)}{\ln\left(\frac{C_{t_2}}{C_{t_1}}\right)}\right)$$

where C_{t1}, C_{t2}, APE_{t1} and APE_{t2} are the concentrations of N (µg N g^{-1} soil dry weight) and atom percent excess of ^{15}N (APE, in %) in the respective target N pool at t_1 and t_2 (days), i.e. the times of IPD sample termination. APE was calculated as the difference between the at% ^{15}N of the given N pool at time t and the background ^{15}N abundance (at% ^{15}N) measured in separate non-labeled samples. Gross rates are therefore expressed in µg N g^{-1} soil dry weight d^{-1}.

Extraction of nucleic acids and quantitative PCR analyses. Nucleic acids were extracted from soil samples (0.5 g fresh weight) using a modification of the method described by Griffiths et al. 2000 [98]. Samples were loaded in 2 mL Lysing Matrix E tubes (MP Biomedicals, Illkirch, France) and cells were disrupted by bead-beating with 0.5 mL of an SDS-based lysis buffer (1% SDS, 0.7 M NaCl, 0.1 M $Na_2(SO_3)$, 0.05 M EDTA pH8 and 0.1 M Tris/HCl pH 7.5) and 0.5 mL of a phenol:-chloroform:isoamyl-alcohol (25:24:1) mixture (pH 7.5). Precipitation of nucleic acids was performed for 1 h at room temperature with 20 µg glycogen (MBI Fermentas, Vienna, Austria), 0.5 mL isopropanol and 40 µL 5 M NaCl. Precipitated nucleic acids were washed twice in 500 µL ice-cold 70% ethanol, re-suspended in 50 µL DEPC-treated water and purified by filtration through PVPP mini-columns centrifuged at 1000×*g* for 2 min at 10°C. Purity and concentration of purified nucleic acids was checked spectrophotometrically using a NanoDrop spectrophotometer (Thermo scientific) and their integrity was determined by agarose gel electrophoresis. A SybrGreen I quantitative PCR approach was used to quantify thaumarchaeal and bacterial *amoA* gene abundance. Thaumarchaeal *amoA* genes were amplified with PCR primers Arch-amoA-104F/Arch-amoA-616R [99] and bacterial *amoA* genes with primers amoA-1F/amoA-2R [100]. Duplicate dilution series (10^2–10^7 copies) of PCR products obtained with M13F and M13R vector primers from Topo® TA cloning® vectors (Invitrogen) containing either a 632 bp fragment of *Nitrososphaera viennensis* EN76 [101] *amoA* gene, obtained with PCR primers Arch-amoA-7F/Arch-amoA-638R [99], or a 491 bp fragment of *Nitrosospira multiformis* ATCC 25196 *amoA1* gene (Nmul_A0799/Nmul_A2325), obtained with primers amoA-1F/amoA-2R, were used as standards for the thaumarchaeal and bacterial *amoA* assays, respectively. Thaumarchaeal *amoA* standards were amplified with an efficiency of 93% and r^2 value of 0.997 and bacterial *amoA* standards with an efficiency of 93% and r^2 value of 0.991. All reactions (20 µL) contained 10 µL of 2× Quantifast™ SYBR® Green PCR Master Mix (Qiagen, Vienna, Austria), 0.2 mg.mL^{-1} BSA, 1 (Archaea) or 0.8 (Bacteria) µM of each primer and for the analysis of soil samples 10 ng of purified nucleic acid extracts. Amplification conditions were 95°C for 15 min, followed by 40 cycles of 95°C for 15 s, 60°C for 30 s (combined annealing and extension step) and plate read at 78°C for thaumarchaeal *amoA* genes and 35 cycles of 95°C for 15 s, 62°C for 30 s and plate read at 82°C for bacterial *amoA* genes, before a final extension step at 60°C for 10 min. Reactions were carried out in a Mastercycler® RealPlex^2 thermocycler (Eppendorf Austria GmbH, Vienna, Austria) and the specificity of the amplification was assessed by melting curve analysis and agarose gel electrophoresis.

Statistical analyses. Mixed between-within subjects ANOVAs (split-plot ANOVA) followed by Tukey HSD post hoc tests were performed in SPSS version 16.0 for Windows (SPSS Inc., USA) to test for the effects of treatment and sampling time on biological and physicochemical soil parameters. If normality (Shapiro-Wilk test) and variance homogeneity (Levene's test) were violated, data were log, square root or reciprocal transformed to meet these assumptions. Mauchly's test was used to test the assumption of sphericity and if significant, the degrees of freedom of the F-distribution were corrected using the Greenhouse-Geisser procedure as the estimated epsilon was always less than 0.75. Biochar effects on pools and processes that were only measured once were tested by oneway ANOVA. Regressions and correlations were calculated with Statgraphics Centurion XVI (StatPoint Technologies Inc., USA). Simple least squares linear regressions were calculated to estimate the amount dependency of biochar effects, with added biochar amount as independent variable. Correlations between microbial community structure parameters, or with nitrification rates were performed based on the Pearson product-moment correlation procedure.

Supporting Information

Figure S1 Structure of beech wood biochar shown by SEM analysis. Cross section across early to late wood transect in beech wood charcoal (top left), longitudinal section through beech charcoal showing open xylem vessels and xylem parenchyma cells (top right), close-up of xylem pits (bottom left) and close-up of a plasmodesmata channel (bottom right).

Figure S2 Scheme of the workflow for soil sample preparations to measure organic and inorganic N transformation rates using ^{15}N isotope pool dilution approaches.

Table S1 Results of ANOVA of soil physicochemical factors and gross nitrification rates for the factors treatment and sampling date for all four treatments.

Table S2 Soil sampling dates, treatments sampled and measurements done on these samples.

Acknowledgments

The setting up and maintenance of the field experiment by the AIT Tulln team is greatly acknowledged. All isotope analyses were run at the SILVER laboratory, Department of Microbiology and Ecosystem Science, University of Vienna, with technical assistance of Margarete Watzka. We further thank the local farmer and landowner, Rudolf Hofmann, Traismauer, Austria for allowing us to setup this biochar project.

Author Contributions

Conceived and designed the experiments: WW RCHN GS. Performed the experiments: JP DT PO FH SS. Analyzed the data: JP WW RCHN DT PO FH SS. Contributed reagents/materials/analysis tools: TU CS BK GS. Wrote the paper: JP WW RCHN PO.

References

1. Bruun EW, Muller-Stover D, Ambus P, Hauggaard-Nielsen H (2011) Application of biochar to soil and N2O emissions: potential effects of blending fast-pyrolysis biochar with anaerobically digested slurry. European Journal of Soil Science 62: 581–589.
2. van Zwieten L, Kimber S, Morris S, Downie A, Berger E, et al. (2010) Influence of biochars on flux of N2O and CO2 from Ferrosol. Australian Journal of Soil Research 48: 555–568.
3. Cayuela ML, Sanchez-Monedero MA, Roig A, Hanley K, Enders A, et al. (2013) Biochar and denitrification in soils: when, how much and why does biochar reduce N2O emissions? Scientific Reports 3.
4. Wang JY, Pan XJ, Liu YL, Zhang XL, Xiong ZQ (2012) Effects of biochar amendment in two soils on greenhouse gas emissions and crop production. Plant and Soil 360: 287–298.
5. Schmidt MWI, Noack AG (2000) Black carbon in soils and sediments: Analysis, distribution, implications, and current challenges. Global Biogeochemical Cycles 14: 777–793.
6. Baldock JA, Smernik RJ (2002) Chemical composition and bioavailability of thermally, altered *Pinus resinosa* (Red Pine) wood. Organic Geochemistry 33: 1093–1109.
7. Kuzyakov Y, Subbotina I, Chen HQ, Bogomolova I, Xu XL (2009) Black carbon decomposition and incorporation into soil microbial biomass estimated by C-14 labeling. Soil Biology & Biochemistry 41: 210–219.
8. Clough TJ, Condron LM, Kammann C, Müller C (2013) A review of biochar and soil nitrogen dynamics. Agronomy 3: 275–293.
9. Joseph SD, Camps-Arbestain M, Lin Y, Munroe P, Chia CH, et al. (2010) An investigation into the reactions of biochar in soil. Australian Journal of Soil Research 48: 501–515.
10. Chen Y, Shinogi Y, Taira M (2010) Influence of biochar use on sugarcane growth, soil parameters, and groundwater quality. Australian Journal of Soil Research 48: 526–530.
11. Ventura M, Sorrenti G, Panzacchi P, George E, Tonon G (2013) Biochar Reduces Short-Term Nitrate Leaching from A Horizon in an Apple Orchard. Journal of Environmental Quality 42: 76–82.
12. Bruun EW, Ambus P, Egsgaard H, Hauggaard-Nielsen H (2012) Effects of slow and fast pyrolysis biochar on soil C and N turnover dynamics. Soil Biology & Biochemistry 46: 73–79.
13. Zheng H, Wang Z, Deng X, Herbert S, Xing BS (2013) Impacts of adding biochar on nitrogen retention and bioavailability in agricultural soil. Geoderma 206: 32–39.
14. Nelissen V, Rütting T, Huygens D, Staelens J, Ruysschaert G, et al. (2012) Maize biochars accelerate short-term soil nitrogen dynamics in a loamy sand soil. Soil Biology & Biochemistry 55: 20–27.
15. Dempster DN, Jones DL, Murphy DV (2012) Organic nitrogen mineralisation in two contrasting agro-ecosystems is unchanged by biochar addition. Soil Biology & Biochemistry 48: 47–50.
16. Anderson CR, Condron LM, Clough TJ, Fiers M, Stewart A, et al. (2011) Biochar induced soil microbial community change: Implications for biogeochemical cycling of carbon, nitrogen and phosphorus. Pedobiologia 54: 309–320.
17. Dempster DN, Jones DL, Murphy DV (2012) Clay and biochar amendments decreased inorganic but not dissolved organic nitrogen leaching in soil. Soil Research 50: 216–221.
18. Taghizadeh-Toosi A, Clough TJ, Sherlock RR, Condron LM (2012) A wood based low-temperature biochar captures NH3-N generated from ruminant urine-N, retaining its bioavailability. Plant and Soil 353: 73–84.
19. Vaccari FP, Baronti S, Lugato E, Genesio L, Castaldi S, et al. (2011) Biochar as a strategy to sequester carbon and increase yield in durum wheat. European Journal of Agronomy 34: 231–238.
20. Jones DL, Rousk J, Edwards-Jones G, DeLuca TH, Murphy DV (2012) Biochar-mediated changes in soil quality and plant growth in a three year field trial. Soil Biology & Biochemistry 45: 113–124.
21. Ma L, Hou Z, Lv N, Ye J, Su S, et al. (2012) Effects of biochar application on wheat growth and nitrogen balance. Xinjiang Agricultural Sciences 49: 589–594.
22. Glaser B, Haumaier L, Guggenberger G, Zech W (2001) The 'Terra Preta' phenomenon: a model for sustainable agriculture in the humid tropics. Naturwissenschaften 88: 37–41.
23. DeLuca TH, MacKenzie MD, Gundale MJ (2009) Biochar Effects on Soil Nutrient Transformations. In: Lehmann J, Joseph S, editors. Biochar for Environmental Management: Science and Technology. London: Earthscan. 251–270.
24. Luo Y, Durenkamp M, De Nobili M, Lin Q, Brookes PC (2011) Short term soil priming effects and the mineralisation of biochar following its incorporation to soils of different pH. Soil Biology & Biochemistry 43: 2304–2314.
25. Fontaine S, Mariotti A, Abbadie L (2003) The priming effect of organic matter: a question of microbial competition? Soil Biology & Biochemistry 35: 837–843.
26. Ball PN, MacKenzie MD, DeLuca TH, Holben WE (2010) Wildfire and Charcoal Enhance Nitrification and Ammonium-Oxidizing Bacterial Abundance in Dry Montane Forest Soils. Journal of Environmental Quality 39: 1243–1253.
27. Liu X, Zhang A, Ji C, Joseph S, Bian R, et al. (2013) Biochar's effect on crop productivity and the dependence on experimental conditions–a meta-analysis of literature data. Plant and Soil DOI 10.1007/s11104-013-1806-x.
28. Jeffery S, Verheijen FGA, van der Velde M, Bastos AC (2011) A quantitative review of the effects of biochar application to soils on crop productivity using meta-analysis. Agriculture Ecosystems & Environment 144: 175–187.
29. Schimel JP, Bennett J (2004) Nitrogen mineralization: Challenges of a changing paradigm. Ecology 85: 591–602.
30. Barraclough D (1991) The Use of Mean Pool Abundances to Interpret N-15 Tracer Experiments.1. Theory. Plant and Soil 131: 89–96.
31. Murphy DV, Recous S, Stockdale EA, Fillery IRP, Jensen LS, et al. (2003) Gross nitrogen fluxes in soil: Theory, measurement and application of N-15 pool dilution techniques. Advances in Agronomy, Vol 79 79: 69–118.
32. Wanek W, Mooshammer M, Blochl A, Hanreich A, Richter A (2010) Determination of gross rates of amino acid production and immobilization in decomposing leaf litter by a novel N-15 isotope pool dilution technique. Soil Biology & Biochemistry 42: 1293–1302.
33. Mooshammer M, Wanek W, Schnecker J, Wild B, Leitner S, et al. (2012) Stoichiometric controls of nitrogen and phosphorus cycling in decomposing beech litter. Ecology 93: 770–782.
34. Gryc V, Vavrcik H, Gomola S (2008) Selected properties of European beech (*Fagus sylvatica* L.). Journal of Forest Science 54: 418–425.
35. Jones DL, Healey JR, Willett VB, Farrar JF, Hodge A (2005) Dissolved organic nitrogen uptake by plants - an important N uptake pathway? Soil Biology & Biochemistry 37: 413–423.
36. Kelleher BP, Simpson MJ, Simpson AJ (2006) Assessing the fate and transformation of plant residues in the terrestrial environment using HR-MAS NMR spectroscopy. Geochimica et Cosmochimica Acta 70: 4080–4094.
37. Farrell M, Hill PW, Farrar J, DeLuca TH, Roberts P, et al. (2013) Oligopeptides represent a preferred source of organic N uptake: A global phenomenon? Ecosystems 16: 133–145.
38. Kasozi GN, Zimmerman AR, Nkedi-Kizza P, Gao B (2010) Catechol and Humic Acid Sorption onto a Range of Laboratory-Produced Black Carbons (Biochars). Environmental Science & Technology 44: 6189–6195.
39. Bergeron SP, Bradley RL, Munson A, Parsons W (2013) Physico-chemical and functional characteristics of soil charcoal produced at five different temperatures. Soil Biology & Biochemistry 58: 140–146.
40. Haslam E (1974) Polyphenol-Protein Interactions. Biochemical Journal 139: 285–288.
41. Mukherjee A, Zimmerman AR, Harris W (2011) Surface chemistry variations among a series of laboratory-produced biochars. Geoderma 163: 247–255.
42. Zimmerman AR, Chorover J, Goyne KW, Brantley SL (2004) Protection of Mesopore-Adsorbed Organic Matter from Enzymatic Degradation. Environmental Science and Technology 38: 4542–4548.
43. Burns RG, DeForest JL, Marxsen J, Sinsabaugh RL, Stromberger ME, et al. (2013) Soil enzymes in a changing environment: Current knowledge and future directions. Soil Biology & Biochemistry 58: 216–234.
44. Bailey VL, Fansler SJ, Smith JL, Bolton H Jr (2011) Reconciling apparent variability in effects of biochar amendment on soil enzyme activities by assay optimization. Soil Biology & Biochemistry 43: 296–301.
45. Mullen CA, Boateng AA, Goldberg NM, Lima IM, Laird DA, et al. (2010) Bio-oil and bio-char production from corn cobs and stover by fast pyrolysis. Biomass & Bioenergy 34: 67–74.
46. Cheng Y, Cai Z-c, Chang SX, Wang J, Zhang J-b (2012) Wheat straw and its biochar have contrasting effects on inorganic N retention and N2O production in a cultivated Black Chernozem. Biology And Fertility Of Soils 48: 941–946.
47. Wang T, Arbestain MC, Hedley M, Bishop P (2012) Chemical and bioassay characterisation of nitrogen availability in biochar produced from dairy manure and biosolids. Organic Geochemistry 51: 45–54.
48. Singh BP, Cowie A, Smernik RJ (2012) Biochar carbon stability in a clayey soil as a function of feedstock and pyrolysis temperature. Environmental Science & Technology 46: 11770–11778.
49. DeLuca TH, MacKenzie MD, Gundale MJ, Holben WE (2006) Wildfire-produced charcoal directly influences nitrogen cycling in ponderosa pine forests. Soil Science Society of America Journal 70: 448–453.
50. Singh B, Singh BP, Cowie AL (2010) Characterisation and evaluation of biochars for their application as a soil amendment. Australian Journal of Soil Research 48: 516–525.
51. Prosser JI, Nicol GW (2012) Archaeal and bacterial ammonia-oxidisers in soil: the quest for niche specialisation and differentiation. Trends in Microbiology 20: 523–531.
52. Stienstra AW, Gunnewiek PK, Laanbroek HJ (1994) Repression of Nitrification in Soils under A Climax Grassland Vegetation. Fems Microbiology Ecology 14: 45–52.
53. Berglund LM, DeLuca TH, Zackrisson O (2004) Activated carbon amendments to soil alters nitrification rates in Scots pine forests. Soil Biology and Biochemistry 36: 2067–2073.
54. Gundale MJ, DeLuca TH (2006) Temperature and source material influence ecological attributes of ponderosa pine and Douglas-fir charcoal. Forest Ecology and Management 231: 86–93.

55. Clough TJ, Condron LM (2010) Biochar and the Nitrogen Cycle: Introduction. Journal of Environmental Quality 39: 1218–1223.
56. Booth MS, Stark JM, Rastetter E (2005) Controls on nitrogen cycling in terrestrial ecosystems: A synthetic analysis of literature data. Ecological Monographs 75: 139–157.
57. Nakajima Y, Ishizuka S, Tsuruta H, Iswandi A, Murdiyarso D (2005) Microbial processes responsible for nitrous oxide production from acid soils in different land-use patterns in Pasirmayang, central Sumatra, Indonesia. Nutrient Cycling in Agroecosystems 71: 33–42.
58. Wrage N, Velthof GL, van Beusichem ML, Oenema O (2001) Role of nitrifier denitrification in the production of nitrous oxide. Soil Biology & Biochemistry 33: 1723–1732.
59. Wang L, Cai Z (2008) Nitrous oxide production at different soil moisture contents in an arable soil in China. Soil Science And Plant Nutrition 54: 786–793.
60. Bateman EJ, Baggs EM (2005) Contributions of nitrification and denitrification to N2O emissions from soils at different water-filled pore space. Biology And Fertility Of Soils 41: 379–388.
61. Offre P, Prosser JI, Nicol GW (2009) Growth of ammonia-oxidizing archaea in soil microcosms is inhibited by acetylene. Fems Microbiology Ecology 70: 99–108.
62. Daum M, Zimmer W, Papen H, Kloos K, Nawrath K, et al. (1998) Physiological and molecular biological characterization of ammonia oxidation of the heterotrophic nitrifier Pseudomonas putida. Current Microbiology 37: 281–288.
63. Hynes RK, Knowles R (1982) Effect of Acetylene on Autotrophic and Heterotrophic Nitrification. Canadian Journal of Microbiology 28: 334–340.
64. Linn DM, Doran JW (1984) Effect of Water-filled Pore-space on Carbon-dioxide and Nitrous-oxide Production in Tilled and Nontilled Soils. Soil Science Society of America Journal 48: 1267–1272.
65. Busscher WJ, Novak JM, Evans DE, Watts DW, Niandou MAS, et al. (2010) Influence of Pecan Biochar on Physical Properties of a Norfolk Loamy Sand. Soil Science 175: 10–14.
66. Case SDC, McNamara NP, Reay DS, Whitaker J (2012) The effect of biochar addition on N2O and CO2 emissions from a sandy loam soil - The role of soil aeration. Soil Biology & Biochemistry 51: 125–134.
67. Schleper C (2010) Ammonia oxidation: different niches for bacteria and archaea? ISME Journal 4: 1092–1094.
68. He J, Shen J, Zhang L, Zhu Y, Zheng Y, et al. (2007) Quantitative analyses of the abundance and composition of ammonia-oxidizing bacteria and ammonia-oxidizing archaea of a Chinese upland red soil under long-term fertilization practices. Environmental Microbiology 9: 2364–2374.
69. Leininger S, Urich T, Schloter M, Schwark L, Qi J, et al. (2006) Archaea predominate among ammonia-oxidizing prokaryotes in soils. Nature 442: 806–809.
70. Dempster DN, Gleeson DB, Solaiman ZM, Jones DL, Murphy DV (2012) Decreased soil microbial biomass and nitrogen mineralisation with Eucalyptus biochar addition to a coarse textured soil. Plant and Soil 354: 311–324.
71. Downie A, Crosky A, Munroe P (2009) Physical properties of biochar. In: Lehmann J, Joseph S, editors. Biochar for Environmental Management: Science and Technology. London, UK: Earthscan.
72. Lehmann J, Rillig MC, Thies J, Masiello CA, Hockaday WC, et al. (2011) Biochar effects on soil biota - A review. Soil Biology & Biochemistry 43: 1812–1836.
73. Mulder C, Cohen JE, Setala H, Bloem J, Breure AM (2005) Bacterial traits, organism mass, and numerical abundance in the detrital soil food web of Dutch agricultural grasslands. Ecology Letters 8: 80–90.
74. Frey SD, Elliott ET, Paustian K (1999) Bacterial and fungal abundance and biomass in conventional and no-tillage agroecosystems along two climatic gradients. Soil Biology & Biochemistry 31: 573–585.
75. Mulder C, Van Wijnen HJ, Van Wezel AP (2005) Numerical abundance and biodiversity of below-ground taxocenes along a pH gradient across the Netherlands. Journal of Biogeography 32: 1775–1790.
76. Strong DT, Sale PWG, Helyar KR (1998) The influence of the soil matrix on nitrogen mineralisation and nitrification. II. The pore system as a framework for mapping the organisation of the soil matrix. Australian Journal of Soil Research 36: 855–872.
77. Baath E, Soderstrom B (1979) Significance of Hyphal Diameter in Calculation of Fungal Bio-volume. Oikos 33: 11–14.
78. Smith SE, Read DJ (2000) Mycorrhizal Symbiosis. Amsterdam, Netherlands: Academic Press.
79. Quilliam RS, Glanville HC, Wade SC, Jones DL (2013) Life in the 'charosphere' - Does biochar in agricultural soil provide a significant habitat for microorganisms? Soil Biology & Biochemistry 65: 287–293.
80. Pietikainen J, Kiikkila O, Fritze H (2000) Charcoal as a habitat for microbes and its effect on the microbial community of the underlying humus. Oikos 89: 231–242.
81. Warnock DD, Lehmann J, Kuyper TW, Rillig MC (2007) Mycorrhizal responses to biochar in soil - concepts and mechanisms. Plant and Soil 300: 9–20.
82. Finlay BJ, Black HIJ, Brown S, Clarke KJ, Esteban GF, et al. (2000) Estimating the growth potential of the soil protozoan community. Protist 151: 69–80.
83. Foissner W (1999) Soil protozoa as bioindicators: pros and cons, methods, diversity, representative examples. Agriculture Ecosystems & Environment 74: 95–112.
84. Coleman DC (1994) The Microbial Loop Concept as used in Terrestrial Soil Ecology Studies. Microbial Ecology 28: 245–250.
85. Moore JC, McCann K, Setala H, De Ruiter PC (2003) Top-down is bottom-up: Does predation in the rhizosphere regulate aboveground dynamics? Ecology 84: 846–857.
86. Knicker H (2011) Soil organic N - An under-rated player for C sequestration in soils? Soil Biology & Biochemistry 43: 1118–1129.
87. Jones DL, Edwards-Jones G, Murphy DV (2011) Biochar mediated alterations in herbicide breakdown and leaching in soil. Soil Biology & Biochemistry 43: 804–813.
88. Hood-Nowotny R, Hinko-Najera Umana N, Inselbacher E, Oswald-Lachouani P, Wanek W (2010) Alternative methods for measuring inorganic, organic, and total dissolved nitrogen in soil. Soil Science Society of America Journal 74: 1018–1027.
89. Jones DL, Owen AG, Farrar JF (2002) Simple method to enable the high resolution determination of total free amino acids in soil solutions and soil extracts. Soil Biology & Biochemistry 34: 1893–1902.
90. Roberts P, Jones DL (2008) Critical evaluation of methods for determining total protein in soil solution. Soil Biology & Biochemistry 40: 1485–1495.
91. Di HJ, Cameron KC, McLaren RG (2000) Isotopic dilution methods to determine the gross transformation rates of nitrogen, phosphorus, and sulfur in soil: a review of the theory, methodologies, and limitations. Australian Journal of Soil Research 38: 213–230.
92. Sorensen P, Jensen ES (1991) Sequential Diffusion of Ammonium and Nitrate from Soil Extracts to a Polytetrafluoroethylene Trap for N-15 Determination. Analytica Chimica Acta 252: 201–203.
93. Lachouani P, Frank AH, Wanek W (2010) A suite of sensitive chemical methods to determine the delta N-15 of ammonium, nitrate and total dissolved N in soil extracts. Rapid Communications in Mass Spectrometry 24: 3615–3623.
94. Khan SA, Mulvaney RL, Hoeft RG (2000) Direct-diffusion methods for inorganic-nitrogen analysis of soil. Soil Science Society of America Journal 64: 1083–1089.
95. Phillips DL, Gregg JW (2001) Uncertainty in source partitioning using stable isotopes. Oecologia 127: 171–179.
96. Doyle A, Weintraub MN, Schimel JP (2004) Persulfate digestion and simultaneous colorimetric analysis of carbon and nitrogen in soil extracts. Soil Science Society of America Journal 68: 669–676.
97. Kirkham D, Bartholomew WV (1954) Equations for following nutrient transformations in soil, utilizing tracer data. Soil Science Society of America Proceedings 18: 33–34.
98. Griffiths RI, Whiteley AS, O'Donnell AG, Bailey MJ (2000) Rapid method for coextraction of DNA and RNA from natural environments for analysis of ribosomal DNA- and rRNA-based microbial community composition. Applied and Environmental Microbiology 66: 5488–5491.
99. Alves RJE, Wanek W, Zappe A, Richter A, Svenning MM, et al. (2013) Nitrification rates in Arctic aoils are associated with functionally distinct populations of ammonia-ixidizing archaea. ISME Journal 7: 1620–1631.
100. Rotthauwe JH, Witzel KP, Liesack W (1997) The ammonia monooxygenase structural gene amoA as a functional marker: Molecular fine-scale analysis of natural ammonia-oxidizing populations. Applied and Environmental Microbiology 63: 4704–4712.
101. Tourna M, Stieglmeier M, Spang A, Konneke M, Schintlmeister A, et al. (2011) Nitrososphaera viennensis, an ammonia oxidizing archaeon from soil. Proceedings of the National Academy of Sciences of the United States of America 108: 8420–8425.

17

Element Pool Changes within a Scrub-Oak Ecosystem after 11 Years of Exposure to Elevated CO_2

Benjamin D. Duval[1,2*¤], Paul Dijkstra[1,2], Bert G. Drake[3], Dale W. Johnson[4], Michael E. Ketterer[5], J. Patrick Megonigal[3], Bruce A. Hungate[1,2]

1 Department of Biological Sciences, Northern Arizona University, Flagstaff, Arizona, United States of America, 2 Merriam-Powell Center for Environmental Research, Northern Arizona University, Flagstaff, Arizona, United States of America, 3 Smithsonian Environmental Research Center, Edgewater, Maryland, United States of America, 4 Department of Natural Resources, University of Nevada-Reno, Reno, Nevada, United States of America, 5 Department of Chemistry and Biochemistry, Northern Arizona University, Flagstaff, Arizona, United States of America

Abstract

The effects of elevated CO_2 on ecosystem element stocks are equivocal, in part because cumulative effects of CO_2 on element pools are difficult to detect. We conducted a complete above and belowground inventory of non-nitrogen macro- and micronutrient stocks in a subtropical woodland exposed to twice-ambient CO_2 concentrations for 11 years. We analyzed a suite of nutrient elements and metals important for nutrient cycling in soils to a depth of ~2 m, in leaves and stems of the dominant oaks, in fine and coarse roots, and in litter. In conjunction with large biomass stimulation, elevated CO_2 increased oak stem stocks of Na, Mg, P, K, V, Zn and Mo, and the aboveground pool of K and S. Elevated CO_2 increased root pools of most elements, except Zn. CO_2-stimulation of plant Ca was larger than the decline in the extractable Ca pool in soils, whereas for other elements, increased plant uptake matched the decline in the extractable pool in soil. We conclude that elevated CO_2 caused a net transfer of a subset of nutrients from soil to plants, suggesting that ecosystems with a positive plant growth response under high CO_2 will likely cause mobilization of elements from soil pools to plant biomass.

Editor: Ben Bond-Lamberty, DOE Pacific Northwest National Laboratory, United States of America

Funding: Research was funded by NSF grants DEB-0092642 and DEB-0543218. The funders had no role in study design, data collection and analysis, decision to publish, or preparation of the manuscript.

Competing Interests: The authors have declared that no competing interests exist.

* E-mail: Benjamin.Duval@ars.usda.gov

¤ Current address: United States Dairy Forage Research Center, United States Department of Agriculture-Agricultural Research Service, Madison, Wisconsin, United States of America

Introduction

Many studies have evaluated the effects of elevated CO_2 on nitrogen cycling, and focused on the hypothesis that tree growth response to elevated CO_2 may be limited by N availability, or change with N use efficiency [1–4], but the impact of CO_2 on elements other than N has been studied less frequently. The nutrients P, K and Ca can also limit plant productivity [5,6], non-N nutrients can limit N_2 fixation and C storage [7], and changes in Al, Mn and Fe concentrations might affect the availability of other mineral nutrients in soils [8]. To fully assess the impact of elevated CO_2 on ecosystem nutrient cycling, it is important to evaluate effects on all elements that are necessary for plant nutrition and elements that control the availability of other nutrients in the soil system.

Photosynthesis and growth are often stimulated by elevated CO_2 in C_3 plants [9,10], often leading to more biomass production. Increased growth increases nutrient demands [11]. It has been suggested that nutrients become more limiting for growth over time and can limit terrestrial C uptake [12]. Increased production of carbohydrates in plants is suggested to reduce element concentrations in plants [13]. Elevated CO_2 generally reduces plant N concentration, but increased growth does not inherently dilute the concentration of other elements in plant tissues [14–18]. In sweet gum (*Liquidambar styraciflua*), Johnson et al. [17] found significant declines in foliar Fe concentration with elevated CO_2. At the POP-EUROFACE CO_2 experiment, there was no change in poplar leaf K or Ca concentrations, while Mg concentration actually increased in those trees [19]. A cross-experiment evaluation of elevated CO_2 by Natali et al. [20] showed significantly lower Fe concentration in sweet gum at the Duke FACE site, decreased Al, V and Fe concentrations in sweet gum at the Oak Ridge FACE site, but increased Mn and Mo concentrations in *Quercus myrtifolia* at the Smithsonian Institution Elevated CO_2 site in Florida. Lastly, a recent meta-analysis of 14 tree species and 10 nutrient elements found that elevated CO_2 lowers Cu, Fe, K, Mg, P and S concentrations, but only at high N availability [15].

Element availability varies by soil type and ecosystem. Soil element availability is a function of soil organic matter content (SOM) and pH, with elements generally less adsorbed to metal oxides and SOM in acid soils [21]. Elevated CO_2 has been shown to increase P availability, possibly a function of decline in SOM in the experimental plots [22]. Elevated CO_2 has also been implicated in reducing leaching of soil N and P from upper soil layers [17]. A recent study found that trace metal concentrations increased in soils exposed to elevated CO_2 at Duke FACE and

Oak Ridge FACE, but decreased under elevated CO_2 for every element surveyed at the Florida SERC experiment [20].

Johnson et al. [23] reported a decrease in foliar N and S concentration of scrub-oak (*Q. myrtifolia*), an increase in oak foliar Mn, and no change in P, K, Cu, B or Zn after 5 years of CO_2 enrichment [23]. However, because biomass was significantly higher under elevated CO_2, total plant pool sizes (on an area basis) for all elements were increased under elevated CO_2. The data reported here were collected at the end of the Florida CO_2 experiment, providing an assessment of the cumulative effect of more than a decade of CO_2 enrichment on soil element pools. We evaluate the impacts of elevated CO2 on a scrub-oak stand in a fire-regenerated ecosystem by quantifying soil, litter and plant tissue (above and below ground tissue) element pools after 11 years, and to determine if elevated CO_2 facilitated nutrient retention, loss or redistribution in this system.

Our overarching hypothesis was that declines in soil nutrients under elevated CO_2 are quantitatively caused by increases in plant pools, in other words, that the cumulative impact of elevated CO_2 is to redistribute elements in the plant-soil system. Specifically, we hypothesized that nutrient cycling in the Florida ecosystem changed under elevated CO_2 in the following ways:

1) Increases in aboveground biomass driven by elevated CO_2 will increase the pool size of elements in those tissues (leaves, stems, litter and roots).

2) Increased plant uptake depletes plant soluble element pools in soils exposed to elevated CO_2.

3) Element retention in this ecosystem will increase under elevated CO_2 because of increased plant element pools, especially in long-lived tissue like wood and coarse roots.

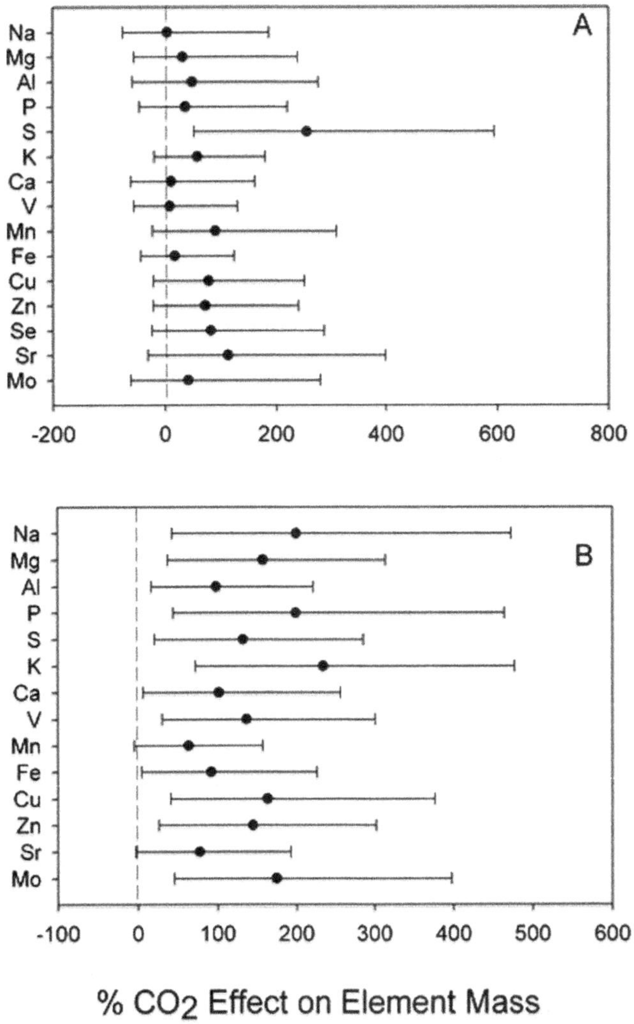

Figure 1. Re-sampled effect size (1000 iterations) of elevated CO_2 compared to ambient CO_2 means and 90% confidence intervals, for element pools in A) *Quercus* spp. leaves, B) *Quercus* spp. stems.

Materials and Methods

Study site

Our study was conducted at the Smithsonian Environmental Research Center's long-term elevated CO_2 experiment at Kennedy Space Center, Cape Canaveral, Florida, USA (28° 38′ N, 80° 42′ W). The experiment consisted of 16 octagonal open-top chambers that were 2.5 m high covering a ground surface area of 9.42 m^2. Eight chambers were kept at ambient atmospheric CO_2 concentration (ambient treatment) and 8 chambers were maintained at ambient +360 µmol mol^{-1} CO_2 (elevated treatment) from May 1996 to June 2007. Soils at the site are acidic sands (Arenic Haplahumods and Spodic Quartzipsamments). The vegetation is Florida coastal scrub-oak palmetto [23,24]. In the experimental chambers, greater than 90% of the aboveground biomass is scrub oak [25].

Field collections

We harvested aboveground biomass in July 2007. We took foliar samples by collecting 5 fully expanded leaves per tree (*Q. myrtifolia*) from 5 distinct trees per plot. Oak stem samples were taken from 5 large branches from the main trunk per plot. We also took samples from the principal symbiotic nitrogen-fixing vine in the system, *Galactia elliottii*. We collected leaves from 5 *G. elliottii* vines per plot.

Soil was collected with a 7 cm diameter soil core at five locations within each chamber. We separated horizons as follows: A horizon (0–10 cm), E horizon (10–30 cm), E2 horizon (30–100 cm) and spodic horizon (Bh), a distinct zone of organic matter accumulation that varied from ~100–250 cm depth. Because the spodic horizon varies in depth and is the deepest soil layer above the vadose zone, not all of the chambers were sampled to the same depth. The cores from each plot were combined per depth into a single composite sample for element analysis.

We collected litter from 1.18 m^2 of the chamber. We collected roots from soil cores by sieving (2 mm mesh). For each soil depth, we separated roots by size into fine (<2 mm diameter) and coarse (>2 mm) fractions. Nutrient concentrations in root tissues were scaled up using root biomass estimates based on minirhizotron photographs (fine roots) and ground penetrating radar imaging [26].

Sample Preparation and Element Analysis

We analyzed soil and plant tissues from the experimental site for the following elements (in order by atomic mass): Na, Mg, Al, P, S, K, Ca, V, Mn, Fe, Cu, Zn, Se, Sr and Mo. All glass wear and plastic containers used for sample extractions and digestions were acid washed in 0.5 M HCl for 48 hours prior to use. All acid reagents were of trace-metal clean purity. Prior to soil digestions and extractions, roots were removed from soil core samples and all soil was passed through a 2 mm sieve and oven dried at 105 °C. An acid digestion was used to prepare samples for measuring total soil element concentrations. Dried soil samples of 100–150 mg were ashed at 600 °C prior to acid digestion in a MARS

Table 1. Element pools in the above ground plant biomass, litter layer and roots (0–100 cm, coarse + fine roots) under ambient and elevated atmospheric CO_2, Kennedy Space Center, Florida.

Element	Tissue	Ambient CO_2	SE	Elevated CO_2	SE
Na (kg · ha^{-1})	Foliar	5.09	2.27	1.89	0.86
	Stems*	0.72	0.31	1.22	0.31
	Litter	45.51	6.60	42.54	6.55
	Roots to 100 cm*	87.45	4.54	146.81	26.50
Mg	Foliar	24.25	10.77	15.55	3.26
	Stems*	0.93	0.25	2.26	0.47
	Litter	32.62	4.87	35.75	5.40
	Roots to 100 cm*	69.81	4.63	97.90	12.17
Al	Foliar	0.16	0.08	0.11	0.02
	Stems	0.30	0.10	0.67	0.17
	Litter	41.60	6.51	47.21	6.40
	Roots to 100 cm	91.17	12.90	145.41	24.73
P	Foliar	12.45	4.55	9.52	1.80
	Stems*	0.02	0.01	0.07	0.02
	Litter	6.13	0.98	7.78	1.04
	Roots to 100 cm	3.04	0.29	4.07	0.56
S (g · ha^{-1})	Foliar	104.08	31.05	299.93	65.71
	Stems	6.26	1.90	13.54	3.11
	Litter	1184.07	157.57	1262.56	135.14
	Roots to 100 cm	168.09	20.35	419.00	130.25
K	Foliar	30.17	8.01	37.87	7.83
	Stems*	78.25	24.26	237.36	51.80
	Litter	80.34	11.40	95.38	14.71
	Roots to 100 cm	382.26	87.86	384.73	72.97
Ca	Foliar	79.76	36.18	48.52	9.10
	Stems	0.32	0.08	0.71	0.16
	Litter	16.96	2.89	20.22	2.53
	Roots to 100 cm	18.47	1.60	26.31	3.20
V (g · ha^{-1})	Foliar	0.44	0.18	0.30	0.06
	Stems*	7.57	3.51	20.81	6.96
	Litter	356.66	44.52	391.05	48.22
	Roots to 100 cm	0.47	0.07	1.04	0.25
Mn	Foliar	0.71	0.22	0.72	0.14
	Stems	0.07	0.02	0.14	0.03
	Litter	5.26	1.20	4.82	1.04
	Roots to 100 cm	1.98	0.39	2.62	0.53
Fe	Foliar	0.22	0.07	0.18	0.03
	Stems	0.09	0.04	0.13	0.04
	Litter	2.10	0.34	11.28	7.87
	Roots to 100 cm	1.34	0.22	1.74	0.24
Cu (g · ha^{-1})	Foliar	43.58	13.43	44.07	9.04
	Stems	7.08	2.00	16.06	3.95
	Litter	298.87	46.66	295.14	37.05
	Roots to 100 cm	678.74	118.14	1215.08	292.35
Zn	Foliar	0.20	0.07	0.22	0.04
	Stems*	0.05	0.01	0.09	0.02
	Litter	2.41	0.23	2.83	0.63
	Roots to 100 cm	2.20	0.18	4.37	1.10
Se (g · ha^{-1})	Foliar	0.14	0.03	0.16	0.06
	Stems	n/a	n/a	n/a	n/a
	Litter	n/a	n/a	n/a	n/a
	Roots to 100 cm	0.02	0.00	0.03	0.01
Sr (g · ha^{-1})	Foliar	235.23	124.75	173.36	41.78
	Stems	55.90	18.00	132.58	36.97
	Litter	2470.91	478.97	2431.55	282.15
	Roots to 100 cm*	3.70	0.43	4.40	0.58
Mo (g · ha^{-1})	Foliar	0.07	0.02	0.05	0.01
	Stems*	0.91	0.31	1.67	0.38
	Litter	241.24	28.68	264.25	31.25
	Roots to 100 cm	0.10	0.02	0.23	0.07

Asterisks denote significant ANOVA results for larger pools under elevated CO_2 compared to ambient CO_2 ,- denotes larger pools in ambient CO_2 plots. All units are kg. ha-1 unless specified differently.

microwave digestor. Microwave digestion was performed for 20 minute runs at 200 °C with trace metal grade, concentrated HF, HNO_3^- and HCl, until all soil was dissolved into solution.

Plant-available element pools in soil were determined using an ammonium oxalate extraction [27]. One (1.0) g of soil was extracted in 15 ml of 0.3 M ammonium oxalate in a 60 ml sample cup, placed on a reciprocal shaker at 180 rpm for 18 h, filtered, diluted 10 times, and re-suspended in 10 ml of 0.32 M trace metal grade HNO_3. We use this extraction as an estimate of plant-available element pools because the yields from this extraction fall within the published values for plant available elements and are consistent with the percent of the plant available element pool compared to the total (in this case acid digest) element pool [8,28].

Plant tissues (leaves, woody biomass and roots) were oven dried at 60 °C for 24 h after collection. Roots were cleaned of excess soil by sonicating ~3–5 g of root tissue in15 ml centrifuge tubes for 30 minutes in ultrapure (18 MΩ) water. The washed roots were again oven-dried for 24 h at 60°C. All plant samples were then ashed at 600 °C, and 500–600 mg of each ashed sample was acid

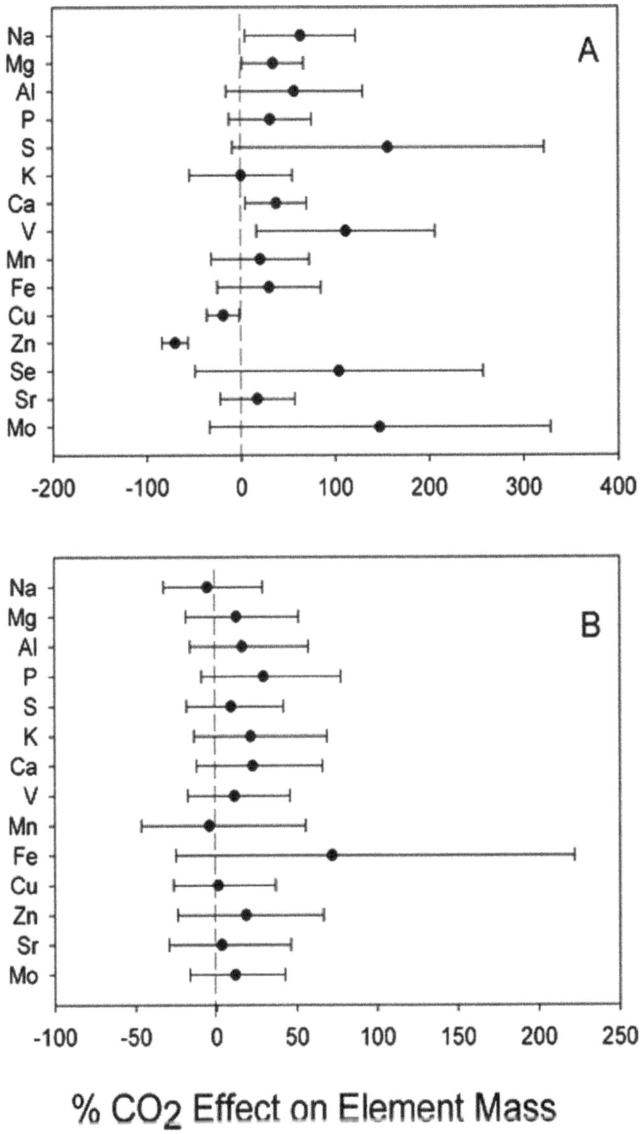

Figure 2. Re-sampled effect size (1000 iterations) of elevated CO_2 compared to ambient CO_2 means and 90% confidence intervals, for element pools in A) all plant roots to a depth of 1 m and B) the litter layer.

digested on the MARS microwave digestor. Element analyses were conducted using Thermo X Series quadrupole ICP-MS.

Estimates of total plant biomass [25], were used to calculate the aboveground pool of elements. We multiplied element concentrations from *Q. myrtifolia* leaves and stems with total leaf and stem biomass to estimate the overall content of nutrients in the three oak species. *G. elliottii* element concentrations were likewise multiplied by biomass. The belowground plant element pool (to 100 cm) was calculated by multiplying the element concentration of roots by their biomass.

Statistical Analyses & Data Availability

We analyzed the effects of CO_2 on *Q. myrtifolia* biomass, element concentration, and element mass with a one-way ANOVA. We also used a two-way ANOVA model to test for CO_2 effects and differences in the element mass of different plant pools (leaves, stems, litter and roots), and interactions between those factors. The effect of elevated CO_2 on soil element mass was analyzed using a repeated measures ANOVA model with CO_2 treatments and soil depth as the repeated measure. We employed the two-tailed Flinger-Killen test to check assumptions of equal variance [29]. Pairwise comparisons of CO_2 effects on soil pools by depth were made using Tukey's HSD test. Due to relatively low sample size ($n=8$ per treatment), we use an alpha of 0.10 to determine significance [20].

To control for family-wise error rates, we used the False Discovery Rate (FDR) test to ensure that using a large number of pair-wise tests for CO_2 effects did not yield a significant number of Type I errors. Our tests of multiple elements within a "group", for example, soil plant available element pools, consisted of 15 individual ANOVAs. In all cases, the FDR expected less than one false discovery per group of multiple tests, justifying our use of multiple ANOVAs.

We also calculated the percent effect of elevated CO_2 on nutrient pools:

$$\%\text{EffectSize} = (\text{ElementmassElevatedCO2} - \text{ElementmassAmbientCO2}) / \text{ElementmassAmbientCO2x100}$$

To ensure that we were able to detect differences due to the CO_2 treatment with relatively low sample size, and reduce our study-level Type II error rate, we also used resampling with replacement to determine the % effect size of elevated CO_2 effects on element pools, an approach complementary to ANOVA for determining differences between treatments [30]. We re-sampled from the sample population of element masses in elevated CO_2 plots and ambient CO_2 plots, with replacement (1000 iterations). This approach enabled us to determine the mean effect of elevated CO_2 compared to the control plots, as well as calculate 90% confidence intervals around the mean effect size. We consider the CO_2 effect meaningful if the confidence intervals do not overlap 0. Statistical analyses were performed in JMP, Microsoft Excel and R [31]. Data used in these analyses are available online via the University of Illinois' Institute for Genomic Biology Public Data Archive [32].

Results

The concentration of V and Ca decreased in *Q. myrtifolia* leaves exposed to elevated CO_2 ($F_{1,14}=3.22$, $P=0.09$; $F_{1,14}=3.20$, $P=0.09$, respectively), but foliar S concentration increased ($F_{1,14}=5.66$, $P=0.03$). Elevated CO_2 reduced the concentration of stem Ca ($F_{1,14}=5.93$, $P=0.03$), Mn ($F_{1,14}=3.17$, $P=0.10$) and Fe ($F_{1,14}=4.18$, $P=0.06$). Elevated CO_2 did not significantly change the concentration of any element measured in root tissue or in the litter layer.

The total aboveground biomass of scrub oak exposed to elevated CO_2 was ~100% higher at the end of the experiment ($F_{1,14}=10.44$, $P=0.01$, from ref. 25). Pools of K and S in total above ground oak biomass were greater under elevated CO_2 ($P<0.05$) compared to ambient CO_2 (Table 1). The effect of CO_2 exposure suggests a greater accumulation of S under elevated CO_2 (Figure 1A). Oak stems under elevated CO_2 hosted significantly greater pools of Na, Mg, P, K, V, Zn and Mo (ANOVA, $P<0.10$; Table 1), and our resample analysis suggests every element other than Mn and Sr was accumulated in woody tissue under elevated CO_2 (Figure 1B).

Table 2. Total pool (acid digest) of soil elements after 11 years exposure to under ambient and elevated atmospheric CO_2, Kennedy Space Center, Florida elevated CO_2, Kennedy Space Center.

Element	Soil Horizon	Ambient CO_2	SEM	Elevated CO_2	SEM
Na (kg · ha^{-1})	A (0–10 cm)	340.5	23.6	287.2	19.9
	E (10–30 cm)	845.6	128.9	756.4	69.2
	E2 (30–100 cm)	231.1	11.1	216.1	13.4
	Spodic	44.6	2.8	53.4	10.7
Mg (kg · ha^{-1})	A (0–10 cm)	46.3	3.1	36.7	1.9
	E (10–30 cm)	64.8	6.6	69	12.8
	E2 (30–100 cm)	7.5	1.5	8.7	1.8
	Spodic	1.5	0.2	1.6	0.2
Al (kg · ha^{-1})	A (0–10 cm)	1687.9	219.9	1284.5	145.3
	E (10–30 cm)	2523.9	501.2	2524	366.6
	E2 (30–100 cm)	515.7	122.9	995.6	545.1
	Spodic	245.6	35.2	247.5	39
P (kg · ha^{-1})	A (0–10 cm)	9.8	2.2	9.1	2
	E (10–30 cm)	131.8	25.9	103.1	16
	E2 (30–100 cm)	10	3.6	13	3.5
	Spodic	4.1	0.7	5	0.9
S (kg · ha^{-1})	A (0–10 cm)	n/a			
	E (10–30 cm)	n/a			
	E2 (30–100 cm)	6.6	1.4	7.6	1.2
	Spodic	1.2	0	1.2	0
K (kg · ha^{-1})	A (0–10 cm)	680.3	62.4	536.7	49.3
	E (10–30 cm)	1527.1	320.1	1346.8	146.5
	E2 (30–100 cm)	126462.7	35285.6	117669.6	31055.1
	Spodic	50398	5619.8	41336.6	4337.6
Ca (kg · ha^{-1})	A (0–10 cm)	30.7	1.6	26.5	1.4
	E (10–30 cm)	65.4	6.2	63.9	3.9
	E2 (30–100 cm)	17.2	5.1	18.4	4.7
	Spodic	4.8	0.1	5	0.2
V (kg · ha^{-1})	A (0–10 cm)	2.50	0.18	1.99	0.13
	E (10–30 cm)	5.01	0.63	4.87	0.60
	E2 (30–100 cm)	2.44	0.20	2.42	0.17
	Spodic	0.48	0.02	0.52	0.03
Mn (kg · ha^{-1})	A (0–10 cm)	19.5	1.5	15.3	1
	E (10–30 cm)	45.8	9.9	40.3	7.9

Table 2. Cont.

Element	Soil Horizon	Ambient CO_2	SEM	Elevated CO_2	SEM
	E2 (30–100 cm)	4.3	0.8	4.1	0.6
	Spodic	0.8	0	0.9	0.1
Fe (kg · ha^{-1})	A (0–10 cm)	72.7	5.8	56.1	3.9
	E (10–30 cm)	137.2	25.5	140	25
	E2 (30–100 cm)	21.9	7.2	19.7	4.5
	Spodic	n/a			
Cu (kg · ha^{-1})	A (0–10 cm)	3.06	0.20	3.11	0.36
	E (10–30 cm)	6.23	0.50	5.47	0.39
	E2 (30–100 cm)	1.94	0.08	1.97	0.07
	Spodic	316.8	9.3	315.7	10
Zn (kg · ha^{-1})	A (0–10 cm)	10.3	1.2	8.1	1
	E (10–30 cm)	16.7	1.5	15.2	1.2
	E2 (30–100 cm)	3.2	0.2	3.7	0.3
	Spodic	0.5	0	0.5	0
Se (g · ha^{-1})	A (0–10 cm)	150.9	6.4	143.4	6.1
	E (10–30 cm)	337	37.4	339.7	21.2
	E2 (30–100 cm)	276.3	264.9	45.3	206.1
	Spodic	54.9	6.5	55.5	8.1
Sr (kg · ha^{-1})	A (0–10 cm)	6.89	0.64	4.93	0.39
	E (10–30 cm)	11.25	2.45	11.20	1.55
	E2 (30–100 cm)	3.24	0.52	3.01	0.45
	Spodic	1.03	0.16	1555.3	0.43
Mo (g · ha^{-1})	A (0–10 cm)	318.2	20.4	289.9	12.9
	E (10–30 cm)	685.7	34.1	681.9	30.4
	E2 (30–100 cm)	1785.3	17.4	1758.9	34.6
	Spodic	307.5	6.2	316.9	9

There was generally a positive effect of CO_2 on the root element pool (Figure 1C). Indeed, our resample analysis showed significant CO_2 effect on Na, Mg, Ca and V, and only a negative effect on Zn (Figure 2A). The mass of litter was not significantly changed by CO_2 exposure, but compared to foliar and stem tissues there were large pools of Na, Mg, S, V, Fe, Cu, Sr and Mo in the litter layer (Table 1). There was at least a slight trend for a positive CO_2 effect on litter pools for every element measured other than Na and Mn (Figure 2B).

Two-way ANOVA including CO_2 treatment and plant pool as predictor variables revealed that there were significant differences among plant pools for every element measured. This difference was driven by larger pools of elements in roots than other plant material for every element other than Al and Ca, which were in

Table 3. Pool of plant available (ammonium oxalate extractable) elements after 11 years exposure to under ambient and elevated atmospheric CO_2, Kennedy Space Center, Florida elevated CO_2, Kennedy Space Center.

Element	Soil Horizon	Ambient CO_2	SEM	Elevated CO_2	SEM
Na (kg · ha^{-1})	A (0–10 cm)	5.12	0.94	4.79	0.65
	E (10–30 cm)	15.20	1.10	15.09	2.15
	E2 (30–100 cm)	11.64	0.29	11.14	0.15
	Spodic	11.94	0.20	10.81	0.60
Mg (kg · ha^{-1})	A (0–10 cm)	8.51	2.13	5.41	0.87
	E (10–30 cm)	7.61	1.03	7.04	1.00
	E2 (30–100 cm)	2.50	0.33	2.22	0.12
	Spodic	4.60	0.19	3.54	0.57
Al (kg · ha^{-1})	A (0–10 cm)	8.61	1.81	5.25	1.90
	E (10–30 cm)	37.95	11.95	14.42	2.45
	E2 (30–100 cm)	312.01	73.30	286.22	85.62
	Spodic	67.92	2.80	53.47	7.65
P (kg · ha^{-1})	A (0–10 cm)	3.50	0.77	3.24	0.70
	E (10–30 cm)	46.99	9.25	36.77	5.69
	E2 (30–100 cm)	7.41	1.55	7.38	1.88
	Spodic	4.81	0.11	3.67	0.77
S (kg · ha^{-1})	A (0–10 cm)	12.03	2.67	11.14	2.39
	E (10–30 cm)	161.66	31.81	126.50	19.56
	E2 (30–100 cm)	25.49	5.32	25.39	6.47
	Spodic	16.57	0.37	12.62	2.65
K (kg · ha^{-1})	A (0–10 cm)	8.20	1.74	7.04	1.06
	E (10–30 cm)	13.98	3.28	10.06	1.33
	E2 (30–100 cm)	5725.41	422.47	5996.16	389.53
	Spodic	18408.19	605.24	14427.58	2228.08
Ca (kg · ha^{-1})	A (0–10 cm)	3.12	1.17	3.11	1.03
	E (10–30 cm)	38.78	3.93	39.55	3.47
	E2 (30–100 cm)	0.63	0.03	0.64	0.03
	Spodic	1.30	0.01	1.01	0.19
V (g · ha^{-1})	A (0–10 cm)	32.97	6.24	14.00	3.05
	E (10–30 cm)	20.98	3.27	15.77	1.05
	E2 (30–100 cm)	152.86	8.68	157.44	12.85
	Spodic	149.04	2.74	133.07	8.72
Mn (g · ha^{-1})	A (0–10 cm)	122.37	19.94	93.03	11.27
	E (10–30 cm)	141.58	11.82	148.75	19.07
	E2 (30–100 cm)	283.02	11.80	281.44	7.11

Table 3. Cont.

Element	Soil Horizon	Ambient CO_2	SEM	Elevated CO_2	SEM
	Spodic	282.94	8.98	239.83	22.77
Fe (kg · ha^{-1})	A (0–10 cm)	7.60	2.16	4.42	1.78
	E (10–30 cm)	27.59	10.71	9.56	1.79
	E2 (30–100 cm)	14.76	3.55	11.53	3.83
	Spodic	3.62	0.17	2.93	0.36
Cu (kg · ha^{-1})	A (0–10 cm)	2.26	0.73	2.64	0.86
	E (10–30 cm)	6.12	1.52	10.66	4.05
	E2 (30–100 cm)	0.32	0.08	0.31	0.08
	Spodic	0.34	0.01	0.29	0.03
Zn (kg · ha^{-1})	A (0–10 cm)	0.58	0.15	0.47	0.12
	E (10–30 cm)	0.97	0.19	1.30	0.32
	E2 (30–100 cm)	0.76	0.17	0.94	0.18
	Spodic	0.87	0.03	0.73	0.07
Se (g · ha^{-1})	A (0–10 cm)	1.00	0.27	0.65	0.22
	E (10–30 cm)	5.07	2.19	1.74	0.37
	E2 (30–100 cm)	n/a			
	Spodic	177.24	12.07	n/a	
Sr (g · ha^{-1})	A (0–10 cm)	31.63	3.55	27.64	4.19
	E (10–30 cm)	85.51	11.01	78.84	10.80
	E2 (30–100 cm)	168.69	7.72	164.63	6.23
	Spodic	142.55	2.77	128.73	6.97
Mo (g · ha^{-1})	A (0–10 cm)	0.99	0.68	0.21	0.02
	E (10–30 cm)	1.25	0.54	0.49	0.03
	E2 (30–100 cm)	97.09	1.11	96.18	1.46
	Spodic	82.78	0.65	80.41	1.19

larger quantity in leaf tissue (Table 1). There were significant, positive main CO_2 effects on the overall above ground plant element pool for Ca ($F_{1,56}=2.76$, $P=0.10$), K ($F_{1,56}=2.99$, $P=0.09$) and Sr ($F_{1,56}=3.00$, $P=0.08$). There was also a significant CO_2 by tissue pool interaction for those three elements (Ca, $F_{3,56}=2.77$, $P=0.05$; K, $F_{3,56}=3.00$, $P=0.04$; Sr, $F_{3,56}=3.09$, $P=0.03$).

Examining the entire soil profile, there was no CO_2 effect on nutrient pools for the entire soil profile (0–100 cm + Bh) for either the total digest or plant available soil elements (Figure 3A, 3B). There were no CO_2 effects when individual horizons were considered independently (Table 2, Table 3), other than plant available Al to a depth of 30 cm ($F_{1,29}=3.33$, $P=0.08$). However, we observed a significant effect of horizon depth for all total digest elements, and a horizon effect for plant-available nutrient pools of Mg, Al, K, Ca, V, Mn, Se and Sr (Table 3). We also observed

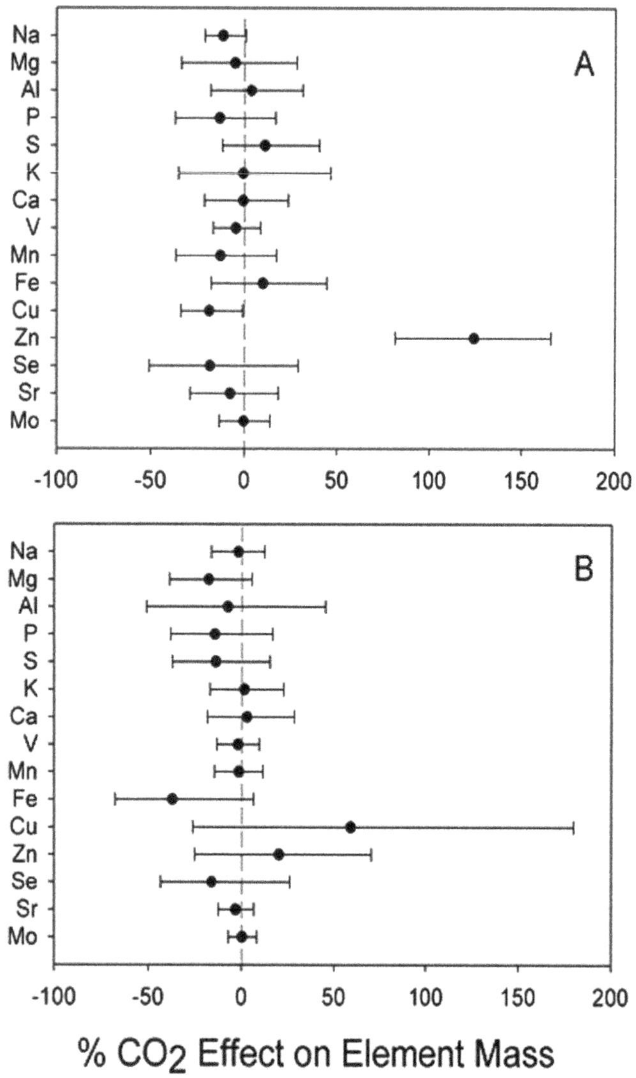

Figure 3. Re-sampled effect size (1000 iterations) of elevated CO_2 compared to ambient CO_2 means and 90% confidence intervals, for elements in A) total acid digest soil pool and B) plant available soil element pools.

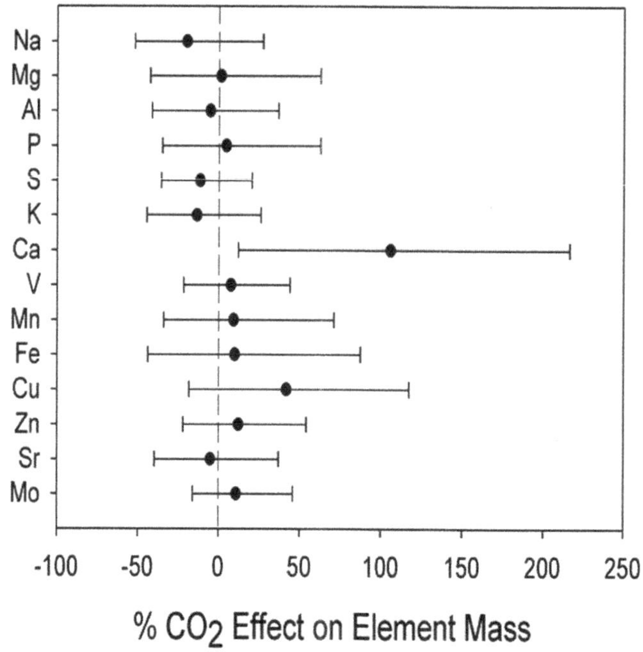

Figure 4. Re-sampled effect size (1000 iterations) of elevated CO_2 compared to ambient CO_2 means and 90% confidence intervals, for the total ecosystem element pool, calculated as the difference between the total plant and plant available element pools in soil under elevated CO_2 compared to ambient CO_2.

trends in the pool size for different elements at different soil depths. Irrespective of CO_2 treatment, there was a large total (acid digest) pool of K in the E2 and Bh (spodic) horizons, and this held for plant available K as well (Table 2, Table 3). Consistent with other spodosols, a large pool of Fe was found in the 10–30 cm portion of the horizon (Table 2). Plant available molybdenum was found in the greatest quantity in the deeper parts of the profile (Table 3).

The effect of CO_2 on the total ecosystem element pool, calculated as the sum of the total plant and plant available element pools in soil, was only significantly higher under elevated CO_2 for Ca (Figure 4).

Discussion

While elevated CO_2 is expected to lower element concentrations in plant leaves [13], element concentrations in oak leaves we measured after 11 years of elevated CO_2 exposure were generally not significantly impacted by high CO_2. Calcium and V concentrations decreased, but S concentrations were higher in oak leaves in the elevated CO_2 treatment. Scrub oaks at the Florida experiment showed significantly greater growth over the 11 years they were exposed to elevated CO_2 [25]. The increased above ground oak biomass under elevated CO_2 was high enough to consistently lead to increases in above ground plant and litter nutrient element pools irrespective of changes in element concentration (Table 1, Figure 1, Figure 2B).

We did not measure significant differences in soil element pools under elevated CO_2 compared to ambient conditions. Combined with the sustained biomass stimulation of oaks under elevated CO_2, there is no evidence that non-nitrogen nutrients are limiting growth after extended CO_2 enrichment. There was a strong signal for a positive effect of elevated CO_2 on root nutrient pools (Figure 1C), and the overall pool of elements in roots for oaks exposed to both elevated and ambient CO_2 was often orders of magnitude higher than the element content of stems and leaves, driven by the large below ground biomass pool [33]. Thus, it is possible that these oaks are liberating nutrient elements from the total element pools into soluble forms by increasing production of root exudates, facilitating mycorrhizal colonization and changes to rhizosphere chemistry that facilitate nutrient uptake [34–37]. At the end of our study, root biomass was significantly higher under elevated CO_2 treatments [33]. If oaks under elevated CO_2 are shifting C allocation belowground, which is in turn providing greater root surface area and potentially more root exudate production, it would explain both the possibility that oaks are mining the soil for elements to meet their nutritional demands, as well as the positive effect of CO_2 on root element pools (Figure 2B).

We do not have direct measurements of *Q. myrtifolia* mining soils for nutrients via root exudation and rhizosphere acidification. However, a source of nutrient liberation (and therefore facilitation of movement from soils to plant biomass) could come from the "priming" effect of elevated CO_2 on organic matter mineralization, which could enhance the release of nutrients like P, Ca and metals bound to SOM. Indeed, we observed that the increase in Ca stocks in oaks was higher than the decline of extractable Ca in the soils (Figure 4). This phenomenon has been observed for N at this site, in the form of increased N mineralization under elevated CO_2 [38], and the Ca result supports the hypothesis that CO_2 induced soil priming increases nutrient availability could be a general phenomenon.

Calculating the CO_2 effect of total ecosystem elements showed that CO_2 enhances Ca retention but not significantly so for other elements (Figure 4). Because of the fire regime of this system, non-volatile elements sequestered in plant biomass will eventually return to the soil, but elements leached from the soil system to the water table are effectively gone from the system [39].

Liu et al. [40] measured increased leaching of Mg (385%), K (223%), Ca (167%) and NO_3^- (108%) under elevated CO_2, and attributed element loss to accelerated mineral weathering and higher soil water content under elevated CO_2. Element loss through leaching is permanent, and we expect that soluble forms of elements that migrate downward through the soil profile will be exported from the system via lateral transfer [28]. However, the total pool of most elements is large relative to the plant available pool (Table 2, Table 3), and soluble forms of elements (especially K, Fe and Mg) can also be replenished in the soil via geochemical processes like chemical weathering, which may be accelerated by exposure to elevated CO_2 [34,40]. Indeed, our observed trend for lower concentrations of amorphous Al-oxides under elevated CO_2 at the Florida site could be a result of accelerated chemical weathering [41].

Our results demonstrate that nutrient cycling is substantially altered after 11 years of exposure to elevated CO_2, but the CO_2 effect is element dependent [15,42]. The strong, positive growth response of oaks to CO_2 [25] led to increased pools of some elements (Na, V, Zn and Mo) in plant biomass and quantifiably lower plant available pools of most elements throughout the soil profile (Table 3). However, because there were only significant changes in the movement of some elements, it is likely that CO_2 effects on element cycles are not easily generalized.

Conclusions

Our results support the hypothesis that increases in oak biomass under elevated CO_2 would increase the pool of nutrient elements in oak tissues. We also observed measurably lower stocks of most nutrients in soils under elevated CO_2. The observation that Ca was retained in this system under elevated CO_2 opens the possibility that some plants actively mine soils under elevated CO_2 for nutrients other than N.

Acknowledgments

We thank David Johnson, Tom Powell and Hans Anderson for their commitment to the upkeep and success of the SERC Elevated CO_2 experiment. We thank Dr. Frank Day and his team at Old Dominion University for invaluable field assistance at the end of the experiment. We also thank Dr. Gwenyth Gordon and Dr. Ariel Anbar of Arizona State University for assistance with macronutrient and trace element analysis with Q-ICP-MS and the use of their facilities.

Author Contributions

Conceived and designed the experiments: BDD BGD DWJ JPM BAH. Performed the experiments: BDD PD MEK. Analyzed the data: BDD BAH. Contributed reagents/materials/analysis tools: BDD PD MEK JPM BAH. Wrote the paper: BDD PD BAH.

References

1. Finzi AC, Moore DJP, DeLucia EH, Lichter J, Hofmockel KS, et al. (2006) Progressive nitrogen limitation of ecosystem processes under elevated CO_2 in a warm-temperate forest. Ecology 87: 15–25.
2. Hungate BA, Johnson DW, Dijkstra P, Hymus G, Stiling P, et al. (2006) Nitrogen cycling during seven years of atmospheric CO_2 enrichment in a scrub oak woodland. Ecology 87: 26–40.
3. Johnson DW (2006) Progressive N limitation in forests: review and implications for long-term responses to elevated CO_2. Ecology 87: 64–75.
4. Norby RJ, Warren JM, Iversen CM, Medlyn BE, McMurtrie RE (2010) CO_2 enhancement of forest productivity constrained by limited nitrogen availability. PNAS 107: 19368–19373.
5. Vitousek PM, Farrington H (1997) Nutrient limitation and soil development: Experimental test of a biogeochemical theory. Biogeochemistry 37: 63–75.
6. Chapin FS, Matson PA, Mooney HA (2002) Principles of terrestrial ecosystem ecology. Springer Press, USA.
7. van Groenigen KJ, Six J, Hungate BA, de Graaff MA, van Breemen N, et al. (2006) Element interactions limit soil carbon storage. PNAS 103: 6571–6574.
8. Kabata-Pendias A (2001) Trace elements in soils and plants. CRC Press, Boca Raton, FL, USA.
9. Ainsworth EA, Long SP (2005) What have we learned from 15 years of free-air CO_2 enrichment (FACE)? A meta-analytic review of the responses of photosynthesis, canopy properties and plant production to rising CO_2. New Phyt 165: 351–372.
10. Norby RJ, DeLucia EH, Gielen B, Calfapietra C, Giardina CP, et al. (2005) Forest response to elevated CO_2 is conserved across a broad range of productivity. PNAS 102: 18052–18056.
11. Sterner RW, Elser JJ (2002) Ecological Stoichiometry: The biology of elements from molecules to the biosphere. Princeton University Press, Princeton, New Jersey.
12. Luo Y, Su B, Currie WS, Dukes JS, Finzi A, et al. (2004) Progressive nitrogen limitation of ecosystem responses to rising atmospheric carbon dioxide. BioSci 54: 731–739.
13. Loladze I (2002) Rising atmospheric CO_2 and human nutrition: toward globally imbalanced plant stoichiometry? Trend Ecol Evol 17: 457–461.
14. Peñuelas J, Filella I, Tognetti R (2001) Leaf mineral concentrations of *Erica arborea*, *Juniperus communis* and *Myrtus communis* growing in the proximity of a natural CO_2 spring. Glob Change Biol 7: 291–301.
15. Duval BD, Blankinship JC, Dijkstra P, Hungate BA (2012) CO2 effects on plant nutrient concentration depend on plant functional group and available nitrogen: a meta-analysis. Plant Ecol 213: 505–521.
16. Cotrufo MF, Ineson P, Scott A (1998) Elevated CO_2 reduces the nitrogen concentration of plant tissues. Glob Change Biol 4: 43–54.
17. Johnson DW, Cheng W, Joslin JD, Norby RJ, Edwards NT, et al. (2004) Effects of elevated CO_2 on nutrient cycling in a sweetgum plantation. Biogeochemistry 69: 379–403.
18. Cotrufo MF, De Angelis P, Polle A (2005) Leaf litter production and decomposition in a poplar short-rotation coppice exposed to free air CO_2 enrichment (POPFACE). Glob Change Biol 11: 971–982.
19. Marinari S, Calfapietra C, De Angelis P, Scarascia Mugnozza G, Grego S (2007) Impact of elevated CO_2 and nitrogen fertilization on foliar elemental composition in a short rotation poplar plantation. Env Pollution 147: 507–515.
20. Natali SM, Sanudo-Wilhelmy SA, Lerdau MT (2009) Plant and soil mediation of elevated CO_2 impacts on trace metals. Ecosystems 12: 715–727. doi: 10.1007/s10021-009-9251-7.
21. Adriano DC (2001) Trace Elements in Terrestrial Environments: Biogeochemistry, Bioavailability, and Risks of Metals. Springer Press.
22. Khan FN, Lukac M, Turner G, Godbold DL (2008) Elevated atmospheric CO_2 changes phosphorus fractions in soils under a short rotation poplar plantation (EuroFACE). Soil Biol Biochem 40: 1716–1723.
23. Johnson DW, Hungate BA, Dijkstra P, Hymus GJ, Hinkle CR, et al. (2003) The effects of elevated CO_2 on nutrient distribution in a fire-adapted scrub oak forest. Ecol Appl 13: 1388–1399.
24. Dijkstra P, Hymus G, Colavito D, Vieglais DA, Cundari CM, et al. (2002) Elevated atmospheric CO_2 stimulates aboveground biomass in a fire-regenerated scrub-oak ecosystem. Glob Change Biol 8: 90–103.
25. Seiler TJ, Rasse DP, Li J, Dijkstra, Anderson HP, et al. (2009) Disturbance, rainfall and contrasting species responses mediated aboveground biomass response to 11 years of CO_2 enrichment in a Florida scrub-oak ecosystem. Glob Change Biol 15: 356-367. doi: 10.1111/j.1365-2486.2008.01740.x.
26. Stover DB, Day FP, Butnor JR, Drake BG (2007) Effect of elevated CO_2 on coarse root biomass in florida scrub detected by ground-penetrating radar. Ecology 88: 1328–1334.
27. Liu D, Clark JD, Crutchfield JD, Sims JL (1996) Effect of pH of ammonium oxalate extracting solutions of prediction of plant available molybdenum in soil. Comm Soil Sci Plant Analy 27: 2511–2541.

28. Brady NC, Weil RR (2002) The nature and properties of soils. Prentice Hall, Upper Saddle River, New Jersey.
29. Crawley MJ (2007) The R Book. John Wiley and Sons, West Sussex, England.
30. Sillen WMA, Dieleman WIJ (2012) Effects of elevated CO2 and N fertilization on plant and soil carbon pools of managed grasslands: a meta-analysis. Biogeosci 9: 2247–2258.
31. R Development Core Team (2007) R: A language and environment for statistical computing. R Foundation for Statistical Computing, Vienna, Austria.
32. Institute for Genomic Biology, DeLucia Laboratory Public Data Archive. Available: http://www.life.illinois.edu/delucia/Public%20Data%20Archive/. Accessed 2013 April 30.
33. Schroeder RE (2011) Effects of 11 years of CO_2 enrichment on root biomass and spatial distribution in a Florida scrub-oak ecosystem. PhD Dissertation, Department of Biological Sciences, Old Dominion University.
34. Oh N-W, Richter DD (2004) Soil acidification induced by elevated atmospheric CO_2. Glob Change Biol 10: 1936–1946.
35. Tang J, Chen J, Chen X (2006) Response of 12 weedy species to elevated CO_2 in low-phosphorus availability soil. Ecol Res 21: 664–670.
36. Lagomarsino A, Moscatelli MC, Hoosbeek MR, De Angelis P, Grego S (2008) Assessment of soil nitrogen and phosphorus availability under elevated CO_2 and N-fertilization in a short rotation poplar plantation. Plant Soil 308: 131–147.
37. Cheng L, Zhu J, Chen G, Zheng X, Oh N-H, et al. (2010) Atmospheric CO_2 enrichment facilitates cation release from soil. Ecol Lett 13: 284-291. doi: 10.1111/j.1461-0248.2009.01421.x.
38. Langley JA, McKinley DC, Wolf AA, Hungate BA, Drake BG, et al. (2009) Priming depletes soil carbon and releases nitrogen in a scrub-oak ecosystem exposed to elevated CO_2. Soil Biol Biochem 41: 54–60.
39. Boerner REG (1982) Fire and nutrient cycling in temperate ecosystems. BioSci 32: 187–192.
40. Liu JX, Zhang DQ, Zhou GY, Faivre-Vuillin B, Deng Q, et al. (2008) CO_2 enrichment increases nutrient leaching from model forest ecosystems in subtropical China. Biogeosci 5: 1783–1795.
41. Kawano M, Tomita K (1996) Amorphous aluminum hydroxide formed at the earliest weathering stages of K-feldspar. Clays Clay Min 44: 671–676.
42. Duval BD, Dijkstra P, Natali S, Megonigal JP, Ketterer ME, et al. (2011) Plant-soil distribution of potentially toxic elements in response to elevated CO2. Env Sci Tech 45: 2570–2574.

Facilitation Drives the Positive Effects of Plant Richness on Trace Metal Removal in a Biodiversity Experiment

Jiang Wang[1], Yuan Ge[2]*, Tong Chen[1], Yi Bai[1], Bao Ying Qian[1], Chong Bang Zhang[1]

1 School of Life Science, Taizhou University, Linhai, China, **2** Bren School of Environmental Science and Management/Earth Research Institute, University of California Santa Barbara, Santa Barbara, California, United States of America

Abstract

Background: Phytoextraction is an environmentally acceptable and inexpensive technique for mine tailing rehabilitation that uses metallophyte plants. These plants reduce the soil trace metal contents to environmentally acceptable levels by accumulating trace metals. Recently, whether more trace metals can be removed by species-rich communities of these plants received great attention, as species richness has been reported having positive effects on ecosystem functions. However, how the species richness affects trace metals removal of plant communities of mine tailing is rarely known.

Methodology/Principal Findings: We examined the effects of species richness on soil trace metal removal in both natural and experimental plant communities. The root lengths and stem heights of each plant species were measured in order to calculate the functional diversity indices. Our results showed that trace metal (Cu, Cd, Pb and Zn) concentrations in mine tailing soil declined as species richness increased in both the natural and experimental plant communities. Species richness, rather than functional diversity, positively affected the mineralomass of the experimental plant communities. The intensity of plant-plant facilitation increased with the species richness of experimental communities. Due to the incremental role of plant-plant facilitation, most of the species had higher biomasses, higher trace metal concentrations in their plant tissues and lower malondialdehyde concentrations in their leaves. Consequently, the positive effects of species richness on mineralomass were mostly attributable to facilitation among plants.

Conclusions/Significance: Our results provide clear evidence that, due to plant-plant facilitation, species richness positively affects the removal of trace metals from mine tailing soil through phytoextraction and provides further information on diversity conservation and environmental remediation in a mine tailing environment.

Editor: Jin Chen, Key Laboratory of Tropical Forest Ecology, Xishuangbanna Tropical Botanical Garden, Chinese Academy of Sciences, China

Funding: This research was financially supported by the National Natural Science Foundation of China (Project 31000256), the Natural Science Foundation of Zhejiang Province, China (Y13C030015) and the Ecology Key Disciplines of Zhejiang Province in Taizhou University (EKD2013-04). The funders had no role in study design, data collection and analysis, decision to publish, or preparation of the manuscript.

Competing Interests: The authors have declared that no competing interests exist.

* E-mail: gyuan@bren.ucsb.edu

Introduction

Research into understanding how species diversity affects ecosystem function has increased over the past two decades [1,2]. Many studies have shown that species richness has positive effects on ecosystem functions and services, particularly in stressful environments [3–9]. Mine tailings are stressful environments characterized by high concentrations of trace metals and low pH, fertility and water retention capacity [10]. Phytoextraction is an environmentally friendly and inexpensive technique for mine tailing rehabilitation that uses metallophyte plants. They accumulate trace metals through their roots, which reduces trace metal content in the soil to environmentally accepted levels [11]. Li et al [12] found that increased algal species richness led to a rise in cadmium removal in an algal microcosm experiment. However, surprisingly few studies have investigated whether species-rich plant communities remove soil trace metals more efficiently than species-poor communities on mine tailing sites.

Previous studies have suggested that the positive effects of species richness on ecosystem functions generally arise through three primary mechanisms: selection effect, niche complementarity and facilitation [13]. Selection effect refers to the increased probability that a randomly assembled species-rich community will contain a species or combinations of species that have the higher than average biomass trait [14]. Niche complementarity and facilitation affect the collective performance of a community through either resource partitioning or positive interactions between species [15]. The two latter mechanisms are called the niche complementarity effect [16]. The selection and complementarity effects are not exclusive and operate simultaneously to bring about changes in ecosystem functions [17]. The principal mechanism behind the positive effects of biodiversity on ecosystem function has been actively debated [18–20].

Severely polluted environments, such as mine tailing sites, put ecosystems under constant stress [21]. The central role of facilitation among higher plants in maintaining ecosystem functions in stressful environments has been widely recognized [22–25]. However, the importance of this facilitative role may depend, to a large degree, on the characteristics of the species being tested, the stress intensity and the ecosystem function measure being considered [23,26–27]. Moreover, recent studies suggest that the role of facilitation may actually decrease in exceptionally severe environments [28–29]. So far, the role played by facilitation in plant communities growing on mine tailings and the underlying mechanism driving it remains unknown.

In a recent study, we found that more diverse natural plant communities led to lower trace metal (Cu, Cd, Pb and Zn) concentrations in mine tailing soil [30]. The results suggested that species-rich plant communities would be more efficient at removing toxic trace metals from mine tailing soils than those with fewer species. Although natural plant communities are more realistic, they hardly allow identifing why the species-rich plant communities having high efficiency as natural plant communities have non-uniform initial conditions, such as different seed pools, soil trace metal concentrations and other soil characteristics. Compared to the natural plant communities, experimental plant communities are more suitable for exploring the underlying mechanism. In this study, we combined both approaches. We aimed to determine: (1) whether plant species richness has a positive effect on the removal of trace metals in mine tailing soil; (2) if a positive effect exists, whether it is mainly attributable to facilitation among plants and (3) what is the mechanism driving the facilitation effects.

Materials and Methods

Ethics Statement

No specific permissions were required for this research. The natural plant community experiment was conducted on Huangyan Pb/Zn mine tailings. The exploitation and processing of minerals from the Huangyan Pb/Zn mine stopped about 20 years ago and now plants spontaneously grow on the mine tailings. Consequently, this site is open and no permissions were required to conduct the experiment. The experimental plant community experiment was located at Taizhou University. Some of the authors of the manuscript are Taizhou University staff and the site used for the study was at the authors' experimental research center. Consequently, no specific permissions were needed to conduct the experimental trial on this site. The plant species used in this study were not endangered or protected species.

Natural Plant Communities

A plant species survey was conducted at the Huangyan Pb/Zn mine tailing site (28°34′23″N, 120°53′44″E) on April 14, 2008 (Table S1 in File S1). Based on the natural species composition data, we designed four species richness levels (1, 2, 4 and 8) for the natural plant communities on the Huangyan Pb/Zn mine tailings. For each species richness level, five plots (1 m×1 m) with different species compositions were randomly selected (Table 1). In each plot, the aboveground and belowground plant material was collected and sorted into species. The plant materials were thoroughly washed with running tap water and rinsed with distilled water to remove any soil particles attached to the plant surfaces. The aboveground and belowground plant materials were separated and oven dried (80°C) to a constant weight and then weighed.

In each plot, five soil samples were randomly collected using a soil core sampler (64 mm in diameter×100 mm in length). Each soil sample was collected at a depth of 0–20 cm, air-dried at room temperature, homogenized and sieved (<2 mm). They were then used to determine the soil trace metal concentrations (Cu, Cd, Pb and Zn) and total N. After digestion with HCl and $HClO_4$ (4:1, v/v) [31], the soil trace metal concentrations were measured using an Inductively Coupled Plasma Optical Emission Spectrometer (ICP/OES, Optima 2100DV, Perkin Elmer, USA). Total N in the filtrates was determined using the Berthelot reaction method [32]. Briefly, 1.0 g of K_2SO_4 catalyst mixture and 5 ml of concentrated H_2SO_4 were added to 0.5 g of air-dried ground tailings in 10 ml digestion tubes. Exactly 20 ml of distilled water was then added to the tubes, which were then filtered. After full digestion, the filtrate and the contents of the tubes were transferred to 50 ml volumetric flasks for the Berthelot reaction.

Table 1. The species composition of each plot at the four different species richness levels for the natural communities growing on the Huangyan Pb/Zn mine tailings.

Species richness			
1	**2**	**4**	**8**
MF	MF+RB	LC+RB+RP+SC	GP+PA+XS+CC+MF+BP+SC+RP
RB	RP+SN	SC+RL+XS+MF	VH+LP+DL+RP+MF+XS+LF+RL
LC	MF+WJ	SC+MF+CP+LC	SC+RL+XS+RP+LF+RC+MF+RB
XS	GP+RB	LC+MF+GP+LF	XS+RP+LJ+AE+LC+MF+RB+CA
DL	SC+GP	MF+RL+GP+XS	SC+LC+WJ+VH+RP+MF+LF+RB

Footnotes: Species abbreviations and their corresponding full names are listed in Table S1.

Experimental Plant Communities

In order to construct the parallel experimental plant communities, eight species (*Bidens pilosa* Linn, *Phytolacca americana* Linn, *Commelina communis* Linn, *Mirabilis jalapa* Linn, *Chenopodium ambrosioides* Linn, *Solanum nigrum* Linn, *Brassica campestris* Linn and *Xanthium sibiricum* Patrin ex Widder) were randomly selected from the species pool of plants found growing on Huangyan Pb/Zn mine tailings (Table S1 in File S1). *P. americana* is a perennial herbaceous plant and the other seven species are annual herbaceous plants. Seeds from the eight species were collected from the plants growing on the Huangyan Pb/Zn mine tailings, then sown in trays between the 5th and 7th April, 2009 and transplanted about two months after germination. Mine tailing soil samples were randomly collected from the Huangyan Pb/Zn mine tailings and then mixed together to make a composite sample (Tailing properties: organic matter, 330±71.2 mg·kg^{-1}; total P, 50.3±10.6 mg·kg^{-1}; total N, 120.7±41.4 mg·kg^{-1}; Cu, 31.29±3.81 mg·kg^{-1}, Cd, 8.92±2.32 mg·kg^{-1}; Pb, 924.57±61.23 mg·kg^{-1} and Zn, 1312.45±52.44 mg·kg^{-1}).

The seedlings from the eight species were transplanted into plastic containers (pots, 80×80×60 cm^3) filled with mine tailing soil between June 2nd and 4th, 2009 and placed outdoors. We constructed 23 different communities and created four species richness levels (Table 2): eight monocultures, five polycultures of two species, five polycultures of four species and five polycultures of eight species, which were replicated five times. Thirty-two plant seedlings were transplanted into each pot. The plant density was similar to that of the natural plant communities growing on Huangyan Pb/Zn mine tailings. The species in the polycultures were planted at equal densities. A random selection of species from the total pool was not used because, with a limited number of communities, equal representation of the species at each species richness level was not guaranteed. Instead, the species compositions were chosen to guarantee that every species would be selected at least once at each of the four species richness levels.

Table 2. The species composition of each pot at the four species richness levels for the experimental plant communities.

Species richness			
1	2	4	8
BP	BP+XS	BP+CC+CA+SN	BP+PA+CC+MJ+CA+SN+BC+XS
PA	PA+MJ	PA+MJ+CA+BC	BP+PA+CC+MJ+CA+SN+BC+XS
CC	CC+CA	XS+PA+SN+MJ	BP+PA+CC+MJ+CA+SN+BC+XS
MJ	BP+PA	PA+BP+CC+BC	BP+PA+CC+MJ+CA+SN+BC+XS
CA	SN+BC	BP+CC+MJ+CA	BP+PA+CC+MJ+CA+SN+BC+XS
SN			
BC			
XS			

Footnotes: Species abbreviations and their corresponding full names are listed in Table S1.

The pots were weeded weekly. In addition to natural rainfall, water was added by spraying the pots during dry periods.

The experimental plant communities were harvested on May 10th, 2010. All the aboveground and belowground biomasses (living plants) were sorted into species, dried and weighed separately. Before biomass harvesting, a water treatment experiment was conducted. The water content of each pot was determined 5 days after it received the same amount of water through artificial spraying (no water was added during the final 5 days before harvesting). Five soil cores, to a depth of 0–20 cm, were randomly collected from each pot and then combined into one sample. Soil water content was calculated by comparing the wet and dry soil masses (mass of wet soil − mass of dry soil)/mass of dry soil). A proportion of each soil sample was air-dried at room temperature, homogenized and sieved (<2 mm) in order to determine the soil trace metal concentrations (Cu, Cd, Pb and Zn) and total N. The soil trace metal concentrations and total N were measured using the same method employed in the natural plant communities experiment.

At each species richness level, five surviving plants were chosen to determine the malondialdehyde (MDA) concentration in the leaves and the trace metal (Cu, Cd, Pb and Zn) concentrations in plant tissues. Generally, one surviving plant of each species was selected in each pot. However, because some species were not present in all the pots, two or three surviving plants were selected from one pot to guarantee five surviving plants per species richness level. Three surviving plants of *X. sibiricum* were selected at species richness levels two and four, as only three plants of *X. sibiricum* survived. *B. campestris* was not involved in the MDA and trace metal analyses because it did not survive in most of the pots. MDA concentration is an indicator of lipid peroxidation, which reflects the damage to plant tissues caused by oxidative stress [33]. The uppermost five leaves of each surviving plant were used to determine the MDA concentration by using the method followed by Heath and Packer [34]. The dried plant tissues were ground and mixed fully in order to determine the trace metal concentrations. After digestion with HNO_3 [31], the trace metal concentrations in the plant tissues were measured using an Inductively Coupled Plasma Optical Emission Spectrometer (ICP/OES, Optima 2100DV, Perkin Elmer, USA). The biomass used to measure the MDA was included in the plant biomass of each pot during data analysis.

Statistical Analysis

The total trace metal content accumulated in the plants (mineralomass) was used to reflect the trace metal removal efficiency by the plants. The mineralomass of a species in one pot was calculated as follows:

$$\text{Mineralomass} = \text{biomass of the species} \times \text{trace metal concentrations in plant tissues}$$

The mineralomass of the plant communities in each pot was the sum of the mineralomasses of all the component species.

Functional diversity indices were represented by the community-weighted mean trait values [35] and the functional diversity Q index [36]. For each species, five plants in monoculture were randomly selected and their root lengths and stem heights were measured with a ruler. The community-weighted mean trait values for each trait were calculated for every pot ($n = 15$) using species trait values and species relative abundances (calculated from the species biomass) following Garnier et al. [35]:

$$CWM = \sum_{i=1}^{n} p_i trait_i$$

where n is the number of species in the pot, p_i is the relative abundance of species i and trait$_i$ is the trait value of species i.

The average values of the root lengths and stem heights for each species were used to calculate the Euclidean distance between species [37]. *B. campestris* disappeared from all the plots so its root lengths and stem heights were not measured. Following the method used by Heemsbergen et al. [38], the trait values of seven species were put into a number of classes (the highest value was defined as the seventh class and the lowest value was defined as the first class). The Euclidean distance was calculated as follows:

$$d_{ij} = \left[\sum_{k=1}^{K} (A_{ki} - A_{kj})^2\right]^{1/2}$$

where A_{ki} and A_{kj} are the class number of species i and j for trait k and$K$ is the number of traits measured ($K = 2$).

The functional diversity Q index was calculated as follows:

$$FD_Q = \sum_{i=1}^{S-1} \sum_{j=i+1}^{S} d_{ij} p_i p_j$$

where S is the number of species observed in the plot and p_i and p_j are the relative abundance of species i and j, respectively.

We used the additive partitioning method to quantify the selection and complementarity effects [16]. The complementarity effect and the selection effect were calculated as follows:

$$\Delta Y = N\overline{\Delta RYM} + N\text{cov}(\Delta RY, M)$$

where ΔY is the net diversity effect, $N\overline{\Delta RYM}$ is the complementarity effect and $N\text{cov}(\Delta RY,M)$ is the selection effect. N is the number of species in the polyculture, $\overline{\Delta RY}$is the average change in relative mineralomass for all species in the mixture and $\overline{M}$ is the average mineralomass of the monocultures. The cov($\Delta RY,M$) term is the covariance between the mineralomass of the species grown in monocultures (*M*) and the change in their relative mineralomass when grown in polycultures.

The relative interaction intensity index (RII) was defined as follows [39]:

$$RII = \frac{B_W - B_O}{B_W + B_O}$$

where *Bo* is the mass potentially achieved by the target plant growing in the absence of inter- or intraspecific interactions, i.e. the expected biomass which is the monoculture biomass divided by the species richness of the polycultures. and *Bw* is the mass observed when the target plant was growing with other plants, i.e. the observed biomass of species in the polycultures. This index represents the ratio of the net mass loss/gain due to the interaction relative to the mass affected by only facilitative interaction and competitive interactions. RII has values ranging from −1 to 1, is symmetrical around zero and is negative for competition and positive for facilitation.

The data of soil trace metal concentration in the natural communities were log transformed to meet the assumptions for linear regression. We first used simple linear regression to examine the effect of species richness on soil trace metal concentrations (Cu, Cd, Pb and Zn) in the natural and experimental plant communities, respectively, and to examine the effect of species richness on the RII. Correlation analysis (Spearman coefficient) was used to determine the dependence of the complementarity effect, the selection effect or the MDA concentrations of each species in the experimental plant communities on the actual species richness, and to determine the dependence of biomass on the actual species richness, FD_Q, CWM_{root} or CWM_{stem}. These analyses were performed using the SPSS 11.5 program (SPSS Inc, Chicago, IL, USA) for Windows and significance levels for all analyses were set at $P<0.05$. The multiple stepwise regressions were performed to examine the effects of actual species richness, FD_Q, CWM_{root} and CWM_{stem} on mineralomass in the polycultures. The optimal model was selected with respect to its value of AIC (Akaike Information Criterion). The model with the lowest value of AIC was selected as the optimal model. For the selected optimal model, the normality of residuals (Shapiro-Wilk test), homosedasticity (Breusch-Pagan test) and multicollinearity among predictors (Variance Inflation Factors) were tested to check for multiple regressions preliminary assumptions. The multiple stepwise regressions were performed using R 2.11.1.

Results

In the natural plant communities growing on the Huangyan Pb/Zn mine tailings, soil trace metal (Cu, Cd, Pb and Zn)

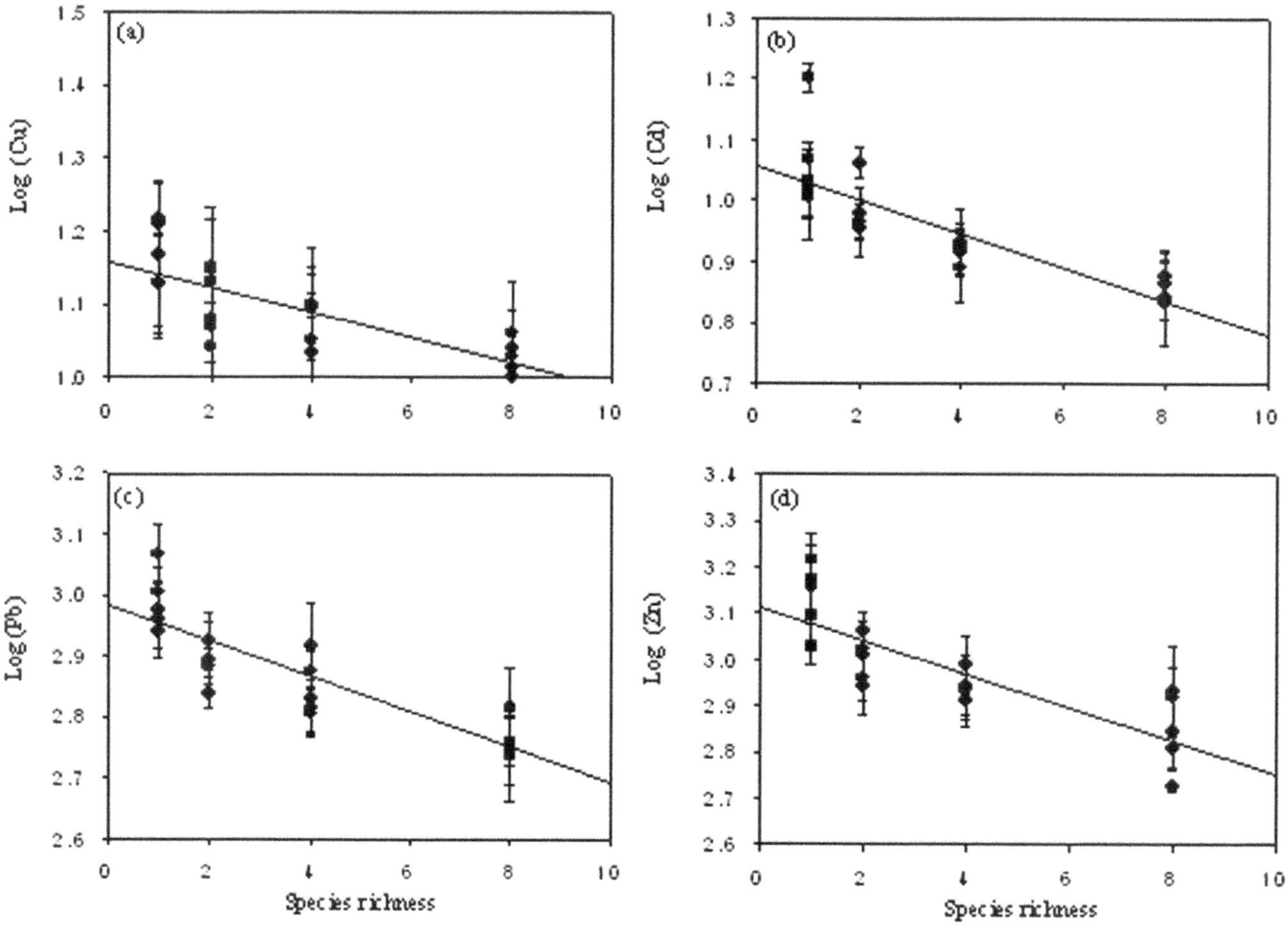

Figure 1. The dependence of soil trace metal concentration on the actual species richness (SR) in the natural plant communities. (a) Fitted line: Log (Cu) = 1.180−0.688 SR, $n=25$, $P=0.001$. (b) Fitted line: Log (Cd) = 1.058−0.832 SR, $n=25$, $P<0.001$. (c) Fitted line: Log (Pb) = 2.981−0.862 SR, $n=25$, $P<0.001$. (d) Fitted line: Log (Zn) = 3.115−0.815 SR, $n=25$, $P<0.001$.

concentrations were negatively correlated with species richness (Fig. 1). In the experimental plant communities, even if some plants disappeared in some pots due to the oxidative stress caused by trace metal toxicity, the actual species richness was still significantly and positively correlated with the initial species richness ($r = 0.966$, $n = 23$, $P<0.001$). The trace metal (Cu, Cd, Pb and Zn) concentrations in the soils from the experimental plant communities were negatively correlated with the actual species richness, which was consistent with the results from the natural plant communities (Fig. 2). These results indicated that more diverse plant communities led to lower soil trace metals in the mine tailings.

Actual species richness was always included in the optimal regression model in the multiple stepwise regressions analysis (Table 3), while FD_Q was always excluded. CWM_{root} was included for Cu, Cd and Zn, and CWM_{stem} was included for Cd and Pb. Moreover, actual species richness was positively correlated with the biomass of the polycultures ($r = 0.876$, $n = 15$, $P<0.001$). However, FD_Q ($r = 0.430$, $n = 15$, $P = 0.110$), CWM_{root} ($r = 0.494$, $n = 15$, $P = 0.061$) and CWM_{stem} ($r = 0.485$, $n = 15$, $P = 0.067$) had no significant relationship with biomass. These results suggested that the significant effect of actual species richness on mineralomass was probably closely linked to biomass production.

We also investigated whether the complementarity effect or the selection effect led to the positive effects that an increase in actual species richness had on the mineralomasses of Cu, Cd, Pb and Zn. The selection effect had no significant relationship with actual species richness in the polycultures (Fig. 3). The selection effect values varied, ranging from positive to negative across all the experimental plant communities. In contrast, the complementarity effect significantly rose with the increase in actual species richness in the polycultures, compared to the other cultures. Moreover, the complementarity effect values were positive across all the experimental plant communities. Our results suggested that the positive effects of species richness on trace metal accumulations by plants were mainly attributable to the complementarity effect.

The relative interaction intensity index (RII) was used to reflect interactions among plants. RII was positive in most polycultures (Fig. 4). Moreover, they significantly increased when actual species richness rose. Moreover, with the exception of *B. campestris*, the increase of survival ratio of other seven plant species increased with the actual species richness (Table 4). The results proved that plant-plant facilitation dominated in the polycultures and increased when the actual species richness rose.

The malondialdehyde (MDA) concentration directly reflects the oxidative damage to plant tissues caused by trace metals. In the

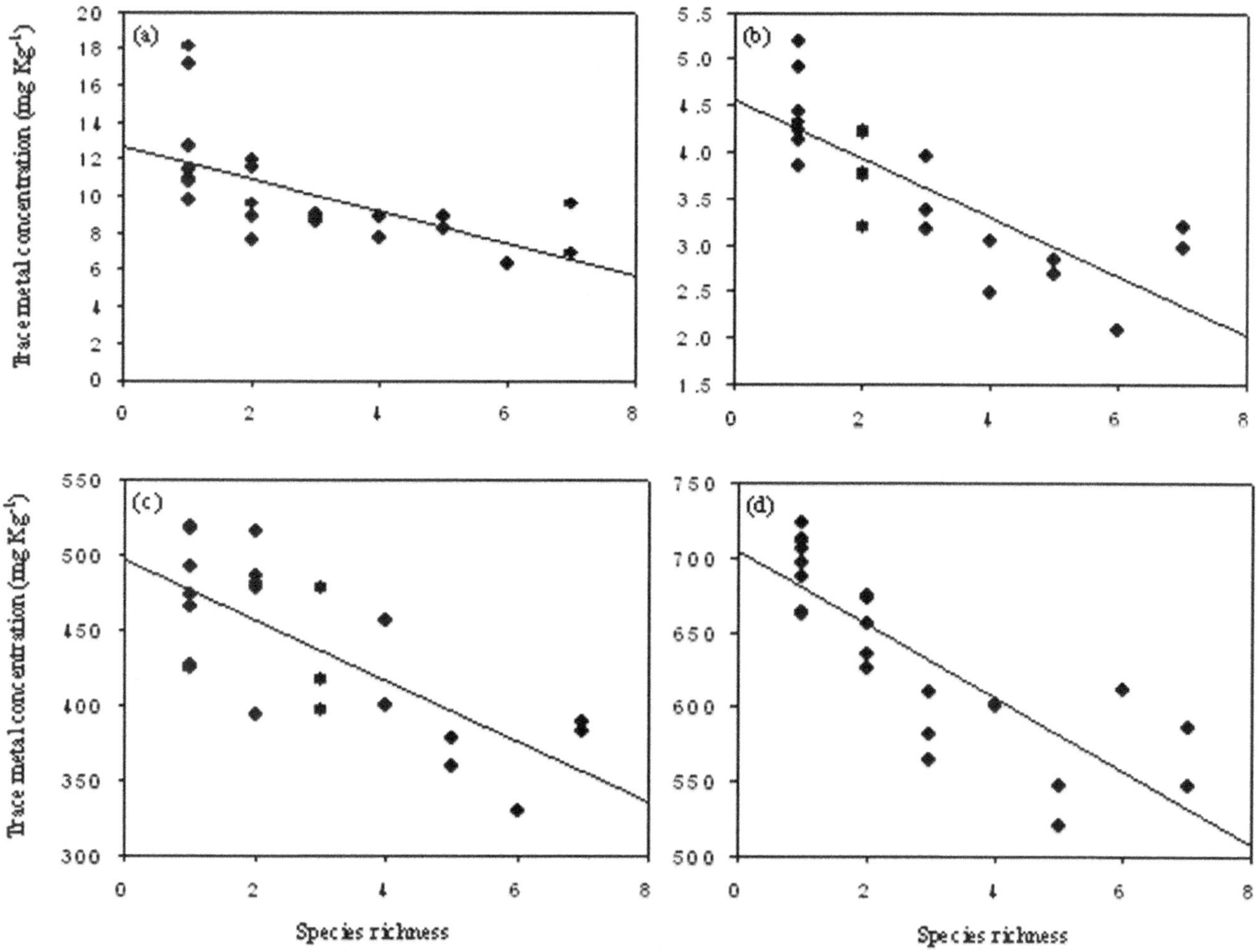

Figure 2. The dependence of soil trace metal concentration (TY) on the actual species richness (SR) in the experimental plant communities. (a) Fitted line: $TY_{Cu} = 12.655 - 0.610$ SR, $n = 23$, $P<0.001$. (b) Fitted line: $TY_{Cd} = 4.547 - 0.796$ SR, $n = 23$, $P<0.001$. (c) Fitted line: $TY_{Pb} = 496.414 - 0.735$ SR, $n = 23$, $P<0.001$. (d) Fitted line $TY_{Zn} = 703.839 - 0.818$ SR, $n = 23$, $P<0.001$.

Table 3. Multiple stepwise regression analyses between mineralomass and explanatory variables from a model that initially contained four variables (actual species richness (SR), FD_Q, CWM_{root} and CWM_{stem} values for Cu, Cd, Pb and Zn).

Mineralomass	Included variables	VIF	AIC	W	BP
$Mineralomass_{Cu}$	SR	1.13	−55.514	0.919(0.188)	3.706(0.157)
	CWM_{root}	1.13			
$Mineralomass_{Cd}$	SR	1.14	−49.798	0.944(0.437)	5.429(0.143)
	CWM_{root}	4.86			
	CWM_{stem}	4.53			
$Mineralomass_{Pb}$	SR	1.13	−52.161	0.912(0.145)	5.902(0.118)
	CWM_{stem}	1.13			
$Mineralomass_{Zn}$	SR	1.13	−52.667	0.903(0.094)	6.032(0.096)
	CWM_{root}	1.13			

Footnotes: the values in the brackets are p-value.

experimental communities, the MDA concentration in the leaves of seven species consistently decreased as actual species richness rose (Table 5). Moreover, the trace metal (Cu, Cd, Pb and Zn) concentrations in the plant tissues consistently increased as actual species richness rose (Figure S1 in File S1). These results suggested that a reduction in oxidative damage to plant tissues probably promoted the accumulation of trace metals.

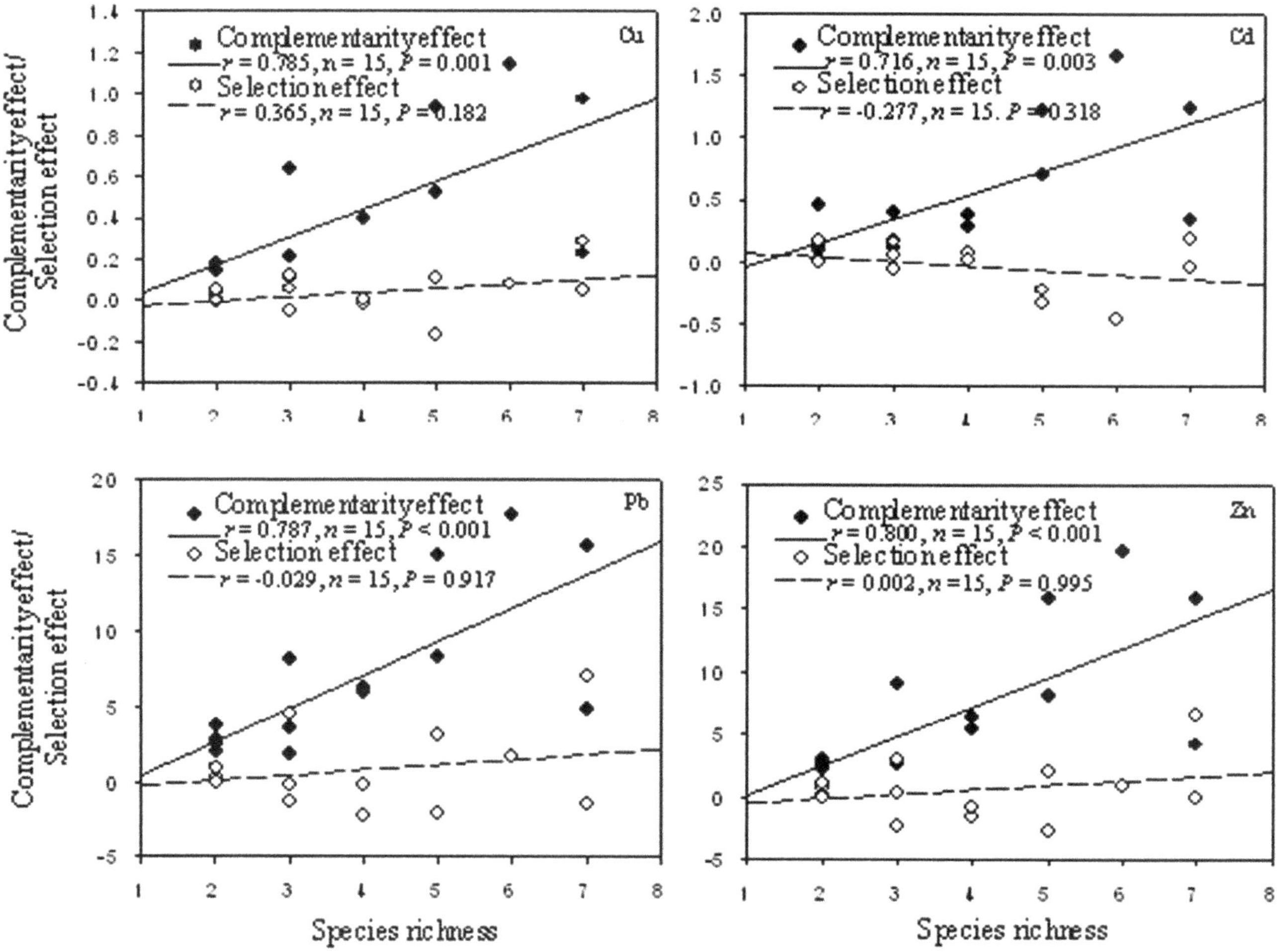

Figure 3. The dependence of the complementarity/selection effect on the actual species richness in the experimental plant communities. a) Cu, b) Cd, c) Pb and d) Zn. The relevant Spearman's correlation coefficient and *P*-value are shown in each panel.

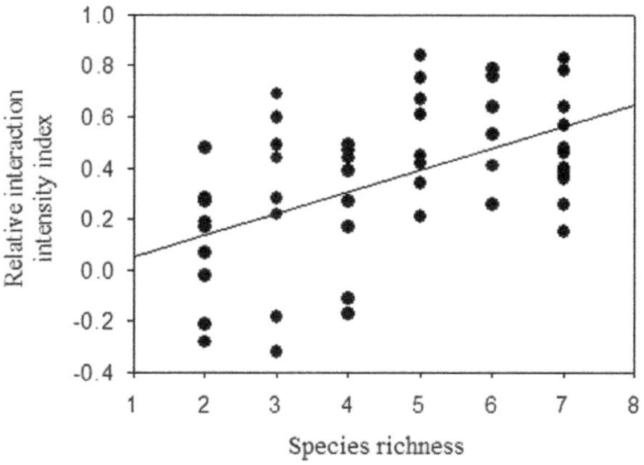

Figure 4. The dependence of relative interaction intensity index (RII) on the actual species richness (SR) in the experimental plant communities. Fitted line: RII = 0.032+0.519 SR, $n = 57$, $P < 0.001$.

Discussion

Our results demonstrated that species-rich plant communities had lower soil trace metal (Cu, Cd, Pb and Zn) concentrations in both the natural and the experimental plant communities than the species-poor communities. More importantly, we found that species richness had significantly positive effects on mineralomass in the experimental plant communities. Consequently, it is almost certain that the positive effects of species richness on mineralomass led to the lower soil trace metal concentrations seen in the more diverse plant communities.

In theory, the observed positive biodiversity effect could be further partitioned into the selection effect and the complementarity effect [16]. Across all the experimental plant communities, the grand mean value of the selection effect was nearly zero and made little contribution to the positive effect. However, the grand mean value of the complementarity effect was significantly positive and positively correlated with actual species richness. Therefore, the positive effect of actual species richness on mineralomass can mainly be attributed to the complementarity effect. Our results suggest that the positive complementarity effect could be attributed to the increase in biomass (see [40] for more details) and the increases in trace metal concentrations in plant tissues (Figure S1 in File S1) could be due to the increase in actual species richness.

Table 4. The survival ratio of each planted species in certain species richness level.

	Planted species richness			
	1	2	4	8
Bidens pilosa Linn	0.531	0.563	0.667	0.800
Phytolacca americana L.	0.156	0.188	0.250	0.500
Commelina communis Linn.	0.438	0.563	0.625	0.900
Mirabilis jalapa L.	0.219	0.438	0.500	0.750
Chenopodium ambrosioides L.	0.344	0.563	0.500	0.700
Solanum nigrum L.	0.156	0.500	0.563	0.650
Brassica campestris L.	0.125	0.125	0.063	0.100
Xanthium sibiricum Patrin ex Widder	0.250	0.188	0.500	0.550

Survival ratio = the survival number of plants in a certain species richness level/ the total planted number in the corresponding species richness level.

Table 5. The correlations between malondialdehyde (MDA) concentration for each species and the actual species richness in the experimental communities.

Species	*r*	*n*	*P*
BP	−0.900	20	$P<0.001$
PA	−0.660	20	$P=0.002$
CC	−0.886	20	$P<0.001$
MJ	−0.829	20	$P<0.001$
CA	−0.744	20	$P<0.001$
SN	−0.935	20	$P<0.001$
XS	−0.892	16	$P<0.001$

Footnotes: Species abbreviations and their corresponding full names are listed in Table S1.

Distinguishing the effects of niche differentiation and facilitation is very difficult in practice [16]. Niche complementarity or facilitation may have important roles in driving the positive complementarity effect seen in the polycultures. Niche differentiation directly linked to interspecific trait differences that are determined by the functional diversity of communities [41–42]. In this study, the pot experimental design limited environmental fluctuations and spatial heterogeneity, which are important for resource partitioning between species [17]. Consequently, in this study, functional diversity probably made only a limited contribution to the increased mineralomass. However, environmental fluctuations and spatial heterogeneity are ubiquitous in natural conditions [17]. Consequently, we speculate that niche complementarity may provide more contribution to the positive effect of species richness on mineralomass in natural communities of mine tailings. In the experimental communities, most of species in the polycultures had positive RII values; Most of the polycultures had a larger biomass than their most productive component species in the monocultures (see [38] for more details); Most of plant species (with the exception of *B. campestris*) had higher survival ratio in the polycultures. All of the above results proved plant-plant facilitation dominated in the polycultures, as overyielding [43] and the increase of survival ratio [44] are strong indicators of facilitation. Moreover, the yield and survival ratio increased with the actual species richness. Consequently, the positive complementarity effect seen in the polycultures should be ascribed to the facilitation.

The plants used in the experimental communities are all the established species of the Huangyan Pb/Zn mine tailing. Moreover, their seeds were also collected from the Huangyan Pb/Zn mine tailings. Consequently, the plants used should have high tolerance to the stress of mine tailing. Although mine tailings are characterized by high concentrations of trace metals and low pH, fertility and water retention capacities [10,30] and mechanically, physically, chemically and biologically restrict plant growth [10,30], the experimental mine tailing environment may not exert the exceptionally severe stress for these plants. Choler et al. [45] also found, due to lack the experiment at the altitude of >3000 m where should be exceptionally severe for nival communities, the facilitation of the nival communities consistent increased with altitude. Consequently, the role of facilitation decreasing in exceptionally severe environments [28–29] was not seen in this experiment.

In this study, the main driving mechanism behind the facilitative processes could be the amelioration of mine tailing soil by the collective performance of all plant species, as soil total N and soil water increased as species richness rose in the experimental communities (see [40] for more details). Plants can improve the soil physical, chemical and biological properties of mine tailings by preventing erosion, adding nutrients and promoting microbial growth and activity [46]. The species-rich experimental communities showed increased biomass production and surface area coverage (data not shown), which likely decreased light penetration and water evaporation and enhance water retention capacity. The root biomass increased as species richness rose (see [40] for more details), which likely increased soil nitrogen due to root exudates. Additionally, soil microbial biomass and activity increased as species richness rose (data not shown), which likely accelerated the decomposition of plant litter and increase the soil nitrogen levels.

Trace metals stress leads to the production of free radicals in plants, which causes lipid peroxidation and tissue damage [47]. MDA is a cytotoxic product of lipid peroxidation [48]. The ameliorated soil properties probably reduced the trace metal stresses, which led to a decrease in MDA concentrations in the plant tissues. Consequently, as actual species richness increased, the plant stress caused by trace metals declined, which may be the reason why trace metal concentrations in plant tissues increased as actual species richness rose.

Taken together, our results provide clear evidence that increased species richness in plant communities exerts a positive effect on soil trace metal removal through increased facilitation among plants present. In comparison to conventional techniques for mine tailing remediation, phytoextraction is cost-effective, safe for humans and environmentally friendly [10,11]. However, the factors limiting the wider application of this technology involve its relatively low trace metal removal efficiency [49] and the small biomass of most of the hyper-accumulators identified so far [50–52]. Our results indicated that diverse plant communities can promote biomass production and trace metal accumulation by plants. However, the selection of species to be used in the construction of plant communities for phytoremediation should be prudent. In this study, *B. campestris* was sensitive to the mine tailing environment and the facilitative effect was not beneficial. Although similar results were found in our natural and experimental plant communities, future studies should be expanded to include realistic plant communities established on different mine tailings in order to determine whether the implications of our results can be extrapolated to realistic phytoremediation programs for mine tailings.

Supporting Information

File S1 This includes Table S1 and Figure S1. Table S1 The name of species identified in the species survey of Huangyan Pb/Zn mine tailing. Figure S1 The heavy metal concentrations in plant tissues of seven species in pots with different actual sepcies richness. a) Cu, b) Cd, c) Pb and d) Zn. Species abbreviations are seen in Table S1. The lines shown in each panel are the regression lines (All the P values of Cu, Cd, Pb and Zn for seven species are lower than 0.05).

Acknowledgments

We thank Xiu Bai and Xing Luo for help in the field and two anonymous referees for their helpful comments.

Author Contributions

Conceived and designed the experiments: JW, YG. Performed the experiments: TC, YB, BYQ, CBZ. Analyzed the data: JW. Wrote the paper: JW, YG.

References

1. Hooper DU, Adair EC, Cardinale BJ, Byrnes JEK, Hungate BA, et al. (2012) A global synthesis reveals biodiversity loss as a major driver of ecosystem change. Nature 486: 105–108.
2. Midgley GF (2012) Biodiversity and ecosystem function. Science 335: 174–175.
3. Kennedy TA, Naeem S, Howe KM, Knops JMH, Tilman D, et al. (2002) Biodiversity as a barrier to ecological invasion. Nature 417: 636–638.
4. Cardinale BJ, Srivastava DS, Duffy JE, Wright JP, Downing AL, et al. (2006) Effects of biodiversity on the functioning of trophic groups and ecosystems. Nature 443: 989–992.
5. Cadotte MW, Cardinale BJ, Oakley TH (2008) Evolutionary history and the effect of biodiversity on plant productivity. PNAS, 105: 17012–17017.
6. Kaneryd L, Borrvall C, Bergl S, Curtsdotter A, Eklöf A, et al. (2012) Species-rich ecosystems are vulnerable to cascading extinctions in an increasingly variable world. Ecol Evol 2: 858–872.
7. Lei PF, Scherer-Lorenzen M, Bauhus J (2012) The effect of tree species diversity on fine-root production in a young temperate forest. Oecologia 169: 1105–1115.
8. Maestre FT, Quero JL, Gotelli NJ, Escudero A, Ochoa V, et al. (2012) Plant species richness and ecosystem multifunctionality in global drylands. Science 335: 214–217.
9. Thompson PL, Shurin JB (2012) Regional zooplankton biodiversity provides limited buffering of pond ecosystems against climate change. J Anim Ecol 81: 251–259.
10. Meeinkuirt W, Pokethitiyook P, Kruatrachue M, Tanhan P, Chaiyarat R (2012) Phytostabilization of a Pb-contaminated mine tailing by various tree species in pot and field trial experiments. Int J Phytoremediat 14: 925–938.
11. Novo LAB, Covelo EF, González L (2013): Phytoremediation of amended copper mine tailings with Brassica juncea, Int J Min, Reclamat & Environ 27: 215–226.
12. Li SP, Li JT, Kuang JL, Duan HN, Zeng Y, et al. (2012) Effects of species richness on cadmium removal efficiencies of algal microcosms. J Appl Ecol 49: 261–267.
13. Hector A, Bazeley-White E, Loreau M, Otway S, Schmid B (2002) Overyielding in grassland communities: testing the sampling effect hypothesis with replicated biodiversity experiments. Ecol Lett 5: 502–511.
14. Cardinale BJ, Gross K, Fritschie K, Flombaum P, Fox JW, et al. (2013) Biodiversity simultaneously enhances the production and stability of community biomass, but the effects are independent. Ecology. In press.
15. Tilman D, Lehman C (2001) Biodiversity, composition, and ecosystem processes: theory and concepts. In: Kinzig AP, Pacala SW, Tilman D, editors. The functional consequences of biodiversity. Princeton: Princeton University Press. pp. 9–41.
16. Loreau M, Hector A (2001) Partitioning selection and complementarity in biodiversity experiments. Nature 412: 72–76.
17. Špaèková I, Lepš J (2001) Procedure for separating the selection effect from other effects in diversity–productivity relationship. Ecol Lett 4: 585–594.
18. Huston MA, Aarssen LW, Austin MP (2000) No consistent effect of plant diversity on productivity. Science 289: 1255a.
19. Naeem S (2000) Reply to Wardle et al. Bull Ecol Soc Am 81: 241–246.
20. Fargione J, Tilman D, Dybzinski R, Lambers JHR, Clark C, et al. (2007) From selection to complementarity: shifts in the causes of biodiversity–productivity relationships in a long-term biodiversity experiment. Proc R Soc B 274: 871–876.
21. Kondratyev KY, Krapivin VF, Phillips GW (2002) Global Environmental Change: Modeling and Monitoring. Springer, Berlin.
22. Baumeister D, Callaway RM (2006) Facilitative effects of *Pinus flexilis* during succession: a hierarchy of mechanisms benefits other plant species. Ecology 87: 1816–1830.
23. Maestre FT, Bowkerl MA, Escolar C, Puche MD, Soliveres S, et al. (2010) Do biotic interactions modulate ecosystem functioning along stress gradients? Insights from semi-arid plant and biological soil crust communities. Phil Trans R Soc B 365: 2057–2070.
24. Odadi WO, Karachi MK, Abdulrazak SA, Young TP (2011) African wild ungulates compete with or facilitate cattle depending on season. Science 333: 1753–1755.
25. He Q, Bertness MD, Altieri AH (2013) Global shifts towards positive species interactions with increasing environmental stress. Ecol Lett 16: 695–706.
26. Maestre FT, Valladares F, Reynolds JF (2005) Is the change of plant-plant interactions with abiotic stress predictable? A meta-analysis of field results in arid environments. J Ecol 93: 748–757.

27. Kawai T, Tokeshi M. (2007) Testing the facilitation-competition paradigm under the stress-gradient hypothesis: decoupling multiple stress factors. Proc Royal Soc B 274: 2503–2508.
28. Bruno JF, Stachowicz JJ, Bertness MD (2003) Inclusion of facilitation into ecological theory. Trends Ecol Evol 18: 119–125.
29. Michalet R, Brooker RW, Cavieres LA, Kikvidze Z, Lortie CJ, et al. (2006) Do biotic interactions shape both sides of the humped-back model of species richness in plant communities? Ecol Lett 9: 767–773.
30. Wang J, Zhang CB, Ke SS, Qian BY (2011) Different spontaneous plant communities in Sanmen Pb/Zn mine tailing and their effects on mine tailing physico-chemical properties. Environ Earth Sci 62: 779–786.
31. Mcgrath SP, Cunliffe CH (1985) A simplified method for the extraction of the metals Fe, Zn, Cu, Ni, Cd, Pb Cr, Co and Mn from soils and sewage sludges. J Sci food Agr 36: 794–798.
32. Page AL, Miller RH, Keeney DR (1982) Methods of Soil Analysis, Chemical and Microbiological Properties. Wisconsin: Madison.
33. Dhir B, Sharmila P, Saradhi PP (2004) Hydrophytes lack potential to exhibit cadmium stress induced enhancement in lipid peroxidation and accumulation of praline. Aquat Toxicol 66: 141–147.
34. Heath RL, Packer L (1968) Photoperoxidation in isolated chloroplasts 1 kinetics and stoichiometry of fatty acid peroxidation. Arch Biochem Biophys 125: 189–198.
35. Garnier E, Cortez J, Billès G, Navas M, Roumet C, et al. (2004). Plant functional markers capture ecosystem properties during secondary succession. Ecology, 85: 2630–2637.
36. Rao CR (1982) Diversity and dissimilarity coefficients – a unified approach. Theor Popul Biol 21: 24–43.
37. Walker B, Kinzig A, Langridge J (1999) Plant attribute diversity, resilience, and ecosystem function: the nature and significance of dominant and minor species. Ecosystems 2: 95–113.
38. Heemsbergen DA, Berg MP, Loreau M, van Hal JR, Faber JH, et al. (2004) Biodiversity effects on soil processes explained by interspecific functional dissimilarity. Science 306: 1019–1020.
39. Armas C, Ordiales R, Pugnaire FI (2004) Measuring plant interactions: anew comparative index. Ecology 85: 2682–2686.
40. Wang J, Zhang CB, Chen T, Li WH (2013) From selection to complementarity: the shift along the abiotic stress gradient in a controlled biodiversity experiment. Oecologia 171: 227–235.
41. Petchey OL, Gaston KJ (2002) Functional diversity (FD), species richness and community composition. Ecol Lett 5: 402–411.
42. Petchey OL, Gaston KJ (2006) Functional diversity: back to basics and looking forward. Ecol Lett 9: 741–758.
43. Drake JM (2003) Why does grassland productivity increase with species richness? Disentangling species richness and composition with tests for overyielding and superyielding in biodiversity experiments. PNAS 270: 1713–1719.
44. Mulder CPH, Uliassi DD, Doak DF (2001) Physical stress and diversity-productivity relationships: the role of positive interactions. Proc Natl Acad Sci USA 98: 6704–6708.
45. Choler P, Michalet R, Callaway RM (2001) Facilitation and competition on gradients in alpine plant communities. Ecology 82: 3295–3308.
46. Gómez-Sagasti M, Alkorta I, Becerril JM, Epelde L, Anza M, et al. (2012) Microbial Monitoring of the Recovery of Soil Quality During Heavy Metal Phytoremediation. Water Air Soil Pollut (2012) 223:3249–3262.
47. Šuran J, Prišć M, Rašić D, Srebočan E, Crnić A P (2013) Malondialdehyde and heavy metal concentrations in tissues of wild boar (*Sus scrofa* L.) from central Croatia. J Environ Sci Heal Part B 48: 147–152.
48. Choudhary M, Jetley UK, Khan MA, Zutshi S, Fatma T (2007) Effect of heavy metal stress on proline, malondialdehyde, and superoxide dismutase activity in the cyanobacterium Spirulina platensis-S5. Ecotoxicology and Environmental Safety 66: 204–209.
49. Cooke SJ, Suski CD (2008) Ecological restoration and physiology: an overdue integration. BioScience 58: 957–968.
50. Parraga-Aguado I, Gonzalez-Alcaraz MN, Alvarez-Rogel J, Jimenez-Carceles FJ, Conesa HM (2013) The importance of edaphic niches and pioneer plant species succession for the phytomanagement of mine tailings. Environ Pollut 176: 134–143.
51. Zhang SJ, Li TX, Huang HG, Zou TJ, Zhang XZ, et al. (2012) Cd accumulation and phytostabilization potential of dominant plants surrounding mining tailings. Environ Sci Pollut Res 19: 3879–3888.
52. Sheng XF, He LY, Wang QY, Ye HS, Jiang CY (2008) Effects of inoculation of biosurfactant-producing *Bacillus* sp. J119 on plant growth and cadmium uptake in cadmium-amended soil. J Hazard Mater 155: 17–22.

19

Soil Carbon Stocks Decrease following Conversion of Secondary Forests to Rubber (*Hevea brasiliensis*) Plantations

Marleen de Blécourt[1]*, Rainer Brumme[1], Jianchu Xu[2], Marife D. Corre[1], Edzo Veldkamp[1]

1 Soil Science of Tropical and Subtropical Ecosystems, Büsgen Institute, Georg-August-Universität Göttingen, Göttingen, Germany, **2** World Agroforestry Centre, East Asia Node, Kunming, China

Abstract

Forest-to-rubber plantation conversion is an important land-use change in the tropical region, for which the impacts on soil carbon stocks have hardly been studied. In montane mainland southeast Asia, monoculture rubber plantations cover 1.5 million ha and the conversion from secondary forests to rubber plantations is predicted to cause a fourfold expansion by 2050. Our study, conducted in southern Yunnan province, China, aimed to quantify the changes in soil carbon stocks following the conversion from secondary forests to rubber plantations. We sampled 11 rubber plantations ranging in age from 5 to 46 years and seven secondary forest plots using a space-for-time substitution approach. We found that forest-to-rubber plantation conversion resulted in losses of soil carbon stocks by an average of 37.4±4.7 (SE) Mg C ha^{-1} in the entire 1.2-m depth over a time period of 46 years, which was equal to 19.3±2.7% of the initial soil carbon stocks in the secondary forests. This decline in soil carbon stocks was much larger than differences between published aboveground carbon stocks of rubber plantations and secondary forests, which range from a loss of 18 Mg C ha^{-1} to an increase of 8 Mg C ha^{-1}. In the topsoil, carbon stocks declined exponentially with years since deforestation and reached a steady state at around 20 years. Although the IPCC tier 1 method assumes that soil carbon changes from forest-to-rubber plantation conversions are zero, our findings show that they need to be included to avoid errors in estimating overall ecosystem carbon fluxes.

Editor: Ben Bond-Lamberty, DOE Pacific Northwest National Laboratory, United States of America

Funding: This study is funded by the German Ministry for Economic Cooperation and Development (http://www.bmz.de) and the Deutsche Gesellschaft für Internationale Zusammenarbeit (GIZ)(http://www.giz.de) for the project "Making the Mekong Connected" (No. 08.7860.3-001.00). The funders had no role in study design, data collection and analysis, decision to publish, or preparation of the manuscript.

Competing Interests: Deutsche Gesellschaft für Internationale Zusammenarbeit (GIZ) partly funded this study. There are no patents, products in development or marketed products to declare. This does not alter the authors' adherence to all the PLOS ONE policies on sharing data and materials.

* E-mail: mblecou@gwdg.de

Introduction

Deforestation and forest degradation in the tropics have been estimated to contribute 12–15% of the global anthropogenic CO_2 emissions [1]. The majority of land-use induced CO_2 emissions arise from the loss of above-ground biomass and to a lesser extent from decomposition of soil organic carbon [2,3]. Currently, the estimations of land-use change effects on above-ground carbon stocks are improving due to remote sensing techniques. However, estimates of land-use change effects on soil carbon stocks remain inconclusive [4,5].

In tropical regions, the magnitude and direction of land-use induced changes in soil carbon stocks are largely determined by mean annual rainfall and clay mineralogy [5,6]. A large number of studies exist on the impact of tropical land-use changes on soil carbon stocks, especially on the conversion from forest to pasture, pasture to secondary forest, and forest to cropland [5]. However, the published field observations are unequally distributed over the tropics with regards to an area-weighted distribution of the above-mentioned biophysical conditions. In addition, little research has been done on currently important land-use changes, one of which is forest-to-rubber (*Hevea brasiliensis*) plantation conversion. These limitations in available field-observations hamper the estimates of land-use change effects on soil carbon stocks in the tropics [5].

Our present study focuses on the land-use change from secondary forests to rubber plantations in Xishuangbanna, the southernmost prefecture of Yunnan Province in the southwest of China. The area of monoculture rubber plantations is rapidly expanding in the montane mainland southeast Asia, spanning southwest China, Laos, Cambodia, Myanmar, northeast Thailand, and northwest Vietnam, causing a large decrease in the region's forest cover [7]. Rubber trees were traditionally not grown in this region, since environmental conditions (low temperatures in winter and a distinctive dry season) were considered marginal for rubber trees [8]. The first rubber plantations were successfully established in Xishuangbanna by the Chinese government in the late 1950s, and the subsequent expansion of rubber plantations resulted in a strong economic development [9]. At present, rubber plantations in montane mainland southeast Asia cover an area of more than 1.5 million ha [8] of which 424,000 ha is in Xishuangbanna [10]. By 2050 the area of rubber plantations is predicted to increase fourfold, mainly by replacing secondary forests, and swidden-related bushes and shrublands [11].

Forest-to-rubber plantation conversion is an important recent land-use change for which environmental impacts have hardly been studied. Meta-analyses of current data have shown that changes of soil carbon stocks after conversion of forests to tree plantations are variable: no effects were reported for tropical tree plantations [5,12] whereas in another review 0–30% decrease in soil carbon stocks were reported for intensified rubber plantations compared to swidden fields in southeast Asia [13]. However, this review was based on studies which did not have a clear reference land-use type for the rubber plantations but merely compared land-use types and therefore any detected difference cannot directly be attributed to changes in land use. To our knowledge there are only three published tropical studies that investigated the effects of forest-to-rubber plantation conversions on soil carbon stocks. Two out of the three studies focused on the conversion from primary forests to rubber plantations in Brazil [14,15] and both reported declines in the soil carbon stocks in 17- and 22-year-old plantations. The other study focused on the conversion from secondary forests to rubber plantations and happened to be conducted in Xishuangbanna [16]. Yang et al. [16] reported a 20% decline in the soil carbon stock in the top 0.6-m depth in a 3-year-old plantation and a 16% decline in a 7-year-old plantation but these estimates were not corrected for changes in soil bulk density with land-use change. It is also important that studies include older rubber plantations as soil carbon stocks may reach equilibrium after several decades, and older plantations will allow us to quantify the long-term effects of this land-use change. Soil carbon losses after deforestation are often related to (i) changes in the quality and quantity of soil carbon input, (ii) accelerated soil carbon decomposition rates driven by changes in microclimatic conditions or the breakdown of soil aggregates, and (iii) enhanced soil surface erosion. The magnitude of soil carbon losses depends furthermore on site-specific conditions such as soil texture, soil mineralogy, topography and climate.

Improved estimates of the effects of this important land-use change on soil carbon stocks are essential to the national greenhouse gas inventories from the Conference of Parties of the United Nations Framework of Climate Change. Although the Intergovernmental Panel of Climate Change (IPCC) provides guidelines for the estimations of the ecosystem carbon fluxes arising from land use changes, the IPCC tier 1 method assumes soil carbon changes to be zero for the conversion from forests to rubber plantations because of limited scientific knowledge [4,17].

We conducted the present study in Xishuangbanna, Yunnan Province, China using a space-for-time substitution approach. Our objectives were: (i) to quantify changes in soil carbon stocks following conversion of secondary forests to rubber plantations over a 46 years' time period, and (ii) to determine the biophysical factors which control soil carbon concentrations, and soil carbon stock changes. We hypothesized that conversion from secondary forests to rubber plantations would result in a decrease of soil carbon stocks. This decrease is expected to be related to management practices commonly employed in rubber plantations such as terrace construction and removal of the vegetation understory.

Materials and Methods

Study Area and Site Characteristics

The study area of 4500 ha was located in Menglong township, Jinghong county of Xishuangbanna prefecture in Yunnan province, China (21°31′17.03″N, 100°37′12.13″E) (Fig. 1). The climate is tropical monsoon and is characterized by a dry season from November to April and a wet season from May to October. The mean annual rainfall is 1377 mm and the mean annual temperature is approximately 22.7°C [9]. The topography is hilly to mountainous, with an elevation that varies between 650 and 1450 m above sea level [9]. The study plots were located between 700 and 830 m above sea level. The soils at the plots are dominated with low activity clays and were classified as Ferralsols having an effective cation exchange capacity (ECEC) of less than 12 $cmol_c\ kg^{-1}$ clay and as (hyper) ferralic Cambisols with an ECEC of less than 24 $cmol_c\ kg^{-1}$ clay [18] (Table 1, Table S1A, Table S1B).

Current and Past Land Use

The major land-use types in Menglong township include rubber plantations, secondary forests, and farmland. According to local plantation owners the dominant land-use change trajectories in Menglong township were: (i) Primary forest–swidden agriculture–secondary forest–rubber plantation, and (ii) primary forest–swidden agriculture–rubber plantation. Swidden agriculture was the dominant land-use type in the region for centuries [19]; this involved cutting and burning of vegetation patches in the forest, thereby creating fields for use as rotation of cropping phases (1–3 years) and fallow periods (5–20 years) during which secondary vegetation regrows [11,20]. The widespread practice of swidden cultivation in the past resulted in loss and degradation of primary forests [21]. Nowadays, almost all swidden fields have been replaced by monoculture rubber plantations. Since primary forest and swidden agriculture are not present anymore, we focused on the more recent land-use change from secondary forest towards rubber plantations. Based on information from local plantation owners, we selected rubber plantations that all went through this land-use change trajectory. Forest clearing was done by hand and no heavy machinery was used. After forest clearing, the sites were

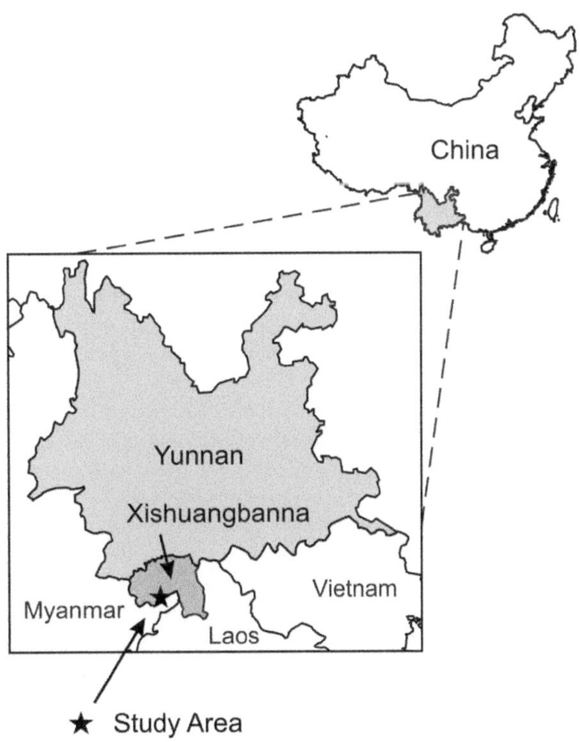

Figure 1. Location of the study area in Xishuangbanna prefecture, Yunnan province, China.

Table 1. Means[1] (± SE) of soil characteristics of land-use types.

Characteristic	Depth (m)	Rubber plantation (n = 11)	Secondary forest (n = 7)
Sand (%)	0–0.15	32.5±3.8	34.7±4.6
	0.15–0.3	30.5±3.9	34.4±5.0
	0.3–1.2	27.8±3.4	30.2±4.4
Silt and clay (%)	0–0.15	67.5±3.8	65.3±4.6
	0.15–0.3	69.5±3.9	65.6±5.0
	0.3–1.2	72.2±3.4	69.8±4.4
Bulk density (g cm^{-3})	0–0.15	1.1±0.1	1.0±0.1
	0.15–0.3	1.1±0.0	1.2±0.0
	0.3–1.2	1.3±0.0	1.3±0.0
C:N ratio	0–0.15	12.2±0.3	12.8±0.5
	0.15–0.3	12.1±0.4	12.5±0.4
	0.3–1.2	9.5±0.4	9.7±0.5
pH (H_2O)[2]	0–0.15	4.8±0.1 a	4.7±0.1 b
	0.15–0.3	4.8±0.1	4.9±0.1
	0.3–1.2	5.0±0.1	4.9±0.1
pH (KCl)	0–0.15	3.9±0.0	3.9±0.0
	0.15–0.3	3.8±0.1	3.9±0.0
	0.3–1.2	4.0±0.0	4.0±0.0
Effective cation exchange capacity[2] ($mmol_c$ kg^{-1} soil)	0–0.15	46.4±1.8 a	55.5±2.4 b
	0.6–0.9	28.8±1.2 a	36.9±3.9 b
Base saturation (%)	0–0.15	24.4±4.2	24.1±6.0
	0.6–0.9	10.9±1.6	11.3±1.5

[1]Means of the 0.3–1.2 m depth interval are means of the 0.3–0.6-m, 0.6–0.9-m and 0.9–1.2-m depth intervals.
[2]Within a row, means followed by different letters differ significantly between rubber plantation and secondary forest (linear mixed effects model at $P \leq 0.05$).

usually burnt. During the first four years after planting, rubber trees may have been intercropped with maize, upland rice, peanuts, and beans. In our study area, four forest remnants remain, including three collective forests and one "watershed protection" forest, which we used as our reference. These broadleaf forests are highly degraded due to the collection of firewood at present and timber extraction in the past. The forests have been cleared from primary forest, after which they were used for swidden agriculture and finally secondary forests were established (information from local farmers). The age of each forest remnant was estimated between 40–60 years. The size of the forest patches ranges from 20–60 ha.

Management Practices in Rubber Plantations

Management practices commonly applied in rubber plantations are terrace establishment and maintenance, fertilizer application, pest control, management of the vegetation understory and rubber tapping. The terrace benches are constructed by hand using a hoe, cutting soil from upper parts of the slope and moving it to lower parts. During terrace maintenance, weeds are removed from the terrace risers by scraping off the soil, which is subsequently evenly distributed over the entire terrace bench. This process is repeated once or twice per year, resulting in older plantations having wider and deeper incised terraces. Rubber trees are planted in a row on the terrace benches and have a tree spacing ranging from 2–3 m. The step height of each individual terrace ranges from 0.3–0.8 m and bench width is typically ~2.5 m. The horizontal distance between two adjacent terrace benches is 5–7 m, depending on the slope of the hill. Between some of the rubber trees pits are dug which have various uses: (i) as a measure to reduce runoff and retain soil moisture, and (ii) to apply fertilizer and collect leaf litter and cut herbs. The dimensions of these pits vary per plantation and range from: 0.4–1.3-m length × 0.2–0.5-m width × 0.2–0.35-m depth. Chemical fertilizers (NPK-compounds) are applied one to two times per year. The management of vegetation understory varies per rubber plantation; some plantation owners use herbicides to control the vegetation understory while others cut the vegetation understory back. Rubber tapping is usually done between April and October and latex collection is done every second day.

Sampling Design

We used the space-for-time substitution approach to quantify changes in soil carbon stocks following conversion from secondary forests to rubber plantations. Soil carbon stocks were measured in clusters consisting of one reference secondary forest plot and one to three plots in rubber plantations. Clusters were established around randomly selected secondary forest plots. To avoid edge effects, the forest plots were selected at least 20 m from the forest edge. Within each cluster, the rubber plantations were chosen based on biophysical conditions, land-use history and distance to the reference plot. We only selected rubber plantations that were established immediately after forest clearing. To keep biophysical conditions within a cluster as similar as possible, we selected rubber plantations with similar altitude, slope, aspect, soil color and soil texture as the reference plot. The maximum distance between plots within a cluster was 3 km. In total we selected seven clusters, with a total of 11 rubber plantations and seven reference forests. The rubber plantations within each cluster differed in age ranging from 5 to 46 years. Selected rubber plantations were both state-owned rubber plantations and plantations belonging to smallholder farmers.

A critical assumption of the space-for-time substitution approach is that plots within a cluster were initially similar with regard to soil characteristics, soil carbon stocks, and land-use history such that measured differences in soil carbon stocks between the reference land-use type and the converted land-use type can be attributed to recent land-use change [22]. To test this assumption we compared land-use independent soil characteristics (i.e. soil texture) of plots within a cluster. Since we did not detect significant differences in soil texture between the secondary forest and rubber plantations within a cluster (Table 1), we assumed that the soils were originally similar and that observed soil carbon changes can be attributed to changes in land use.

Fieldwork Permission

Our research was part of the project "Making the Mekong Connected (MMC)". This project had been officially registered at the Kunming Institute of Botany, Chinese Academy of Sciences, which provided rights for access to field research in China. We received permission from the owners and managers of the rubber plantations to conduct the described fieldwork in their plantations. The secondary forests are part of local collective forests belonging to the villages. The local forestry station of Menglong Township, Jinghong County, has been informed a priori about our fieldwork in the secondary forests. No official permits were required for the described fieldwork since the secondary forests are not part of a national nature reserve. The fieldwork did not involve endangered or protected species.

Plot Layout and Soil and Litter Sampling

In each land-use type, we established a plot with a size of 20×20 m, corrected for slope. Within each plot we established five parallel transects with 5-m spacing in between. Transects had a fixed north-south orientation. We randomly positioned four sampling points along each transect, resulting in 20 sampling points per plot.

Soil samples were collected down to 1.2-m depth from five depth intervals: 0–0.15 m, 0.15–0.3 m, 0.3–0.6 m, 0.6–0.9 m, and 0.9–1.2 m. The upper three depth intervals were sampled using a Dutch auger at the 20 sampling points. The 20 soil samples were pooled in the field to form one composite sample for each depth interval. Soil samples for the 0.6–0.9-m and 0.9–1.2-m depth were sampled in a soil pit which was positioned on the slope between two adjacent terraces in the middle of each plot. The soil samples were air dried for five days and sieved through a 2-mm sieve prior to laboratory analyses. Bulk-density samples were taken in the soil pit for each of the five depth intervals, using the core method [23]. Very few of the bulk-density samples contained stones or rock fragments and thus we did not correct for the gravel content.

Litter-layer samples were taken from every second sampling point, resulting in 10 litter samples per plot. Leaf litter and organic material (including seeds and twigs) were collected from a 0.04-m^2 quadrant sample frame. The collected material was oven dried at 60°C for 48 hours and weighed. Subsamples of each sample were pooled by plot and analyzed for total carbon and nitrogen concentration. The carbon stock of the litter layer was calculated with the carbon concentration (%), the mass of the litter layer, and the sample-frame area.

Tree Inventory, Topographical and Land use Data

In the rubber-plantation plots, we measured for all the trees the diameter at breast height (DBH) at 1.3 m above the soil surface. In the forest plots, we measured the DBH for trees with a DBH >4 cm, and the DBH of bamboos. For bamboos, we measured one stem DBH per clump and we recorded the number of stems per clump. Here we report both the tree basal area and total basal area, which is the sum of the basal area of trees and bamboos. Other site characteristics that were collected of each plot included: slope, aspect, altitude, and GPS coordinates. Information on current and past land use and their management was collected through interviews with land-owners and elders in the villages.

Laboratory Analyses and Soil Carbon Stock Calculations

Total carbon and nitrogen concentrations were measured from ground soil and litter samples by dry combustion using CNS Elemental analyzer (Elementar Vario EL, Hanau, Germany). As soil pH was below 5.5, carbonates were not expected in these soils and we made no attempt to remove them. Soil pH (H_2O) and pH (KCl) were measured on air-dried soil for all individual soil samples in a 1:2.5 soil-to-solution ratio. ECEC was measured on soil samples of the 0–0.15-m and 0.6–0.9-m depth. The soil samples were percolated with un-buffered 1 M NH_4Cl and the percolates were analyzed for exchangeable cations using ICP-AES (Spectroflame, Spectro Analytical Instruments, Kleve, Germany) [24]. Soil texture analyses were determined for all depth intervals with the pipette method, distinguishing the fractions: clay (<0.002 mm), silt (0.002–0.063 mm), and sand (0.063–2 mm).

Soil carbon stocks (Mg C ha^{-1}) in each depth interval were calculated using the following equation:

$$SOC\ (Mg\ C\ ha^{-1}) = \frac{\%C}{100} \times BD\ (Mg\ m^{-3}) \times D\ (m)\ \times\ 10{,}000\ m^2\ ha^{-1} \quad (1)$$

where, BD is the bulk density and D refers to the thickness of the depth interval. Total soil carbon stocks down to 1.2-m depth were calculated as the sum over all depth intervals. Land-use changes often coincide with changes in bulk density due to management practices which may compact or loosen the soil. In order to be able to compare the same soil mass and to avoid the interference of bulk density changes with soil carbon stocks changes, we used the bulk density data of the reference plots to calculate the soil carbon stock of the rubber plantation plots [6].

Statistical Analyses

All statistical analyses were done using the open source statistical software R version 2.15.0 [25]. To make statistical inferences on the differences in soil carbon stocks and soil characteristics between rubber plantations and secondary forest, we applied linear mixed effects models (LME) using the nlme package [26]. Response variables were the soil carbon stocks and soil characteristics and we included land-use type, depth interval, and the interaction between land-use type and depth interval as fixed effects. Cluster was included as a random factor. Comparisons of soil carbon stocks and soil characteristics between land-use types at each depth interval were obtained by defining and testing contrasts with the generalized linear hypothesis test in the multcomp package [27]. For the multiple comparisons of soil carbon stock changes between depth intervals, the P values were adjusted using Holm's correction. For each LME, assumptions on normality and homogeneity of variance were checked by visual inspection of plots of residuals against fitted values. In cases of unequal variances of residuals, we included a variance function that allows for unequal variances [28].

To examine monotonic trends of soil carbon concentrations and relative soil carbon stock differences with potential explanatory variables, we did spearman rank correlation tests. Relative soil carbon stock differences were calculated as carbon stock in rubber plantation minus carbon stock of the reference secondary forest divided by carbon stock of the reference secondary forest multiplied by 100. Relative soil carbon stock differences were correlated with explanatory variables of the rubber plantations. As potential explanatory variables we included litter carbon stock, litter C:N ratio, total basal area, sum of silt and clay content, slope, and altitude. Correlation tests were done for each depth interval.

The trend between soil carbon and rubber plantation age was examined using non-linear regression. We tested the fit of both a mono-exponential model and a bi-exponential model according to Lobe et al. [29]. The mono-exponential model assumes a single soil carbon pool which following land-use change tends towards a new equilibrium:

$$X_t = X_e + (X_0 - X_e) \times \exp(-k \times t) \quad (2)$$

where, X_0 is the initial soil carbon stock of the secondary forest plots ($t = 0$), X_t is the soil carbon stock in the rubber plantation plots at age t, X_e is the soil carbon stock at steady state, k is the decay rate per year, and t is year since land-use change. The age t at which steady state was reached was calculated as the point where the proportion of carbon remaining in the soil (X_t) did not differ more than 1% of the calculated steady state value X_e [29]. The bi-exponential model considers both labile and stable soil carbon pools:

$$X_t = X_1 \times \exp(-k_1 \times t) + X_2 \times \exp(-k_2 \times t) \quad (3)$$

where, X_1 is the proportion of carbon in the labile pool, and X2 is the proportion of carbon in stable pool ($X_2 = 100 - X_1$), k_1 is the decay rate per year of the labile pool, k_2 is the decay rate per year of the stable pool. We expressed soil carbon as the proportion of the soil carbon stock in the rubber plantation to the initial amount in the reference secondary forest. The exponential models were fitted to the data using nonlinear least-squares estimations. The goodness of the fit was assessed by Pearson's correlation coefficient (r) showing the relationship between the observed and fitted values.

Results

Soil Characteristics, Litter Layer, and Tree Basal Area in Rubber Plantations and Secondary Forests

Soil texture, bulk density, soil C:N ratio, pH (KCl), and base saturation did not differ between rubber plantations and secondary forests (Table 1). The pH (H_2O) in the top 0.15-m depth was higher in rubber plantations than in secondary forests. The ECEC in all depth intervals was lower in rubber plantations than in secondary forest. Litter carbon concentration, litter C:N ratio, and litter carbon stock did not differ between rubber plantations and secondary forests (Table 2). The tree basal area in rubber plantations ranged from 3.2 to 42.4 and was positively correlated with plantation age (spearman's rho = 0.93, $p \leq 0.001$); the mean tree basal area was twice that of the secondary forests (Table 2). However, the total basal area (sum of trees and bamboos) did not differ between rubber plantations and secondary forests.

Soil Carbon Concentrations and Stocks in Rubber Plantations and Secondary Forests

All rubber plantations had a lower soil carbon stock in the total soil profile (0–1.2-m depth) than secondary forests ($P \leq 0.01$) (Table 3). The differences in soil carbon stocks between rubber plantations and secondary forests ranged from −15.4 to −59.4 Mg C ha^{-1} with a mean of −37.4±4.7 Mg C ha^{-1}, equivalent to a 19.3±2.7% loss of the initial soil carbon stock. The soil carbon losses were depth dependent as was shown by a significant interaction between land-use type and soil depth ($P \leq 0.001$). The decrease in soil carbon concentrations and soil carbon stocks was significant for the three depth intervals in the top 0.6-m depth (Table 3). The largest decrease was found in the top 0.15-m of the soil ($P \leq 0.01$) accounting for 32% of soil carbon losses.

For the top 0.15-m depth, the proportion of carbon remaining in the soil exponentially decreased with the years since land-use change, as described by the mono-exponential model (Equation 2) (Fig. 2A). The largest decrease could be observed in the first 5 years following land-use change, when the soil carbon stocks had declined to approximately 80% of the original amount. A steady state was reached after approximately 20 years, when soil carbon stocks had declined to 68% of the original amount. At 0.15–0.3-m depth, soil carbon had the tendency to exponentially decrease with time but the estimated decay rate of the mono-exponential model was not significant; a steady state after approximately 20 years showed a soil carbon stock decline of 75% of the original amount (Fig. 2B). At 0.3–0.6-m depth (Fig. 2C), a mono-exponential trend was not detectable. Bi-exponential model (Equation 3) fitting resulted in insignificant decay rates for both the labile and stable soil carbon pool for all soil depths (data not shown). Furthermore, the fitted curves of the bi-exponential model and mono-exponential model were identical. Together these results indicate that the mono-exponential model was most suitable to describe the observed soil carbon changes in relation to the years since land-use change.

Correlations of Soil Carbon Concentrations and Soil Carbon Stock Changes with Environmental Factors

In rubber plantations, soil carbon concentrations in the top 0.6-m of the soil showed positive correlations with altitude and with the sum of clay and silt content (and a negative correlation with sand content) (Table 4). However, at 0.15–0.3-m depth the correlation with the sum of silt and clay content was only marginally significant ($P = 0.1$). Rubber plantation age was not correlated to soil carbon concentrations in the top 0.6-m depth. However, for 0.9–1.2-m depth a positive correlation was observed between soil carbon concentrations and plantation age (spearman's rho = 0.66, $P \leq 0.05$). In secondary forests, soil carbon concentrations at 0.15–0.3-m depth were positively correlated with the sum of clay and silt content and at 0.3–0.6-m depth with the total basal area of the forest (Table 4). The trends with soil texture and total basal area were also apparent at 0.6–0.9-m depth (data not shown). Relative differences in soil carbon stocks between rubber plantations and secondary forests in the top 0.15-m of the soil were negatively correlated with total basal area and rubber plantation age. In the top 0.6-m of the soil, relative differences in soil carbon stocks were positively correlated with altitude, but for the top 0.15-m of the soil this correlation was marginally significant ($P = 0.06$).

Table 2. Means (± SE) of litter and tree characteristics of land-use types.

Characteristic	Rubber plantation (n = 11)	Secondary forest (n = 7)
Litter carbon concentration (%)	41.1±0.7	40.0±0.7
Litter C : N ratio	46.1±3.8	44.9±3.6
Litter carbon stock (Mg C ha^{-1})	2.1±0.2	2.7±0.4
Tree basal area[1] (m^2 ha^{-1})	18.6±3.8 a	9.7±2.4 b
Total basal area[2] (m^2 ha^{-1})	18.6±3.8	15.3±1.7

[1]Within a row, means followed by different letters differ significantly between rubber plantation and secondary forest (linear mixed effects model at P≤0.05).
[2]Total basal area is calculated as the sum of the basal area of trees and bamboos.

Discussion

Impact of Land-use Changes on Soil Carbon Stocks

Our findings of decreased soil carbon stocks under rubber plantations, not only in the top 0.6-m depth but also when considering the whole 1.2-m depth, are consistent with published studies of paired comparisons and chronosequences that all reported soil carbon losses following the conversion from primary or secondary forests to rubber plantations [14–16].

To estimate the effects of land-use changes on soil carbon stocks it is crucial to account for changes in soil bulk density. The importance of correction for bulk density changes has been emphasized by many authors [2,6,22,30], but it was not applied in the published studies on forest-to-rubber plantation conversions [14–16]. We examined the effects of bulk density changes on the estimated land-use change effects for our own data and for the cited studies that reported data on depth, bulk density, and soil carbon concentrations [15,16]. Comparison of corrected and uncorrected estimates showed that in these studies, not accounting for bulk density changes resulted in overestimations up to 5% and underestimations as high as 18% of the relative soil carbon stock difference (Table S2). Errors were greatest for the top 0.3-m depth. This comparison illustrates again that ignoring bulk density changes potentially causes large errors in land-use induced soil carbon stocks changes. For the following discussion of forest-to-rubber plantation conversions, we used the corrected values of relative soil carbon stock changes.

The only published study on secondary forest-to-rubber plantation conversion was conducted in Xishuangbanna on an Udic Ferrisol (Chinese classification system) [16]. Yang et al. [16] observed soil carbon losses (corrected for bulk density changes) of 24% in a 3-year-old rubber plantation and 21% in a 7-year-old rubber plantation in the top 0.6-m depth. The soil carbon stocks in rubber plantations and secondary forests reported by Yang et al. [16] were comparable to our estimated soil carbon stocks (Table 3). In Brazil, the conversion from primary forests to rubber plantations resulted in soil carbon losses of 21% down to 0.5-m depth in a 22-year-old rubber plantation on an Ultisol [14], and of 48% down to 1.0-m depth in a 17-year-old rubber plantation on an Oxisol [15]. The soil carbon losses observed by Araujo et al. [14] and Yang et al. [16] correspond well with our observed declines of 24% at 0–0.3-m depth and 21% at 0–0.6-m depth (Table 3). However, the 48% decline down to 1-m depth reported by Salimon et al. [15] is much larger than the 19% we observed down to 1.2-m depth.

Important methodological differences exist between our study and the cited studies. The large decrease in soil carbon stocks reported by Salimon et al. [15] should be interpreted with care, since in their study soil samples below 0.10-m depth were taken from the middle of each depth interval instead of sampling the entire depth interval, which can lead to inaccurate estimations of the soil carbon stocks. Furthermore, the studies from Araujo et al. [14] and Salimon et al. [15] consisted of only one forest and one rubber plantation and the study from Yang et al. [16] consisted of two rubber plantations and two forests. Results from such case studies with no or insufficient replications should not be extrapolated to a large scale. Finally, Yang et al. [16] and Salimon et al. [15] established in each land-use type one plot where they took no more than three replicate soil samples per depth interval.

At first sight, our observed decline in soil carbon stocks of 24±2.5% in the top 0.3-m depth seems to be much larger than the insignificant effects reported for the compiled studies of the whole

Table 3. Means (± SE) of soil carbon stocks and absolute[1] and relative[2] differences between land-use types.

	Rubber plantation (n = 11)		Secondary forest (n = 7)		Difference (n = 7)	
Depth (m)	C (%)	C (Mg ha^{-1})	C (%)	C (Mg ha^{-1})	Absolute (Mg C ha^{-1})	Relative (C %)
0–0.15	2.1±0.1	30.3±1.9	2.9±0.1	43.9±2.6	−11.8±1.1***	−26.9±2.8***
0.15–0.3	1.7±0.1	29.8±1.6	2.2±0.1	38.9±1.5	−8.2±1.4 ***	−21.4±3.2***
0.3–0.6	1.2±0.1	43.6±2.6	1.4±0.1	52±1.6	−8.0±3.0*	−15.4±5.6*
0.6–0.9	0.7±0.1	28.0±1.9	0.9±0.1	35.2±3.7	−6.5±3.6	−16.0±8.0
0.9–1.2	0.6±0.0	23.2±1.3	0.7±0.0	26.0±1.0	−2.9±1.8	−11.2±7.0
Total	-	154.9±6.2	-	196.0±3.5	−37.4±4.7**	−19.3±2.7**

Significant at *P≤0.05, **P≤0.01, and ***P≤0.001, (linear mixed effects model).
[1]Absolute differences in stocks were calculated as means of rubber plantations within a cluster minus reference forest.
[2]Relative differences in stocks were calculated as means of rubber plantations within a cluster minus reference forest divided by reference forest multiplied by 100.

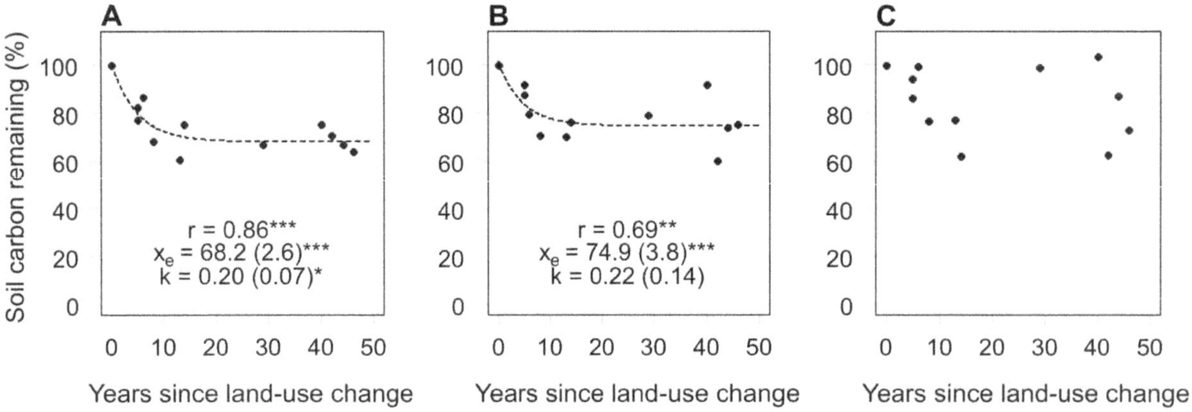

Figure 2. Soil carbon remaining after land-use change at (A) 0–0.15-m, (B) 0.15–0.3-m, and (C) 0.3–0.6-m depth. Soil carbon remaining is expressed as the proportion of soil carbon in rubber plantations relative to the secondary forest. The dashed lines represent fitted mono-exponential models (see Equation 2). r = Pearson's correlation coefficient between observed and fitted values; k = decay rate ($year^{-1}$) and X_e = equilibrium ratio (%), and values in brackets are SE. Pearson's r and parameter estimates are significant at $^*P \leq 0.05$, $^{**}P \leq 0.01$, and $^{***}P \leq 0.001$.

tropics for forest-to-tropical tree plantation conversions [5]. However, analysis of the supplementary dataset from Powers et al. [5] revealed that 12 out of the 15 tree plantation types included in their meta-analysis showed soil carbon losses of 19±4.4% down to 0.3-m depth. The overall insignificant effect reported in their study resulted from the large number of observations from the other three plantation types (comprising 36% of all observations) that showed soil carbon accumulation, thereby offsetting the soil carbon losses. Our observed decline was thus within the same magnitude as carbon losses reported for many other tropical tree plantations. Taken as a whole, tropical tree plantations established after forest conversion appear to be more prone to soil carbon losses than previously reported.

The space-for-time substitution approach used in our study has as advantage that long-term soil carbon stock dynamics can be studied within a relatively short time period. However, this approach has also disadvantages: (i) It requires the untestable assumption that the land-use changes were random in the landscape regarding forest soil carbon stocks, and (ii) the spatial variation in other biophysical conditions of plots within a cluster (i.e. each cluster included a reference secondary forest and one to three rubber plantations) may interfere with land-use change effects and time trends. To deal with these limitations we carefully selected plots for comparison within a cluster and the clusters were replicated spatially. We included seven replicated clusters, which reduces the chance that soil carbon stock differences between rubber plantations and secondary forests are not due to land use.

Soil Carbon Losses Related to Years since Land-use Change

The observed exponential decrease in soil carbon stocks in the top 0.15-m depth in rubber plantations with years since land-use change is similar to trends often reported for forest-to-agriculture conversions [31–33]. For our study, the potential drivers of the rapid soil carbon losses during the first five years following land-use change are: (i) soil disturbances during site preparation and terrace construction, which may accelerate soil surface erosion and increase soil carbon decomposition, and (ii) the sparse vegetation cover in young plantations, which reduces the soil carbon input and may change the microclimatic conditions, the latter could in turn result in enhanced soil carbon decomposition rates. The reduced soil carbon losses in older plantations may be explained by the denser vegetation cover, thereby increasing soil carbon input and soil stability. Management practices that are likely to

Table 4. Correlation coefficients[1] of soil carbon concentrations and relative soil carbon stock differences[2] with explanatory variables at three depths.

	Rubber plantation C (%) (n = 11)			Secondary forest C (%) (n = 7)			Relative soil C differences (%) (n = 9)		
Explanatory variable	**0–0.15 m**	**0.15–0.3 m**	**0.3–0.6 m**	**0–0.15 m**	**0.15–0.3 m**	**0.3–0.6 m**	**0–0.15 m**	**0.15–0.3 m**	**0.3–0.6 m**
Litter C stock (Mg ha^{-1})	0.21	0.39	0.40	−0.07	−0.18	0.36	−0.10	0.52	0.49
Litter C : N ratio	−0.40	−0.23	0.15	−0.43	−0.57	−0.11	−0.35	−0.25	0.04
Total basal area (m^2 ha^{-1})	−0.39	−0.47	−0.24	0.18	0.57	0.89**	−0.64*	$-0.55^{\dagger}$	−0.48
Silt and clay (%)	0.72*	$0.53^{\dagger}$	0.66*	0.36	0.93**	0.61	0.13	0.20	0.49
Rubber plantation age (year)	−0.26	−0.34	−0.03	–	–	–	−0.65*	−0.46	−0.24
Slope (%)	−0.08	0.07	0.14	−0.46	−0.04	0.11	0.15	0.35	0.08
Altitude (m)	0.75**	0.76**	0.71*	−0.50	0.00	0.21	$0.59^{\dagger}$	0.70*	0.70*

[1]Spearman rank correlation test, marginally significant at $^{\dagger}P \leq 0.1$, and significant at $^*P \leq 0.05$, and $^{**}P \leq 0.01$.
[2]Relative soil carbon stock differences were calculated as carbon stock in rubber plantation minus carbon stock of the reference secondary forest divided by carbon stock of the reference secondary forest multiplied by 100. Relative soil carbon stock differences were correlated with explanatory variables of the rubber plantations.

affect the soil carbon balance during the entire rotation period of the rubber plantation are terrace maintenance, rubber tapping, fertilization and the removal of the vegetation understory.

Although a soil carbon steady state was reached approximately 20 years after conversion, we expect the land-use change induced soil carbon losses in this region to continue for a longer period of time. This is because the lifespan of rubber plantations in this region ranges between 30–50 years, and it is a common practice to establish new rubber plantations after felling the previous plantation. Site preparation for the new rubber trees involves the establishment of new terraces, which would be accompanied with a further decline in the soil carbon stock.

Environmental Controls on Soil Carbon Concentrations and Soil Carbon Losses

In rubber plantations and secondary forests, the positive correlation of soil carbon concentrations with silt and clay content (Table 4) is consistent with literature and can be explained by the chemical and physical stabilization mechanisms of clay and silt particles [34]. The correlation between altitude and soil carbon concentrations, as we observed for rubber plantations, is often related to temperature effects. However, a temperature gradient is probably not the cause of the observed correlation, considering the relatively small altitude range of the sampled rubber plantations (700–830 m). Most likely the observed correlation reflects the marginally significant relationship between altitude and the sum of silt and clay content in the rubber plantations (data not shown). In secondary forests, the positive correlation of soil carbon concentrations with total basal area (which reflects forest productivity) suggests that increases of above-ground biomass could increase soil carbon input through increased input of leaf litter and root residues. This implies that in this region, restoration of the above-ground biomass in degraded secondary forests may increase soil carbon concentrations. The negative correlation between soil carbon stock differences and plantation age in 0–0.15-m depth attests that soil carbon stocks in rubber plantations progressively declined with increasing plantation age. This trend was also reflected in the negative correlation between soil carbon stock differences and total basal area due to the collinearity between plantation age and total basal area. The positive correlation between soil carbon stock differences and altitude may reflect the previously described soil texture gradient in rubber plantations with altitude.

Changes in Above-ground Carbon Stocks versus Soil Carbon Losses

It is often assumed that soil carbon emissions arising from deforestation and forest degradation in the tropics are relatively small compared to above-ground carbon losses [2,3,35]. In our case, comparison of the observed soil carbon losses with the estimates of above-ground carbon changes based on regional data [36] reveals that forest conversion to rubber plantations had a much larger effect on soil carbon stocks than on above-ground carbon stocks. Above-ground carbon stock estimates for forests outside nature reserves range from 32.2–71.0 Mg C ha^{-1} with a mean of 53.2±2.1 Mg C ha^{-1} [36]. We assumed that these forests may reflect the conditions of our sampled forests, which were also situated outside nature reserves. For rubber plantations, means of above-ground carbon stock estimates range from 61.4 Mg C ha^{-1} for plantations <800-m altitude to 35.1 Mg C ha^{-1} for plantations between 800–1000-m altitude [36]. We used these estimates as our sampled plantations were situated between 700 and 830-m altitude. Together, this indicates that land-use change may results in above-ground carbon stock changes ranging from a loss of 18 Mg C ha^{-1} to an increase of 8 Mg C ha^{-1}. These estimates are in agreement with our data on total basal area (Table 2), for which we did not detect differences between the total basal area of forests and rubber plantations. Such changes in above-ground carbon stocks were much lower than the soil carbon loss of 37.4±4.7 Mg C ha^{-1} for the entire 1.2-m depth.

Consequences of Observed Soil Carbon Losses

We showed that the conversion from secondary forests to rubber plantations leads to diminishing soil carbon stocks and that this decline is much larger than the changes in above-ground carbon stocks. Given the clear pattern of our locally collected data, it is likely that in montane mainland southeast Asia on comparable soils, this land-use change may cause declines in the soil carbon stock, the magnitude of which will depend on site-specific biophysical conditions and management practices. The size of the observed losses has implications for the estimates of ecosystem carbon fluxes arising from land-use changes in the national inventories based on the IPCC guidelines. The IPCC tier 1 method does not include soil carbon fluxes for the forest-to-rubber plantation conversions [4,17]. Our findings show that soil carbon changes need to be included to avoid possibly large errors in the estimates of the overall ecosystem carbon fluxes.

Supporting Information

Table S1 S1A. Site and soil characteristics of the secondary forest plots. S1B. Site and soil characteristics of the rubber plantation plots.

Table S2 The effect[1] of correction for bulk density changes on estimates of soil carbon stock changes. Corrections for bulk density changes were applied for estimates of the soil carbon stocks in rubber plantations and for the estimates of absolute[2] and relative[3] differences in soil carbon stocks between rubber plantations and secondary forests for our own data[4] (means ± SE), and for the cited studies that reported soil carbon concentration and bulk density data.

Acknowledgments

We are grateful for the valuable support in China, by Rhett D. Harrisson and Yan Zhuang Xiang (Xishuangbanna Tropical Botanical garden, Menglun), the staff of World Agroforestry Centre (ICRAF, East Asia office, Kunming), Maria Hänsel (University of Bayreuth), and the owners of the rubber plantations. We also thank Andreas Keuter (University of Goettingen, Germany) and Harmen P. Hendriksma (B. Triwaks Bee Research Center, Rehovot, Israel) for their feedback on statistics.

Author Contributions

Conceived and designed the experiments: MdB RB JX EV. Performed the experiments: MdB. Analyzed the data: MdB MDC. Wrote the paper: MdB RB JX MDC EV.

References

1. Werf GR van der, Morton DC, DeFries RS, Olivier JGJ, Kasibhatla PS, et al. (2009) CO_2 emissions from forest loss. Nature Geoscience 2: 737–738.
2. Detwiler RP (1986) Land use change and the global carbon cycle: the role of tropical soils. Biogeochemistry 2: 67–93.

3. Houghton RA, Hackler JL (1999) Emissions of carbon from forestry and land-use change in tropical Asia. Global Change Biology: 481–492.
4. Aalde H, Gonzalez P, Gytarsky M, Krug T, Kurz WA, et al. (2006) Forest land. In: Eggleston HS, Buendia L, Miwa K, Ngara T, Tanabe K, editors. 2006 IPCC guidelines for national greenhouse gas inventories, prepared by the national greenhouse gas inventories programme. Hayama, Japan: IGES. 4.1–4.83.
5. Powers JS, Corre MD, Twine TE, Veldkamp E (2011) Geographic bias of field observations of soil carbon stocks with tropical land-use changes precludes spatial extrapolation. Proceedings of the National Academy of Sciences of the United States of America 108: 6318–6322.
6. de Koning GHJ, Veldkamp E, Lopez-Ulloa M (2003) Quantification of carbon sequestration in soils following pasture to forest conversion in northwestern Ecuador. Global Biogeochemical Cycles 17: 1–12.
7. Ziegler AD, Fox JM, Xu J (2009) The rubber juggernaut. Science 324: 1024–1025.
8. Li Z, Fox JM (2012) Mapping rubber tree growth in mainland Southeast Asia using time-series MODIS 250 m NDVI and statistical data. Applied Geography 32: 420–432.
9. Xu J, Fox J, Vogler J, Yongshou P, Lixin Y, et al. (2005) Land-use and land-cover change and farmer vulnerability in Xishuangbanna prefecture in southwestern China. Environmental Management 36: 404–413.
10. Xu J, Grumbine RE, Beckschäfer P (2012) Landscape transformation through the use of ecological and socioeconomic indicators in Xishuangbanna, Southwest China, Mekong Region. Ecological Indicators: In press.
11. Fox J, Vogler JB, Sen OL, Giambelluca TW, Ziegler AD (2012) Simulating land-cover change in montane mainland southeast Asia. Environmental management 49: 968–979.
12. Marín-Spiotta E, Sharma S (2013) Carbon storage in successional and plantation forest soils: a tropical analysis. Global Ecology and Biogeography 22: 105–117.
13. Bruun TB, de Neergaard A, Lawrence D, Ziegler AD (2009) Environmental consequences of the demise in swidden cultivation in southeast Asia: carbon storage and soil quality. Human Ecology 37: 375–388.
14. Araujo QR, Comerford NB, Ogram AV, Al-Agely A, Santos Filho LP, et al. (2004) Soil carbon and physical property changes in Brazilian coastal tableland soils with land use following deforestation. Agroforestry Systems 63: 193–198.
15. Salimon C, Wadt PG, Alves S (2009) Decrease in carbon stocks in an oxisol due to land use and cover change in southwestern Amazon. Ambiente e Agua - An Interdisciplinary Journal of Applied Science 4: 57–65.
16. Yang JC, Huang JH, Pan QM, Tang JW, Han XG (2004) Long-term impacts of land-use change on dynamics of tropical soil carbon and nitrogen pools. Journal of Environmental Sciences-China 16: 256–261.
17. Lasco RD, Ogle S, Raison J, Verschot L WR (2006) Cropland. In: Eggleston HS, Buendia L, Miwa K, Ngara T TK, editor. 2006 IPCC guidelines for national greenhouse gas inventories, prepared by the national greenhouse gas inventories programme. Hayama, Japan: IGES. 5.1–5.66.
18. IUSS Working Group WRB (2006) World reference base for soil resources 2006. World Soil Resources Report No. 103. Rome: FAO.
19. Xu J (2006) The political, social, and ecological transformation of a landscape. Mountain Research and Development 26: 254–262.
20. Mertz O (2009) Trends in shifting cultivation and the REDD mechanism. Current Opinion in Environmental Sustainability 1: 156–160.
21. Fox J, Vogler J (2005) Land-use and land-cover change in montane mainland southeast Asia. Environmental Management 36: 394–403.
22. Veldkamp E (1994) Organic carbon turnover in three tropical soils under pasture after deforestation. Soil Science Society of America Journal 58: 175–180.
23. Blake GR, Hartge KH (1986) Bulk Density. In: Klute A, editor. Methods of Soil Analysis, Part 1:Physical and Mineralogical Methods. Madison, Wisconsin: Soil Science Society of America. 363–375.
24. König N, Fortmann H (1996) Probenvorbereitungs-, Untersuchungs- und Elementbestimmungs- Methoden des Umweltanalytik-Labor der Niedersächsischen Forstlichen Versuchsanstalt und des Zentrallabor II des Forschungszentrums Waldökosysteme.Teil 1: Elementbestimmungsmethoden A-M. Berichte des Forschungszentrums Waldökosysteme de Universität Göttingen.
25. R Development Core Team (2012) R: A language and environment for statistical computing: ISBN 3–900051–07–0. Available: http://www.r-project.org. Accessed 30 January 2013.
26. Pinheiro J, Bates D, DebRoy S, Sarker D, R Development Core Team (2012) nlme: Linear and Nonlinear Mixed Effects Models. R package version 3.1–108, Available: http://cran.r-project.org/web/packages/nlme/index.html. Accessed 30 January 2013.
27. Hothorn T, Bretz F, Westfall P (2008) Simultaneous inference in general parametric models. Biometrical Journal 50: 346–363.
28. Zuur AF, Ieno EN, Walker NJ, Saveliev AA, Smith GM (2009) Mixed Effects Models and Extensions in Ecology with R. New York: Springer. 574 p.
29. Lobe I, Amelung W, Du Preez CC (2001) Losses of carbon and nitrogen with prolonged arable cropping from sandy soils of the South African Highveld. European Journal of Soil Science 52: 93–101.
30. Davidson EA, Ackerman IL (1993) Changes in soil carbon inventories following cultivation of previously untilled soils. Biogeochemistry 20: 161–193.
31. Zingore S, Manyame C, Nyamugafata P, Giller KE (2005) Long-term changes in organic matter of woodland soils cleared for arable cropping in Zimbabwe. European Journal of Soil Science 56: 727–736.
32. Dalal RC, Mayer RJ (1986) Long-term trends in fertility of soils under continuous cultivation and cereal total organic carbon and its rate of loss from the soil profile. Australian Journal of Soil Research 24: 281–292.
33. Jenny H (1941) Factors of soil formation: A system of quantitative pedology. New York: McGraw-Hill. 281 p.
34. Schimel DS, Braswell BH, Holland EA, McKeown R, Ojima DS, et al. (1994) Climatic, edaphic and biotic controls over storage and turnover of carbon in soils. Global Biogeochemical Cycles 8: 279–293.
35. Fearnside PM, Imbrozio Barbosa R (1998) Soil carbon changes from conversion of forest to pasture in Brazilian Amazonia. Forest Ecology and Management 108: 147–166.
36. Li H, Ma Y, Aide TM, Liu W (2008) Past, present and future land-use in Xishuangbanna, China and the implications for carbon dynamics. Forest Ecology and Management 255: 16–24.

20

Non-Native Plant Litter Enhances Soil Carbon Dioxide Emissions in an Invaded Annual Grassland

Ling Zhang[1,2], Hong Wang[1], Jianwen Zou[1,3]*, William E. Rogers[4], Evan Siemann[1,2]*

1 College of Resources and Environmental Sciences, Nanjing Agricultural University, Nanjing, China, **2** Department of Ecology and Evolutionary Biology, Rice University, Houston, Texas, United States of America, **3** Department of Horticulture, Cornell University, Ithaca, New York, United States of America, **4** Department of Ecosystem Science and Management, Texas A & M University, College Station, Texas, United States of America

Abstract

Litter decomposition is a fundamental ecosystem process in which breakdown and decay of plant detritus releases carbon and nutrients. Invasive exotic plants may produce litter that differs from native plant litter in quality and quantity. Such differences may impact litter decomposition and soil respiration in ways that depend on whether exotic and native plant litters decompose in mixtures. However, few field experiments have examined how exotic plants affect soil respiration via litter decomposition. Here, we conducted an *in situ* study of litter decomposition of an annual native grass (*Eragrostis pilosa*), a perennial exotic forb (*Alternanthera philoxeroides*), and their mixtures in an annual grassland in China to examine potential invasion effects on soil respiration. *Alternanthera* litter decomposed faster than *Eragrostis* litter when each was incubated separately. Mass loss in litter mixes was more rapid than predicted from rates in single species bags (only 35% of predicted mass remained at 8 months) showing synergistic effects. Notably, exotic plant litter decomposition rate was unchanged but native plant litter decomposition rate was accelerated in mixtures (decay constant $k = 0.20$ month^{-1}) compared to in isolation ($k = 0.10$ month^{-1}). On average, every litter type increased soil respiration compared to bare soil from which litter was removed. However, the increases were larger for mixed litter (1.82 times) than for *Alternanthera* litter (1.58 times) or *Eragrostis* litter (1.30 times). Carbon released as CO_2 relative to litter carbon input was also higher for mixed litter (3.34) than for *Alternathera* litter (2.29) or *Eragrostis* litter (1.19). Our results indicated that exotic *Alternanthera* produces rapidly decomposing litter which also accelerates the decomposition of native plant litter in litter mixtures and enhances soil respiration rates. Thus, this exotic invasive plant species will likely accelerate carbon cycling and increase soil respiration even at intermediate stages of invasion in these annual grasslands.

Editor: Harald Auge, Helmholtz Centre for Environmental Research – UFZ, Germany

Funding: This work was supported by the NSFC (41225003), the Ministry of Education 111 project (B12009), US-NSF (DEB 0820560), the Priority Academic Program Development of Jiangsu Higher Education Institutions (PAPD), and the China Scholarship Council (for providing funding support to LZ). The funders had no role in study design, data collection and analysis, decision to publish, or preparation of the manuscript.

Competing Interests: The authors have declared that no competing interests exist.

* E-mail: jwzou21@njau.edu.cn (JWZ); siemann@rice.edu (ES)

Introduction

Litter decomposition has been shown to be an important ecosystem process associated with carbon (C) cycling and nutrient release [1,2]. Exotic plant invasions can enhance plant litter decomposition and C cycling of invaded ecosystems [3–5]. During decomposition, more C may go belowground as substrate for soil microbial activities [6]. In addition, invasive plant species usually have larger net primary production (NPP) relative to the native species via higher C fixing efficiency [3,7]. Thus, exotic plant invasions are commonly accompanied by higher litter inputs and more C input belowground.

Native ecosystems are generally encroached or invaded over a prolonged time period. Consequently, invasive plant litter mixes with litter of native species and may not be decomposing alone over the course of a typical invasion. Mixing invasive plant litter with native plant litter of different quality may enhance decomposition of the litter mixture via nutrient transfers between component plant litters, leading to non-additive effects [8–13]. Thus, enhanced decomposition rates of litter mixtures containing both native and invasive plant detritus may be common in invaded ecosystems, even though results might vary among different invasions [14]. Alterations of litter decomposition and C cycling following exotic plant invasions may be underestimated based on assessments of decomposition using litter of only a single species [9,11]. Consequently, empirical studies measuring the effects of litter mixing will help to better predict the effects of plant invasions on nutrient dynamics.

Litter decomposition processes involve C transfers between plant litter and the soil ecosystem at the litter-soil interface [6,15]. Carbon cycling during litter decomposition includes C loss to the atmosphere as CO_2 emissions and input to belowground organic pools [6]. However, while higher litter decomposition rates may lead to more C input belowground, this does not necessarily mean equivalent C sequestration in soil [6,16,17]. In general, soil C stocks are a balance of photosynthetic C input and output via CO_2 emissions [18]. Hence, CO_2 emissions at the soil-atmosphere interface play an important role in C loss from terrestrial C pools [19,20]. These emissions have been estimated to be the largest source of CO_2 released to the atmosphere, most notably being an order of magnitude larger than the combined sum of C emissions by anthropogenic fossil fuel combustion and deforestation [20].

Thus, even a slight shift in soil CO_2 emissions may lead to large changes in atmospheric composition and increase rates of global climate change. Carbon released belowground by accelerated litter decomposition may provide fresh C substrate for soil microbes to decompose but may also enhance decomposition of recalcitrant forms of C. This can lead to high levels of additional CO_2 release through this positive priming effect [6,21–23]. However, few studies have been conducted *in situ* to examine how soil CO_2 emissions are affected by litter composition [24,25], especially native ecosystems that are experiencing exotic plant encroachment or invasions.

Eragrostis pilosa (Poaceae, Indian love grass, hereafter referred to as "*Eragrostis*") is a native annual grass in China with a wide global distribution [26]. The grassland dominated by *Eragrostis* in our experimental site is being invaded by the perennial exotic forb *Alternanthera philoxeroides* (Amaranthaceae, alligator weed, hereafter referred to as "*Alternanthera*") introduced from South America via Japan [27]. Both plant species occurred in this invaded grassland and have coexisted for ~5 years. During the invasion by *Alternanthera*, litter produced by the native and the invasive plant have been decomposing both alone and in mixture. While a number of studies have been conducted to examine native and invasive plant litter decomposition individually, few studies have tried to examine litter mixing effects on litter decomposition and soil respiration rates in invaded areas. Here, we conducted an *in situ* experiment to examine the effect of litter addition and litter mixing on soil CO_2 emissions using the native plant *Eragrostis* and invasive plant *Alternanthera* as model species. By examining decomposition rates of single- and mixed-species litter, as well as CO_2 emissions from soils with different litter compositions (i.e., single-species litters, mixed-species litter, and litter-free controls), we addressed the following questions: 1) Do native and invasive plant litter differ in litter quality and decomposition rates? 2) Does litter mixing have non-additive effects on decomposition; 3) Do CO_2 emissions from litter-soil systems vary between litter types and species?

Materials and Methods

Study Site

The experiment was conducted in 2010–2011 in an unmanaged annual grassland at the Nanjing Agricultural University experimental station (6 m elevation, 32°01′N 118°37′E) in China. Mean monthly temperatures in this area range from −10°C in January to 32.5°C in July with a mean annual temperature of 25°C. Annual mean precipitation of this area is 980 mm with up to 90% of precipitation occurring between March and November. Soils are classified as hydromorphic soils in a Chinese soil taxonomic classification and are high in clay content (6% sand, 40% silt and 54% clay) with an initial pH (H_2O) of 6.8 and organic carbon content of 8.3 kg C m^{-2}.

Focal Species

Eragrostis pilosa is an annual bunchgrass native to China and is distributed throughout southeast China. The field station was dominated by naturally occurring *Eragrostis* before the introduction of exotic invasive plants. Presently, remnants of the native *Eragrostis* dominated grassland are threatened by *Alternanthera* which is expanding its range. The co-existence of *Eragrostis* with this invasive species and its broad distribution make *Eragrostis* a useful model plant to study ecosystem C cycling changes after exotic plant invasions.

The most abundant invasive plant at the field station is the perennial forb *Alternanthera philoxeroides* that covers more than 50% of the grassland during peak growing time. *Alternanthera* originates from South America, was introduced to China via Japan in the 1930s, and has been reported to be invasive in at least 19 provinces [27]. Moreover, it is also invasive in Australia and North America. Basset et al. [28] reported faster litter decomposition rates of invasive *Alternanthera* in New Zealand, indicating it may potentially facilitate ecosystem element cycling in areas it invades.

Litter Decomposition

An *in situ* litterbag approach was used to test the litter decomposition rate of both focal species. At the end of the 2010 growing season, we hand collected newly produced litter from standing *Eragrostis* and *Alternanthera* at the Nanjing Agricultural University experimental station. Neither plant species is endangered or protected. All litter samples were air dried and carefully processed (cut into a size of litter length at 3~6 cm) to fit into the litterbag. For each litter treatment, 10 g of litter was put into 10×15 cm litterbags (mesh size of 1 mm^2) [29,30]. Single species bags received 10 g of a single litter type. For the mixed-species treatment, 5 g of each litter type was put in the bag. All litterbags were deployed simultaneously in invaded areas where both species co-existed in March 2011 and were retrieved after *in situ* decomposition for 3, 6 or 8 months. A subsample of litter of each species was oven dried for water, C and N measurement. Samples used for chemical properties measurement were ground to pass through a 2 mm sieve, decarbonized with HCl, and then analyzed for C and N content by a CNS elemental analyzer (Variomax CNS Analyzer, Elementar GmbH, Hanau, Germany) [31].

In situ soil CO_2 emission measurements

In the invaded areas where both plants coexisted, 12 plots (2×2 m and 5 m apart) were set up in December 2010, prior to the 2011 growing season by thoroughly removing all existing plant materials. The shallow root depth of both plant species enabled the thorough clearance of belowground biomass. Soil CO_2 emissions were measured weekly for 8 months from plots that had existing litter removed and then received different litter compositions (i.e. litter of *Alternanthera*, *Eragrostis* or mixtures of both plant) and from blank control plots with litter removed and no subsequent litter addition. CO_2 was measured with the static-opaque chamber method, a commonly used, well-established method [32]. Within each plot, circular grooved clay collars (height = 25 cm, inside diameter = 20 cm) were installed flush with the soil surface for permanent use. The area within the clay collars received litter treatments of 10 g litter of *Alternanthera*, *Eragrostis* or 1:1 mixture. Mesh cloth was placed above the clay collars to prevent new litterfall and to fix the litterbags to the ground. While soil CO_2 emissions were measured, an open-bottomed cylindrical PVC gas sampling chamber (100 cm high) was fit into the circular groove. Each groove was filled with water to seal the rim of the opaque cylindrical chamber. The gas sampling chambers were wrapped by foam and aluminum foil to minimize temperature variations during measurements. Battery-powered fans mixed the air inside the chambers. Three samples of headspace gas were collected using single-use syringes. The first sample was collected when chambers were installed (0 minutes) followed by collections 10 and 20 minutes after installation. A gas chromatograph (Agilent 7890) with a flame ionization detector (FID) was used to measure CO_2 concentrations. Soil CO_2 emission rates were determined from the slope of the mixing ratio change of the three samples using the following equation:

$$F = P \times V \times \frac{dCO_2}{dt} \times \frac{1}{RT} \times M \times \frac{1}{A} \times \frac{M_c}{M}$$

where F refers to rate of soil CO_2 emission rate (mg CO_2–C m^{-2} h^{-1}), P is the standard atmospheric pressure (Pa), V and A are the volume (m^3) and interior bottom area (m^2) of the cylindrical chamber, R stands for universal gas constant, T is the absolute air temperature (K) when the gas sample was aspirated and Mc and M are the molecular masses of carbon and CO_2 (g mol^{-1}), respectively. Cumulative CO_2 emissions within a given time period were calculated by multiplying average CO_2 emission rates and the associated time span assuming that measurements were representative. Since the C content of the plant litter differed, cumulative CO_2 emissions are also reported normalized to litter C input (g CO_2-C per g litter-C input) and emissions from bare soil were subtracted when discussing effects of the different litter quality on CO_2 emissions.

Statistics

Effects of litter mixing and decomposition time on *Alternanthera* and *Eragrostis* litter mass were tested by two-way analysis of variance (ANOVA). Two-way ANOVA was also used to test the effects of plant species and decomposition time on litter mass remaining within each litter type (in isolation or in mixtures). Post-hoc student's t tests were used to examine significant differences among treatment means for factors with more than 2 levels.

Litter decay constants (k, $month^{-1}$) were calculated to examine litter mixing effects on litter decomposition rates. The k values were calculated from a single negative exponential model as following [33,34]:

$$\frac{M_t}{M_0} = e^{-kt}$$

where M_t is litter mass remaining at time t and M_0 is initial litter biomass at time $t = 0$, and k is a first-order litter decay constant. The mean k values of litter decomposing in isolation and in mixture of both species were also calculated by the model. Litter mass remaining was examined using all combinations of the three single-species or mixed-species results at each time point (9 possible combinations).

Effects of litter mixing on overall litter decomposition rates were examined by comparing observed mass remaining in mixed-species bags and expected mass remaining of litter mixtures. Expected mass remaining was obtained based on decomposition rate of single plant litter samples according to the following equation [35,36]:

$$\text{Expected mass remaining} = \frac{M_{Alt}}{M_{Alt} + M_{Era}} \times R_{Alt} + \frac{M_{Era}}{M_{Alt} + M_{Era}} \times R_{Era}$$

where R refers to the remaining litter mass of a species (*Alt* = *Alternathera*, *Era* = *Eragrostis*) in the single-species litterbag and M refers to the initial litter mass of a species in the litter mixture. A randomization approach was used in calculating expected mass remaining. Specifically, the set of all combinations of the three single-species results of each species (9 possible combinations) were used to generate expected mass remaining. Post-hoc student's t tests were used to examine differences between observed and expected litter mass remaining at each time point.

Statistical results at $\alpha = 0.05$ were considered significant. All statistical analysis was conducted in JMP 9.0 (SAS Inc., Cary, NC, USA).

Results

Litter Chemistry and Litter Mass Loss

Initial litter quality of invasive *Alternanthera* (68.74±5.90) did not differ from that of native *Eragrostis* (80.63±3.99; $F_{1,\ 4} = 2.79$, $P = 0.1702$; Fig. 1) in terms of C:N ratio, but initial N concentration of *Alternanthera* litter (9.16±0.27 g kg^{-1}) was significantly higher than that of *Eragrostis* litter (6.69±0.08 g kg^{-1}; $F_{1,\ 4} = 15.73$, $P = 0.0166$). However, after 8 months of decomposition in single species bags, litter C:N ratio of *Alternanthera* (31.63±0.66) was significantly lower than that of *Eragrostis* (47.76±2.73; $F_{1,\ 4} = 32.95$, $P = 0.0046$) and was similar after 8 months of decomposition in mixture (Fig. 1).

When each was decomposing alone, *Alternanthera* litter decomposed faster than *Eragrostis* litter (Table 1; Fig. 2a). The litter decay constant (k) of *Alternanthera* was approximately three times that of *Eragrostis* when each was decomposing alone (Table 2). Litter mass loss rate of *Eragrostis* was significantly accelerated by litter mixing (Table 1; Fig. 2cd) with k values doubled but mixing did not change the mass loss rate of *Alternanthera* litter (Table 2). This decreased the difference in mass loss rate between the two species in mixtures (Fig. 2b). Synergistic non-additive effects on litter mixture decomposition were observed after 6 and 8 months of decomposition (Table 3).

Soil CO_2 Emissions

CO_2 emission rates were highest from soils covered by litter mixtures, followed by soils with only *Alternanthera* litter and only *Eragrostis* litter, respectively (Fig. 3a). Soils with all litter removed (bare soil) were characterized by the lowest CO_2 emission rates (Fig. 3a). Finally, cumulative CO_2 emissions were consistent with the CO_2 emission rates of soils with different litter coverage (Fig. 3b). The net increases in cumulative soil CO_2 emissions from soils covered with litter compared to bare soils were 544 g CO_2–C m^{-2} for mixtures, 388 g CO_2–C m^{-2} for *Alternathera* litter, and 199 g CO_2–C m^{-2} for *Eragrostis* litter. Net increases in carbon

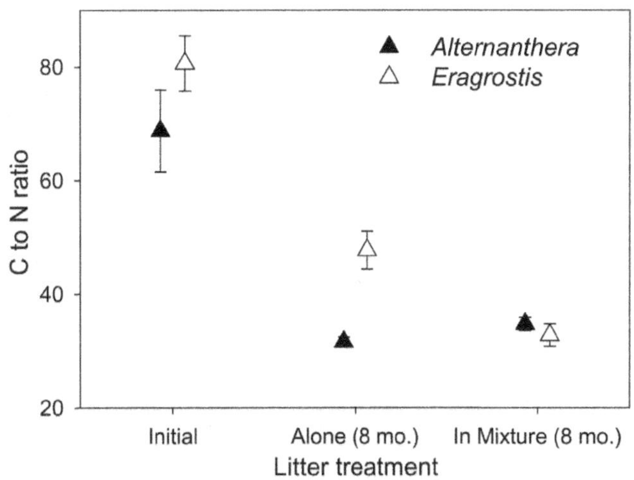

Figure 1. Mean initial and final C:N ratio of *Alternanthera* and *Eragrostis* litter decomposing both alone and in mixture. Means ±1 SE are shown.

Table 1. ANOVAs for remaining litter mass as affected by species and litter mixing with time in the field incubation study.

		Single species		Mixed species	
Factor	*df*	*F value*	*P value*	*F value*	*P value*
Species	1,12	**128.1**	**<0.0001**	**70.9**	**<0.0001**
Time	2,12	**9.0**	**0.0049**	**118.0**	**<0.0001**
Interaction	2,12	2.0	0.1804	**60.8**	**<0.0001**
Model	5,12	**32.0**	**<0.0001**	**85.7**	**<0.0001**
		Alternanthera		*Eragrostis*	
Factor	*df*	*F value*	*P value*	*F value*	*P value*
Litter mixing	1,12	0.4	0.5219	**32.0**	**0.0001**
Time	2,12	**11.1**	**0.0023**	**71.2**	**<0.0001**
Interaction	2,12	0.5	0.6034	**14.3**	**0.0007**
Model	5,12	**5.0**	**0.0127**	**40.6**	**<0.0001**

Significant results are shown in bold.

released as CO_2 relative to carbon input as litter were also higher for mixed litter (3.34 g CO_2–C per g litter-C) than for *Alternathera* litter (2.29) or *Eragrostis* litter (1.19).

Discussion

Litter decomposition is an important ecosystem process releasing C and nutrients bound up in plant litter. Thus, changes in litter decomposition might profoundly influence ecosystem C

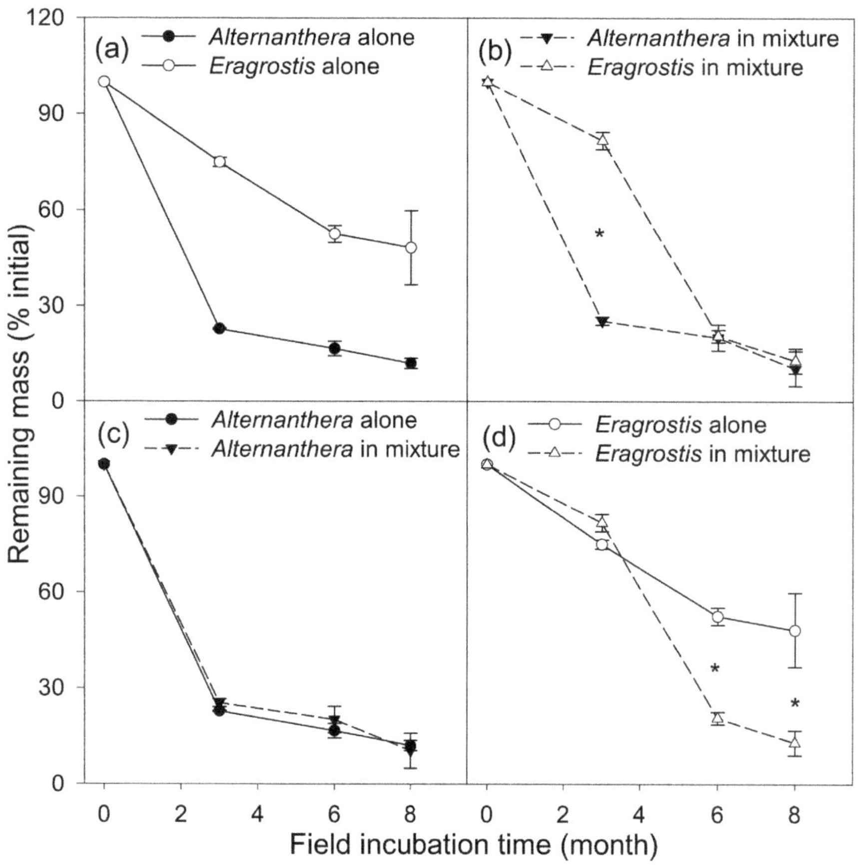

Figure 2. Remaining mass (% initial) of native and invasive plant litter when each of them was decomposing alone (a) or in mixtures (b). Dynamics of litter mass remaining between single-species and litter mixtures within *Alternanthera* and *Eragrostis* were presented in (c) and (d), respectively. Asterisks indicate time points when means were significantly different at $\alpha = 0.05$.

Table 2. Litter mass decay constants (k, month^{-1}) during litter decomposition in single- (k_{alone}) or mixed-species litterbags (k_{mixed}) after 8 months of decomposition in the field.

	K_{alone}					K_{mixed}				
Species	**Estimate**	**S.E.**	***t* ratio**	***P* value**	**R^2**	**Estimate**	**S.E.**	***t* ratio**	***P* value**	**R^2**
Alternanthera	0.37	0.04	9.45	<0.0001	0.97	0.36	0.04	9.12	<0.0001	0.95
Eragrostis	0.10	0.01	8.67	<0.0001	0.89	0.20	0.03	6.00	0.0001	0.86
Mean	0.18	0.01	21.59	<0.0001	0.95	0.24	0.01	19.65	<0.0001	0.98

Values were calculated from a first order negative exponential model.

cycling. In fact, as evidenced by our study, litter of the invasive plant *Alternanthera* had higher initial N concentration and decomposed faster compared to that of native *Eragrostis* when each species decomposed in isolation. Moreover, mixing the litter of both native and invasive species significantly enhanced litter mass loss rate of *Eragrostis*. Observed litter mass remaining in mixtures was lower than expected based on single species results, indicating synergistic non-additive effects of litter mixing on litter mixture decomposition. In turn, both CO_2 emission rate and cumulative CO_2 emissions were higher from soils under litter mixtures composed of both native and invasive plant litter relative to those with a single litter type, especially *Eragrostis*, or bare soils, thereby indicating that the faster decomposition rate of invasive plant litter and accelerated decomposition rate of litter mixtures can substantially contribute to higher soil CO_2 emissions.

Faster Decomposition Rate of Invasive Plant Litter

Litter of the invasive plant *Alternanthera* had higher N concentration and a faster decomposition rate which was consistent with other studies of invasive plants [3,4,28]. In general, the contact area of decomposing litter with soil could be an important factor enhancing litter decomposition via effects on environmental conditions such as soil water content and soil microbial communities. In this study, the litter samples were processed to be similar length and size and litterbags were fixed to the ground by mesh cloth. Therefore, the differences in litter contact area with soil were likely minimal. Along with environmental conditions, litter quality has been considered an important factor controlling litter decomposition rate. Indeed, litter of the invasive plant *Alternanthera* had higher N concentration and lower initial C to N ratio compared to *Eragrostis* litter (Fig. 1), indicating high quality litter for microbes to metabolize. Since invasive plant species may have greater ability to uptake soil nutrients than native plants [37], in some cases, higher soil nutrient uptake may lead to higher litter quality for the invasive plant [7]. *Alternanthera* litter was reported to decompose faster than litter from native sedge species (*Schoenoplectus tabernaemontani*, *Isolepis prolifer*) in a litterbag experiment in New Zealand [28]. However, in our study, the dominant native plant *Eragrostis* produced litter with lower N concentration relative to that of the invasive *Alternanthera* and hence had a slower mass loss rate. Consequently, there may be effects of *Alternanthera* invasions on litter decomposition and element cycling both through displacement of native species but also through litter mixing during the invasion process.

Synergistic Effects of Litter Mixing on Litter Mixture Decomposition

Recent studies have shown that effects of litter mixing on decomposition may be non-additive when there is variation in component litter properties [8,11,12,38]. In this study, synergistic non-additive effects of litter mixing on litter mixture mass loss were also observed (Table 3; Fig. 2) [11,24,39]. Decomposition of the lower quality *Eragrostis* litter was accelerated by mixing with *Alternanthera* litter (Table 2; Fig. 2). Moreover, mass loss rate of litter mixture was faster after 6 and 8 months of decomposition than expected based on each litter type decomposing alone (Table 3). Several mechanisms have been proposed to explain effects of litter mixing including microclimate conditions, secondary chemical release, and nutrient transfer [40–42]. In this study, litterbags were deployed among similar plant and soil environmental conditions, so it is likely that variations in microclimate were minimal. Moreover, even though a recent study reported allelopathic effects of *Alternanthera* on algal growth [43], releasing of secondary compounds would have been expected to reduce the

Table 3. Expected and observed litter mass remaining (g) in mixed-species litter bags after field incubation for 3, 6 and 8 months.

	Litter mass remaining (g)			
Time	**Expected**	**Observed**	***t* ratio**	***P* value**
3 months	5.99±0.46	5.38±0.07	1.39	0.2028
6 months	3.30±0.05	2.08±0.10	**9.23**	**<0.0001**
8 months	3.33±0.25	1.17±0.14	**6.83**	**0.0001**

Predicted remaining mass was calculated from litter mass measured in single-species litter bags. Means ± SE. Differences between predicted and observed values were examined by Student's *t* tests. Significant results are shown in bold.

rate of decomposition of *Eragrostis*, not accelerate it (Tables 1 and 2; Fig. 2) [41]. Thus, nutrient transfer may be the most likely cause of synergistic effects of litter mixing in this study [8,9]. When decomposed in isolation, *Alternanthera* litter had significantly lower C to N ratio than *Eragrostis* litter after 8 months. When decomposed in mixture, however, both had similar C to N ratio, indicating potential N transfer between component species (Fig. 1). Together these results suggest that nutrient transfer likely caused the non-additive effects of litter mixing for these two species [8,11].

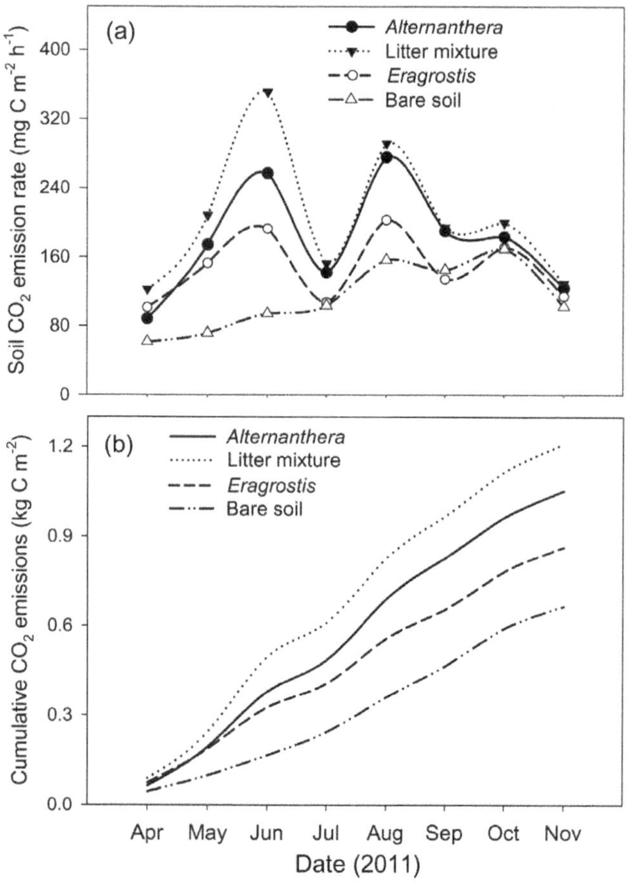

Figure 3. Monthly mean soil CO_2 emission rates based on weekly measurements (a) and accumulated soil CO_2 emissions (b) of soils with different litter types and soils without litter coverage (bare soil).

Higher Soil CO_2 Emissions as Affected by Litter Addition

Soil CO_2 emissions were enhanced by both invasive and native plant litter additions, especially when litter was a mixture of both component species (Fig. 3). Isotope labeling studies have shown that C input belowground is often the major C flux associated with litter biomass loss [6] so it is possible to have rapidly decomposing litter that increases soil C but not necessarily CO_2 emissions. Here, *Eragrostis* litter had low rates of mass loss and low rates of soil CO_2 emissions. The mass loss of *Alternanthera* and the litter mixtures were each much larger (Fig. 2). Moreover, with emissions from bare soils subtracted, soil CO_2 emissions were much higher for the mixtures compared to either *Alternanthera* litter or *Eragrostis* litter added alone (Fig. 3). However, litter mixing that had synergistic non-additive effects on litter mass loss may lead to more C loss as CO_2 emissions. Indeed, a similar result was reported by a study on the combined effect of beech-spruce litter on CO_2 emissions. The highest CO_2 emissions were found from soils covered by litter mixtures that had faster litter mass loss rate, even though results varied between bedrocks [24].Therefore, the invasion process of *Alternanthera* might be associated with higher soil CO_2 emissions due to faster litter decomposition. In addition, invasive plant species often have greater C fixing capacity relative to the native species [7]. In another study at this research site, we found *Alternanthera* produced 1.6 times as much aboveground litter as native *Eragrostis* (data in review). Thus, in *Alternanthera* invaded areas, soil CO_2 emissions and belowground C input may increase due to both greater litter production as well as higher mass specific input [23] with the degree of mixing influencing the relative magnitude of these two effects [16]. More studies considering both aboveground and belowground litter decomposition would help to obtain a thorough understanding of exotic plant invasion effects on soil C dynamics.

Litter Effects on Invasions

Differences in litter composition and decomposition may influence the process of invasion [4,44]. High quality, rapidly decomposing litter may indirectly enhance the survival and performance of invasive species by elevating soil nutrient availability [4,45,46]. On the other hand, a recent study on invasive plant *Phragmites australis* in New England wetlands reported positive litter legacies on competitive ability of its offspring through buildup of slow decomposing litter that suppressed native plant recruitment [44]. Even though litter legacy effects on offspring performance were not examined in this study, the high nitrogen litter of *Alternanthera* suggests that positive effects through increasing soil nitrogen may occur [23].

Conclusions

Our study showed that the invasive plant *Alternanthera philoxeroides* produced litter with faster decomposition rate than the dominant native species, *Eragrostis pilosa*. Mixing litter of invasive *Alternanthera* with that of native *Eragrostis* leads to synergistic non-additive effects on litter mixture decomposition and soil CO_2 emissions. Differences in litter quality and nutrient transfers between component species might be factors contributing to the non-additive effects. Soil CO_2 flux of areas with both native and invasive plant litter was higher than areas with equivalent invasive plant litter, native plant litter, and litter removal. The accelerated litter decomposition rate and enhanced soil CO_2 emissions following invasion have implications for C sequestration in invaded ecosystems and the potential range expansion of invasive plant. The effects of litter mixing suggest that these effects may depend on the spatial scale and extent of incomplete or ongoing invasions. It should be noted that our study included only one native and one invasive species that co-occurred in an invaded ecosystem and may not characterize other invasions. More studies that investigate additional invasive plant species effects on ecosystem processes should be conducted to generalize our results to other kind of invasions associated with various species or invaded ecosystems.

Acknowledgments

We thank: Shuwei Liu, Yaojun Zhang for their help in collecting and analyzing air, plant and soil samples; Shuijin Hu and two reviewers for their comments on an earlier version of the manuscript.

Author Contributions

Conceived and designed the experiments: LZ JZ ES. Performed the experiments: LZ HW. Analyzed the data: LZ ES. Wrote the paper: LZ JZ WER ES.

References

1. Aerts R, de Caluwe H (1997) Nutritional and plant-mediated controls on leaf litter decomposition of *Carex* species. Ecology 78: 244–260.
2. Raich JW, Schlesinger WH (1992) The global carbon dioxide flux in soil respiration and its relationship to vegetation and climate. Tellus B 44: 81–99.
3. Liao CZ, Peng RH, Luo YQ, Zhou XH, Wu XW, et al. (2008) Altered ecosystem carbon and nitrogen cycles by plant invasion: a meta-analysis. New Phytol 177: 706–714.
4. Allison SD, Vitousek PM (2004) Rapid nutrient cycling in leaf litter from invasive plants in Hawai'i. Oecologia 141: 612–619.
5. Ehrenfeld JG, Kourtev P, Huang WZ (2001) Changes in soil functions following invasions of exotic understory plants in deciduous forests. Ecol Appl 11: 1287–1300.
6. Rubino M, Dungait JAJ, Evershed RP, Bertolini T, De Angelis P, et al. (2010) Carbon input belowground is the major C flux contributing to leaf litter mass loss: Evidences from a ^{13}C labelled-leaf litter experiment. Soil Biol Biochem 42: 1009–1016.
7. Feng YL, Auge H, Ebeling SK (2007) Invasive *Buddleja davidii* allocates more nitrogen to its photosynthetic machinery than five native woody species. Oecologia 153: 501–510.
8. Gartner TB, Cardon ZG (2004) Decomposition dynamics in mixed-species leaf litter. Oikos 104: 230–246.
9. Hättenschwiler S, Tiunov AV, Scheu S (2005) Biodiversity and litter decomposition in terrestrial ecosystems. Annu Rev Ecol Evol Syst 36: 191–218.
10. Lecerf A, Marie G, Kominoski JS, LeRoy CJ, Bernadet C, et al. (2011) Incubation time, functional litter diversity, and habitat characteristics predict litter-mixing effects on decomposition. Ecology 92: 160–169.
11. Berglund SL, Ågren GI (2012) When will litter mixtures decompose faster or slower than individual litters? A model for two litters. Oikos 121: 1112–1120.
12. Chen BM, Peng SL, D'Antonio CM, Li DJ, Ren WT (2013) Non-additive effects on decomposition from mixing litter of the invasive *Mikania micrantha* HBK with native plants. PloS ONE 8: e66289.
13. McLaren JR, Turkington R (2011) Plant identity influences decomposition through more than one mechanism. PLoS ONE 6: e23702.
14. Hickman JE, Ashton IW, Howe KM, Lerdau MT (2013) The native-invasive balance: implications for nutrient cycling in ecosystems. Oecologia 173: 319–328.
15. Frey SD, Six J, Elliott ET (2003) Reciprocal transfer of carbon and nitrogen by decomposer fungi at the soil-litter interface. Soil Biol Biochem 35: 1001–1004.
16. Fontaine S, Bardoux G, Abbadie L, Mariotti A (2004) Carbon input to soil may decrease soil carbon content. Ecol Lett 7: 314–320.
17. Sayer EJ, Powers JS, Tanner EVJ (2007) Increased litterfall in tropical forests boosts the transfer of soil CO_2 to the atmosphere. PLoS ONE 2(12): e1299.
18. Amundson R (2001) The carbon budget in soils. Annu Rev Earth Pl Sc 29: 535–562.
19. Raich JW, Potter CS, Bhagawati D (2002) Interannual variability in global soil respiration, 1980–94. Global Change Biol 8: 800–812.
20. Schlesinger WH, Andrews JA (2000) Soil respiration and the global carbon cycle. Biogeochemistry 48: 7–20.
21. Kuzyakov Y, Friedel JK, Stahr K (2000) Review of mechanisms and quantification of priming effects. Soil Biol Biochem 32: 1485–1498.
22. Fontaine S, Mariotti A, Abbadie L (2003) The priming effect of organic matter: a question of microbial competition? Soil Biol Biochem 35: 837–843.
23. Meisner A, de Boer W, Cornelissen JHC, van der Putten WH (2012) Reciprocal effects of litter from exotic and congeneric native plant species via soil nutrients. PLoS ONE 7: e31596.
24. Berger TW, Inselsbacher E, Zechmeister-Boltenstern S (2010) Carbon dioxide emissions of soils under pure and mixed stands of beech and spruce, affected by decomposing foliage litter mixtures. Soil Biol Biochem 42: 986–997.
25. Sayer EJ, Heard MS, Grant HK, Marthews TR, Tanner EVJ (2011) Soil carbon release enhanced by increased tropical forest litterfall. Nat Clim Change 1: 304–307.
26. Zeid M, Echenique V, Díaz M, Pessino S, Sorrells M (2011) Eragrostis. In: Kole C, editor. Wild Crop relatives: Genomic and Breeding Resources. Berlin Heidelberg: Springer. pp. 135–151.
27. Weber E, Sun SG, Li B (2008) Invasive alien plants in China: diversity and ecological insights. Biol Invasions 10: 1411–1429.
28. Bassett IE, Beggs JR, Paynter Q (2010) Decomposition dynamics of invasive alligator weed compared with native sedges in a Northland lake. N Z J Ecol 34: 324–331.
29. Bärlocher F (2005) Leaf mass loss estimated by litter bag technique. In: Graça MS, Bärlocher F, Gessner M, editors. Methods to Study Litter Decomposition.Netherlands: Springer. pp. 37–42.
30. Crossley DA, Hoglund MP (1962) A litter-bag method for the study of microarthropods inhabiting leaf litter. Ecology 43: 571–573.
31. Shang QY, Yang XX, Gao CM, Wu PP, Liu JJ, et al. (2011) Net annual global warming potential and greenhouse gas intensity in Chinese double rice-cropping systems: a 3-year field measurement in long-term fertilizer experiments. Global Change Biol 17: 2196-2210.
32. Zou JW, Huang Y, Zheng XH, Wang YS, Chen YQ (2004) Static opaque chamber-based technique for determination of net exchange of CO_2 between terrestrial ecosystem and atmosphere. Chin Sci Bull 49: 381–388.
33. Wieder RK, Lang GE (1982) A critique of the analytical methods used in examining decomposition data obtained from litter bags. Ecology 63: 1636–1642.
34. Olson JS (1963) Energy storage and the balance of producers and decomposers in ecological systems. Ecology 44: 322–331.
35. Briones MJI, Ineson P (1996) Decomposition of eucalyptus leaves in litter mixtures. Soil Biol Biochem 28: 1381–1388.
36. McTiernan KB, Ineson P, Coward PA (1997) Respiration and nutrient release from tree leaf litter mixtures. Oikos 78: 527–538.
37. Kourtev PS, Ehrenfeld JG, Haggblom M (2003) Experimental analysis of the effect of exotic and native plant species on the structure and function of soil microbial communities. Soil Biol Biochem 35: 895–905.
38. Chapman SK, Newman GS, Hart SC, Schweitzer JA, Koch GW (2013) Leaf litter mixtures alter microbial community development: mechanisms for non-additive effects in litter decomposition. PLoS ONE 8: e62671.
39. Ashton IW, Hyatt LA, Howe KM, Gurevitch J, Lerdau MT (2005) Invasive species accelerate decomposition and litter nitrogen loss in a mixed deciduous forest. Ecol Appl 15: 1263–1272.
40. Chapman K, Whittaker JB, Heal OW (1988) Metabolic and faunal activity in litters of tree mixtures compared with pure stands. Agr Ecosyst Environ 24: 33–40.
41. Horner JD, Gosz JR, Cates RG (1988) The role of carbon-based plant secondary metabolites in decomposition in terrestrial ecosystems. Am Nat 132: 869–883.
42. Kainulainen P, Holopainen JK (2002) Concentrations of secondary compounds in Scots pine needles at different stages of decomposition. Soil Biol Biochem 34: 37–42.
43. Zuo SP, Ma YQ, Shinobu I (2012) Differences in ecological and allelopathic traits among *Alternanthera philoxeroides* populations. Weed Biol Manag 12: 123–130.

44. Holdredge C, Bertness MD (2011) Litter legacy increases the competitive advantage of invasive *Phragmites australis* in New England wetlands. Biol Invasions 13: 423–433.
45. Rodgers VL, Wolfe BE, Werden LK, Finzi AC (2008) The invasive species *Alliaria petiolata* (garlic mustard) increases soil nutrient availability in northern hardwood-conifer forests. Oecologia 157: 459–471.
46. Siemann E, Rogers WE (2003) Changes in light and nitrogen availability under pioneer trees may indirectly facilitate tree invasions of grasslands. J Ecol 91: 923–931.

21

Simulated Nitrogen Deposition Reduces CH_4 Uptake and Increases N_2O Emission from a Subtropical Plantation Forest Soil in Southern China

Yongsheng Wang[1,2], Shulan Cheng[2], Huajun Fang[1]*, Guirui Yu[1], Minjie Xu[2], Xusheng Dang[1,2], Linsen Li[1], Lei Wang[1,2]

1 Key Laboratory of Ecosystem Network Observation and Modeling, Institute of Geographical Sciences and Natural Resources Research, Chinese Academy of Sciences, Beijing, China, **2** University of Chinese Academy of Sciences, Beijing, China

Abstract

To date, few studies are conducted to quantify the effects of reduced ammonium (NH_4^+) and oxidized nitrate (NO_3^-) on soil CH_4 uptake and N_2O emission in the subtropical forests. In this study, NH_4Cl and $NaNO_3$ fertilizers were applied at three rates: 0, 40 and 120 kg N ha^{-1} yr^{-1}. Soil CH_4 and N_2O fluxes were determined twice a week using the static chamber technique and gas chromatography. Soil temperature and moisture were simultaneously measured. Soil dissolved N concentration in 0–20 cm depth was measured weekly to examine the regulation to soil CH_4 and N_2O fluxes. Our results showed that one year of N addition did not affect soil temperature, soil moisture, soil total dissolved N (TDN) and NH_4^+-N concentrations, but high levels of applied NH_4Cl and $NaNO_3$ fertilizers significantly increased soil NO_3^--N concentration by 124% and 157%, respectively. Nitrogen addition tended to inhibit soil CH_4 uptake, but significantly promoted soil N_2O emission by 403% to 762%. Furthermore, NH_4^+-N fertilizer application had a stronger inhibition to soil CH_4 uptake and a stronger promotion to soil N_2O emission than NO_3^--N application. Also, both soil CH_4 and N_2O fluxes were driven by soil temperature and moisture, but soil inorganic N availability was a key integrator of soil CH_4 uptake and N_2O emission. These results suggest that the subtropical plantation soil sensitively responses to atmospheric N deposition, and inorganic N rather than organic N is the regulator to soil CH_4 uptake and N_2O emission.

Editor: Dafeng Hui, Tennessee State University, United States of America

Funding: This project was supported by National Natural Science Foundation of China (No. 31290222, 31130009, 31290221, 31070435, and 41071166), Development Plan Program of State Key Basic Research (No. 2012CB41710, 32010CB833502, and 2010CB833501), Bingwei's Funds for Young Talents from Institute of Geographical Sciences and Natural Resources Research, CAS, (No. 2011RC202) and CAS Strategic Priority Program (No. XDA 05050600). The funders had no role in study design, data collection and analysis, decision to publish, or preparation of the manuscript.

Competing Interests: The authors have declared that no competing interests exist.

* E-mail: fanghj@igsnrr.ac.cn

Introduction

Humid tropical biome stores approximately 10% of global soil carbon (C) [1] and plays a vital role in the budget of ecosystem C and nitrogen (N) fluxes. The amount of nitrous oxide (N_2O) emission from the subtropical and tropical forest soils is estimated at 0.9–3.6 T g yr^{-1}, accounting for 14% to 23% of the global N_2O budget [2]. Simultaneously, well-aerated soils in the subtropical and tropical forests potentially function as a significant sink of atmospheric methane (CH_4) during the dry season [3–6]. The uptake of CH_4 from the subtropical and tropical forest soils is estimated to be 6.2 T g yr^{-1}, accounting for 28% of the global CH_4 sink [7]. Although the importance of the subtropical and tropical forest soils as atmospheric CH_4 sink and N_2O source is well understood, few observations can be available in this region [8–10]. Moreover, low-frequency measurement of gas fluxes in the few studies is unable to accurately estimate the annual amount of soil CH_4 uptake and N_2O emission, which leads to a high uncertainty in the budget of global soil CH_4 and N_2O fluxes.

Chronic N deposition input into terrestrial ecosystems alters plant physiology and soil microbial community, thereby changes the soil biogenic CH_4 and N_2O fluxes [11–13]. Based on a meta-analysis of N addition experimental data in globe, Liu and Greaver [14] concluded that N addition reduced CH_4 uptake by 38% and increased N_2O emission by 216%. In general, chronic N deposition will increase $NH_4^+$and NO_3^- availability in the forest ecosystems, thereby affects CH_4 uptake from the forest soils through changing the activity and composition of methanotrophic community [15–17]. Soil NH_4^+ accumulation can decrease, increase or have no effects on soil CH_4 uptake in the forest ecosystems, depending on forest types [5], duration of N application [18], and N fertilizer types and doses [19]. Three potential mechanisms have been proposed to clarify the inhibition of NH_4^+ accumulation to soil CH_4 uptake: (1) the competition of soil NH_3 to use CH_4 monooxygense with soil CH_4 [20], (2) the toxic inhibition of hydroxylamine and nitrite produced during soil NH_4^+ oxidation [21], and (3) the indirect effects of applied N and associated salt ions through osmotic stress [22]. On the contrary, elevated soil NH_4^+ availability can increase soil CH_4 uptake, which is related to the increase in the quantity of soil ammonia-oxidizing microorganisms [23]. Soil NO_3^- accumulation can also decrease or increase soil CH_4 uptake [18,24]. Osmotic stress caused by NO_3^--N fertilizer-associated salts [22] and anaerobic

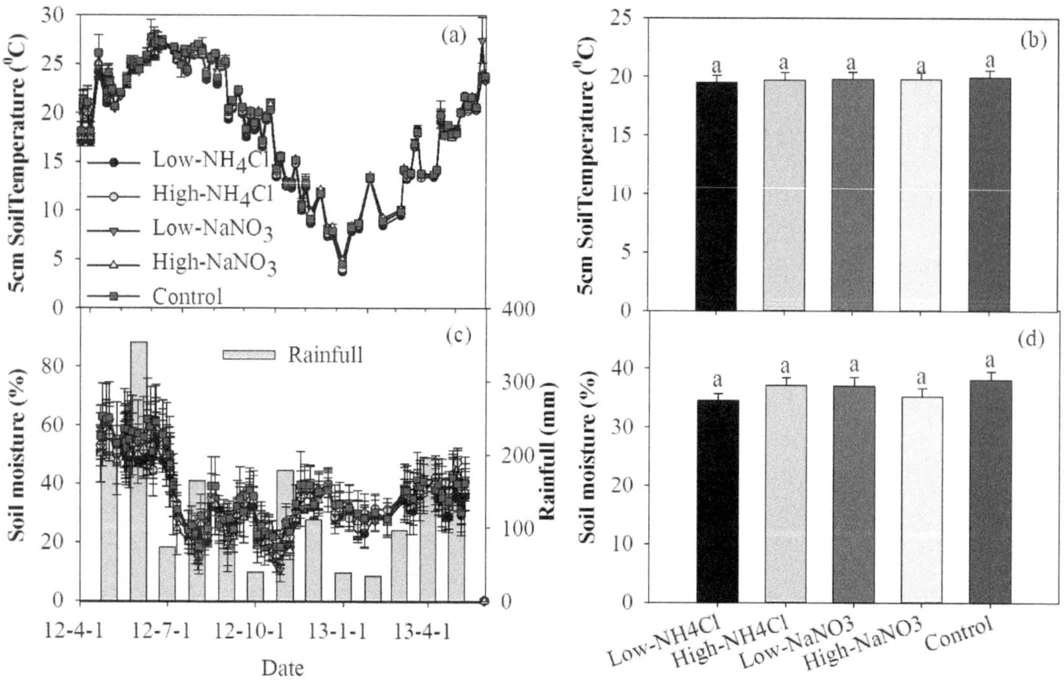

Figure 1. Temporal variations of 5 cm soil temperature and 10 cm soil moisture and their responses to N addition. Different letters above the columns mean significant differences between experimental treatments.

'microsites' produced by NO_3^- reduction [25] are toxic to CH_4-oxidizing bacteria. The mechanism responsible for the promotion of NO_3^- addition to soil CH_4 uptake is still unclear and need a number of experimental evidences to support [26]. A positive relationship between the amount of N addition and N_2O fluxes from the subtropical forest soils is mainly attributed to the promotion of soil nitrification or/and denitrification rates caused by increased N availability [27,28]. Some studies reported that denitrification was the main source of N_2O emission from the subtropical forest soils [29–31], whereas other studies claimed that nitrification dominated soil N_2O fluxes [32]. To date, single N fertilizer (i.e., NH_4NO_3) is widely used to simulate the effects of N deposition in all N manipulative experiments of subtropical forests in China [8,33,34]. The above studies have not evaluated the relative contributions of the deposited N ions (NH_4^+ and NO_3^-) to soil CH_4 uptake and N_2O emission. Moreover, most of soil CH_4 and N_2O fluxes are measured by low-frequency sampling over the short term, which is difficult to accurately assess the budget of soil CH_4 and N_2O fluxes and leads to great uncertainty.

In China, the plantations cover an area of 6.2×10^7 ha, accounting for 31.8% of China's forest area and ranking first in the world [35]. Approximately 63% of plantations concentrate in the subtropical region of southern China [36]. Meanwhile, the southern China is the most economically developed regions with high population density, and plantations, cities and farmlands are interspersed. Because a number of reactive N originated from fossil fuel combustion and fertilizer use is emitted to atmosphere, the forests in this region are receiving a high level of anthropogenic N deposition, mostly as ammonium [37]. Atmospheric N deposition rate via precipitation in southern China has been reported and ranges from 30 to 73 kg N ha^{-1} yr^{-1} [8]. So far, few studies are conducted to examine the effects of N deposition on CH_4 uptake and N_2O emission from the plantation of this region [33,34].

Humid subtropical forest soils are generally characterized by high N availability and high N turnover [38]. Therefore, we hypothesize that increased NH_4^+ and NO_3^- availability via experimental N deposition will inhibit soil CH_4 uptake and promote N_2O emissions from the subtropical plantation. Furthermore, NH_4^+-N fertilizer addition will decrease CH_4 uptake and increase N_2O emission due to soil NH_4^+-N accumulation. In contrast, the effects of NO_3^--N fertilizer addition on soil CH_4 uptake and N_2O emission depend on the concentration of soil NO_3^--N as well as associated salt ions. Our objectives were (1) to quantify the effects of NH_4^+-N and NO_3^--N fertilizer application on soil CH_4 and N_2O fluxes and soil variables in the subtropical plantation; (2) to examine the relationships between soil CH_4 and N_2O fluxes and the relevant soil properties.

Materials and Methods

Site description

This study was conducted in a subtropical slash pine plantation at the Qianyanzhou Ecological Station (QYZ, 26°44′39″N, 115°03′33″E) in southern China. The station belongs to the Institute of Geographic Sciences and Natural Resources Research, Chinese Academy of Sciences. All necessary permits were obtained for this field study. The field study did not involve endangered or protected species. According to local climate records from 1989 to 2008, mean temperature of QYZ site varies between 17 and 19°C. Mean annual precipitation ranges from 945 to 2145 mm, of which 24%, 41%, 23% and 12% occurs in four quarters in turn. The rainfall scarcity and high temperature in late summer often result in seasonal drought [39]. The exotic slash pine plantation was established in 1985. Mean tree height, diameter at breast height, stand basal area, and leaf area index were 12.0 m, 15.8 cm, 35 m^2 ha^{-1}, and 4.5 m^2 m^{-2}, respectively [40]. The main understory and midstory species are *Woodwardia japonica* (L.f.) Sm., *Dicranopteris dichotoma* (Thunb) Bernh, *Loropetalum*

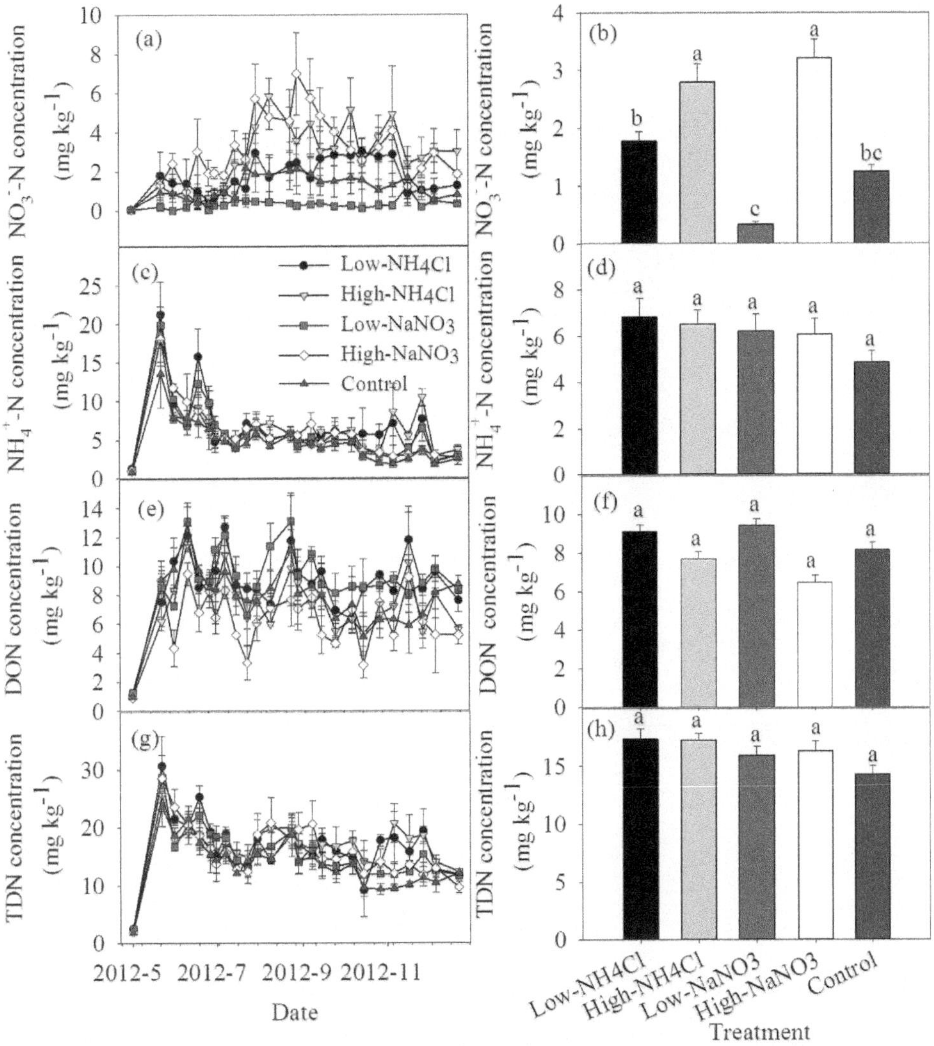

Figure 2. Temporal variations of soil NO_3^--N, NH_4^+-N, DON, and TDN concentrations and their responses to N addition. Different letters above the columns mean significant differences between experimental treatments.

chinense (R.Br.) Oliv, and *Quercus fabric* Hance. The red soil is weathered from red sand rock, and soil texture is divided into 2.0-0.05 mm (17%), 0.05-0.002 mm (68%), and <0.002 mm (15%) [39].

Experimental design

The N addition experiment is a randomized block design. In May 1, 2012, two N fertilizers (NH_4Cl and $NaNO_3$) were used to simulate the effects of deposited NH_4^+ and NO_3^- on ecosystem processes and functions. According to the level of atmospheric N deposition at the QYZ site, two levels referred to as low N (40 kg N ha^{-1} yr^{-1}) and high N (120 kg N ha^{-1} yr^{-1}) were used to simulate a future increase in the atmospheric N deposition by 1-, and 3-fold. A control treatment was designed at each block to calculate the net effect of N addition. Each N treatment was replicated three times, and a total of 15 plots were included. Each plot with 20 m×20 m was divided into four subplots with 5 m×5 m, and the plots were separated by 10 m wide buffer strips. Three subplots were used to measure soil CH_4 and N_2O fluxes, and the other one was used to investigate aboveground biomass and diversity. N fertilizer solutions were sprayed on the plots once a month in 12 equal applications over the entire year, and the control plots received equivalent deionized water only.

Measurement of soil CH_4 and N_2O fluxes

Flux measurements of soil CH_4 and N_2O fluxes were performed by using a static opaque chamber and gas chromatography method [41]. The static chambers were made of stainless steel and consisted of two parts: a square base frame (length×width×height = 50 cm×50 cm×10 cm) and a removable top (length×width×height = 50 cm×50 cm×15 cm). The installed equipments on the static chambers were detailed by Fang et al. [42]. The frames were inserted directly into the soil to a depth of 10 cm and remained intact during the entire observation period. To assess the spatial heterogeneity of soil C and N fluxes, a pre-experiment was conducted to examine the difference of CH_4 and N_2O fluxes among the three subplots of each plot before N addition. No significant difference of CH_4 and N_2O fluxes among the three subplots was found during the observation, suggesting a negligible effect of soil heterogeneity. Considering the practical reasons such as high labor intensity, we collected gas samples through changing the subplots within a month. The soil CH_4 and

Table 1. Results of repeated-measures ANOVAs on the effects of experimental treatment, month and their interaction on soil dissolved N concentrations.

Source of variation	Soil NO_3^--N		Soil NH_4^+-N		Soil DON		Soil TDN	
	F	p	F	p	F	p	F	p
Month	11.49	0.016	38.30	<0.001	16.71	<0.001	37.97	0.002
Treatment	4.40	0.026	1.62	0.24	1.81	0.20	1.51	0.27
Month×Treatment	1.31	0.24	1.41	0.13	1.33	0.17	0.75	0.77

N_2O fluxes were measured twice a week and conducted between 9:00 and 11:00 am (China Standard Time, CST). Five gas samples were taken using 100 ml plastic syringes at intervals of 0, 10, 20, 30, and 40 min after closing the chambers. CH_4 and N_2O concentrations of gas samples were analyzed within 24 h with a gas chromatography (Agilent 7890A, USA) equipped with an electron capture detector (ECD) for N_2O analysis and a flame ionization detector (FID) for CH_4 analysis. The high purity N_2 and H_2 were used as carrier gas and fuel gas, respectively. The ECD and FID were heated to 350°C and 200°C, respectively, and the column oven was kept at 55°C. The soil fluxes were calculated based on their rate of concentration change within the chamber, which was estimated as the slope of linear or nonlinear regression between concentration and time [41]. All the coefficients of determination (r^2) of the regression were greater than 0.90 in our study.

Measurements of soil temperature and moisture

Simultaneously, soil temperature at 5 cm (T_s) and soil moisture at 10 cm below soil surface (SM) were monitored at each chamber. Soil temperature was measured using a portable temperature probes (JM624 digital thermometer, Living-Jinming Ltd., China). Volumetric soil moisture ($m^3\ m^{-3}$) was measured using a moisture probe meter (TDR100, Spectrum, USA).

Soil sampling and mineral N analysis

During the measurement of soil CH_4 and N_2O fluxes, soil samples were collected weekly nearby the static chambers from a depth of 0–20 cm using an auger (2.5 cm in diameter). Five soils were collected and were pooled to one composite sample for each soil layer at each plot. Soils were immediately passed through a 2 mm sieve to remove roots, gravel and stones. Soil samples were extracted in 1.0 M KCl solution (soil: water = 1:10) and shaken for 1 h. The soil suspension was subsequently filtered through Whatman No. 40 filter papers for NH_4^+-N, NO_3^--N, and total dissolved nitrogen (TDN) determination on a continuous-flow autoanalyzer (Seal AA3, Germany). Dissolved organic nitrogen (DON) concentration was calculated as the difference between TDN and total inorganic nitrogen (NH_4^+-N and NO_3^--N).

Statistical analyses

Repeated measures analysis of variance (AVOVA) with Duncan test was applied to examine the differences of soil temperature, soil moisture, soil dissolved N, and soil CH_4 and N_2O fluxes between control and N addition plots. Experimental treatments were set as factors of between-subjects and measurement date was selected as a variable of within-subjects. Linear and nonlinear regression analyses were used to examine the relationships between soil CH_4 and N_2O fluxes and the measured soil variables in monthly scale. All statistical analyses were conducted using the SPSS software package (version 16.0), and statistical significant differences were set with P values<0.05 unless otherwise stated. All figures were drawn using the Sigmaplot software package (version 10.0).

Results

Soil temperature, moisture and precipitation

During the whole observation period, soil temperature at 5 cm depth fluctuated greatly, which correlated with the weather condition. Soil temperature varied as a single-peak and single-sink curve, i.e. temperature was the highest in early July, gradually reached the lowest value in early January, and then increased (Fig. 1a). There was no significant difference in surface temperature among various treatments (Fig. 1b).

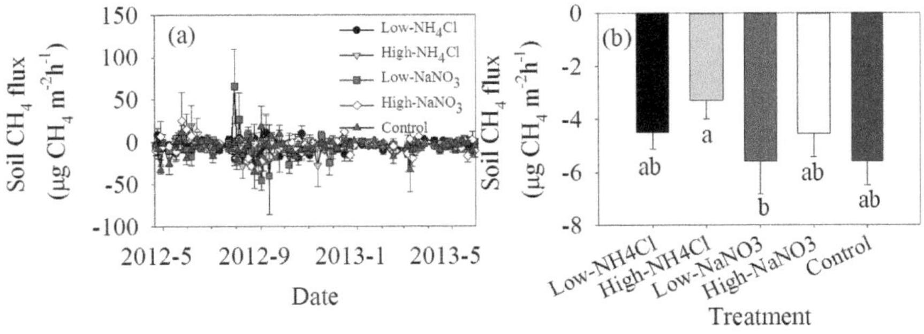

Figure 3. Temporal variations of soil CH_4 fluxes and their responses to N addition. Different letters below the columns mean significant differences between experimental treatments.

Soil moisture at 10 cm depth behaved as significant seasonal variation, with the maximum occurred in May and June and the minimum occurred in August and October (Fig. 1c). The seasonality of soil moisture was well consistent with that of precipitation (Fig. 1c). Similar to surface temperature, no significant difference in soil moisture was found among various treatments (Fig. 1d).

Soil dissolved N concentrations

Soil NO_3^--N concentration showed significant seasonal variation, with the minimum and maximum occurring in May and August (Fig. 2a, Table 1, $P=0.016$). In the control, the concentration of soil NO_3^--N ranged from 0.06 to 2.19 mg kg^{-1}, with an average of 1.25 mg kg^{-1} (Fig. 2a–b). N addition tended to alter soil NO_3^--N concentration, and the difference was significant among five experimental treatments (Table 1, $P=0.026$). Compared with the control, high level of $NaNO_3$ addition tended to increase soil NO_3^--N concentration, while an opposite pattern was found in the low level of $NaNO_3$ addition treatment (Fig. 2b). Furthermore, the promotion of high level of NH_4Cl addition to soil NO_3^--N concentration seemed to be stronger than that of low level of NH_4Cl addition (Fig. 2b).

Soil NH_4^+-N concentration peaked in the middle of May, and then continued to decrease (Fig. 2c). The seasonal variation of soil NH_4^+-N concentration was significant (Table 1, $P<0.001$). In the control, soil NH_4^+-N concentration ranged from 1.91 to 10.80 mg kg^{-1}, with an average of 4.84 mg kg^{-1} (Fig. 2d). Overall, although N addition treatments tended to increase soil NH_4^+-N concentration, the difference between N addition treatments and control was not significant (Fig. 2d, Table 1, $P=0.244$).

Soil DON concentration exhibited a significant seasonal variation (Fig. 2e, Table 1, $P<0.001$), and its seasonality was the same as that of soil NO_3^--N concentration (Fig. 2a and Fig. 2e). In the control, soil DON concentration ranged from 5.30 to 14.11 mg kg^{-1}, with an average of 8.18 mg kg^{-1} (Fig. 2e–f). Low level of N addition tended to increase the concentration of soil DON, while high level of N addition tended to reduce the concentration of soil DON (Fig. 2f). However, N addition did not change soil DON concentration at the level of 0.05 (Table 1, $P=0.203$).

The seasonal variation of soil TDN concentration was consistent with that of soil NH_4^+-N concentration, dramatically decreasing from May to December (Fig. 2c and Fig. 2g). The seasonal variation of soil TDN concentration was significant (Table 1, $P=0.002$). N addition tended to increase soil TDN concentration; moreover, the promotion of NH_4Cl application to soil TDN concentration was slightly higher than that of $NaNO_3$ addition (Fig. 2h). However, the difference of soil TDN concentration among the five experimental treatments was not significant (Table 1, $P=0.273$).

Soil CH_4 and N_2O fluxes

Soil CH_4 fluxes showed a significant seasonal pattern (Table 2, $P=0.008$). We observed both soil CH_4 uptake and emission in the control plots, ranging from -34.9 to 17.9 μg CH_4 m^{-2} h^{-1}, with an average of -5.56 μg CH_4 m^{-2} h^{-1} (Fig. 3a–b). A weak interaction between measurement date and treatment was found (Table 2, $P=0.079$). Significant differences in CH_4 fluxes between the control and N addition treatments were only found in July and September (Fig. 3a). For the same level of N addition, NH_4Cl fertilizer exhibited a higher inhibition to soil CH_4 uptake than $NaNO_3$ fertilizer. However, there was no significant difference in soil CH_4 fluxes between the control and N addition treatments (Fig. 3b).

Soil N_2O fluxes also showed a significant seasonality with the minimum occurring from early October to March next year (Fig. 4 a, Table 2, $P<0.001$). In the control, Soil N_2O flux ranged from −

Table 2. Results of repeated-measures ANOVAs on effects of experimental treatment, month and their interaction on soil CH_4 and N_2O fluxes.

Source of variation	CH_4 flux		N_2O flux	
	F	p	F	p
Month	2.34	0.008	23.83	<0.001
Treatment	0.66	0.064	2.47	0.011
Month × Treatment	1.37	0.079	2.35	<0.001

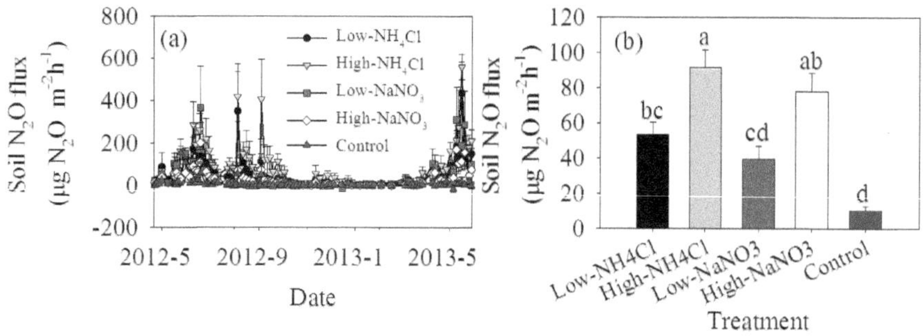

Figure 4. Temporal variations of soil N_2O fluxes and their responses to N addition. Different letters above the columns mean significant differences between experimental treatments.

15.26 to 559.30 μg N_2O m^{-2} h^{-1}, with an average of 10.60 μg N_2O m^{-2} h^{-1} (Fig. 4b). Nitrogen addition produced obvious peaks of soil N_2O emission, which was detected within one week after N addition (Fig. 4a). Soil N_2O fluxes positively responded to N addition, and the promotion increased with the levels of N addition (Fig. 4b). In addition, there was a significant interaction between month and N treatment in the entire observation period (Table 2, $P<0.001$). For the same level of N addition, NH_4Cl fertilizer had a higher promotion to soil N_2O emission than $NaNO_3$ fertilizer (Fig. 4b).

Relationships between soil CH_4 and N_2O fluxes and soil properties

Both soil CH_4 and N_2O fluxes were positively correlated with soil temperature at 5 cm depth and soil moisture at 10 cm depth (Fig. 5, Table 3). The relationships between soil CH_4 fluxes and soil temperature and between soil CH_4 fluxes and soil moisture could be well fitted with quadratic and linear equations, respectively (Fig. 5a–b, Table 3). Similarly, soil N_2O fluxes were linearly correlated with soil temperature and soil moisture (Fig. 5c–d, Table 3).

Soil CH_4 fluxes were positively correlated with soil NO_3^--N concentrations, whereas no significant correlations between soil CH_4 fluxes and other dissolved N species were found (Fig. 6a–d, Table 3). Soil N_2O fluxes were linearly correlated with soil NO_3^--N and TDN concentrations (Fig. 6e, Fig. 6h, Table 3), and the relationship between soil N_2O fluxes and soil NH_4^+-N concentrations was well fitted with a logarithm equation (Fig. 6f, Table 3).

Discussion

Effects of N addition on soil CH_4 fluxes

The subtropical plantation soils can act as a sink of atmospheric CH_4. The mean annual soil CH_4 uptake in the control (0.49 kg CH_4 ha^{-1} yr^{-1}) was lower than those of subtropical plantation in Pingxiang (3.84 kg CH_4 ha^{-1} yr^{-1}) and Dinghushan of southern China (1.34 kg CH_4 ha^{-1} yr^{-1}) [5,6] as well as that of subtropical rainforest in Australia (3.13 kg CH_4 ha^{-1} yr^{-1}) [3]. Except low level of $NaNO_3$ treatment, the other three treatments decreased, on average, the rates of soil CH_4 uptake by 18.38% to 41.04% relative to control (Fig. 3). The decrease in soil CH_4 uptake caused by N addition in our site was higher than those of plantations in Dinghushan [5] and Heshan stations of southern China [8], despite the levels of N addition are similar in the three forest sites (120 vs. 150 kg ha^{-1} yr^{-1}). This result indicated that the response of soil CH_4 uptake to N addition was more sensitive in the northern subtropical plantations than in the southern subtropical

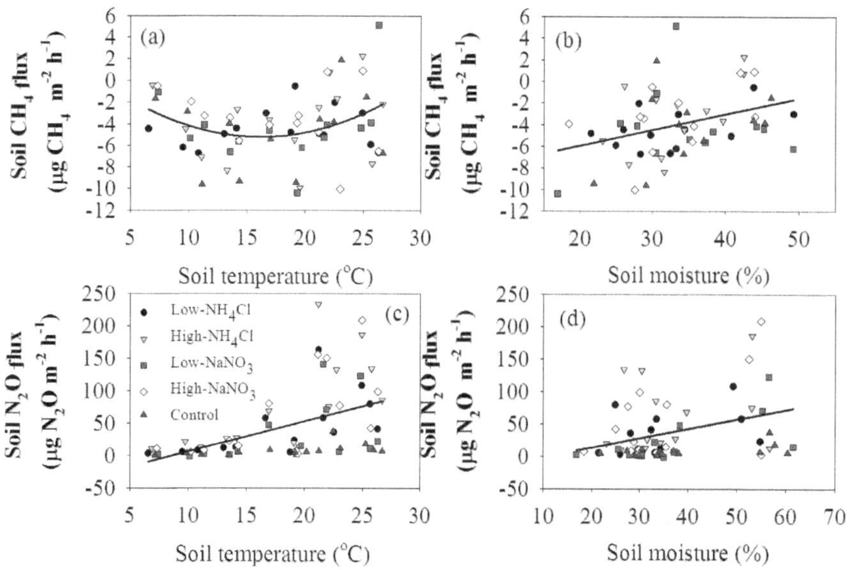

Figure 5. Relationships between soil CH_4 and N_2O fluxes, 5 cm soil temperature, and 10 cm soil moisture (n = 70).

Table 3. Regression models between soil CH_4 and N_2O fluxes and soil properties.

Flux	Soil variables[a]	Regression equation	R^2	*P* value
CH_4	T_s	$y=0.028\ T_S^2-0.90\ T_S+2.02$	0.094	0.044
	M_S	$y=0.15\ M_S-8.84$	0.123	0.006
	NO_3^-	$y=-1.28\ NO_3^-\ -1.77$	0.21	0.004
	Combined	$y=0.17\ M_S-10.66$	0.21	0.003
N_2O	T_S	$y=4.67\ T_S-40.11$	0.25	<0.0001
	M_S	$y=1.99\ M_S-29.76$	0.10	0.001
	NO_3^-	$y=12.29\ NO_3^-+10.58$	0.22	0.005
	NH_4^+	$y=81.52\ \ln(NH_4^+)-85.34$	0.37	<0.0001
	TDN	$y=9.26\ TDN-93.50$	0.17	0.008
	Combined	$y=0.01\ NH_4^++0.013\ NO_3^-\ -0.041$	0.50	<0.0001

[a]: T_S is soil temperature at 5 cm depth, M_S is soil moisture at 10 cm depth, NH_4^+, NO_3^-, and TDN are the concentrations of soil NH_4^+, NO_3^-, and TDN at 20 cm depth.

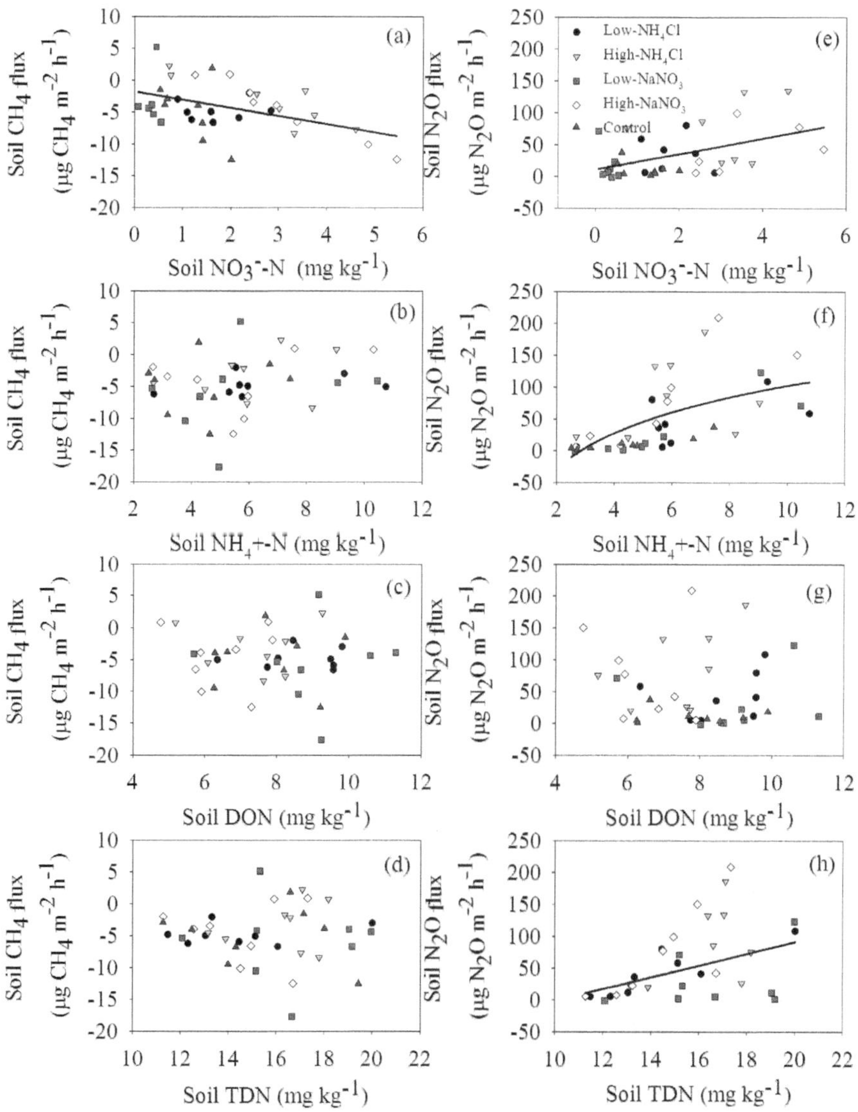

Figure 6. Relationships between soil CH_4 and N_2O fluxes and soil dissolved N concentrations (n = 40).

plantations. This could be attributed to the lower soil N availability, lower atmospheric deposition rate, and the shorter duration of N application in QYZ, compared with the southern subtropical plantations [5,8]. Furthermore, the subsurface mineral soils generally have higher capacity of oxidizing CH_4 than surface litter layer [22,24]. In our site, exogenous N input would directly affect the soil methanotrophic community as well as the amount of CH_4 oxidation due to the lacking of litter layer.

Generally, atmospheric N deposition increases NH_4^+ accumulation and thereby inhibits CH_4 uptake in the well-drained forest soils [8,12,43], despite contrasting results such as promotion and no effect have also been documented [44,45]. In this study, we found that various levels and forms of N addition did not significantly change soil CH_4 uptake over one year (Fig. 3). This could be related to the following third aspects. First, the short-term N fertilizers application did not significantly lead to soil NH_4^+-N accumulation (Fig. 2b), and no significant relationship between soil CH_4 fluxes and soil NH_4^+-N concentrations was found (Fig. 6b). Whalen and Reeburgh [46] also concluded that N inputs did not influence CH_4 uptake until they significantly increased soil NH_4^+ availability in the boreal forest soils. Although an inhibitory trend of soil CH_4 uptake under the NH_4^+-N addition treatments was found, the competition and toxic inhibition of accumulated NH_4^+ did not occur over the short term. Second, N addition enhances the availability of NH_4^+ to soil nitrifiers, which will accordingly decrease the extent to which CH_4 consumers are exposed to NH_4^+ [20]. A slight accumulation in soil NO_3^--N concentration under NH_4^+-N fertilizer treatments indirectly supported our deduction (Fig. 3a). Third, we also found that soil NO_3^--N accumulation could significantly promote soil CH_4 uptake (Fig. 6a), which had been documented in the subtropical plantations of southern China [6]. Especially, the low level of $NaNO_3$ treatment tented to reduce soil NO_3^--N concentration, and thereby it slightly stimulated soil CH_4 uptake (Fig. 2b, Fig. 3b). Moreover, stronger relationships were found between soil CH_4 fluxes and soil NO_3^--N concentrations than between soil CH_4 fluxes and other soil dissolved N concentrations (Fig. 6), suggesting that soil NO_3^- played a more important role in soil CH_4 uptake than other soil dissolved N species in the subtropical plantation.

Soil CH_4 flux is controlled by methanogens operating at anaerobic conditions and methanotrophs taking oxygen as a terminal electron acceptor [47]. Activities and population sizes of these microbial communities depend on a series of soil factors, including soil temperature, moisture, pH, substrate availability, and aeration of soil profile [19,48,49]. Soil CH_4 uptake is dominated by an optimal soil temperature [50]. In our study, the optimal soil temperature was about 15°C (Table 3), and the capacity of soil methanotrophs to oxidize CH_4 would decline when soil temperature was lower or higher than the threshold [51]. Also, soil moisture controls the mass flow of air and diffusion of atmospheric CH_4 into the soil by altering the water filled pore space (WFPS) of soils [52]. We also found that soil CH_4 fluxes under the N addition and control treatments were significantly related to soil moisture (Table 3). Based on the result of stepwise regression analysis, we found that the variation in soil CH_4 uptake was less affected by soil moisture (Table 3). Because N addition did not change soil moisture (Fig. 1), we reasonably deduced that the variation of CH_4 uptake elicited by N treatments was mainly attributed to the change in soil N availability.

Effects of N addition on soil N_2O fluxes

Our result showed that the subtropical slash pine plantation in QYZ exhibited a source of atmospheric N_2O under natural conditions. The average soil N_2O flux in the control (0.93 kg N_2O ha^{-1} yr^{-1}) was comparable with that of Heshengqiao station in Hubei province (0.71 kg N_2O ha^{-1} yr^{-1}) [53], but lower than that of Dinghushan station in South China (2.11 kg N_2O ha^{-1} yr^{-1}) [33]. In our study, NH_4Cl and $NaNO_3$ addition at rates of 40 and 120 kg N ha^{-1} yr^{-1} increased soil N_2O emission by 403% to 762%. On the contrary, in the pine, mixed and evergreen broadleaved forests of Dinghushan station, NH_4NO_3 addition at rates of 50, 100 and 150 kg N ha^{-1} yr^{-1} only increased soil N_2O fluxes by 38% to 58% [33]. These results indicated that the subtropical plantation had high turnover rates of soil N and sensitively responded to increased N deposition. The potential reasons include that the optimal hydrothermal conditions [54], low soil pH [55], and high clay content [39], which favor both soil nitrification and denitrification as well as soil N_2O emission.

Except soil DON concentration, soil N_2O fluxes were significantly correlated with concentrations of soil NH_4^+, NO_3^-, and TDN (Fig. 6), suggesting soil N_2O flux was dominated by both soil nitrification and denitrification processes. Furthermore, the promotion of $NaNO_3$ addition to N_2O emission was slightly lower than that of NH_4Cl addition (Fig. 4). Two potential mechanisms can be responsible for this phenomenon: (1) the high rates of NO_3^- immobilization and nitrification [38], and the low denitrification potential are found in the same type of subtropical plantations [56]; and (2) temperature regulates soil N_2O flux through influencing soil N_2O-producing microorganisms, such as nitrifers and denitrfiers [57]. Soil moisture effects on soil N_2O fluxes are a result of the limitation of O_2 diffusion into the soil and the expansion of soil anaerobic microbial community [58]. The relatively high temperature in wet season was benefit for soil nitrifers and denitrfiers activities, which partly explained the seasonal variation of soil N_2O fluxes with maximum occurring in between May and June (Fig. 4a). Because N addition did not change soil temperature and soil moisture (Fig. 1), the changes in soil N_2O emission under N addition treatments were unlikely to be caused by the changes in soil temperature and soil moisture. Therefore, soil NH_4^+-N and NO_3^--N concentrations were the dominant factors controlling soil N_2O emission in our study, and could explain 49.9% of the temporal variability of soil N_2O fluxes (Table 3).

Conclusions

This study emphasizes the contrasting effects of oxidized NO_3^- and reduced NH_4^+ inputs on the fluxes of CH_4 uptake and N_2O emission from a subtropical plantation soil based on high frequency observations. We found that N addition tended to inhibit soil CH_4 uptake, and dramatically promoted soil N_2O emission. Compared with NO_3^--N fertilizer application, NH_4^+-N fertilizer application had a stronger inhibition to soil CH_4 uptake and a stronger promotion to soil N_2O emission. Also, both soil CH_4 and N_2O fluxes were driven by soil moisture and temperature, but soil inorganic N availability was a key integrator of soil CH_4 uptake and N_2O emission. Overall, short-term N addition has already changed soil CH_4 and N_2O fluxes, which indicated that the subtropical plantation soil was sensitive to N deposition input. In the future, the long-term observation of soil fluxes and the measurement of key microbial functional groups are necessary to clarify the mechanisms responsible for the coupling between soil CH4 and N2O fluxes.

Author Contributions

Conceived and designed the experiments: YW HF SC GY. Performed the experiments: YW MX XD LL LW. Analyzed the data: YW HF SC. Wrote the paper: YW HF SC.

References

1. Post WM, Emanuel WR, Zinke PJ, Stangenberger AG (1982) Soil carbon pools and world life zones. Nature 298: 156–159.
2. IPCC (2001) Climate change 2001: the scientific basis. Contribution of working group I of intergovernmental panel on climate change. Cambridge University Press, Cambridge.
3. Rowlings DW, Grace PR, Kiese R, Weier KL (2012) Environmental factors controlling temporal and spatial variability in the soil-atmosphere exchange of CO_2, CH_4 and N_2O from an Australian subtropical rainforest. Global Change Biology 18: 726–738.
4. Tang XL, Liu SG, Zhou GY, Zhang DQ, Zhou CY (2006) Soil-atmospheric exchange of CO_2, CH_4, and N_2O in three subtropical forest ecosystems in southern China. Global Change Biology 12: 546–560.
5. Zhang W, Mo JM, Zhou GY, Gundersen P, Fang YT, et al. (2008) Methane uptake responses to nitrogen deposition in three tropical forests in southern China. Journal of Geophysical Research-Atmospheres 113, D11116. doi:10.1029/2007JD009195.
6. Wang H, Liu SR, Wang JX, Shi ZM, Lu LH, et al. (2013) Effects of tree species mixture on soil organic carbon stocks and greenhouse gas fluxes in subtropical plantations in China. Forest Ecology and Management 300: 4–13.
7. Dutaur L, Verchot LV (2007) A global inventory of the soil CH_4 sink. Global Biogeochemical Cycles 21: GB4013. doi: 10.1029/2006GB002734.
8. Zhang W, Zhu XM, Liu L, Fu SL, Chen H, et al. (2012) Large difference of inhibitive effect of nitrogen deposition on soil methane oxidation between plantations with N-fixing tree species and non-N-fixing tree species. Journal of Geophysical Research-Biogeosciences 117. doi:10.1029/2012JG002094.
9. Martinson GO, Corre MD, Veldkamp E (2013) Responses of nitrous oxide fluxes and soil nitrogen cycling to nutrient additions in montane forests along an elevation gradient in southern Ecuador. Biogeochemistry 112: 625–636.
10. Steudler PA, Garcia-Montiel DC, Piccolo MC, Neill C, Melillo JM, et al. (2002) Trace gas responses of tropical forest and pasture soils to N and P fertilization. Global Biogeochemical Cycles 16.
11. Butterbach-Bahl K, Gasche R, Huber C, Kreutzer K, Papen H (1998) Impact of N-input by wet deposition on N-trace gas fluxes and CH_4-oxidation in spruce forest ecosystems of the temperate zone in Europe. Atmospheric Environment 32: 559–564.
12. Kim YS, Imori M, Watanabe M, Hatano R, Yi MJ, et al. (2012) Simulated nitrogen inputs influence methane and nitrous oxide fluxes from a young larch plantation in northern Japan. Atmospheric Environment 46: 36–44.
13. Jassal RS, Black TA, Roy R, Ethier G (2011) Effect of nitrogen fertilization on soil CH_4 and N_2O fluxes, and soil and bole respiration. Geoderma 162: 182–186.
14. Liu LL, Greaver TL (2009) A review of nitrogen enrichment effects on three biogenic GHGs: the CO_2 sink may be largely offset by stimulated N_2O and CH_4 emission. Ecology Letters 12: 1103–1117.
15. Castro MS, Peterjohn WT, Melillo JM, Steudler PA, Gholz HL, et al. (1994) Effects of Nitrogen-Fertilization on the Fluxes of N_2O, CH_4, and CO_2 from Soils in a Florida Slash Pine Plantation. Canadian Journal of Forest Research-Revue Canadienne De Recherche Forestiere 24: 9–13.
16. Jang I, Lee S, Zoh KD, Kang H (2011) Methane concentrations and methanotrophic community structure influence the response of soil methane oxidation to nitrogen content in a temperate forest. Soil Biology & Biochemistry 43: 620–627.
17. Mohanty SR, Bodelier PLE, Floris V, Conrad R (2006) Differential effects of nitrogenous fertilizers on methane-consuming microbes in rice field and forest soils. Applied and Environmental Microbiology 72: 1346–1354.
18. Aronson EL, Helliker BR (2010) Methane flux in non-wetland soils in response to nitrogen addition: a meta-analysis. Ecology 91: 3242–3251.
19. Reay DS, Nedwell DB (2004) Methane oxidation in temperate soils: effects of inorganic N. Soil Biology & Biochemistry 36: 2059–2065.
20. Chan ASK, Steudler PA (2006) Carbon monoxide uptake kinetics in unamended and long-term nitrogen-amended temperate forest soils. Fems Microbiology Ecology 57: 343–354.
21. Nyerges G, Stein LY (2009) Ammonia cometabolism and product inhibition vary considerably among species of methanotrophic bacteria. Fems Microbiology Letters 297: 131–136.
22. Bodelier PLE, Laanbroek HJ (2004) Nitrogen as a regulatory factor of methane oxidation in soils and sediments. Fems Microbiology Ecology 47: 265–277.
23. King GM, Schnell S (1994) Effect of increasing atmospheric methane concentration on ammonium inhibition of soil methane consumption. Nature 370: 282–284.
24. Fang H, Cheng S, Yu G, Cooch J, Wang Y, et al. (2014) Low-level nitrogen deposition significantly inhibits methane uptake from an alpine meadow soil on the Qinghai–Tibetan Plateau. Geoderma 213: 444–452.
25. Xu XK, Inubushi K (2004) Effects of N sources and methane concentrations on methane uptake potential of a typical coniferous forest and its adjacent orchard soil. Biology and Fertility of Soils 40: 215–221.
26. Bodelier PLE (2011) Interactions between nitrogenous fertilizers and methane cycling in wetland and upland soils. Current Opinion in Environmental Sustainability 3: 379–388.
27. Matson P, Lohse KA, Hall SJ (2002) The globalization of nitrogen deposition: Consequences for terrestrial ecosystems. Ambio 31: 113–119.
28. Venterea RT, Groffman PM, Verchot LV, Magill AH, Aber JD, et al. (2003) Nitrogen oxide gas emissions from temperate forest soils receiving long-term nitrogen inputs. Global Change Biology 9: 346–357.
29. Chen GC, Tam NFY, Ye Y (2012) Spatial and seasonal variations of atmospheric N_2O and CO_2 fluxes from a subtropical mangrove swamp and their relationships with soil characteristics. Soil Biology & Biochemistry 48: 175–181.
30. Zhang JB, Cai ZC, Zhu TB (2011) N_2O production pathways in the subtropical acid forest soils in China. Environmental Research 111: 643–649.
31. Zhu J, Mulder J, Solheimslid SO, Dorsch P (2013) Functional traits of denitrification in a subtropical forest catchment in China with high atmogenic N deposition. Soil Biology & Biochemistry 57: 577–586.
32. Wang LF, Cai ZC (2008) Nitrous oxide production at different soil moisture contents in an arable soil in China. Soil Science and Plant Nutrition 54: 786–793.
33. Zhang W, Mo JM, Yu GR, Fang YT, Li DJ, et al. (2008) Emissions of nitrous oxide from three tropical forests in Southern China in response to simulated nitrogen deposition. Plant and Soil 306: 221–236.
34. Hu ZH, Zhang H, Cheng ST, Li T, Sheng SH (2011) Effects of simulated nitrogen deposition on N_2O and CH_4 fluxes of soils in forest belt. China Environmental Science 31: 892–897.
35. Department of Forest Resources Management S (2010) The 7th National forest inventory and status of forest resources. Forest Resource and Management 1: 3–10.
36. SFA (2007) China's Forestry 1999–2005. China Forestry Publishing House, Beijing.
37. Liu XJ, Duan L, Mo JM, Du EZ, Shen JL, et al. (2011) Nitrogen deposition and its ecological impact in China: An overview. Environmental Pollution 159: 2251–2264.
38. Zhang JB, Cai.Z.C,Zhu TB, Yang WY, Christoph M (2013) Mechanisms for retention of inorganic N in acid forest soils in southern China. Nature 3: 1–8.
39. Wen XF, Wang HM, Wang JL, Yu GR, Sun XM (2010) Ecosystem carbon exchanges of a subtropical evergreen coniferous plantation subjected to seasonal drought, 2003–2007. Biogeosciences 7: 357–369.
40. Wang YD, Wang ZL, Wang HM, Guo CC, Bao WK (2012) Rainfall pulse primarily drives litterfall respiration and its contribution to soil respiration in a young exotic pine plantation in subtropical China. Canadian Journal of Forest Research-Revue Canadienne De Recherche Forestiere 42: 657–666.
41. Wang YS, Wang YH (2003) Quick measurement of CH_4, CO_2 and N_2O emissions from a short-plant ecosystem. Advances in Atmospheric Sciences 20: 842–844.
42. Fang HJ, Cheng SL, Yu GR, Zheng JJ, Zhang PL, et al. (2012) Responses of CO_2 efflux from an alpine meadow soil on the Qinghai Tibetan Plateau to multi-form and low-level N addition. Plant and Soil 351: 177–190.
43. Le Mer J, Roger P (2001) Production, oxidation, emission and consumption of methane by soils: A review. European Journal of Soil Biology 37: 25–50.
44. Basiliko N, Khan A, Prescott CE, Roy R, Grayston SJ (2009) Soil greenhouse gas and nutrient dynamics in fertilized western Canadian plantation forests. Canadian Journal of Forest Research-Revue Canadienne De Recherche Forestiere 39: 1220–1235.
45. Borken W, Beese F, Brumme R, Lamersdorf N (2002) Long-term reduction in nitrogen and proton inputs did not affect atmospheric methane uptake and nitrous oxide emission from a German spruce forest soil. Soil Biology & Biochemistry 34: 1815–1819.
46. Whalen SC, Reeburgh WS (2000) Methane oxidation, production, and emission at contrasting sites in a boreal bog. Geomicrobiology Journal 17: 237–251.
47. Topp E, Pattey E (1997) Soils as sources and sinks for atmospheric methane. Canadian Journal of Soil Science 77: 167–178.
48. Werner C, Kiese R, Butterbach-Bahl K (2007) Soil-atmosphere exchange of N_2O, CH_4, and CO_2 and controlling environmental factors for tropical rain forest sites in western Kenya. Journal of Geophysical Research-Atmospheres 112.
49. Merino A, Perez-Batallon P, Macias F (2004) Responses of soil organic matter and greenhouse gas fluxes to soil management and land use changes in a humid temperate region of southern Europe. Soil Biology & Biochemistry 36: 917–925.
50. Fang HJ, Yu GR, Cheng SL, Zhu TH, Wang YS, et al. (2010) Effects of multiple environmental factors on CO_2 emission and CH_4 uptake from old-growth forest soils. Biogeosciences 7: 395–407.
51. Steinkamp R, Butterbach-Bahl K, Papen H (2001) Methane oxidation by soils of an N limited and N fertilized spruce forest in the Black Forest, Germany. Soil Biology & Biochemistry 33: 145–153.
52. Lin XW, Wang SP, Ma XZ, Xu GP, Luo CY, et al. (2009) Fluxes of CO_2, CH_4, and N_2O in an alpine meadow affected by yak excreta on the Qinghai-Tibetan plateau during summer grazing periods. Soil Biology & Biochemistry 41: 718–725.
53. Lin S, Iqbal J, Hu RG, Ruan LL, Wu JS, et al. (2012) Differences in nitrous oxide fluxes from red soil under different land uses in mid-subtropical China. Agriculture Ecosystems & Environment 146: 168–178.
54. Xu YB, Xu ZH, Cai ZC, Reverchon F (2013) Review of denitrification in tropical and subtropical soils of terrestrial ecosystems. Journal of Soils and Sediments 13: 699–710.

55. Xu YB, Cai ZC (2007) Denitrification characteristics of subtropical soils in China affected by soil parent material and land use. European Journal of Soil Science 58: 1293–1303.
56. Zhang JB, Cai ZC, Cheng Y, Zhu TB (2009) Denitrification and total nitrogen gas production from forest soils of Eastern China. Soil Biology & Biochemistry 41: 2551–2557.
57. Bijoor NS, Czimczik CI, Pataki DE, Billings SA (2008) Effects of temperature and fertilization on nitrogen cycling and community composition of an urban lawn. Global Change Biology 14: 2119–2131.
58. Luo GJ, Kiese R, Wolf B, Butterbach-Bahl K (2013) Effects of soil temperature and moisture on methane uptake and nitrous oxide emissions across three different ecosystem types. Biogeosciences 10: 3205–3219.

22

Competition Increases Sensitivity of Wheat (*Triticum aestivum*) to Biotic Plant-Soil Feedback

W. H. Gera Hol[1]*, Wietse de Boer[2,3], Freddy ten Hooven[1], Wim H. van der Putten[1,4]

1 Netherlands Institute of Ecology, Department of Terrestrial Ecology, Wageningen, The Netherlands, **2** Netherlands Institute of Ecology, Department of Microbial Ecology, Wageningen, The Netherlands, **3** Wageningen University, Department of Soil Quality, Wageningen, The Netherlands, **4** Wageningen University, Laboratory of Nematology, Wageningen, The Netherlands

Abstract

Plant-soil feedback (PSF) and plant competition play an important role in structuring vegetation composition, but their interaction remains unclear. Recent studies suggest that competing plants could dilute pathogenic effects, whereas the standing view is that competition may increase the sensitivity of the focal plant to PSF. In agro-ecosystems each of these two options would yield contrasting outcomes: reduced versus enhanced effects of weeds on crop biomass production. To test the effect of competition on sensitivity to PSF, we grew *Triticum aestivum* (Common wheat) with and without competition from a weed community composed of *Vicia villosa*, *Chenopodium album* and *Myosotis arvensis*. Plants were grown in sterilized soil, with or without living field inoculum from 4 farms in the UK. In the conditioning phase, field inocula had both positive and negative effects on *T. aestivum* shoot biomass, depending on farm. In the feedback phase the differences between shoot biomass in *T. aestivum* monoculture on non-inoculated and inoculated soils had mostly disappeared. However, *T. aestivum* plants growing in mixtures in the feedback phase were larger on non-inoculated soil than on inoculated soil. Hence, *T. aestivum* was more sensitive to competition when the field soil biota was present. This was supported by the statistically significant negative correlation between shoot biomass of weeds and *T. aestivum*, which was absent on sterilized soil. In conclusion, competition in cereal crop-weed systems appears to increase cereal crop sensitivity to soil biota.

Editor: James F. Cahill, University of Alberta, Canada

Funding: The work was funded by the EU project SOILSERVICE. SOILSERVICE is a collaborative, medium scale focussed research project, which is funded by EU's Seventh Framework Programme. (http://www4.lu.se/soil-ecology-group/research/soilservice.) The funders had no role in study design, data collection and analysis, decision to publish, or preparation of the manuscript.

Competing Interests: The authors have declared that no competing interests exist.

* E-mail: g.hol@nioo.knaw.nl

Introduction

Plants have selective effects on soil biota, with negative, neutral or positive consequences for future occupants of the same location [1,2]. These biotic plant-soil feedback (PSF) effects are caused by net effects of soil-borne mutualists and pathogens that develop in the rhizosphere during plant growth [3]. In addition, there are abiotic feedbacks such as changes in nutrient or water availability. If the performance of conspecific plants in the next growing cycle is stimulated, this is called a positive PSF. If conspecific plants perform worse in the next growing cycle, this is called a negative PSF. This process is important for plant succession [4,5] and diversity-productivity relationships [6,7], but it also plays an important role in applied areas, such as agriculture [8], restoration ecology [9,10] and invasion management [11]. The success of later successional plants traditionally has been attributed to their ability to constrain growth of predecessors [12], but more recently it has been acknowledged that species-specific PSF can influence predecessor-successor interactions [1,13,14].

In agriculture, crop species are being grown in rotation in order to avoid the development of negative PSF, as crop rotation prevents the accumulation of crop-specific soil pathogens [8]. Negative intraspecific PSF effects can be avoided by growing crops in successional cycles with other crops that are not sensitive to the pathogens from previous crops. The crop cycles can also be related to using positive interspecific PSF, for example green manures that increase nutrient availability and nitrogen-fixing plants that enhance the pools of available nitrogen in soil. Intercropping with legumes has been used to both reduce pathogen pressure (negative PSF) and simultaneously increase soil nitrogen (positive PSF) [15].

Empirical work on PSF has been done both with single plant species and with communities [2], mostly composed of wild plants. Shifts in relative abundance of plant species within a community indicate that PSF does affect competitive relations, e.g [5], but only a few studies have addressed the role of competition in PSF studies explicitly [1,13].

There are two mechanisms in relation to feedback in plant mixtures which would lead to opposite effects. Plants that are stressed due to competition can be more vulnerable to pathogens [16]. Successful competitors may need to invest in longer stems to compete for light or larger roots to compete for nutrients and these investments may come at the expenses of plant defense [17]. On the other hand, pathogen effects can be density-dependent and since the relative abundance of species in mixtures is lower than in monocultures less pathogenic effects can be expected and observed in mixtures [6,7].

Thus far there is limited evidence that intraspecific competition increases PSFs [2] or that interspecific competition decreased PSF

[18–20]; but see [1,21–23]. We analyze the effect of competition on plant soil feedback of wheat in the feedback phase. Competition can be studied with replacement (constant density) or additive designs (increasing density) [24]. We choose an additive design without the option to distinguish intra- from interspecific competition since this resembles the agricultural situation best. The outcome of the interaction between feedback and competition may depend on species identity [25], and thus it is important to use species which co-occur in order to provide relevant information for a particular system. All soils were conditioned with *T. aestivum* (common wheat), and nutrients were added to minimize the role of nutrient availability on plant growth responses and to focus on biotic feedback [23,26]. We determined how the presence of a weed community affects responses of common wheat (*Triticum aestivum*) to PSF, by performing a feedback experiment with monoculture and plant mixtures in the feedback phase. For our experiment soil was collected from 4 different locations to determine whether the effect of competition on PSF is general or rather depends on the specific soil biota. Given the high density of crop plants in agriculture, we expect that increased sensitivity to pathogens will overrule the reduction in pathogen density dependence effects by weeds. Thus we hypothesize that negative plant-soil feedback effects of *T. aestivum* plants will be enhanced when *T. aestivum* grows in plant mixtures. Our results show that plants with competition are more vulnerable to negative plant soil feedback effects.

Materials and Methods

The effect of plant mixtures on sensitivity to plant soil feedback was tested by growing *Triticum aestivum* L. plants for two growing cycles in the same soil, in the presence or absence of arable weeds in the second growing cycle. The first growing cycle consisted of wheat plants growing in sterilized soil with or without field soil inoculum. The aim of this phase was to standardize abiotic conditions and determine the variation in effects of soil biota from different farms on wheat growth. Since all soils were conditioned with *T. aestivum*, we minimized the role of soil nutrient feedback and focused on the biotic feedback.

Soils

Soil for the experiment was collected with approval from the owners of 4 farms from the Chilterns area (United Kingdom). The soil-type is chromic luvisol/leptosol with pH in the range of 7.2–7.6. At each farm, soil from three fields was collected. Soil was passed through a 1 cm sieve to remove stones and reduce the larger clumps of clay. For each farm 66 kg soil (a mix from the 3 fields) was sterilized by gamma radiation (25 kGray, Isotron, Ede, The Netherlands) in order to create a sterilized background soil. A small subsample from each field was kept apart and remained non-sterilized in order to serve as an inoculum for soil biota. The sterilized soil was used to fill square pots (11×11×12 cm) with 1 kg soil, including 90 g fresh living soil inoculum (9% inoculum). Eight pots per field were filled, resulting in 4 (farms) * 3 (fields) * 8 (pots) = 96 pots in total. In addition to the aforementioned treatments of sterilized soil with non-sterilised soil inoculum, a control treatment was included which consisted of 16 pots per farm filled with sterilized soil only. This added 4 (farms) * 16 (pots) = 64 pots to the experiment. The amount of replicates for the control pots was doubled from 8 to 16 to increase the balance when comparing inoculated vs non-inoculated pots.

Conditioning phase

For the first phase of the experiment, all pots were sown with 7 seeds of common wheat (*Triticum aestivum*) in a row in the middle of the pots. There were 24 inoculated (3 fields * 8 replicates) and 16 non-inoculated control pots per farm. Plants were kept under controlled conditions in the greenhouse (60% RH; 16 h L, 8 h D, 21°C/16°C) and placed in random order on a bench. Additional illumination was provided by 400W growing bulbs (Philips SON-T Agro, Philips, Eindhoven, The Netherlands). Light intensity at plant level was 225 µmol PAR. Pots were watered regularly with demineralized water and received 60 ml week^{-1} Hoagland solution (half strength, [27]). Nutrient addition may alter feedback effects [28], but was chosen to reduce possible nutrient feedback effects and focus on the biotic component. In addition, fertilization is common practice in agricultural systems. A thrips infestation was controlled by the predatory mite *Amblyseius cucumeris* (Koppert Biological Systems, Berkel en Rodenrijs, The Netherlands). Plants were harvested after 60 days; aboveground biomass was clipped and put in paper bags. Shoot dry weight was determined after 72 h drying at 70°C.

Feedback phase

For the second phase of the experiment, soil with roots was left in the pots and re-growing shoots were removed. After two weeks no more re-growth was observed and the pots were sown again with 7 seeds of *T. aestivum*. Seeds were sown perpendicular to the previous sown seeds, to be able to discriminate between potential re-growth and germination, but this precaution proved unnecessary. Half of the pots was planted with one seedling from each of the three weed species: hairy vetch (*Vicia villosa* Roth), lamb's quarters (*Chenopodium album* L.) and field forget-me-not (*Myosotis arvensis* (L.) Hill.). Seeds from *C. album* and *V. villosa* were collected in the wild in the Netherlands, while *M. arvensis* was ordered from a commercial supplier (Rieger-Hofmann, Blaufelden-Raboldshausen, Germany). Seeds were germinated on glass pearls by 16 h L, 8 h D, 23°C/15°C. *Chenopodium album* and *M. arvensis* are common arable weeds [29]; *Vicia villosa* is often applied as winter cover crop and might return as weed in follow-up crops. The weeds resulted in 2 treatments in the second phase: Mono and Mix (n = 12 for inoculated pots and n = 8 for non-inoculated control pots). The pots were kept in the same greenhouse as before, under the same light and temperature conditions. Pots were watered regularly with demineralized water and received 60 ml week^{-1} Hoagland solution (half strength, [27]). After 55 days the aboveground biomass was harvested as described above for the conditioning phase. Shoots from all pots were separated per species and dry weight per species per pot was determined after 72 h drying at 70°C. For a subset of plants (farm 3) the shoot material was ground in a Retsch mill and carbon and nitrogen was measured in three milligrams of the leaves by combustion with an elemental autoanalyzer Flash EA 1112 NC analyzer (Interscience, Breda, the Netherlands) to determine whether inoculation affected nitrogen and carbon content.

Statistical analyses

Differences in shoot biomass of *T. aestivum* on inoculated and non-inoculated soils in the conditioning phase were tested with a linear model: shoot biomass ~ Inoculum * Farm. Farm was included as fixed effect since we were interested in the generality of the tested responses. Normality of the residuals was tested with the non-parametric Kolmogorov-Smirnov test and Homogeneity of variances with the non-parametric Fligner test. Residuals were not normally distributed and this could not be improved by transformation or removal of outliers. However, non-parametric tests (Wilcoxon test per farm) gave the same results as the linear model and thus violation of the assumptions appeared not too influential. For consistency with the analyses in the feedback phase, we decided to present the linear model results. In the feedback phase we compared shoot biomass of *T. aestivum* between inoculated and non-inoculated soils with a linear model using Inoculation, Farm and Competition as fixed categorical factors. Model check plots were made to inspect behavior of the residuals and normality of errors; no transformations were necessary. We also run an additional model where weed biomass was included as a covariate to verify whether this could explain possible Inoculation effects. When covariates were significant, correlations were tested with Spearman's rank correlation tests. This was done separately for the two treatments in the feedback phase (inoculated and non- inoculated). Difference in nitrogen percentage in the shoots of plants growing on inoculated and non-inoculated soils was tested with a linear model nitrogen ~ Inoculum * Competition and verified non-parametrically with a Wilcoxon test. Since the interaction Inoculum * Competition was not significant ($P = 0.96$) we only present the main fixed effects for the model nitrogen ~ Inoculum + Competition. All statistical analyses were carried out with R 2.14.1 statistical package [30]. The data has been archived in the Data and Information Portal of the NIOO-KNAW http://data.nioo.knaw.nl/index.php and is available on request.

Results

Conditioning phase

The effect of inoculation with field soil on *T. aestivum* shoot biomass varied between farms (Fig. 1, Table 1). For two farms inoculated soil yielded more *T. aestivum* shoot biomass (10%±3) than non-inoculated soil, whereas the other two farms showed the opposite pattern (−12%±0.5).

Effect of crop-weed mixtures on vulnerability to PSF

In the feedback phase the effect of inoculation interacted with farm and competition (significant 3-way interaction, Table 2). The presence of weeds reduced *T. aestivum* shoot biomass in all cases (Fig. 2A, 2B), but the strength of the effect is modified by farm and inoculation. On farms 1, 2 and 3 the growth reduction by competition was smaller on sterilized soil (7–10%, Fig. 2A) than on

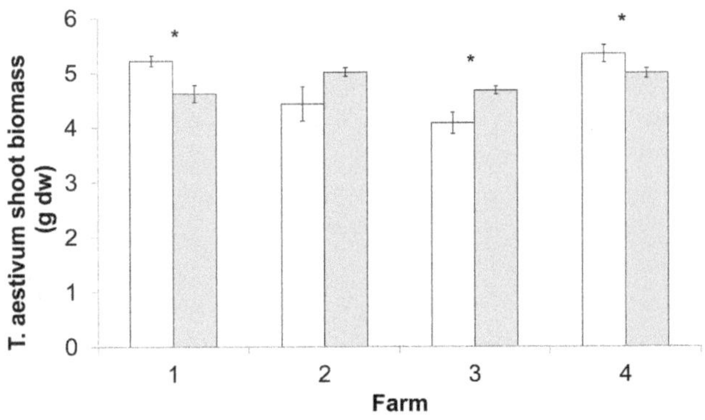

Figure 1. Soil inoculation and plant biomass. Shoot biomass of *Triticum aestivum* (mean ± SE) on sterilized soil without inoculum or with 9% field soil inoculum from 4 different farms in the conditioning phase. The asterisks indicate significant differences ($P<0.05$) between non-inoculated and inoculated soil.

Table 1. Linear model testing effect of inoculation and farm on shoot biomass of *Triticum aestivum* in the conditioning phase.

	Estimate	Std. Error	*t*-value	*P*-value
Intercept	4.61	0.13	34.45	<0.01
Inoculation	0.61	0.21	2.86	<0.01
Farm 2	0.40	0.19	2.10	0.04
Farm 3	0.06	0.19	0.31	0.76
Farm 4	0.37	0.19	1.95	0.05
Inoculation*Farm 2	−1.19	0.30	−3.97	<0.01
Inoculation*Farm 3	−1.20	0.30	−4.01	<0.01
Inoculation*Farm 4	−0.25	0.30	−0.82	0.41

inoculated soil (10–27%, Fig. 2B). Farm 4 is the exception, with stronger growth reduction by competition on the sterilized soil. Since total weed biomass was not significantly affected by inoculation, the difference in *T. aestivum* shoot biomass between sterilized and inoculated soils is probably not due to weeds being smaller on non-inoculated soil. When weed biomass was included as covariate, Inoculum is retained as a, albeit marginally, significant factor in the model. Apparently *T. aestivum* is more sensitive to competition when the field soil biota was present. Further support for this hypothesis can be found when considering the relation between shoot biomass of weeds and *T. aestivum*. For sterilized soil there is no correlation between those two (Fig. 3A), while for inoculated soil this correlation is significantly negative (Fig. 3B).

The presence of biota appeared to have increased either nutrient availability or uptake, since nitrogen percentage in the shoot during the feedback phase was slightly higher for inoculated soils (1.60±0.05% for inoculated soils, n = 23, 1.42±0.06% for non-inoculated soils, n = 16, lm Inoculum $P = 0.02$, Competition $P = 0.31$, no significant interaction). Legumes are known for increasing nitrogen, but in our experiment the presence of weeds including one legume species did not significantly affect nitrogen or carbon percentages in the shoot of *T. aestivum*.

Discussion

Effect of crop-weed mixtures on vulnerability to PSF

Plant mixtures in the feedback phase showed reduced wheat focal shoot biomass, which might be a combination of interspecific competition and enhanced sensitivity to plant soil feedback. In monoculture, however, there was very little evidence of plant soil feedback deviating from neutral. Negative effects of soil inoculation on shoot biomass in the conditioning phase may represent net pathogenic effects of soil biota, which might be even stronger in the feedback phase since the pathogens accumulated during conditioning. The same could be expected for the positive effects, unless positive biota need more time to establish, but in fact soils from only 1 out of 4 farms showed a repeated pattern of the conditioning phase in the feedback phase. Plants in soil from Farm 4 always grew better in sterilized soil compared to inoculated soil. Plants in soil from Farms 1, 2 and 3 did equally well on inoculated and non-inoculated soil in the feedback phase, despite the differences in seen in the conditioning phase. Therefore, regardless of the legacy effects in the conditioning phase, there was little difference between inoculated and non-inoculated soils for the

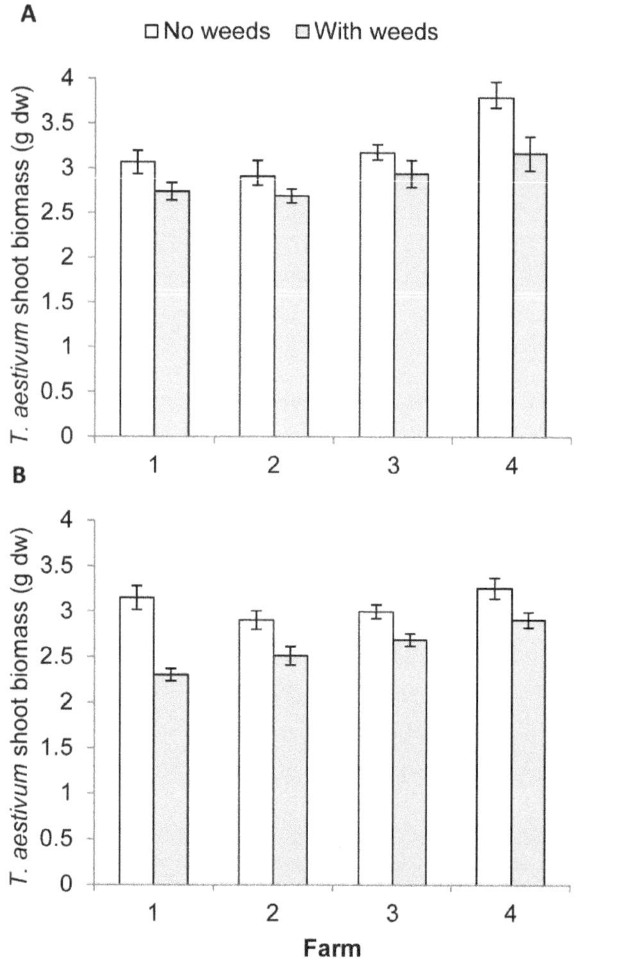

Figure 2. Competition interacts with inoculation effects. Shoot biomass of *Triticum aestivum* (mean ± SE) growing in soil with or without weeds. **A** growing in sterilized soil, **B** growing in inoculated soil.

Table 2. Linear model testing effect of inoculation, competition and farm on shoot biomass of *Triticum aestivum* in the feedback phase.

	Estimate	Std. Error	*t*-value	*P*-value
Intercept	3.15	0.10	30.32	<0.01
Inoculation	−0.08	0.16	−0.51	0.61
Farm 2	−0.25	0.15	−1.69	0.09
Farm 3	−0.15	0.15	−1.02	0.31
Farm 4	0.11	0.15	0.72	0.47
Competition	−0.85	0.15	−5.76	<0.01
Inoculation*Farm 2	0.09	0.23	0.38	0.70
Inoculation*Farm 3	0.25	0.23	1.08	0.28
Inoculation*Farm 4	0.61	0.23	2.63	<0.01
Competition*Farm2	0.46	0.21	2.20	0.03
Competition*Farm3	0.54	0.21	2.58	0.01
Competition*Farm4	0.50	0.21	2.40	0.02
Inoculation*Competition	0.52	0.23	2.22	0.03
Inoculation*Farm 2*Competition	−0.35	0.33	−1.06	0.29
Inoculation*Farm 3* Competition	−0.44	0.33	−1.33	0.19
Inoculation*Farm 4* Competition	−0.79	0.33	−2.40	0.02

monoculture treatment in terms of biomass and only the nitrogen data indicated that there is some biotic feedback. Feedback effects became more apparent under competition; addition of weed seedlings to half of the pots in the feedback phase led to consistently smaller *T. aestivum* plants on inoculated soils as compared to the non-inoculated ones. Likewise, Callaway et al. [22] only found effects of soil biota on *Centaurea maculosa* in greenhouse experiments when competitors were present. Stronger negative feedback effects under competition have been found before, in experiments with replacement designs [13,18,19], where total plant density is kept constant. Manipulation of soil communities independent of plant communities showed that

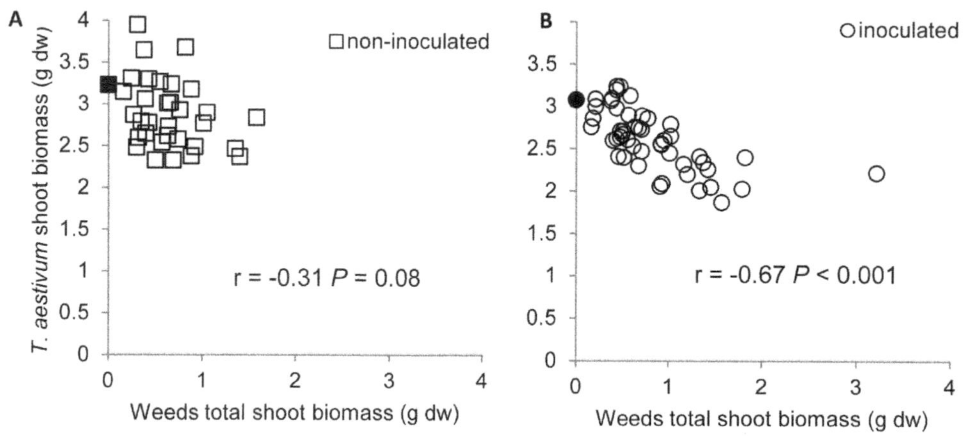

Figure 3. Correlation between crop and weeds depend on soil inoculation. Correlation between *T. aestivum* shoot biomass and weeds shoot biomass within the same pot on sterilized **A** non-inoculated soil and **B** inoculated soil in the feedback phase. Spearman rank correlation coefficients and *P*-values are shown in the graphs. Filled symbols indicate the average *T. aestivum* shoot biomass in the absence of weeds; n = 32 for non-inoculated soils and n = 48 for inoculated soils.

overyielding was due to plants inhibited by their own soil biota and unrelated to nutrient availability [31]. Shannon et al. [20] also used an additive design and found no evidence for feedback when plants were grown separately; only in competitive mixtures did feedback become apparent. Yet there is no consistency in the relationship between competition and feedback; feedback effects might also disappear under conditions of competition [22] and in some circumstances feedback effects might override competition effects [23]. The variety in effects might be related to soil origin and to plant species identity [22], although a study with 24 species showed that the majority of species suffered stronger negative feedback when grown in competition [19]. Even without competition plant soil feedbacks have been described as idiosyncratic [32,33] often meaning unpredictable [34] and the challenge is to find generalities in plant-soil feedback. The added value of the present study is the demonstration that biotic feedback effects were increased when co-occurring species compete and this was found across a wide range of farms and fields.

On non-inoculated soils *T. aestivum* plants had larger shoots and appeared to be less affected by competing weeds than on the inoculated soil. One possible explanation would be that the weeds did not grow well on the non-inoculated soil. However, weed biomass was not significantly different between non-inoculated and inoculated soils and including weed biomass as a covariate did not affect the significant difference between control soil and the inoculated soils. The degradation of wheat roots might also have differed between inoculated and non-inoculated soils, with consequences for plant nutrition. Since we made no measurements of plant growth during the experiment, we cannot exclude the possibility that there were initially differences in *T. aestivum* monocultures between sterilized non-inoculated and inoculated soil which disappeared in the later growth phase. This has been found before [13] and in plant mixtures the weeds could have taken advantage of the temporary growth delay of the wheat plants, thus effectively fixating the difference. This would also result in larger wheat plants in the feedback phase on sterilized control soils than on the inoculated soils. However, if weeds used the window of opportunity created in inoculated soil, then larger weed biomass would be expected for inoculated soil than for the sterilized control soils, but this was not found. Thus the most parsimonious interpretation is that *T. aestivum* plants growing in mixtures were more sensitive to their biotic PSF, e.g. due to a trade-off between resources allocated to either competitive or defense traits.

A possible mechanism for increased effects of PSF under competition could be the increased amount of roots in pots with plant mixtures. Root density is an important factor in development of fungal diseases [35]. However, the question is whether this would also work for specific pathogens. Generally, increasing plant diversity is thought to dilute species-specific pathogens [6,7]. The fact that the current study reveals that *T. aestivum* growing in plant mixtures was more sensitive to PSF might depend on the dominance of *T. aestivum*. This should be tested in a PSF experiment where the focal plant density is kept constant and the competitors are added in a large range to test whether there is a tipping point from increased negative effects of PSF in mixtures towards a dilution of plant-specific pathogens. Aguilera [36] modeled two-species competition under a range of competitive abilities and feedback scenario's and demonstrated that feedbacks can reverse the outcome of competition. The existence of such a tipping point will depend on the strength of the PSF versus the competitive effects. Knowledge of such a point could be used to determine optimal planting densities in mixed cropping systems.

Acknowledgments

We thank Simon Mortimer and Stephanie Harris for providing soil from farms in the UK. This is publication 5443 of the NIOO-KNAW.

Author Contributions

Conceived and designed the experiments: WHGH WdB WvdP. Performed the experiments: WHGH FtH. Analyzed the data: WHGH. Wrote the paper: WHGH WvdP WdB FtH.

References

1. Bever JD (1994) Feedback between plants and their soil communities in an old field community. Ecology 75: 1965–1977.
2. Kulmatiski A, Beard KH, Stevens JR, Cobbold SM (2008) Plant-soil feedbacks: a meta-analytical review. Ecol Lett 11: 980–992.
3. Bever JD, Westover KM, Antonovics J (1997) Incorporating the soil community into plant population dynamics: the utility of the feedback approach. J Ecol 85: 5261–573.
4. Reynolds HL, Packer A, Bever JD, Clay K (2003) Grassroots ecology: plant-microbe-soil interactions as drivers of plant community structure and dynamics. Ecology 84: 2281–2291.
5. Kardol P, Bezemer TM, Van der Putten WH (2006) Temporal variation in plant-soil feedback controls succession. Ecol Lett 9: 1080–1088.
6. Maron JL, Marler M, Klironomos JN, Cleveland CC (2011) Soil fungal pathogens and the relationship between plant diversity and productivity. Ecol Lett 14: 36–41.
7. Schnitzer SA, Klironomos JN, HilleRisLambers J, Kinkel LL, Reich PB, et al. (2011) Soil microbes drive the classic plant diversity-productivity pattern. Ecology 92: 296–303.
8. Bullock DG (1992) Crop rotation. Crit Rev Plant Sci 11: 309–326.
9. Eviner VT, Hawkes CV (2008) Embracing variability in the application of plant-soil interactions to the restoration of communities and ecosystems. Restor Ecol 16: 713–729.
10. Jordan NR, Larson DL, Huerd SC (2011) Evidence of qualitative differences between soil-occupancy effects of invasive vs. native grassland plant species. Invasive Plant Sci Management 4: 11–21.
11. Inderjit, Van der Putten WH (2010) Impacts of soil microbial communities on exotic plant invasions. TREE 25: 512–519.
12. Tilman D (1982) Resource competition and community structure. Princeton University Press, Princeton, NJ.
13. van der Putten WH, Peters BA (1997) How soil-borne pathogens may affect plant competition. Ecology 78: 1785–1795.
14. Harrison KA, Bardgett RD (2010) Influence of plant species and soil condition on plant-soil feedback in mixed grassland communities. J Ecol 98: 384–395.
15. Hauggaard-Nielsen H, Gooding M, Ambus P, Corre-Hellou G, Crozat Y, et al. (2009) Pea-barley intercropping and short-term subsequent crop effects across European organic cropping conditions. Nutr Cycl Agroecosyst 85: 141–155.
16. Gilbert GS (2002) Evolutionary ecology of plant diseases in natural ecosystems. Ann Rev Phytopathol 40: 13–43.
17. Fine PVA, Miller ZJ, Mesones I, Irazuzta S, Appel HM, et al. (2006) The growth-defense trade-off and habitat specialization by plants in Amazonian forests. Ecology 87: S150–S162.
18. Kardol P, Cornips NJ, van Kempen MML, Bakx-Schotman JMT, Van der Putten WH (2007) Microbe-mediated plant-soil feedback causes historical contingency effects in plant community assembly. Ecol Monogr 77: 147–162.
19. Petermann JS, Fergus AJF, Turnbull LA, Schmid B (2008) Janzen-Connell effects are widespread and strong enough to maintain diversity in grasslands. Ecology 89: 2399–2406.
20. Shannon S, Flory SL, Reynolds H (2012) Competitive context alters plant-soil feedback in an experimental woodland community. Oecologia 169: 235–243.
21. Callaway RM, Thelen GC, Barth S, Ramsey PW, Gannon JE (2004) Soil fungi alter interactions between the invader *Centaurea maculosa* and North American natives. Ecology 85: 1062–1071.
22. Callaway RM, Thelen GC, Rodriguez A, Holben WE (2004) Soil biota and exotic plant invasion. Nature 427: 731–733.
23. Casper BB, Castelli JP (2007) Evaluating plant-soil feedback together with competition in a serpentine grassland. Ecol Lett 10: 394–400.
24. Snaydon RW (1991) Replacement or additive designs for competition studies. J Appl Ecol 28: 930–946.
25. Callaway RM, Mahall BE, Wicks C, Pankey J, Zabinski C (2003) Soil fungi and the effects of an invasive form on grasses: neighbor identity matters. Ecology 84: 129–135.

26. Brinkman EP, van der Putten WH, Bakker E-J, Verhoeven KJF (2010) Plant-soil feedback: experimental approaches, statistical analysis and ecological interpretations. J Ecol 98: 1063–1073.
27. Hoagland DR, Arnon DI (1950) The water-culture method for growing plants without soil. Calif Agricultural Experiment Stn Circular 347:1–32.
28. Manning P, Morrison SA, Bonkowski M, Bardgett RD (2008) Nitrogen enrichment modifies plant community structure via changes to plant-soil feedback. Oecologia 157: 661–673.
29. Lundkvist A, Salomonsson L, Karlsson L, Gustavssone AMD (2008) Effect of organic farming on weed flora composition in a long term perspective. Eur J Agron 28: 570–578.
30. R development Core Team (2008) R: a Language and Environment for Statistical Computing. R website. Available: http://www.R-project.org Accessed 2012 Oct 8.
31. Hendriks M, Mommer L, de Caluwe H, Smit-Tiekstra AE, van der Putten WH, et al. (2013) Independent variation of plant and soil mixtures reveal soil feedback effects on plant community overyielding. J Ecol 101: 287–297.
32. Brandt AJ, Seabloom EW, Hosseini PR (2009) Phylogeny and provenance affect plant-soil feedbacks in invaded California grasslands. Ecology 90: 1063–1072.
33. Jiang LL, Han X, Zhang GM, Kardol P (2010) The role of plant-soil feedbacks and land-use legacies in restoration of a temperate steppe in northern China. Ecol Res 25: 1101–1111.
34. Hol WHG, Meyer KM, Van der Putten (2011) Idiosyncrasy in ecology – what's in a word? Front Ecol Environ 9: 431–432.
35. Otten W, Gilligan CA (2006) Soil structure and soil-borne diseases: using epidemiological concepts to scale from frungal spread to plant epidemics. Eur J Soil Sci 57: 26–37.
36. Aguilera AG (2011) The influence of soil community density on plant-soil feedbacks: an important unknown in plant invasion. Ecol Model 222: 3413–3420.

23

Contributions of Understory and/or Overstory Vegetations to Soil Microbial PLFA and Nematode Diversities in Eucalyptus Monocultures

Jie Zhao[1,3⁹], Songze Wan[1⁹], Chenlu Zhang[2⁹], Zhanfeng Liu[1], Lixia Zhou[1], Shenglei Fu[1]*

1 Key Laboratory of Vegetation Restoration and Management of Degraded Ecosystems, South China Botanical Garden, Chinese Academy of Sciences, Guangzhou, China, 2 State Key Laboratory of Cotton Biology, Key Laboratory of Plant Stress Biology, College of Life Sciences, Henan University, Kaifeng, China, 3 Key Laboratory of Agro-Ecological Processes in Subtropical Region, Institute of Subtropical Agriculture, Chinese Academy of Sciences, Changsha, China

Abstract

Ecological interactions between aboveground and belowground biodiversity have received many attentions in the recent decades. Although soil biodiversity declined with the decrease of plant diversity, many previous studies found plant species identities were more important than plant diversity in controlling soil biodiversity. This study focused on the responses of soil biodiversity to the altering of plant functional groups, namely overstory and understory vegetations, rather than plant diversity gradient. We conducted an experiment by removing overstory and/or understory vegetation to compare their effects on soil microbial phospholipid fatty acid (PLFA) and nematode diversities in eucalyptus monocultures. Our results indicated that both overstory and understory vegetations could affect soil microbial PLFA and nematode diversities, which manifested as the decrease in Shannon–Wiener diversity index (H') and Pielou evenness index (J) and the increase in Simpson dominance index (λ) after vegetation removal. Soil microclimate change explained part of variance of soil biodiversity indices. Both overstory and understory vegetations positively correlated with soil microbial PLFA and nematode diversities. In addition, the alteration of soil biodiversity might be due to a mixing effect of bottom-up control and soil microclimate change after vegetation removal in the studied plantations. Given the studied ecosystem is common in humid subtropical and tropical region of the world, our findings might have great potential to extrapolate to large scales and could be conducive to ecosystem management and service.

Editor: Dafeng Hui, Tennessee State University, United States of America

Funding: This study was financially supported by the Knowledge Innovation Program of the Chinese Academy of Sciences (KSCZ2-EW-Z-6) and Natural Science Fund of China (31210103920). The funders had no role in study design, data collection and analysis, decision to publish, or preparation of the manuscript.

Competing Interests: The authors have declared that no competing interests exist.

* E-mail: sfu@scbg.ac.cn

⁹ These authors contributed equally to this work.

Introduction

Biodiversity plays a crucial role in ecosystem function and processes not only because of its importance in production of food, fiber and fuel, but also because of its roles in ground water replenishment, flooding controls, prevention of soil erosion and ecological invasion, influences on restoration succession, and so on [1,2]. Although the relationship between biodiversity and ecosystem functions and processes has been the focuses of quite a few studies over the last two decades, much controversy still exists [3]. The impact of plant diversity on ecosystem functions and processes has been explored mainly in grasslands (e.g. [4,5]); and primary productivity is frequently used to assess the ecosystem stability [4]. In general, ecosystems with high biodiversity level are more stable, sustainable, reliable and predictable [1,4,5,6]. By contrast, several studies reported no significant effect of plant diversity on plant productivity [7,8].

Soil contains the most diverse communities of organisms and soil biodiversity provides a number of ecosystem services [2,9]. Ecological interactions between aboveground and belowground biodiversity have received many attentions in the recent decades [10]. In one hand, plants play a key role in determining soil biota because plants are the main food sources (i.e. leaf litter, dead roots and root exudates) to soil biota, which may exert a bottom-up control on soil organisms. In another hand, soil biota may affect aboveground plant diversity and productivity through influencing nutrient cycling and distribution, and damaging plants by parasites, pathogens and herbivores [11,12].

Plant functional groups (PFG) or plant functional types (PFT) are assemblages of species with similar roles in ecosystem processes by responding in similar ways to multiple environmental factors [13]. They are necessary devices for reducing the complexity of plant community and often uncharted characteristics of species diversity in function and structure when attempting to project the nature and function of species assemblages into future environments [14]. However, due to the diverse aims of different researches, diverse plant functional groups are chosen to simplify the plant community, i.e. legumes, forbs/herbs, C3/C4-plants, overstory vegetation, and understory vegetation.

Vegetation removal has been frequently used to explore how specific plant species or functional groups affect ecosystem processes and functions. Overstory removal (timber harvest) has been a common forest management practice which also provides opportunities for soil ecologists to study relationships between overstory tree species and soil biota. In general, overstory removal suppresses soil biota such as bacteria, fungus, nematodes, and microarthropods in the long run [15,16,17]. However, there are studies which reported overstory removal had slight or non-significant impacts on soil biota, especially in the short term [18,19]. To date, less attention has been paid to understory vegetation in terms of their impacts on soil biota, probably because the biomass and economic value of the understory are less than overstory tree species. Understory vegetation is an important component in many forest ecosystems worldwide [20,21] and drives ecological processes and functions, such as biodiversity, stand productivity, tree-seedling regeneration, forest succession, litter decomposition, soil nutrient cycling and soil water conservation [19,22,23,24]. However, how understory vegetation affects the soil biodiversity has never been studied before.

In the present study, we monitored the soil microbial phospholipid fatty acid (PLFA) and nematode diversity after overstory and/or understory removal in eucalyptus monocultures in South China. The eucalyptus species is *Eucalyptus urophylla*, which is a fast-grow species and has been widely planted for reforestation in South China [19]. The understory vegetation was dominated by *Dicranopteris dichotoma*, a common fern species in humid subtropical and tropical region [20]. Other common understory species included *Miscanthus sinensis* and *Rhodomyrtus tomentosa*. Thus, the diversity of overstory and understory vegetations is not high in the studied ecosystems. Therefore, we did not calculate the plant diversity. In this experiment, overstory and understory vegetations were removed separately or in combination to assess their impacts on soil microbial PLFA profile and nematode community. Both *E. urophylla* and *D. dichotoma* are high in biomass and primary productivity [20,25]. Previous study, conducted at the same site as this study, reported that understory removal showed stronger effects on soil microbial community structure, and abundance of nematode and microarthropod than overstory removal [20]. Therefore, the hypothesis of this study was that the effects of understory removal on soil microbial and nematode diversity were greater than overstory removal. The primary goal of this study was to evaluate the contributions of the understory and/or overstory vegetation to soil microbial and nematode diversity. Our analyses also intend to address the question of which factors (i.e. resource and soil microclimate) affected soil biodiversity after altering of aboveground vegetations.

Materials and Methods

Site Description

This study was conducted at the Heshan Hilly Land Interdisciplinary Experimental Station (112°50′E, 22°34′N), property of the Chinese Academy of Sciences (CAS) in Guangdong Province of China. The climate is subtropical monsoon with a distinct wet and dry Season. The mean annual temperature and precipitation are 21.7°C and 1,700 mm, respectively. The soil is an acrisol [26]. Vegetation at the experimental site consisted of three (replications) 4-year-old *E. urophylla* plantations. The seedlings were planted with a spacing of 3×2 m.

Experimental Design

Our experiment was conducted in three replicated plots of 4-year-old *E. urophylla* monoculture plantations. In September 2009, we established an experiment with a split-plot design. Within each plantation, main plots (15×15-m) were overstory trees treatment (present or removed) and subplots were understory treatment (present or removed). In total, there were four treatments in each plantation, namely, control with intact overstory and understory (CK), understory removal (UR), overstory tree removal (TR) and all-plant removal (PR). The trees were cut with an electrical saw leaving 5 cm stumps aboveground. All woody debris was removed from the plots immediately after vegetation was removed. The shoots of all understory plants were removed manually with the aid of a machete. Trenches were created around each main-plot and subplot to prevent the interactions of the roots from other subplots and to prevent the exchange of nutrients among subplots. Germination of the stumps (TR plots) and understory vegetation (UR and PR subplots) were monthly removed.

Soil Sampling and Analysis

Soils were sampled on 20 September, 1 October and 6 November 2009, 11 March, 15 June and 14 October 2010, corresponding to 10 days before and 7, 37, 160, 256, 376 days after treatments were implemented. Soil cores (2.5 cm in diameter, 5 cm in length) were taken at 0–5 cm and 5–10 cm depths from eight locations selected randomly in each subplot within each plantation. Eight cores of the same depth from each subplot were combined to form one composite sample; there were three replicate samples for each treatment. The surface litter was removed carefully when soil sample was taken.

Soil water content (SWC %, g of water per 100 g dry soil) was measured by oven-drying for 48 h at 105°C. Soil temperature was recorded every 2 h with the DS1922L temperature logger iButtons (Dallas Semiconductor Corp., Dallas, TX) from September 2009 to October 2010. Soil microbial PLFAs were analyzed according to [27]. Briefly, the soil was extracted in a one-phase extraction mixture containing chloroform:methanol:phosphate buffer (1:2:0.8 v/v/v), with the amount of phosphate buffer corrected to account for existing soil water content mixture of chloroform:-methanol:citrate buffer (1:2:0.8 v/v/v). After extraction the phospholipids were separated from neutral lipids and glycolipids on solid phase extraction columns, 0.50 g Si (Supelco, Inc., Bellefonte, Penn). Polar lipids were eluted and were then subjected to mild alkaline methanolysis. Resulting fatty acid methyl esters (FAMEs) were separated, quantified, and identified using capillary gas chromatography (GC). Qualitative and quantitative fatty acid analyses were performed using an Agilent 6890 gas chromatograph (Agilent Technologies, Palo Alto, CA, USA) and the MIDI Sherlock Microbial Identification System (MIDI Inc., Newark, DE, USA). Concentrations of each PLFA were calculated relative to 19:0 internal standard concentrations. The following PLFAs were considered to be of soil microbial origin: i15:0, a15:0, 15:0, i16:0, 16:1ω7, i17:0, a17:0, 17:0, cy17:0, cy19:0 18:2ω6,9, 16:1ω5, 18:1ω7, and 18:1ω9 [28,29,30].

Nematodes were extracted from 50 g of fresh soil using the Baermann funnel method. After fixation in 4% formalin solution, nematodes were counted with a differential interference contrast (DIC) microscope (ECLIPSE 80 i, Nikon), and the first 100 individuals encountered were identified to genus level. All nematodes were identified to genus level when the sample contained fewer than 100 individuals.

Statistical Analysis

Shannon–Wiener diversity index, Pielou evenness index, Margalef richness index, and Simpson dominance index were employed to assess the soil microbial PLFA and nematode diversity.

$$\text{Shannon}-\text{Wiener diversity index, } H' = \sum_{i=1}^{S} Pi * \ln Pi \quad (1)$$

$$\text{Pielou evenness index, } J = \frac{H'}{H' \max} \quad (2)$$

$$H' \max = \ln S \quad (3)$$

$$\text{Margalef richness index, } SR = \frac{S-1}{\ln N} \quad (4)$$

$$\text{Simpson dominance index, } \lambda = \sum Pi^2 \quad (5)$$

Where '*Pi*' is the proportion of the individuals of "*i*th" group in the community; '*S*' is the total number of microbial PLFAs or nematode genera in the community; and '*N*' is the amount of microbial PLFAs or number of total nematodes in the community.

Repeated-measure ANOVA was employed to determine the time effect and treatment effect through the whole experimental period with TR and UR as between-subjects factors. To address the relationship between soil microclimates (i.e. soil water content and soil temperature) and soil biodiversity, linear and quadratic regression analysis were performed. Repeated-measure ANOVA and regression model were performed using SPSS software (SPSS Inc., Chicago, IL). Statistical significance was determined at $p<0.05$.

Results

Microbial PLFA Diversity

There was a significant time effect on all the microbial variables, namely, Shannon–Wiener diversity index (H'), Pielou evenness index (J), Margalef richness index (SR), and Simpson dominance index (λ) ($p<0.05$) (Fig. 1). There were trends that understory removal and overstory removal decreased soil microbial PLFA diversity at 0–5 cm soil depth by 0.024 ($p=0.08$) and 0.026 ($p=0.055$) units, respectively (Fig. 1a). Overstoty removal significantly reduced soil microbial PLFA diversity at 5–10 cm soil depth by 0.048 units ($p=0.034$) (Fig. 1b). Repeated measure ANOVA showed that understory removal and overstory removal apparently increased dominance index at 0–5 cm soil depth by 0.007 ($p=0.038$) and 0.008 ($p=0.035$) units, respectively (Fig. 1g); and overstory removal significantly increased dominance index at 5–10 cm soil depth by 0.011 units ($p=0.032$) (Fig. 1h). There was no remarkable effect of vegetation removal on evenness index and richness index of microbial PLFA profile at both 0–5 cm and 5–10 cm soil depths during the study (Fig. 1c, d, e, and f). No interaction effect of understory removal and overstory removal on soil microbial variables was found during the study.

Nematode Diversity

There was a significant time effect on all the nematode variables, namely, Shannon–Wiener diversity index (H'), Pielou evenness index (J), Margalef richness index (SR), and Simpson dominance index (λ) ($p<0.05$) (Fig. 2). There were trends that understory removal decreased soil nematode diversity index by 0.086 units ($p=0.096$) (Fig. 2a) and evenness index by 0.022 units ($p=0.058$) (Fig. 2c) at 0–5 cm soil depth, respectively. Overstory removal tended to increase the soil nematode evenness at 0–5 cm soil depth by 0.022 units ($p=0.062$) (Fig. 2c), and significantly increased the soil nematode evenness at 5–10 cm soil depth by 0.036 units ($p=0.004$) (Fig. 2d). There was a trend that interaction of understory removal and overstory removal tended to affect soil nematode diversity index at 0–5 cm soil depth during the study ($p=0.072$) (Fig. 2a). Neither soil nematode Margalef richness index (SR) nor Simpson dominance index (λ) were significantly affected by understory removal or overstory removal at either 0–5 or 5–10 cm depth (Fig. 2e, f, g, and h).

Correlations between Soil Biodiversity and Soil Microclimates

Regression analysis showed that soil water content and soil biodiversity were not well fitting in either linear or quadratic model (Appendix S1). Coefficients of determination (R^2) of both linear and quadratic regressions between soil water content and biodiversity indices (H', J, SR, and λ) of soil microbial PLFAs and nematodes were very low (Appendix S1). Shannon–Wiener diversity index (H') ($R^2=0.15$, $p=0.006$) and Margalef richness index (SR) ($R^2=0.02$, $p=0.001$) of microbial PLFA profile were quadratic correlated with soil temperature (Fig. 3a and c). Pielou evenness index (J) of nematode community was quadratic correlated with soil temperature ($R^2=0.111$, $p=0.024$) (Fig. 4b).

Discussion

Contributions of Understory or Overstory Vegetations to Soil Biodiversity

Apparently, all plant removal showed strongest suppression on soil microbial and nematode diversity in the present study. In addition, both understory removal and overstory removal reduced soil microbial PLFA and nematode diversities, which was inconsistent with our hypothesis that the effects of understory removal on soil biodiversity might be more powerful than overstory removal. Consistent with our finding, many studies reported that all plant removal had significant negative influence on soil biota in forest and grassland ecosystem [17,31,32]. Our previous study at the same site reported that understory removal had stronger negative effects on abundances and compositions of soil biota (microbes and nematodes) than overstory removal [20]. That was why our hypotheses stated as understory had greater contributions on soil biodiversity than overstory. Contrary to our hypotheses, our analyses clearly showed that the overstory vegetation impacted soil biodiversity too. In addition, overstory removal rather than understory removal significantly reduced amount of total PLFAs in 5–10 cm soil depth during the study (Appendix S2). There are many studies focused on the effects of overstory removal on soil biota because canopy tree harvest is a common forest management practice. For example, fifteen years after harvesting, diversity and structure of soil bacterial and fungal communities remained apparently altered by harvesting disturbances; and three fungi taxa (i.e. ectomycorrhizal fungi, saprobic taxa, and ascomycetes) and actinomycetes were most sensitive to harvesting disturbance in sub-boreal spruce forest in BC, Canada [15]. Both stem-only and whole-tree harvesting decreased microarthropod abundance in a temperate mixed conifer–hardwood stand [33] and in sub-boreal spruce stand [34]. However, these studies usually focused on size of populations and structure of communities of soil biota; there is no mention of how soil biodiversity responses to overstory removal in these studies. Furthermore, most of the studies did not clearly state

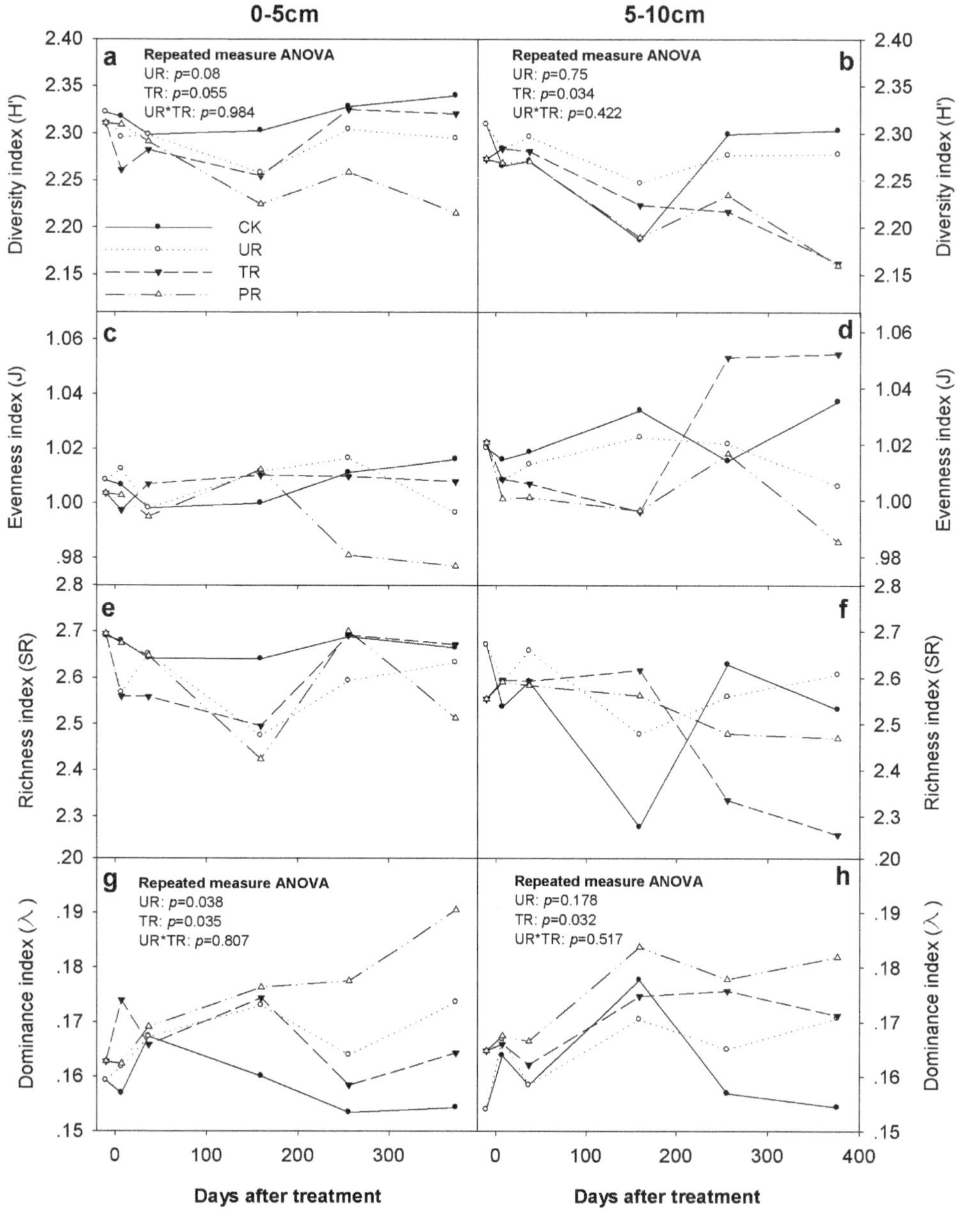

Figure 1. Microbial PLFA diversity as affected by vegetation removal during the study. Shannon–Wiener diversity index (H'), Pielou evenness index (J), Margalef richness index (SR) and Simpson dominance index (λ) under control (CK), understory removal (UR), tree removal (TR), and all-plant removal (PR) in each sampling event at 0–5 cm (a, c, e, g) and 5–10 cm (b, d, f, h) soil depths.

whether understory was remained un-harvesting or not when overstory was harvested. In addition, understory vegetation is important ecological component that plays important roles in driving forest ecosystem functions and processes [20,21]; however, it is usually overlooked in the previous studies. Several studies reported the relationship between understory vegetation and aboveground biodiversity (e.g., birds, reptiles, arthropods) [35,36,37]. Interactions between understory vegetation and soil biodiversity have not been documented yet. This is the first study reported the linkages between understory vegetation and soil microbial PLFA and nematode diversities.

Shannon–Wiener diversity index (H') gives weight to rarer taxa, whereas Simpson dominance index (λ) gives weight to more abundant taxa [38,39]. The suppressive effect of overstory removal on H' of soil microbial PLFA profile might indicate the linkage between overstory vegetation and soil rare microbial taxa. The promotive effect of overstory removal on λ of soil microbial PLFA profile might indicate the linkage between overstory vegetation and soil abundant microbial taxa. Therefore, overstory vegetation might link to both the rare and abundant taxa of soil microbial organisms. Although no significant effect of overstory removal on H' and λ of nematode community, overstory removal increased the Pielou evenness index (J) of nematode community, which implied that the abundance of nematode taxa did not differ from each other as much as their abundance before overstory removal. The likely reason might be that the rare species disappeared and were not collected after overstory removal, which eliminated the weight of the rare taxa to Pielou evenness

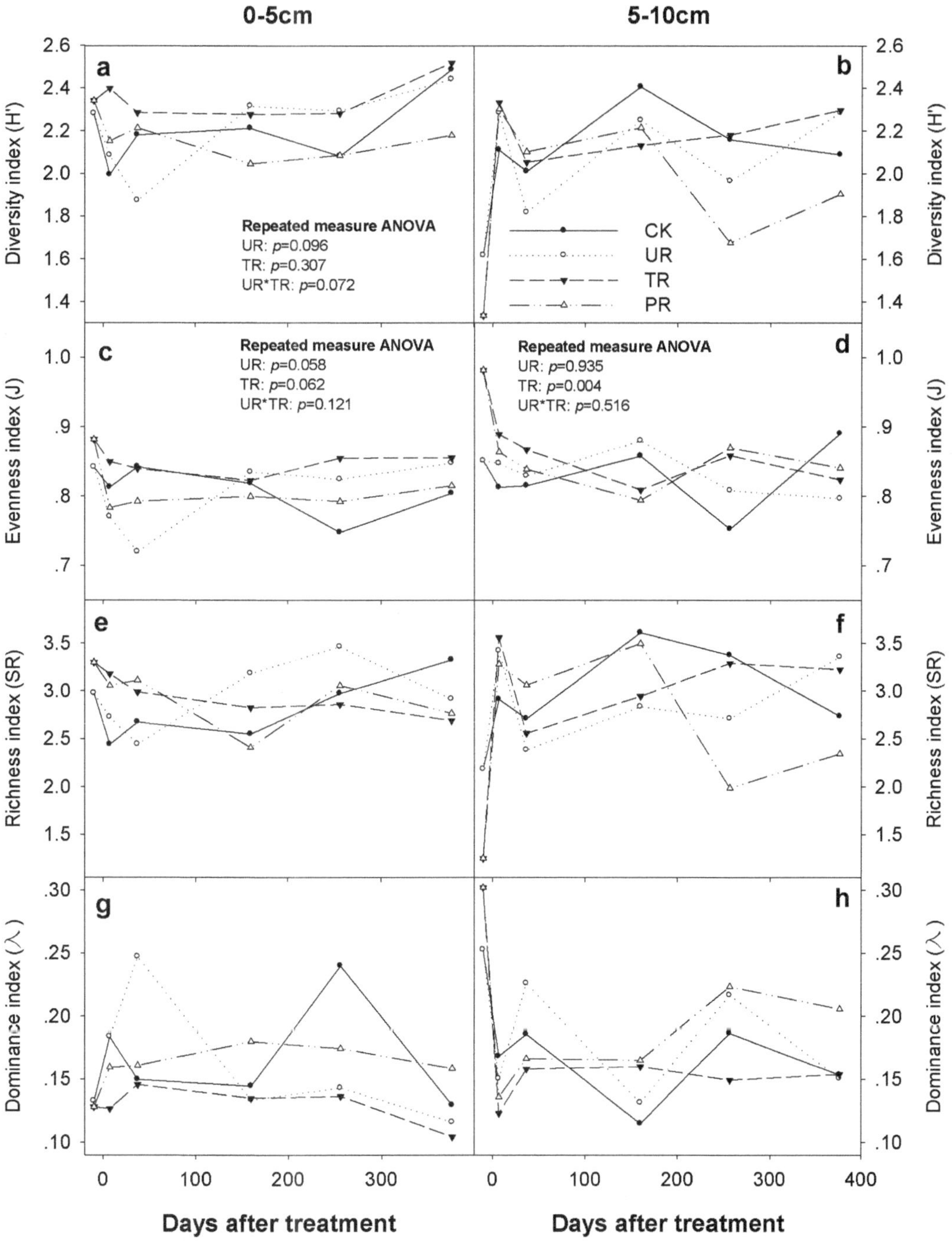

Figure 2. Nematode diversity as affected by vegetation removal during the study. Shannon-Wiener diversity index (*H'*), Pielou evenness index (J), Margalef richness index (SR) and Simpson dominance index (λ) under control (CK), understory removal (UR), tree removal (TR), and all-plant removal (PR) in each sampling event at 0–5 cm (a, c, e, g) and 5–10 cm (b, d, f, h) soil depths.

index. In the similar way, understory vegetation linked to both abundant and rare taxa at 0–5 cm soil depth.

Factor(s) that Affect Soil Biodiversity

Plant diversity may influence soil biodiversity via two main ways. High plant diversity may (1) enhance net primary productivity and (2) lead to high food resources (e.g. litter and root exudates) diversity, and subsequently supports high soil biodiversity [40]. However, many studies found plant species identities were more important than plant diversity in controlling soil biodiversity [41,42,43]; and sometimes the effect of plant identities can overrule the effect of plant diversity [41,44,45].

The bottom-up control might explain the alterations of soil biodiversity after vegetation removal [19,20,32]. *E. urophylla* is the only overstory species and *D. dichotoma* is the dominant understory species in the studied plantations. The productivity of both species are much high. The annual carbon increment of *E. urophylla* can reach 1960 gC m^{-2} yr^{-1} in the developmental stage in the studied

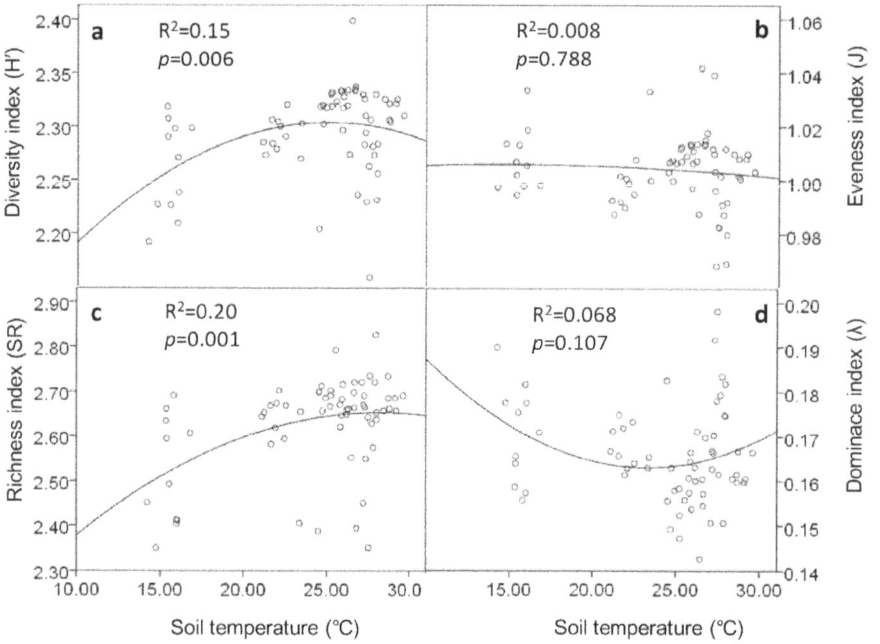

Figure 3. Relationships between indices of microbial PLFAs and soil temperature. Bivariate plots of the microbial PLFA diversity (a), evenness (b), richness (c) and dominance (d) indices versus soil temperature. Points represent results from all six sampling events at 0–5 cm soil depth. Soil temperatures used here were the mean daily temperatures of the six days that soil samplings were carried out.

area [25], while the biomass of *D. dichotoma* may contribute to 22%~48% of total live biomass in *E. urophylla* plantations in the studied area (unpublished data). Both *E. urophylla* and *D. dichotoma* removal resulted in the reduction of carbon input to soil food web [19], which in turn may change soil biodiversity. The bottom-up control opinion was also supported by the differential responses of soil biodiversity at different depths to overstory and understory vegetation. Overstory removal decreased soil biodiversity at both 0–5 cm and 5–10 cm soil depths while understory removal mainly suppressed soil biodiversity at 0–5 cm soil layer. The reason might be that the overstory *E. urophylla* is a deep-rooted species and the dominant understory *D. dichotoma* is a shallow-rooted fern species

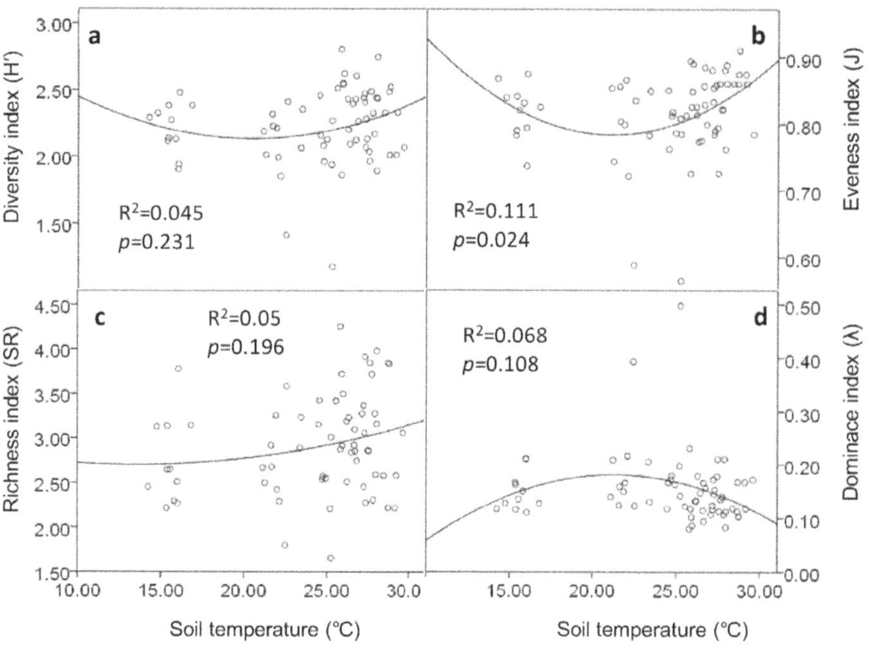

Figure 4. Relationships between indices of nematode diversity and soil temperature. Bivariate plots of the soil nematode diversity (a), evenness (b), richness (c) and dominance (d) indices versus soil temperature. Points represent results from all six sampling events at 0–5 cm soil depth. Soil temperatures used here were the mean daily temperatures of the six days that soil samplings were carried out.

with fibrous root on stolons. The deep-rooted species removal might reduce the availability of soil carbon and nutrients for soil biota at deeper soil depth. Previous studies reported that vegetation removal reduced fine root biomass in eucalyptus plantations in the studied area [19,25]. Unfortunately, none of these studies differentiated either overstory or understory roots from the total plant roots and they did not report the composition of root at different soil depths.

In the present study, soil microclimate (especially soil temperature) explained a certain proportion of variability of soil biodiversity indices. Therefore, soil microclimate change that induced by vegetation removal probably contributed to the alteration of soil biodiversity. Supporting this hypothesis, growth temperature affected the cellular fatty acids composition of sulphate-reducing bacteria that were of marine origin [46]. In addition, our own data revealed that vegetation removal increased soil temperature and decreased soil water content in the later stage of the study, and the effect of understory removal on soil microclimate was greater than overstory removal.

The negative effects of vegetation removal on soil biodiversity also gradually increased over time, which indicated that alterations in bottom-up control and soil microclimate induced by vegetation removal were time-dependent. Firstly, vegetation removal significantly reduced the food resource for soil biota. However, the decomposition of the remained soil residuals (e.g., roots and litter) is time consuming. Therefore, the soil biodiversity declined with the exhaustion of soil residuals. Secondly, the soil microclimate changes that induced by vegetation removal were from March to October (mainly in wet season) in 2010, which was overlapped with the significant declines of soil biodiversity in this study. Thus, the seasonal variation of soil microclimate might be another reason for the time-dependent response of soil biodiversity to vegetation removal.

Conclusions

Here we reported the first study designed to explore the relationship between overstory and understory vegetations and soil biodiversity. Our results showed that both overstory and understory vegetations correlated with soil biodiversity of Eucalyptus plantations in southern China. Interaction between overstory vegetation and soil biodiversity might be due to its high-biomass property that exerts a bottom-up control on soil biodiversity in the studied plantations. Interaction between understory vegetation and soil biodiversity might due to the mixing effect of vegetation identity and soil microclimate change after understory removal that drives soil biodiversity in the studied plantations. Eucalyptus has been extensively planted in many tropical and subtropical regions [25] and *Dicranopteris* is a common fern distributed through humid subtropical and tropical regions of the world [20]. Therefore, our findings might have great potential to extrapolate to large scales, especially to the humid subtropical and tropical region of the world.

Acknowledgments

This study was conducted at Heshan Hilly Land Interdisciplinary Experimental Station, CAS and Heshan National Field Research Station of Forest Ecosystem. We thank Dejun Li for refine the language of the manuscript. We also thank the editor Dafeng Hui and two anonymous reviewers for their comments.

Author Contributions

Conceived and designed the experiments: SF. Performed the experiments: JZ SW CZ ZL LZ. Analyzed the data: JZ SW CZ. Contributed reagents/materials/analysis tools: SF. Wrote the paper: JZ SF.

References

1. Naeem S, Li S (1997) Biodiversity enhances ecosystem reliability. Nature 390: 507–509.
2. Bardgett RD, Usher MB, Hopkins DW (2005) Biological diversity and function in soils. Cambridge, UK: Cambridge University Press.
3. Loreau M, Naeem S, Inchausti P, Bengtsson J, Grime JP, et al. (2001) Biodiversity and ecosystem functioning: current knowledge and future challenges. Science 294: 804–808.
4. Tilman D, Reich PB, Knops J, Wedin D, Mielke T, et al. (2001) Diversity and productivity in a long-term grassland experiment. Science 294: 843–845.
5. Hector A, Schmid B, Beierkuhnlein C, Caldeira MC, Diemer M, et al. (1999) Plant diversity and productivity experiments in European grasslands. Science 286: 1123–1127.
6. McGrady-Steed J, Harris PM, Morin PJ (1997) Biodiversity regulates ecosystem predictability. Nature 390: 162–165.
7. Huston MA (1997) Hidden treatments in ecological experiments: re-evaluating the ecosystem function of biodiversity. Oecologia 110: 449–460.
8. Huston MA, Aarssen LW, Austin MP, Cade BS, Fridley JD, et al. (2000) No consistent effect of plant diversity on productivity. Science 289: 1255.
9. Brussaard L (1997) Biodiversity and ecosystem functioning in soil. Ambio 26: 563–570.
10. Wardle DA, Bardgett RD, Klironomos JN, Setala H, van der Putten WH, et al. (2004) Ecological linkages between aboveground and belowground biota. Science 304: 1629–1633.
11. van der Heijden M, Klironomos J, Ursic M, Moutoglis P, Streitwolf-Engel R, et al. (1998) Mycorrhizal fungal diversity determines plant biodiversity, ecosystem variability and productivity. Nature 396: 69–72.
12. Van Der Heijden MGA, Bardgett RD, Van Straalen NM (2008) The unseen majority: soil microbes as drivers of plant diversity and productivity in terrestrial ecosystems. Ecology Letters 11: 296–310.
13. Lavorel S, McIntyre S, Landsberg J, Forbes TDA (1997) Plant functional classifications: from general groups to specific groups based on response to disturbance. Trends in Ecology & Evolution 12: 474–478.
14. Woodward FI, Cramer W (1996) Plant functional types and climatic change: Introduction. Journal of Vegetation Science 7: 306–308.
15. Hartmann M, Howes CG, VanInsberghe D, Yu H, Bachar D, et al. (2012) Significant and persistent impact of timber harvesting on soil microbial communities in Northern coniferous forests. The ISME Journal.
16. Seastedt T, Crossley Jr D (1981) Microarthropod response following cable logging and clear-cutting in the southern Appalachians. Ecology: 126–135.
17. Forge T, Simard S (2001) Structure of nematode communities in forest soils of southern British Columbia: relationships to nitrogen mineralization and effects of clearcut harvesting and fertilization. Biology and Fertility of Soils 34: 170–178.
18. Panesar TS, Marshall VG, Barclay HJ (2000) The impact of clearcutting and partial harvesting systems on population dynamics of soil nematodes in coastal Douglas-fir forests. Pedobiologia 44: 641–665.
19. Wu J, Liu Z, Wang X, Sun Y, Zhou L, et al. (2011) Effects of understory removal and tree girdling on soil microbial community composition and litter decomposition in two Eucalyptus plantations in South China. Functional Ecology 25: 921–931.
20. Zhao J, Wan S, Li Z, Shao Y, Xu G, et al. (2012) *Dicranopteris*-dominated understory as major driver of intensive forest ecosystem in humid subtropical and tropical region. Soil Biology and Biochemistry 49: 78–87.
21. Nilsson MC, Wardle DA (2005) Understory vegetation as a forest ecosystem driver: evidence from the northern Swedish boreal forest. Frontiers in Ecology and the Environment 3: 421–428.
22. Barrett J, Cochran P, Busse M (1996) Changes in ponderosa pine site productivity following removal of understory vegetation. Soil Science Society of America Journal 60: 1614–1621.

23. Zhao J, Shao Y, Wang X, Neher DA, Xu G, et al. (2013) Sentinel soil invertebrate taxa as bioindicators for forest management practices. Ecological Indicators 24: 236–239.
24. Zhao J, Wan S, Fu S, Wang X, Wang M, et al. (2013) Effects of understory removal and nitrogen fertilization on soil microbial communities in *Eucalyptus* plantations. Forest Ecology and Management 310: 80–86.
25. Chen D, Zhang C, Wu J, Zhou L, Lin Y, et al. (2011) Subtropical plantations are large carbon sinks: Evidence from two monoculture plantations in South China. Agricultural and Forest Meteorology 151: 1214–1225.
26. FAO (2006) World reference base for soil resources 2006. World Soil Resources Report 103 FAO, Rome.
27. Bossio DA, Scow KM (1998) Impacts of carbon and flooding on soil microbial communities: Phospholipid fatty acid profiles and substrate utilization patterns. Microbial Ecology 35: 265–278.
28. Frostegård A, Bååth E (1996) The use of phospholipid fatty acid analysis to estimate bacterial and fungal biomass in soil. Biology and Fertility of Soils 22: 59–65.
29. Joergensen RG, Wichern F (2008) Quantitative assessment of the fungal contribution to microbial tissue in soil. Soil Biology & Biochemistry 40: 2977–2991.
30. Frostegård Å, Tunlid A, Bååth E (2011) Use and misuse of PLFA measurements in soils. Soil Biology and Biochemistry 43: 1621–1625.
31. Wardle DA, Bonner KI, Barker GM, Yeates GW, Nicholson KS, et al. (1999) Plant removals in perennial grassland: vegetation dynamics, decomposers, soil biodiversity, and ecosystem properties. Ecological Monographs 69: 535–568.
32. Zhao J, Wang X, Shao Y, Xu G, Fu S (2011) Effects of vegetation removal on soil properties and decomposer organisms. Soil Biology and Biochemistry 43: 954–960.
33. Bird G, Chatarpaul L (1988) Effect of forest harvest on decomposition and colonization of maple leaf litter by soil microarthropods. Canadian Journal of Soil Science 68: 29–40.
34. Battigelli JP, Spence JR, Langor DW, Berch SM (2004) Short-term impact of forest soil compaction and organic matter removal on soil mesofauna density and oribatid mite diversity. Canadian Journal of Forest Research 34: 1136–1149.
35. Atwell RC, Schulte LA, Palik BJ (2008) Songbird response to experimental retention harvesting in red pine (Pinus resinosa) forests. Forest Ecology and Management 255: 3621–3631.
36. López G, Moro M (1997) Birds of Aleppo pine plantations in south-east Spain in relation to vegetation composition and structure. Journal of Applied Ecology: 1257–1272.
37. Humphrey J, Hawes C, Peace A, Ferris-Kaan R, Jukes M (1999) Relationships between insect diversity and habitat characteristics in plantation forests. Forest Ecology and Management 113: 11–21.
38. Neher DA, Lee Campbell C (1994) Nematode communities and microbial biomass in soils with annual and perennial crops. Applied Soil Ecology 1: 17–28.
39. Zhao J, Zhang W, Wang K, Song T, Du H (2013) Responses of the soil nematode community to management of hybrid napiergrass: The trade-off between positive and negative effects. Applied Soil Ecology. DOI: 10.1016/j.apsoil.2013.10.011.
40. Wardle DA (2006) The influence of biotic interactions on soil biodiversity. Ecology Letters 9: 870–886.
41. De Deyn GB, Raaijmakers CE, Van Ruijven J, Berendse F, Van Der Putten WH (2004) Plant species identity and diversity effects on different trophic levels of nematodes in the soil food web. Oikos 106: 576–586.
42. Viketoft M, Palmborg C, Sohlenius B, Huss-Danell K, Bengtsson J (2005) Plant species effects on soil nematode communities in experimental grasslands. Applied Soil Ecology 30: 90–103.
43. De Deyn GB, Quirk H, Bardgett RD (2011) Plant species richness, identity and productivity differentially influence key groups of microbes in grassland soils of contrasting fertility. Biology Letters 7: 75–78.
44. Wardle DA, Yeates GW, Williamson W, Bonner KI (2003) The response of a three trophic level soil food web to the identity and diversity of plant species and functional groups. Oikos 102: 45–56.
45. Wardle DA, Yeates GW, Barker GM, Bonner KI (2006) The influence of plant litter diversity on decomposer abundance and diversity. Soil Biology and Biochemistry 38: 1052–1062.
46. Könneke M, Widdel F (2003) Effect of growth temperature on cellular fatty acids in sulphate-reducing bacteria. Environmental Microbiology 5: 1064–1070.

24

Effects of Straw Incorporation on Soil Organic Matter and Soil Water-Stable Aggregates Content in Semiarid Regions of Northwest China

Peng Zhang[1,2], Ting Wei[1,2], Zhikuan Jia[1,2]*, Qingfang Han[1,2], Xiaolong Ren[1,2], Yongping Li[3]

1 The Chinese Institute of Water-Saving Agriculture, Northwest A&F University, Yangling, Shaanxi, China, **2** Key Laboratory of Crop Physi-Ecology and Tillage Science in Northwestern Loess Plateau, Ministry of Agriculture, Northwest A&F University, Yangling, Shaanxi, China, **3** Guyuan Institute of Agricultural Sciences, Guyuan, Ningxia, China

Abstract

The soil degradation caused by conventional tillage in rain-fed areas of northwest China is known to reduce the water-use efficiency and crop yield because of reduced soil porosity and the decreased availability of soil water and nutrients. Thus, we investigated the effects of straw incorporation on soil aggregates with different straw incorporation rates in semiarid areas of southern Ningxia for a three-year period (2008–2010). Four treatments were tested: (i) no straw incorporation (CK); (ii) incorporation of maize straw at a low rate of 4 500 kg ha^{-1} (L); (iii) incorporation of maize straw at a medium rate of 9000 kg ha^{-1} (M); (iv) incorporation of maize straw at a high rate of 13 500 kg ha^{-1} (H). The results in the final year of treatments (2010) showed that the mean soil organic carbon storage of the 0–60 cm soil layers were significantly ($P<0.05$) increased with H, M, and L, by 21.40%, 20.38% and 8.21% compared with CK, respectively. Straw incorporation increased >0.25 mm water-stable macroaggregates level, geometric mean diameter, mean weight diameter and the aggregate stability, which were ranked in order of increasing straw incorporation rates: H/M > L > CK. Straw incorporation significantly ($P<0.05$) reduced the fractal dimension in the 0–40 cm soil layers compared with CK. Our results suggest that straw incorporation is an effective practice for improving the soil aggregate structure and stability.

Editor: Raffaella Balestrini, Institute for Plant Protection (IPP), CNR, Italy

Funding: This work was supported by the China Support Program (2012BAD09B03, 2011AA100504, and 2011BAD29B09) for Dryland Farming in the 12th 5-year plan period, the Youth project of National Natural Science Fund (31201156), and the Basic Science Research Fund in Northwest A&F University (QN2013005). The funders had no role in study design, data collection and analysis, decision to publish, or preparation of the manuscript.

Competing Interests: The authors have declared that no competing interests exist.

* E-mail: jiazhk@126.com

These authors contributed equally to this work.

Introduction

Soil infertility [1], soil erosion, and water deficiency [2] are the major factors that limited crop growth in semiarid areas of northwest China. The rates of crop straw use for fuel and forage have declined significantly since the 1980s and crop straw is increasingly burned after the harvest, which leads to high losses of soil organic substances [3,4], and increased emission of CO_2 that pollute the environment [5]. Furthermore, this practice has led to the degradation of the agricultural ecological environment [6].

The soil organic matter (SOM) content is one of the major factors that affects soil properties and functions including a range of physical characteristics such as the water-holding capacity [7], water infiltration [8], and aggregate stability [9]. SOM is considered to be a major binding agent that stabilizes soil aggregates [10,11]. Soil aggregates are the basic units of the soil structure [12], which are composed of primary particles and binding agents that determine the microbial biomass and mineral nutrient reserves [13–15]. These soil properties are also affected by soil organic matter decomposition processes [16,17].

Many studies have shown that crop straw is rich in organic material and soil nutrients, so it is increasingly considered to be an important natural organic fertilizer [18–20]. Straw can be incorporated to soil either directly or indirectly, which can promote the production of a favorable soil environment. Straw also maintains the physicochemical condition of the soil and improves the overall ecological balance of the crop production system [20,21]. Nelson [22] and Wilhelm *et al.* [23] showed that the incorporation of crop residues into soil significantly prevented soil erosion and enhanced the soil quality. Sonnleitner *et al.* [24] found that straw incorporation also improved the aggregate stability and other soil properties compared with farmyard manure. Mulumba and Lal [25] also reported that the addition of crop residues to cultivated soil had positive effects on the soil porosity, available water content, soil aggregation, and bulk density. Bhagat and Verma [26] showed that the incorporation of crop straws for five years significantly increased the crop yield and improved the soil properties.

The soil improvement effect of straw incorporation has been recognized widely [23–25] but information is still limited on the responses of the SOC and water-stable aggregates under different rates of straw incorporation, particularly in the loessal soil in semiarid areas of northwest China. The theory and technique of straw incorporation in this region have also not been reported. Thus, the present study investigated the effects of different crop straw application rates combined with conventional planting on SOC, the >0.25 mm water-stable macroaggregate rate, and various soil properties in the southern Ningxia region of China.

Materials and Methods

Ethics Statements

The study was carried out on the private land, we rent the farmland from the local farmers, and contracts and deeds are signed. No specific permissions were required in this area to run the experiment as the study sites are farming area without any protection zone, and the farming activities won't hurt the local animals. And we only plant the grain crop in the field, so the field studies did not involve endangered or protected species.

Site description

The experiment was conducted between 2008 and 2010 at the Dryland Agricultural Research Station, Pengyang County, Ningxia, China (106°45'N, 35°79'E and 1800 m a.s.l.). The experimental area was in a hilly and gully region of the Loess Plateau, which was characterized by a semiarid, warm temperate, continental monsoon climate. The average annual rainfall was 435 mm, which fell mainly from June to September. The annual mean evaporation was 1050 mm and the annual temperature average was 8.1°C with a frost-free period of 155 days.

Rainfall during the experimental period was measured using an automatic weather station (WS-STD1, England) at the experimental site. Monthly precipitation distributions during the experimental period are shown in Fig. 1. The total precipitation for 2008, 2009, and 2010 was 390.9, 335.2, and 537 mm, while the precipitation during the maize-growing season was 362, 298.2, and 476.1 mm, respectively.

The soil at the experimental site was a loessal soil with a pH of 8.5. In the 0–40 cm soil layer, the organic matter, total N, P, and K were 8.32 g kg^{-1}, 0.61 g kg^{-1}, 0.58 g kg^{-1}, and 5.4 g kg^{-1}, respectively, while the available N, P, and K were 46.25 mg kg^{-1}, 10.41 mg kg^{-1}, and 104.82 mg kg^{-1}. In 2007, the site was planted with maize prior to the experiment

The experimental field was flat and, according to the FAO/UNESCO Soil Classification [27], the soil was a Calcic Cambisol (sand 14%, silt 26%, and clay 60%) with low fertility. The key physical properties of the soil layers (0–40 cm depth) are shown in Table 1.

Experimental design and field management

The experiment used a randomized block design with three replicates. Each plot was 3 m wide and 6 m long. The experiment included four straw incorporation rate treatments: (i) no straw incorporation (CK); (ii) incorporation of maize straw at a low rate of 4 500 kg ha^{-1} (L); (iii) incorporation of maize straw at a medium rate of 9000 kg ha^{-1} (M); (iv) incorporation of maize straw at a high rate of 13 500 kg ha^{-1} (H).

The maize straws were mixed manually with the top 25 cm of soil in the field. Before mixing with the soil, the maize straws were chopped into 5 cm pieces and then applied to the soil six months before the crop was planted to facilitate decomposition of the straw. The straw was incorporated into the soil layer on 15 October 2007 and after the crop harvests during 2008–2010.

Ten days before sowing, a basis fertilizer containing 102 kg N ha^{-1} and 90 kg P ha^{-1}, was spread evenly over the each plot and plowed into soil layer. Maize (cv. Shendan 16) was sown at a rate of 5.25 seed m^{-2} on 18 April 2008, 15 April 2009 and 20 April 2010 using a holesowing (3 cm in diameter) machine. An additional 102 kg N ha^{-1} was applied as a top dressing in late June. And on 7 October 2008, 5 October 2008, and 10 October 2010. No irrigation was provided during the experimental years. Manual weeding was performed throughout the experiment.

Sampling and measurement

Rainfall data were recorded using a standard weather station located at the experimental site. After the maize harvest in 2008 and 2010, soil samples were collected for the four incorporation treatments. A soil sample was collected from each plot at depths of 0–20 cm, 20–40 cm and 40–60 cm to determine the soil organic matter. A similar soil sample was collected at depths of 0–10 cm, 10–20 cm, 20–30 cm and 30–40 cm to determine the aggregate stability. The soil samples were collected from four points in each plot replicate and mixed to produce a composite sample. Each soil samples was passed through an 8 mm sieve by gently breaking the soil clods, whereas pebbles and stable clods >8 mm were discarded. Soil samples were air-dried for 24 h in the laboratory before analysis.

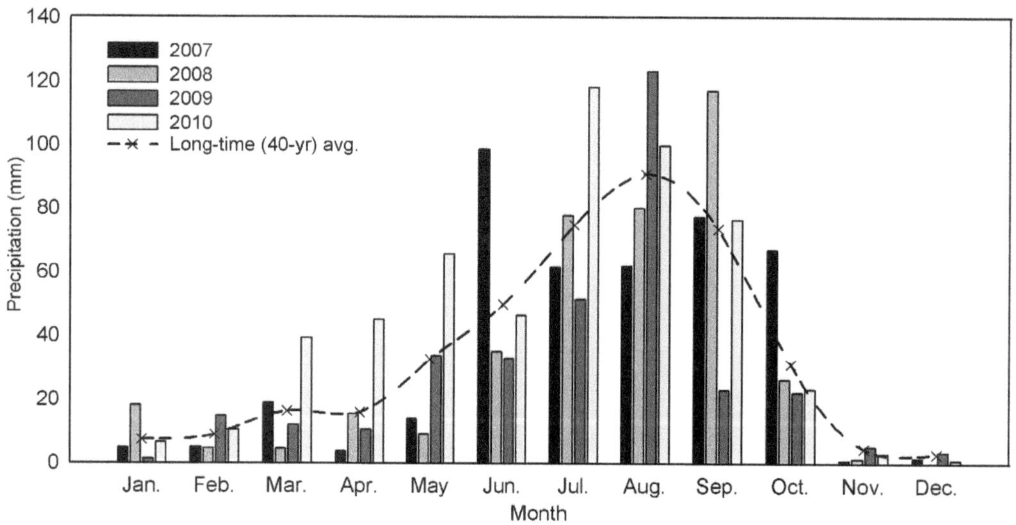

Figure 1. Distribution of mean monthly precipitation at the experimental site during 2007–2010.

Table 1. Physical properties of the tilth soil (0–40 cm depth) in the experimental site.

Depth (cm)	Bulk density (g cm^{-3})	Aggregate size (%)					
		>5 mm	5-2 mm	2-1 mm	1-0.5 mm	0.5-0.25 mm	<0.25 mm
0–10	1.33	0.1	0.27	2.25	4.5	4.47	88.41
10–20	1.33	0.15	0.19	1.38	4.2	4.15	89.93
20–30	1.36	0.1	0.28	1.41	4.02	3.36	90.83
30–40	1.38	0.1	0.22	1.07	3.29	3.8	91.52

Soil organic carbon was determined by the $K_2Cr_2O_7$–H_2SO_4 digestion method, and SOM content was calculated as a portion of SOC which has been described by Wang *et al.* [28].

$$OM_i = O_i \times A \tag{1}$$

$$C_i = \frac{d_i \times \rho_i \times O_i}{10} \tag{2}$$

where OM_i is the soil organic matter content (g kg^{-1}), O_i is the soil organic carbon content (g kg^{-1}), C_i is the soil organic carbon storage (Mg C ha^{-1}), d_i is the soil depth (cm), ρ_i is the soil bulk density (g cm^{-3}), A is Van Bemmelen coefficient (A = 2) [29].

The size distribution of water-stable aggregates was determined by placing a soil sample on a stack of sieves (5,2, 1, 0.5 and 0.25 mm) fitted with a soil aggregate analyzer (Japan, QD24–DIK–2001). The stacked sieves were immersed in water and moved up and down by 3.5 cm at a frequency of 30 cycles $60s^{-1}$ for 15 min. The proportions of aggregates that measured >5, 5–2, 2–0.5, 0.5–0.25 and <0.25 mm were calculated [30].

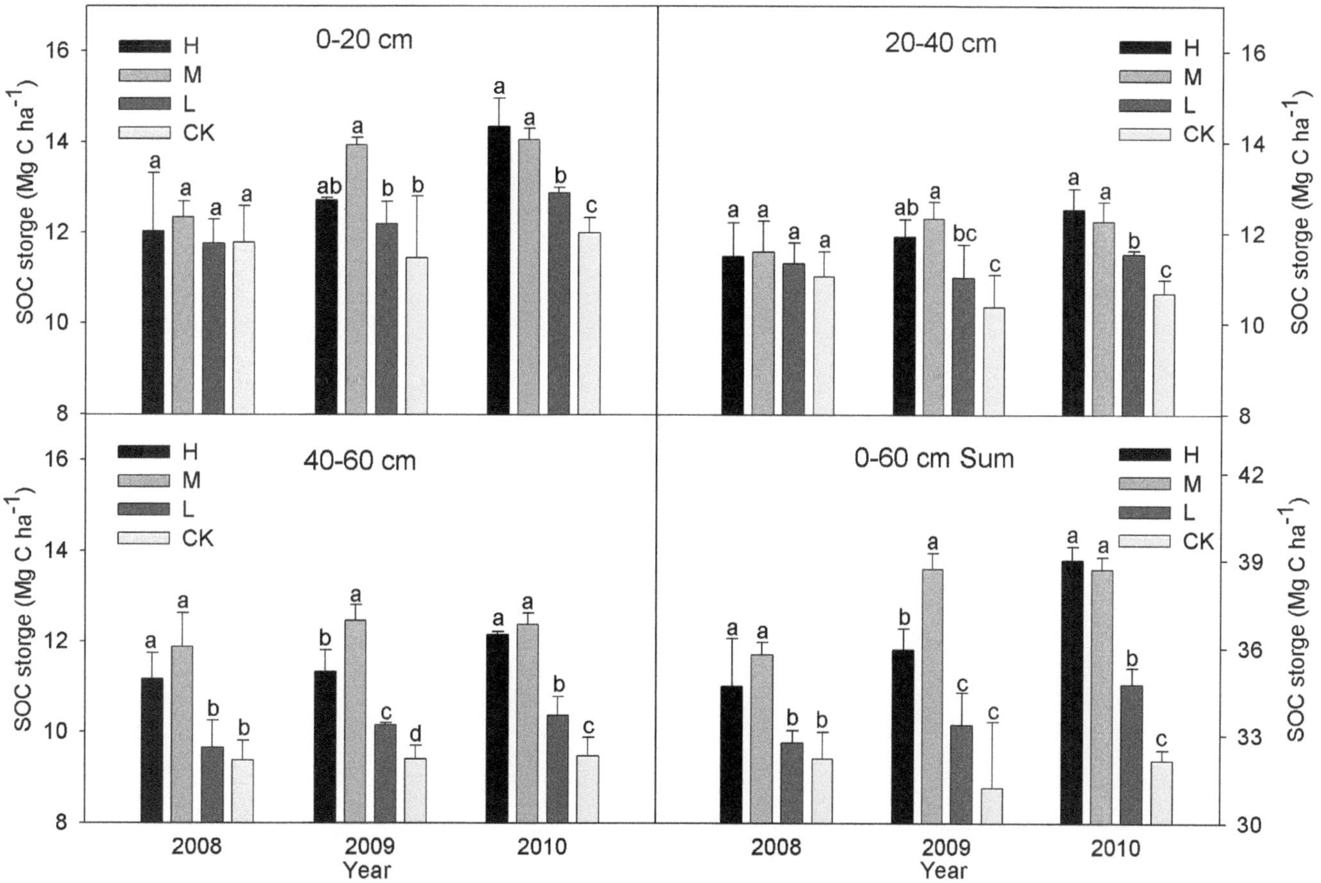

Figure 2. Soil organic carbon storage with different straw incorporation treatments (Mg C ha^{-1}). Note: CK, no straw incorporation; L, incorporation of straw at a low rate of 4 500 kg ha^{-1} maize straw; M, incorporation of straw at a medium rate of 9 000 kg ha^{-1} maize straw; H, incorporation of straw at a high rate of 13 500 kg ha^{-1} maize straw; Sum, the sum value of the 0–60 cm soil layers; Bars with different lower case letters indicate significant differences at $P<0.05$.

Table 2. $WR_{0.25}$ with different straw incorporation treatments (%).

Year	Treatment	Soil depth (cm) 0–10	10–20	20–30	30–40	0–40 AVG
2008	H	$13.37a_1 \pm 0.33^{B2}$	$10.57a \pm 1.28_3{}^{A}$	$7.72ab \pm 0.27^{B}$	$7.33a \pm 0.51^{A}$	$9.75a \pm 0.31^{B}$
	M	$11.79b \pm 1.05^{A}$	$9.28a \pm 0.60^{B}$	$7.97a \pm 0.65^{A}$	$6.62ab \pm 0.85^{B}$	$8.91b \pm 0.26^{B}$
	L	$10.10c \pm 0.46^{A}$	$7.39b \pm 0.58^{A}$	$6.53bc \pm 1.08^{A}$	$6.07ab \pm 0.81^{A}$	$7.52c \pm 0.16^{A}$
	CK	$9.12c \pm 0.70^{A}$	$6.93b \pm 0.78^{A}$	$5.89c \pm 0.57^{A}$	$5.68b \pm 0.63^{A}$	$6.91d \pm 0.44^{A}$
2009	H	$14.64a \pm 0.49^{AB}$	$11.46a \pm 0.17^{A}$	$9.51a \pm 0.26^{AB}$	$7.85a \pm 1.00^{A}$	$10.86a \pm 0.19^{AB}$
	M	$14.93a \pm 0.89^{A}$	$11.22a \pm 0.14^{AB}$	$9.34a \pm 0.32^{A}$	$7.06ab \pm 1.46^{AB}$	$10.64a \pm 0.27^{AB}$
	L	$10.39b \pm 0.26^{A}$	$9.48b \pm 1.49^{A}$	$7.02b \pm 1.18^{A}$	$6.48ab \pm 1.10^{A}$	$8.34b \pm 0.32^{A}$
	CK	$9.75b \pm 0.26^{A}$	$7.18c \pm 0.73^{A}$	$6.40b \pm 1.00^{A}$	$5.65b \pm 0.35^{A}$	$7.25c \pm 0.10^{A}$
2010	H	$15.81a \pm 1.20^{A}$	$12.76a \pm 0.44^{A}$	$11.50a \pm 0.20^{A}$	$8.57a \pm 0.05^{A}$	$12.16a \pm 0.59^{A}$
	M	$16.72a \pm 1.36^{A}$	$12.81a \pm 0.05^{A}$	$11.30a \pm 1.10^{A}$	$8.81a \pm 0.44^{A}$	$12.41a \pm 0.72^{A}$
	L	$11.00b \pm 0.78^{A}$	$9.06b \pm 0.81^{A}$	$8.03b \pm 2.14^{A}$	$6.96b \pm 0.81^{A}$	$8.76b \pm 0.66^{A}$
	CK	$10.23b \pm 0.46^{A}$	$7.79b \pm 0.23^{A}$	$6.07b \pm 0.46^{A}$	$5.71c \pm 0.23^{A}$	$7.45c \pm 0.25^{A}$

Note: CK, no straw incorporation; L, incorporation of straw at a low rate of 4 500 kg ha^{-1} maize straw; M, incorporation of straw at a medium rate of 9 000 kg ha^{-1} maize straw; H, incorporation of straw at a high rate of 13 500 kg ha^{-1} maize straw; AVG, the mean value of the 0-40 cm soil layers.
[1]Values followed by the same lowercase letter in the same line are not significantly different according to Duncan's multiple range test ($P<0.05$) between the four straw incorporation treatments in the same year.
[2]Values followed by the same uppercase letter in the same line are not significantly different according to Duncan's multiple range test ($P<0.05$) between the different years of the same straw incorporation treatment.
[3]Means ± standard deviations.

And the proportions of aggregates were used to calculate the water-stable macroaggregates content with a diameter of >0.25 mm [31], mean weight diameter (MWD) [32,33], geometric mean diameter (GMD) [34], and the soil aggregate stability (WSAR) [35]. These parameters were calculated as follows:

$$macroaggregates(>0.25mm)=\sum_{i=1}^{n} W_i \quad (3)$$

$$MWD=\sum_{i=1}^{n} \overline{X_i} W_i \quad (4)$$

$$GMD=\exp\left(\frac{\sum_{i=1}^{n} W_i \log \overline{X_i}}{\sum_{i=1}^{n} W_i}\right) \quad (5)$$

$$WSAR=WSA/A\times 100\% \quad (6)$$

where $macroaggregates(>0.25mm)$ is the volume of soil particles with a diameter of >0.25 mm, $W_i$is weight of the aggregates in that size range as a fraction of total dry weight of the sample analysed, and n is number of sieves, $\overline{X_i}$ is the mean diameter of aggregates over each sieve size, WSA is the mass of $macroaggregates(>0.25mm)$, and A is the mass of the soil aggregates with a diameter of >0.25 mm.

Fractal dimension D were then obtained to describe the characteristics of soil aggregate size distribution. As suggested by Tyler and Wheatcraft [36] and Zhang *et al.*, [37], the volume of soil with particle diameter $>d_i(d_i>d_{i+1}, i=1,2,\ldots)$ is defined as:

$$V(\delta>d_i)=A\left[1-\left(d_i/k\right)^{3-D}\right)], \quad (7)$$

where δ is yard measure, k and A are the constants representing size and shape, respectively, and D is the fractal dimension. For a given soil, d_i represents the average value of soil particles between d_i and d_{i+1}.Generally, variations of particle density ρ among different soil particles could be ignored. And hence ρ is a constant. Therefore, another expression (Eq. (8)) is derived from Eq. (7):

$$W(\delta>d_i)=V(\delta>d_i)\rho=\rho A\left[1-\left(d_i/k\right)^{3-D}\right)], \quad (8)$$

where $W(\delta>d_i)$ is the cumulative mass of particles with sizes $\delta>d_i$, and W_0 is the total mass of any sizes of soil particles. The fractal equation, reflecting the relationship between the mass distributions of soil particles and average particle diameter, can be obtained as follows:

$$\frac{W(\delta>d_i)}{W_0}=1-\left(\frac{\overline{d_i}}{d_{max}}\right)^{3-D} \quad or \quad \frac{W(\delta>d_i)}{W_0}=\left(\frac{\overline{d_i}}{d_{max}}\right)^{3-D} \quad (9)$$

Then after regression analysis between $\log(\overline{d_i}/d_{max})$ and $\log((W(\delta<\overline{d_i}))/W_0)$, the fractal dimension D can be calculated.

Statistical analysis

Statistical analysis was carried out by using the SPSS 13.0 (Statistical Package for the Social Sciences) package. The effects of treatments on the measured parameters were evaluated using a one-way ANOVA. Duncan's new multiple range test was used to calculate the least significant difference (LSD) between means

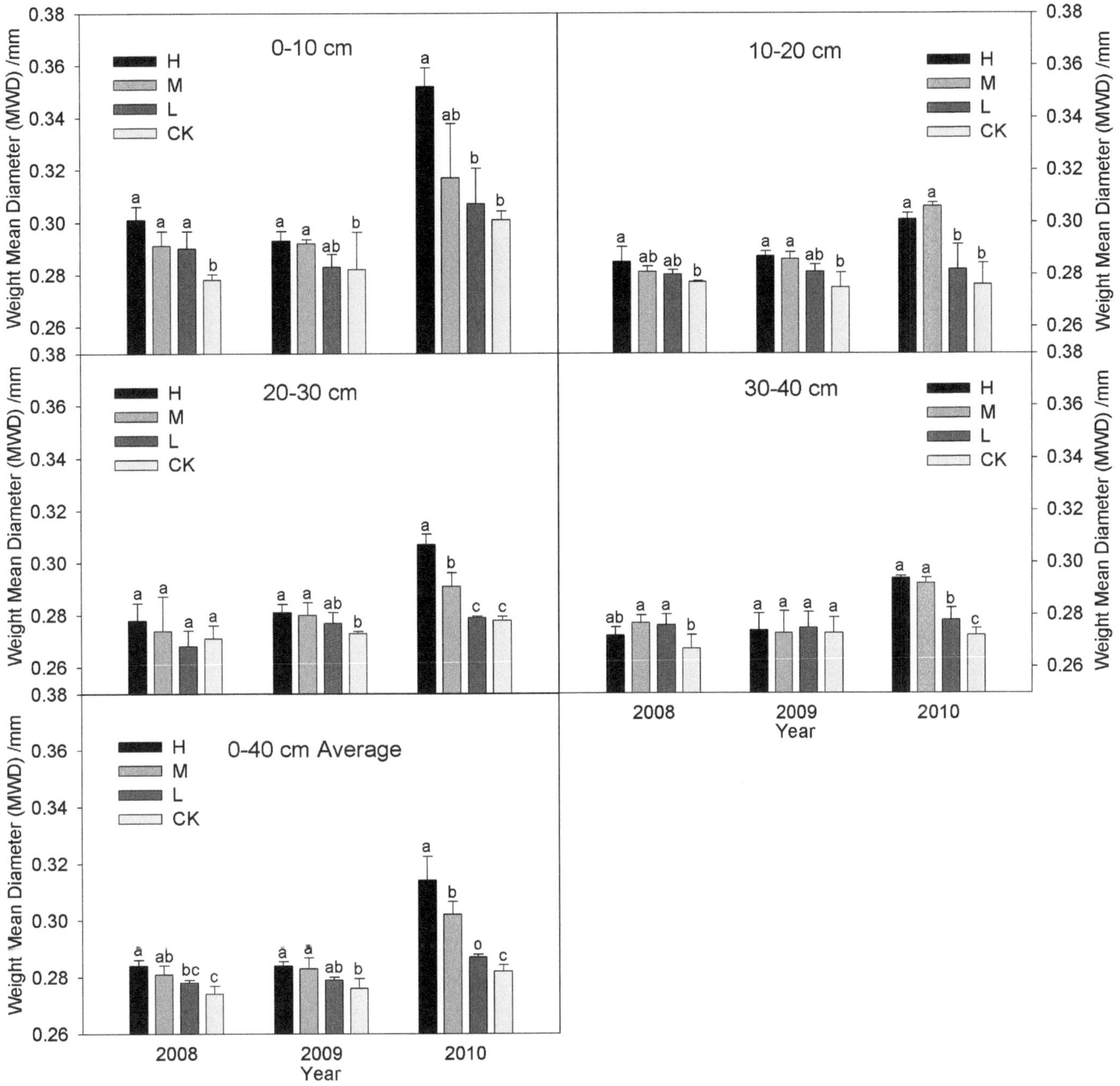

Figure 3. MWD values with wet sieving under the different straw incorporation treatments. Note: MWD, weight mean diameter; CK, no straw incorporation; L, incorporation of straw at a low rate of 4 500 kg ha^{-1} maize straw; M, incorporation of straw at a medium rate of 9 000 kg ha^{-1} maize straw; H, incorporation of straw at a high rate of 13 500 kg ha^{-1} maize straw; Average, the mean value of the 0–40 cm soil layers. Bars with different lower case letters indicate significant differences at $P<0.05$.

when F-values were significant. In all cases, differences were deemed to be significant if $P<0.05$.

Results

Soil organic carbon storage (SOC)

The effect of straw incorporation on SOC storage is shown in Fig. 2, where the soil organic carbon storage increased with the straw incorporation. The sum of SOC storage in 0–60 cm layers with the three incorporation treatments were higher than CK, i.e., 7.71% ($P<0.05$), 11.14% ($P<0.05$) and 1.70% in 2008, 15.15% ($P<0.05$), 24.00% ($P<0.05$) and 6.86% in 2009, and 21.40% ($P<0.05$), 20.38% ($P<0.05$) and 8.21% ($P<0.05$) in 2010, respectively. The SOC storage increased with the number of years of incorporation, i.e., the SOC storage (0–60 cm depth) in 2010 had increased by 6.19–12.48% compared with 2008, and decreased with the soil layer depth, i.e., by 3.75–25.68% in 2008, 11.85–21.70% in 2009 and 13.51–26.64% in 2010. The SOC storage of H and M was slightly higher than CK in 2008, although the difference was significant in 40–60 cm layer only. In 2009, compared with CK, H and M significantly increased the SOC storage by 11.01% and 21.74% ($P<0.05$) in 0–20 cm layer,

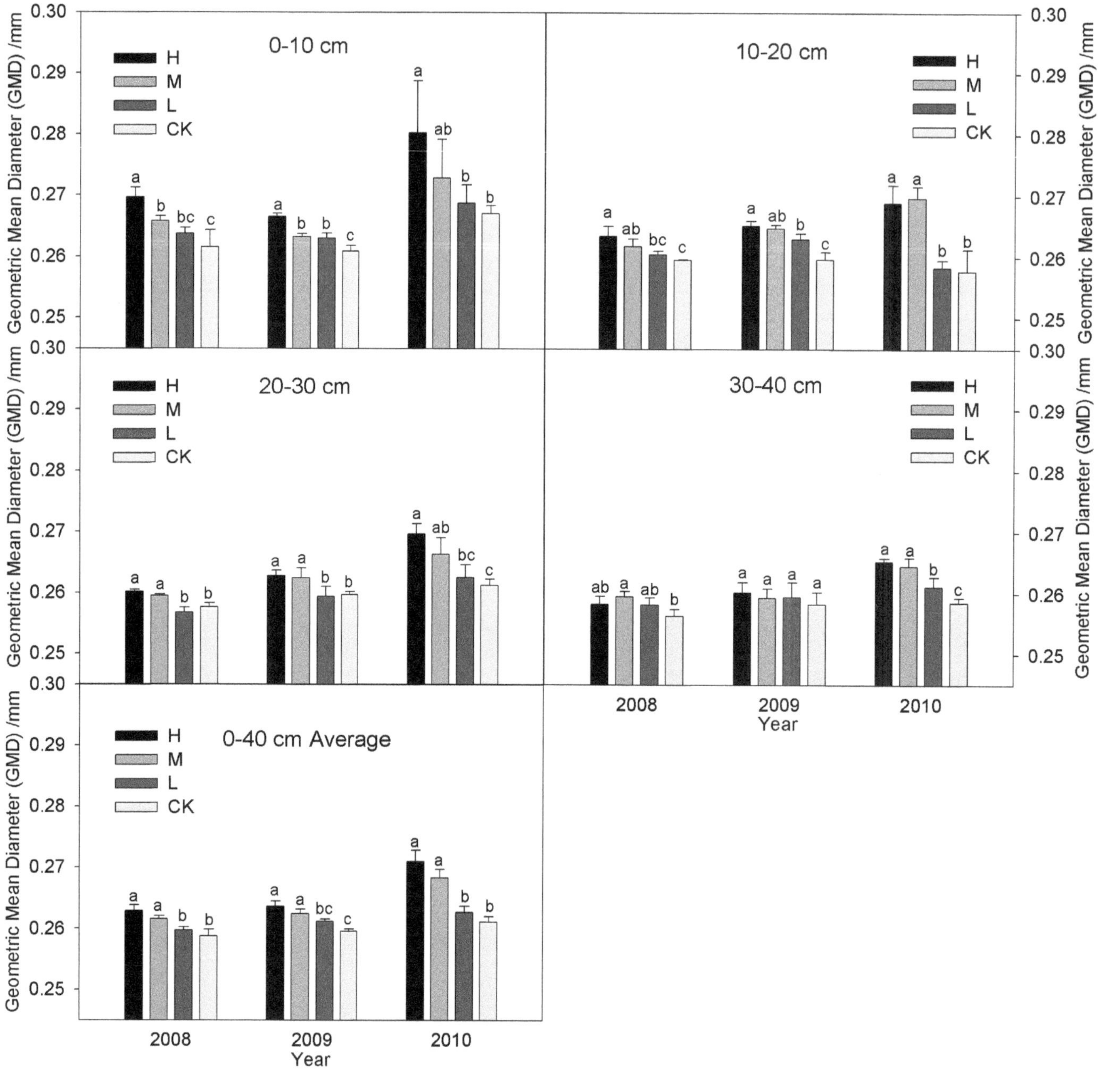

Figure 4. GMD values with wet sieving under the different straw incorporation treatments. Note: GMD, geometric mean diameter; CK, no straw incorporation; L, incorporation of straw at a low rate of 4 500 kg ha^{-1} maize straw; M, incorporation of straw at a medium rate of 9 000 kg ha^{-1} maize straw; H, incorporation of straw at a high rate of 13 500 kg ha^{-1} maize straw; Average, the mean value of the 0–40 cm soil layers. Bars with different lower case letters indicate significant differences at $P<0.05$.

14.94% and 18.81% ($P<0.05$) in 20–40 cm layer, and 20.40% and 32.47% ($P<0.05$) in 40–60 cm layer, respectively, and L was significantly increased by 7.89% ($P<0.05$) in 40–60 cm layer only. There was a significant difference between straw incorporation treatments and CK for each of the three soil layers in 2010, but there were no significant differences between H and M.

Water-stable macroaggregates (>0.25 mm)

The effects of straw incorporation on >0.25 mm macroaggregates are shown in Table 2. The mean >0.25 mm macroaggregates contents of the 0–40 cm layers of incorporation treatments were significantly ($P<0.05$) higher than CK during 2008–2010, i.e., 8.91–41.14%, 15.13–49.94%, and 17.64–66.58%, respectively. The >0.25 mm macroaggregates content also decreased with the soil layer depth, and increased with the number of years of incorporation,, i.e., the H, M, L and CK in 2010 was 24.72% ($P<0.05$), 39.28% ($P<0.05$), 16.49% and 7.81% higher than that in 2008. Compared with CK, the >0.25 mm macroaggregates content of three straw incorporation treatments in the 0–10 cm layer was increased by 10.75–46.60% in 2008, 6.56–53.13% in 2009, and 7.53–63.44% in 2010, respectively, while the differences between L and CK were not significant ($P>0.05$) during 2008–2010. The

Table 3. Effects on the soil aggregate stability rate with different straw incorporation treatments (%).

Year	Treatments	Soil Depth (cm)				0–40 AVG
		0–10	10–20	20–30	30–40	
2008	H	21.86a[1]±0.89[A2]	16.29a±0.41[3][A]	10.83a±0.54[B]	9.19a±0.65[A]	14.60a±0.33[A]
	M	19.62ab±1.09[A]	15.31b±0.62[A]	11.16a±0.82[A]	8.47b±0.21[A]	13.70b±0.62[A]
	L	17.83bc±1.56[A]	12.88c±0.24[A]	9.70b±0.64[A]	8.00bc±0.42[A]	11.83c±0.18[A]
	CK	16.45c±2.42[A]	12.52c±0.58[A]	9.39b±0.80[A]	7.53c±0.37[A]	11.79c±0.45[A]
2009	H	20.32a±0.27[A]	16.14a±0.54[A]	13.15a±0.34[AB]	9.74a±0.99[A]	14.84a±0.23[A]
	M	20.02a±0.46[A]	15.41a±0.50[A]	12.61a±0.41[A]	8.66ab±1.28[A]	14.18b±0.22[A]
	L	15.76b±0.49[A]	14.45a±1.78[A]	10.16b±1.36[A]	8.28ab±1.01[A]	12.16c±0.35[A]
	CK	15.47b±0.30[A]	11.81b±0.83[A]	9.59b±1.10[A]	7.50b±0.41[A]	11.09d±0.05[A]
2010	H	20.19a±1.93[A]	16.48a±1.73[A]	15.24a±0.12[A]	10.45a±0.11[A]	15.59a±0.26[A]
	M	20.87a±0.68[A]	15.78a±1.32[A]	13.99a±1.23[A]	10.08a±0.43[A]	15.18a±0.64[A]
	L	15.72b±0.72[A]	13.39a±1.81[A]	12.53b±0.46[A]	8.66b±0.77[A]	12.58b±1.00[A]
	CK	15.11b±0.12[A]	12.79a±1.70[A]	12.00b±0.76[A]	7.52c±0.26[A]	11.85b±1.32[A]

Note: CK, no straw incorporation; L, incorporation of straw at a low rate of 4 500 kg ha^{-1} maize straw; M, incorporation of straw at a medium rate of 9 000 kg ha^{-1} maize straw; H, incorporation of straw at a high rate of 13 500 kg ha^{-1} maize straw; AVG, the mean value of the 0-40 cm soil layers.
[1]Values followed by the same lowercase letter in the same line are not significantly different according to Duncan's multiple range test (P<0.05) between the four straw incorporation treatments in the same year.
[2]Values followed by the same uppercase letter in the same line are not significantly different according to Duncan's multiple range test (P<0.05) between the different years of the same straw incorporation treatment.
[3]Means ± standard deviations.

>0.25 mm macroaggregates content showed the same trends in the 10–20 and 20–30 cm layers, and the H and M treatments had the significant difference (P<0.05) with CK. The H, M and L increased by 29.05% (P<0.05), 16.55% and 6.87% in 2008, 38.94% (P<0.05), 24.96% and 14.69% in 2009, and 50.09% (P<0.05), 54.29% (P<0.05), 21.89% (P<0.05) in 2010, respectively. A linear correlation was found between the >0.25 mm macroaggregates and soil organic carbon (R^2>0.64, P<0.01).

Mean weight diameter (*MWD*) and Geometric mean diameter (*GMD*)

Fig. 3 & 4 show that the *MWD* and *GMD* values with the three incorporation treatments increased significantly throughout the three-year study. The average *MWD* and *GMD* values under the incorporation treatments were higher in the 0–40 cm layers than CK during 2008–2010, i.e., 1.46–3.65% and 0.39–1.54% in 2008, 1.09–2.90% and 0.62–1.55% in 2009, 1.77–11.35% and 0.77–3.83% in 2010, respectively. There was no significant difference (P>0.05) between L and CK during the study period. The *MWD* and *GMD* values decreased with the soil layer depth and increased with the number of years of incorporation, i.e., the H, M, L and CK in 2010 was 10.56% (P<0.05) and 3.04% (P<0.05), 7.47% (P<0.05) and 2.29% (P<0.05), 3.24% and 1.15%, and 2.92% and 0.77% higher than that in 2008. The *MWD* and *GMD* values of treatments in the 0–10 and 10–20 cm layers were ranked in the order: H > M > L > CK. The H and M levels were significantly higher than CK while H and M levels were similar during 2008–2010 (Fig. 3–4). The *MWD* and *GMD* exhibited the same trends in 20–30 cm layers and there were no significant (P>0.05) differences among incorporation treatments during 2008–2009, the *MWD* values with the H were significantly (P<0.05) higher than the other three treatments during 2010, i.e., by 5.50%, 10.04%, and 10.43%, respectively. The *GMD* did not differ significantly among the four treatments, with the exceptions of H and L, and H and CK in 2010. The *MWD* and *GMD* values in the 30–40 cm layers under the incorporation treatments were increased by 1.87%, 3.75% (P<0.05), 3.37% (P<0.05) and 0.76%, 1.76% (P<0.05), 0.78% in 2008, 0.37%, 0.01%, 0.73% and 0.73%, 0.45%, 0.45% in 2009, 8.09% (P<0.05), 7.35% (P<0.05), 2.21% (P<0.05) and 2.32% (P<0.05), 1.93% (P<0.05), 0.77% (P<0.05) in 2010, compared with CK, respectively. There was no significant difference (P>0.05) between H and M during the study period.

Soil aggregate stability (*WSAR*)

The straw incorporation significantly affected the *WSAR* after three years (Table 3). The *WSAR* values in the 0–40 cm layers of incorporation treatments was significantly higher than CK during 2008–2010, i.e., 0.34–19.80%, 9.65–33.79%, and 6.16–35.08%, respectively. The *WSAR* also decreased with the soil layer depth, while it increased with the number of years of incorporation. The *WSAR* in the 0–10 cm layers increased significantly with the amount of straw and the *WSAR* was highest with H throughout 2008–2010, i.e., the *WSAR* significantly (P<0.05) increased by 28.35%, 28.93% and 25.78% compared with L, respectively, and by 33.62%, 31.35% and 32.89% (P<0.05) compared with CK, respectively. Compared with L and CK, the WSAR values with M increased significantly by 20.08–24.67% and 26.87–29.78% during 2008–2010, respectively, but there were no significant (P>0.05) differences between H and M, and L and CK. With the four treatments, the *WSAR* values in the 10–20 and 20–30 cm layers were lower than the 0–10 cm layers, which were ranked in the order: H > M > L > CK. There were no significant differences (P>0.05) among the four treatments in the 30–40 cm layers throughout the three years, and the *WSAR* maintained at 7.50–10.45%.

Fractal dimension (D)

Fig. 5 shows that the fractal dimensions with the three incorporation treatments decreased significantly (P<0.05) after three years of straw incorporation (the R^2 of fitting curve is:

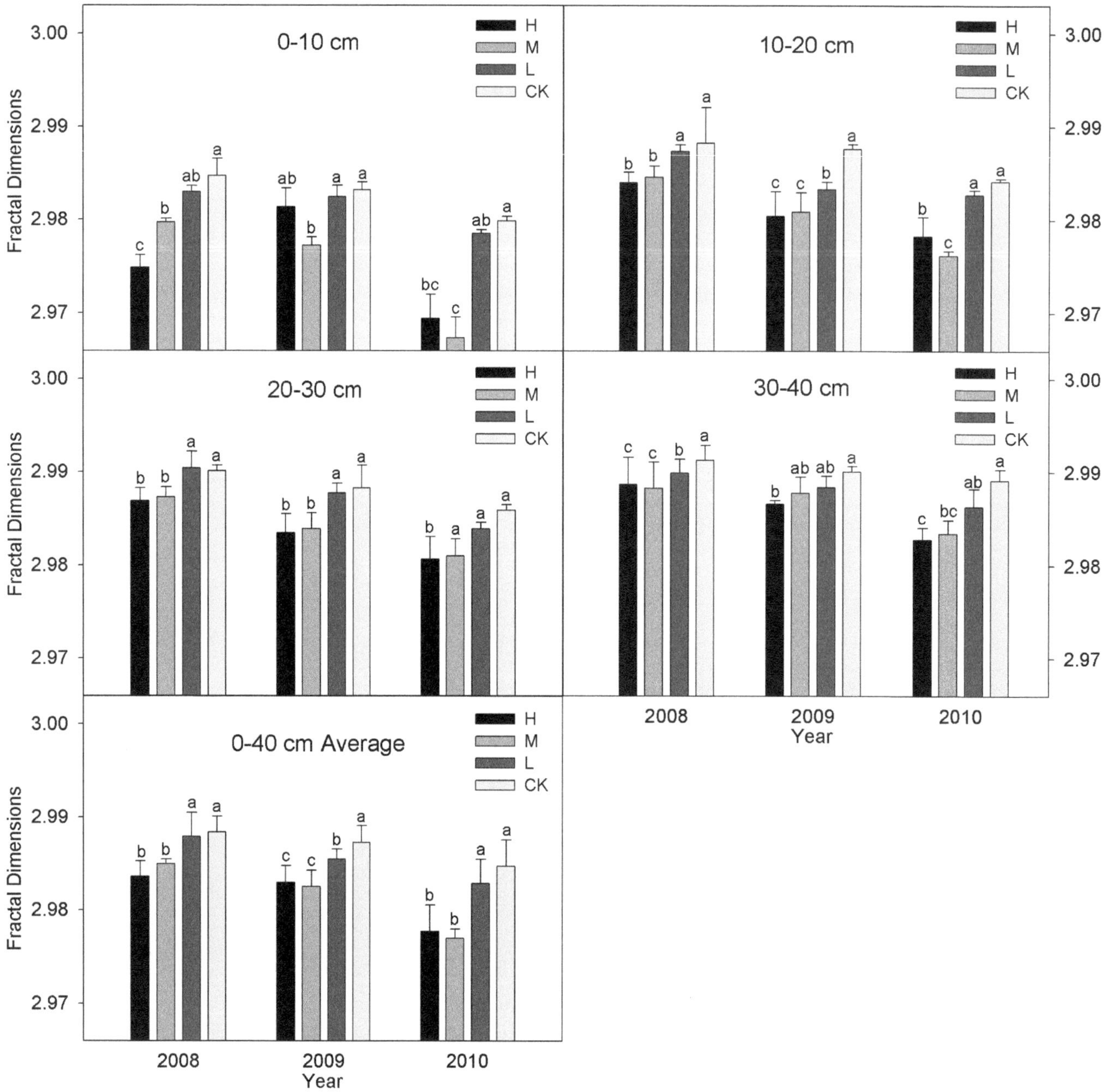

Figure 5. Fractal dimensions (D) of soil aggregates with different straw incorporation treatments. Note: CK, no straw incorporation; L, incorporation of straw at a low rate of 4 500 kg ha^{-1} maize straw; M, incorporation of straw at a medium rate of 9 000 kg ha^{-1} maize straw; H, incorporation of straw at a high rate of 13 500 kg ha^{-1} maize straw; Average, the mean value of the 0–60 cm soil layers; Bars with different lower case letters indicate significant differences at $P<0.05$; $0.92<R^2<0.99$.

0.92~0.99). The fractal dimensions (0–40 cm layer) with the incorporation treatments were lower than those of CK during 2008–2010, i.e., 0.16% ($P<0.05$), 0.11% ($P<0.05$) and 0.02% in 2008, 0.14% ($P<0.05$), 0.16% ($P<0.05$) and 0.06% ($P<0.05$) in 2009, and 0.24% ($P<0.05$), 0.26% ($P<0.05$) and 0.06% in 2010, respectively. The fractal dimensions also increased with the soil layer depth and decreased with the number years of incorporation, i.e., the H, M, L and CK in 2010 was 0.20% ($P<0.05$), 0.27% ($P<0.05$), 0.17% ($P<0.05$) and 0.12% higher than that in 2008. The fractal dimensions in the 0–10 cm layers with the four treatments were ranked in the order: H<M<L<CK in 2008. The fractal dimensions with H and M were significantly ($P<0.05$) lower than CK, i.e., 0.33%, 0.17% and 0.06%, respectively. However, the ranking was M<H<L<CK during 2009–2010, i.e., 0.06%, 0.20% ($P<0.05$), 0.02% in 2009, 0.35% ($P<0.05$), 0.42% ($P<0.05$) and 0.04% in 2010, respectively. While the differences between L and CK were not significant ($P>0.05$) during 2008–2010. Compared with L and CK, the fractal dimensions in the 10–20 cm layers under H was significantly ($P<0.05$) reduced the fractal dimensions by 0.11% and 0.14% in 2008, and 0.10% and 0.24% in 2009,, while M significantly ($P<0.05$) reduced by 0.09% and 0.12% in 2008, and 0.08% and 0.23% in 2009, respectively.

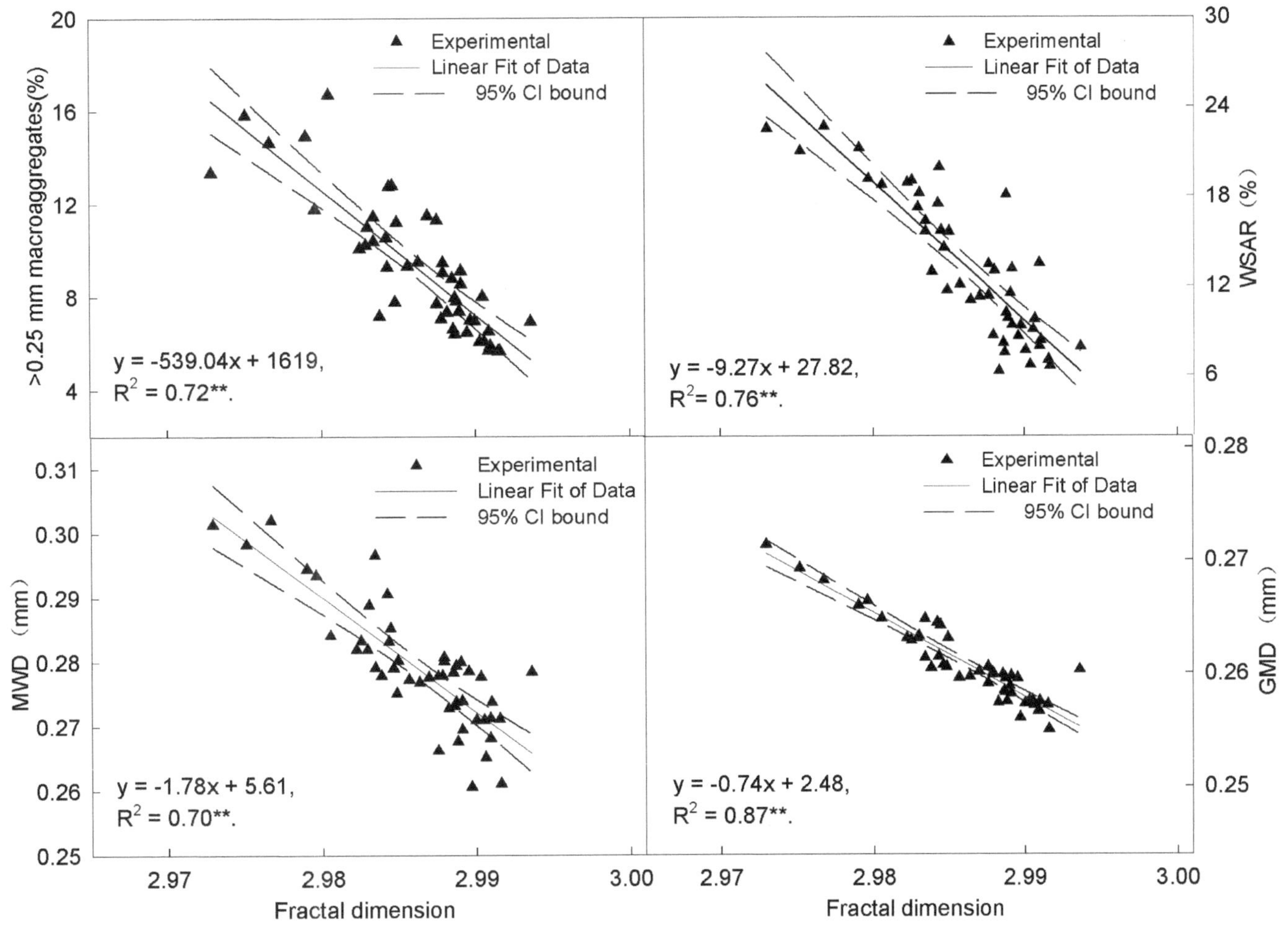

Figure 6. Correlations between D and >0.25 mm macroaggregates, MWD, GMD, WSAR of the soil aggregates. Note: MWD, weight mean diameter of soil aggregates; GMD, geometric mean diameter of soil aggregates; WSAR, soil aggregate stablility rate; D, Fractal dimension of soil aggregates. 95% CI, 95% confidence interval.** $P<0.01$.

The fractal dimension was lowest with M during 2010, i.e., it was decreased by 0.22% ($P<0.05$) compared with L, and by 0.27% ($P<0.05$) compared with CK. The differences among the four treatments decreased gradually in the 20–40 cm layer and the fractal dimension remained at 2.987~2.991.

Data fitting detected a linear correlation between the indexes of soil aggregates and the fractal dimension (Fig. 6).

Discussion

Results of this study demonstrated that application of crop straws had positive effects on the soil physico-chemical properties. On the other hand, there are Loess plateau regions at northwest China with intensive cultivation systems and poor soil management strategies. Therefore, crop residual management is very important for preserving natural ecosystems [38]. Another problem for this area is conventional tillage that reduces the soil water storage and destroys the aggregates and soil structure [39] which prompts deterioration of crop yield.

The results indicated that the SOC storage increased significantly in the 0–60 cm layers after three years of straw incorporation (Fig. 2). Rasmussen and Collins [40] reported that the soil organic matter content was strongly related to the amount of residues added and only weakly related to the type of residue applied. The H and M treatments were significantly different from the L throughout the three years and the soil organic matter level increased as the straw incorporated and decomposed [20], which effectively mitigated the loss of soil organic carbon from in the agroecosystem caused by intensive cropping [41]. The SOC storage decreased gradually in all the treatments with the soil layer depth because the degree of straw incorporation was lower in the deeper layers compared with the surface layers (0–20 cm) of the soil [42]. This was because the amount of straw incorporated in the topsoil was greater than that in the deeper layers [40,43].

Soil aggregates are the basic units of the soil structure and they are composed of primary particles and binding agents [12]. They are also necessary soil conditions for high crop yields [44]. Conventional tillage disturbs the soil and increase the effects of drying–rewetting and freezing–thawing, which increases the susceptibility of the macroaggregate (>0.25 mm) to disruption [45–47]. Pinheiro *et al.* [48] showed that soil exposure with tillage and the lack of residue inputs led to a decline in aggregation and organic carbon, both of which made the soil susceptible to erosion. Our study showed that the straw incorporation of straws determined significantly more and larger soil aggregates than CK, thus indicating an improvement of soil physical quality. This may have been attributable to the significant increase in the SOC storage (an average increase by 8.21–21.40% in 2010), the lower soil bulk density (an average decreased by 1.80–4.13% in 2010, data no shown) [49], and increased soil porosity (an average increased by 1.70–3.90% in 2010, data no shown) [50] after straw incorporation. It also stimulated the activity of soil microorganisms [51] and an abundance of polyose metabolites were produced during the straw decay process [52]. These soil physical and chemical conditions may have accelerated the incorporated SOM decomposing process and increased soil aggregation [24].

Soil aggregate structural stability is widely recognized as a key indicator of soil quality, which is closely related to a number of soil properties, processes, and functions, e.g., the quantity and composition of SOM [48], infiltration capacity [53], soil biotic activity [54] and the resistance to erosion [55,56]. Wei *et al.* [57] showed that the addition of crop residues was the most effective measure for increasing the rhizosphere aggregate stability. Sonnleitner *et al.* [24] and Karami *et al.* [41] also found that straw application improved the aggregate stability and other soil properties. In our study, the soil aggregate stability of the straw incorporation treatments were significantly higher than CK in 2010, and it was decreased with the soil layer depth. These results agreed with studies by Tripathy and Singh [42] and Karami *et al.* [41]. Our results also indicated that straw incorporation was positively related to the physical protection of organic matter [20] and an increased aggregate quantity [58], but it also improved the soil aggregate stability [20] and reduced soil degeneration [22,23].

Many studies have shown that soil is a porous medium with fractal characteristics [59–60]. Thus, fractal theory can be used to describe the complex characteristic of soil structure [61]. Castrignanò and Stelluti [62] reported that a higher fractal dimension indicated the heavier texture of a soil and its inferior permeation properties. This showed that fractal theory is an effective method to describing the soil aggregate distribution [37] and changed with different levels of straw incorporation [31]. The fractal dimensions of the 0–40 cm layers with the four treatments after three years were ranked in the order: H<M<L<CK and the three straw incorporation treatments were significantly different from CK. These results agreed with Zhang *et al.* [37] and Zhang *et al.* [31]. The low values of *D* indicated a size distribution dominated by a large number of macroaggregates (>0.25 mm) [59,63]. This improvement in the fractal dimension may have been accelerated by the incorporation of straw, which improved the soil structure, increased the SOM content and microbial activity [64], and significantly increased the mount and size of soil aggregates [37]. Our results indicated there were significant improvements in the soil macroaggregates and the aggregate structure after straw incorporation [20].

Conclusion

The incorporation of different amount of straw significant increased the *SOC* storage, >0.25 mm macroaggregates, *MWD* and *GMD* in a semiarid soil. The *SOC* storage, >0.25 mm macroaggregates, *MWD* and *GMD* also increased with higher straw incorporation rates. The fractal dimension decreased with increasing straw incorporation rates. Therefore, the incorporation of straw into the soil in semiarid areas is an effective practice for improving the soil aggregate content and stability.

Acknowledgments

We are grateful to Mr Li Yongping and Liu Shixin for managing the field experiments and professional English editor Jackson who is from UK, and kind help with the language of this manuscript.

Author Contributions

Conceived and designed the experiments: ZJ QH. Performed the experiments: PZ TW. Analyzed the data: PZ. Contributed reagents/materials/analysis tools: XR YL. Wrote the paper: PZ.

References

1. Hu KL, Li BG, Lin QM, Li GT, Chen DL (1999) Spatial variabi lity of soil nutrient in wheat field. Transactions of the Chinese Society of Agricultural Engineering 15:33–38. (in Chinese with English abstract)
2. Mupangwa W, Twomlow S, Walker S (2008) The influence of conservation tillage methods on soil water regimes in semi-arid southern Zimbabwe. Phys Chem Earth, Parts A/B/C 33: 762–767.
3. Biederbeck V, Campbell C, Bowren K, Schnitzer M, McIver R (1980) Effect of burning cereal straw on soil properties and grain yields in Saskatchewan. Soil Sci Soc Am J 44: 103–111.
4. Wuest SB, Caesar-TonThat T, Wright SF, Williams JD (2005) Organic matter addition, N, and residue burning effects on infiltration, biological, and physical properties of an intensively tilled silt-loam soil. Soil Tillage Res 84: 154–167.
5. Duan F, Liu X, Yu T, Cachier H (2004) Identification and estimate of biomass burning contribution to the urban aerosol organic carbon concentrations in Beijing. Atmos Environ 38: 1275–1282.
6. Mandal KG, Misra AK, Hati KM, Bandyopadhyay KK, Ghosh PK, et al. (2004) Rice residue-management options and effects on soil properties and crop productivity. J Food Agric Environ 2: 224–231.
7. Carter MR (2002) Soil quality for sustainable land management. Agron J 94: 38–47.
8. Hillel D (2004) Encyclopedia of Soils in the Environment: London, Elsevier, v. 4, p. 295–303.
9. Six J, Bossuyt H, Degryze S, Denef K (2004) A history of research on the link between (micro) aggregates, soil biota, and soil organic matter dynamics. Soil Tillage Res 79: 7–31.
10. Tisdall J, Oades J (1982) Organic matter and water-stable aggregates in soils. J Soil Sci 33:141–163. doi:10.1111/j.1365-2389.1982.tb01755.x
11. Haynes R, Beare M (1997) Influence of six crop species on aggregate stability and some labile organic matter fractions. Soil Biol Biochem 29: 1647–1653.
12. Scanlon BR, Andraski BJ, Bilskie J (2002) Methods of Soil Analysis: Part 4 Physical Methods: Soil Science Society of America. 643–668 p.
13. Hernández-Hernández R, López-Hernández D (2002) Microbial biomass, mineral nitrogen and carbon content in savanna soil aggregates under conventional and no-tillage. Soil Biol Biochem 34: 1563–1570.
14. Villar M, Petrikova V, Dıaz-Ravina M, Carballas T (2004) Changes in soil microbial biomass and aggregate stability following burning and soil rehabilitation. Geoderma 122: 73–82.
15. Ashagrie Y, Zech W, Guggenberger G (2005) Transformation of a *Podocarpus falcatu* dominated natural forest into a monoculture *Eucalyptus globulus* plantation at Munesa, Ethiopia: soil organic C, N and S dynamics in primary particle and aggregate-size fractions. Agric Ecosyst Environ 106: 89–98.
16. Jastrow J (1996) Soil aggregate formation and the accrual of particulate and mineral-associated organic matter. Soil Biol Biochem 28: 665–676.
17. Chevallier T, Blanchart E, Albrecht A, Feller C (2004) The physical protection of soil organic carbon in aggregates: a mechanism of carbon storage in a Vertisol under pasture and market gardening (Martinique, West Indies). Agric Ecosyst Environ 103: 375–387.
18. Duiker S, Lal R (1999) Crop residue and tillage effects on carbon sequestration in a Luvisol in central Ohio. Soil Tillage Res 52: 73–81.
19. Saroa G, Lal R (2003) Soil restorative effects of mulching on aggregation and carbon sequestration in a Miamian soil in central Ohio. Land Degrad Dev 14: 481–493.
20. Tan DS, Jin JY, H SW, Li ST, He P (2007) Effect of long-term application of K fertilizer and wheat straw to soil on crop yield and soil K under different planting systems. Agric Sci in China 6: 200–207.
21. Li Q, Chen X, Zhang F, Römheld V (2002) Study on balance of phosphorus and potassium in winter wheat and summer maize rotation system. Plant Nutr Fert Sci 8: 152–156. (in Chinese with English abstract)
22. Nelson RG (2002) Resource assessment and removal analysis for corn stover and wheat straw in the Eastern and Midwestern United States—rainfall and wind-induced soil erosion methodology. Biomass Bioenergy 22: 349–363.
23. Wilhelm W, Johnson JMF, Karlen DL, Lightle DT (2007) Corn stover to sustain soil organic carbon further constrains biomass supply. Agron J 99: 1665–1667.
24. Sonnleitner R, Lorbeer E, Schinner F (2003) Effects of straw, vegetable oil and whey on physical and microbiological properties of a chernozem. App Soil Ecol 22: 195–204.
25. Mulumba LN, Lal R (2008) Mulching effects on selected soil physical properties. Soil Tillage Res 98: 106–111.
26. Bhagat R, Verma T (1991) Impact of rice straw management on soil physical properties and wheat yield. Soil Sci 152: 108–115.

27. FAO/UNESCO (1993) 'World soil resources. An explanatory note on the FAO World Soil Resource Map at 1 : 25,000,000 scales.' (FAO: Rome)
28. Wang S, Tian H, Liu J, Pan S (2003) Pattern and change of soil organic carbon storage in China: 1960s–1980s. Tellus B 55: 416–427.
29. Pribyl DW (2010) A critical review of the conventional SOC to SOM conversion factor. Geoderma 156: 75–83.
30. Oades J, Waters A (1991) Aggregate hierarchy in soils. Soil Res 29: 815–828.
31. Zhang P, Jia ZK, Wang W, Lu WT, Gao F, et al. (2012) Effects of straw returning on characteristics of soil aggregates in semi-arid areas in southern Ningxia of China. Sci Agric Sin 45:1513-1520. (in Chinese with English abstract)
32. Van Bavel C (1949) Mean weight diameter of soil aggregates as a statistical index of aggregation. Soil Sci Soc Am Proc. pp. 20–23.
33. Youker R, McGuinness J (1957) A short method of obtaining mean weight-diameter values of aggregate analyses of soils. Soil Sci 83: 291–294.
34. Mazurak AP (1950) Effect of gaseous phase on water-stable synthetic aggregates. Soil Sci 69: 135–148.
35. Hou XQ, Li R, Jia ZK, Han QF, Wang W, et al. (2012) Effects of rotational tillage practices on soil properties, winter wheat yields and water-use efficiency in semi-arid areas of north-west China. Field Crops Res 129: 7–13.
36. Tyler SW, Wheatcraft SW (1992) Fractal scaling of soil particle-size distributions: analysis and limitations. Soil Sci Soc Am J 56: 362–369.
37. Zhang Z, Wei C, Xie D, Gao M, Zeng X (2008) Effects of land use patterns on soil aggregate stability in Sichuan Basin, China. Particuology 6: 157–166.
38. Zhu H, Wu J, Huang D, Zhu Q, Liu S, et al. (2010) Improving fertility and productivity of a highly-weathered upland soil in subtropical China by incorporating rice straw. Plant Soil 331: 427–437.
39. Mele PM, Crowley DE (2008) Application of self-organizing maps for assessing soil biological quality. Agric Ecosyst Environ 126: 139–152.
40. Rasmussen PE, Collins HP (1991) Long-term impacts of tillage, fertilizer, and crop residue on soil organic matter in temperate semiarid regions. Adv Agron 45: 93–134.
41. Karami A, Homaee M, Afzalinia S, Ruhipour H, Basirat S (2012) Organic resource management: Impacts on soil aggregate stability and other soil physico-chemical properties. Agric Ecosyst Environ 148: 22–28.
42. Tripathy R, Singh AK (2004) Effect of water and nitrogen management on aggregate size and carbon enrichment of soil in rice-wheat cropping system. J Plant Nutr Soil Sci 167: 216–228.
43. Prasad R, Power J (1991) Crop residue management. Adv Soil Sci 15: 205–251.
44. Limon-Ortega A, Govaerts B, Sayre KD (2009) Crop rotation, wheat straw management, and chicken manure effects on soil quality. Agron J 101: 600–606.
45. Beare M, Hendrix P, Cabrera M, Coleman D (1994) Aggregate-protected and unprotected organic matter pools in conventional-and no-tillage soils. Soil Sci Soc Am J 58: 787–795.
46. Paustian K, Levine E, Post WM, Ryzhova IM (1997) The use of models to integrate information and understanding of soil C at the regional scale. Geoderma 79: 227–260.
47. Mikha MM, Rice CW (2004) Tillage and manure effects on soil and aggregate-associated carbon and nitrogen. Soil Sci Soc Am J 68: 809–816.
48. Pinheiro E, Pereira M, Anjos L (2004) Aggregate distribution and soil organic matter under different tillage systems for vegetable crops in a Red Latosol from Brazil. Soil Tillage Res 77: 79–84.
49. Lal R (2000) Physical management of soils of the tropics: priorities for the 21st century. Soil Sci 165: 191–207.
50. Pagliai M, Vignozzi N, Pellegrini S (2004) Soil structure and the effect of management practices. Soil Tillage Res 79: 131–143.
51. Kasteel R, Garnier P, Vachier P, Coquet Y (2007) Dye tracer infiltration in the plough layer after straw incorporation. Geoderma 137: 360–369.
52. Pascual JA, García C, Hernandez T (1999) Comparison of fresh and composted organic waste in their efficacy for the improvement of arid soil quality. Bioresour Technol 68: 255–264.
53. Abu-Sharar T, Salameh A (1995) Reductions in hydraulic conductivity and infiltration rate in relation to aggregate stability and irrigation water turbidity. Agric Water Manage 29: 53–62.
54. Roldán A, Salinas-García J, Alguacil M, Caravaca F (2005) Changes in soil enzyme activity, fertility, aggregation and C sequestration mediated by conservation tillage practices and water regime in a maize field. Appl Soil Ecol 30: 11–20.
55. Barthes B, Roose E (2002) Aggregate stability as an indicator of soil susceptibility to runoff and erosion; validation at several levels. Catena 47: 133–149.
56. Ramos M, Nacci S, Pla I (2003) Effect of raindrop impact and its relationship with aggregate stability to different disaggregation forces. Catena 53: 365–376.
57. Wei CF, Gao M, Shao JG, Xie DT, Pan GX (2006) Soil aggregate and its response to land management practices. China Particuology 4: 211–219.
58. Tarafdar JC, Meena SC, Kathju S (2001) Influence of straw size on activity and biomass of soil microorganisms during decomposition. Eur J Soil Biol 37: 157–160.
59. Perfect E, Kay B (1991) Fractal theory applied to soil aggregation. Soil Sci Soc Am J 55: 1552–1558.
60. Rieu M, Sposito G (1991) Fractal fragmentation, soil porosity, and soil water properties: II. Applications. Soil Sci Soc Am J 55: 1239–1244.
61. Bird N, Bartoli F, Dexter A (2005) Water retention models for fractal soil structures. Eur J Soil Sci 47: 1–6.
62. Castrignanò A, Stelluti M (1999) Fractal geometry and geostatistics for describing the field variability of soil aggregation. J Agric Eng Res 73: 13–18.
63. Ding Q, Ding W (2007) Comparing stress wavelets with fragment fractals for soil structure quantification. Soil Tillage Res 93: 316–323.
64. Liu JF, Hong W, Wu CZ (2002) Fractal features of soil clusters under some precious hardwood stands in the central subtropical region, China. Acta Ecol. Sin. 22, 197. (in Chinese with English abstract).

Impacts of Agricultural Management and Climate Change on Future Soil Organic Carbon Dynamics in North China Plain

Guocheng Wang, Tingting Li*, Wen Zhang, Yongqiang Yu

State Key Laboratory of Atmospheric Boundary Layer Physics and Atmospheric Chemistry, Institute of Atmospheric Physics, Chinese Academy of Sciences, Beijing, China

Abstract

Dynamics of cropland soil organic carbon (SOC) in response to different management practices and environmental conditions across North China Plain (NCP) were studied using a modeling approach. We identified the key variables driving SOC changes at a high spatial resolution (10 km×10 km) and long time scale (90 years). The model used future climatic data from the FGOALS model based on four future greenhouse gas (GHG) concentration scenarios. Agricultural practices included different rates of nitrogen (N) fertilization, manure application, and stubble retention. We found that SOC change was significantly influenced by the management practices of stubble retention (linearly positive), manure application (linearly positive) and nitrogen fertilization (nonlinearly positive) – and the edaphic variable of initial SOC content (linearly negative). Temperature had weakly positive effects, while precipitation had negligible impacts on SOC dynamics under current irrigation management. The effects of increased N fertilization on SOC changes were most significant between the rates of 0 and 300 kg ha^{-1} yr^{-1}. With a moderate rate of manure application (i.e., 2000 kg ha^{-1} yr^{-1}), stubble retention (i.e., 50%), and an optimal rate of nitrogen fertilization (i.e., 300 kg ha^{-1} yr^{-1}), more than 60% of the study area showed an increase in SOC, and the average SOC density across NCP was relatively steady during the study period. If the rates of manure application and stubble retention doubled (i.e., manure application rate of 4000 kg ha^{-1} yr^{-1} and stubble retention rate of 100%), soils across more than 90% of the study area would act as a net C sink, and the average SOC density kept increasing from 40 Mg ha^{-1} during 2010s to the current worldwide average of ~55 Mg ha^{-1} during 2060s. The results can help target agricultural management practices for effectively mitigating climate change through soil C sequestration.

Editor: Ben Bond-Lamberty, DOE Pacific Northwest National Laboratory, United States of America

Funding: This work was jointly supported by the Ministry of Science and Technology of China (Grant No. 2010CB950603), the CAS Strategic Priority Research Program (Grant No. XDA05050507), the National Natural Science Foundation of China (Grant No. 41021004), and the State Key Program of National Natural Science of China (Grant No. 41130104). The funders had no role in study design, data collection and analysis, decision to publish, or preparation of the manuscript.

Competing Interests: The authors have declared that no competing interests exist.

* E-mail: litingting@mail.iap.ac.cn

Introduction

Cultivation generally leads to dramatic changes in soil organic carbon (SOC) by influencing the processes involved with soil C production and decomposition [1,2]. Changes in agricultural SOC are characterized by dynamic exchange processes which is strongly linked to environmental conditions such as air temperature, precipitation, soil pH and texture; and agronomic management such as crop rotation, stubble retention, tillage regimes, and application of chemical fertilizers and animal manure [3,4]. The complex interactions between management practices and environmental conditions, as well as the lack of spatiotemporal continuity in SOC monitoring data over meaningfully large areas, hamper our ability to accurately predict the regional cropland SOC change and identify the factors that control the SOC dynamics.

A modeling approach has advantages in estimating spatiotemporal changes in SOC under various management practices and environmental conditions [5]. A number of process-based carbon models such as DNDC [6], RothC [7], and CENTURY [8] have been used to simulate SOC dynamics in agricultural systems at both national and continental scales [9–11]. Recently, Huang et al. [12] developed a biogeophysical model, Agro-C, which has been validated and further used to assess the long-term agricultural SOC changes on a national scale in China [5] and Australia [2].

The North China Plain (NCP) is one of the most important agricultural production areas in China, covering a total cropland area of ~18 Mha, with a typical continuous winter wheat-summer maize cropping system (Figure 1). During the past several decades, significant changes in both climate and agricultural management practices have taken place in NCP. For example, Ding et al. [13] reported that both temperature and precipitation had shown an increasing trend across the northern part of China during the past 50 years. Improved management practices such as crop residue retention and application of mineral fertilizers and/or organic manure have been promoted as strategies to both increase crop productivity and optimize soil fertility [14]. Although an overall net increase in SOC across the croplands of NCP during the past 30 years has been reported [5], there remains large uncertainties in future cropland SOC dynamics due to the unclear future changes in both climate and management practices. For instance, it is generally recognized that the current global warming will increase the rates of organic matter decomposition, thereby accelerating climate change through C cycle feedbacks [15]. However, in areas with relatively low mean annual temperature and without water and nutrient deficiencies (e.g., the North China Plain, typically with irrigation and N fertilization to support crop production), increasing air temperature could potentially increase crop productivity and result in more input C into soils, thereby offsetting the loss in SOC through soil respiration. More recently, Yu et al. [16] simulated the Chinese cropland SOC changes during the next 40 years, by simply designing three levels of carbon input based on the historic changing trend in annual amounts of crop residues and manure. However, the quantity of input C is strongly linked to management such as stubble retention and N fertilization, which needs to be considered during a comprehensive assessment of long-term SOC dynamics, and determined to provide strategies in effectively managing the farming systems. Moreover, since it takes several decades or even a century to reach a new equilibrium of SOC following a change in management practice [17,18], the role of soils to sequester C (source or sink) as affected by environmental variables and management practices needs to be identified on a long time scale, e.g., around 100 years.

In this study, we quantified the influence of both climatic variables and management practices on SOC across croplands of North China Plain. The SOC was simulated under 900 combinations of climate, N fertilization, stubble retention, and manure application scenarios at a daily time step, from 2011 to 2100, and at a high spatial resolution, using the pre-calibrated and validated Agro-C model. The impacts of management practices and climatic variables on SOC dynamics were then statistically analyzed. This would not only provide a better understanding of the factors and processes regulating C cycling and balance in the agro-ecosystems, but would also provide insight into the effectiveness of methods for enhancing soil quality and mitigating climate change.

Materials and Methods

1.1 Agro-C

Agro-C [12] is a biogeophysical model, and it consists of two submodels, Crop-C and Soil-C. The Crop-C submodel simulates processes involved with crop photosynthesis, autotrophic respiration, and net primary production (NPP); and the Soil-C submodel simulates soil heterotrophic respiration via the decomposition of both input C (e.g., crop residues, roots and manure) and SOC. Changes in SOC are then determined by balancing the loss of SOC with the gain of input C. The input C is split into two components of labile-C and resistant-C, and the SOC pool is divided into two sub-pools named light-C and heavy-C. The light-C sub-pool is hypothesized to be more biologically reactive, while the heavy-C sub-pool is much more resistant to decomposition. Decomposition of each component and sub-pool is described by first-order kinetics at predefined potential rates (d^{-1}) of 2.6×10^{-2}, 8.4×10^{-4}, 2.5×10^{-4} and 1.8×10^{-5} for labile-C, resistant-C, light-C and heavy-C, respectively. The actual decomposition rate of each component and sub-pool is further modified by soil parameters such as temperature, moisture, texture and pH [12]. The C flow between different pools was determined based on the following assumptions: (i) the decomposition of both the labile-C and the resistant-C converts a fraction of the C into the light-C pool, (ii) a fraction of the decomposed light-C is transferred into

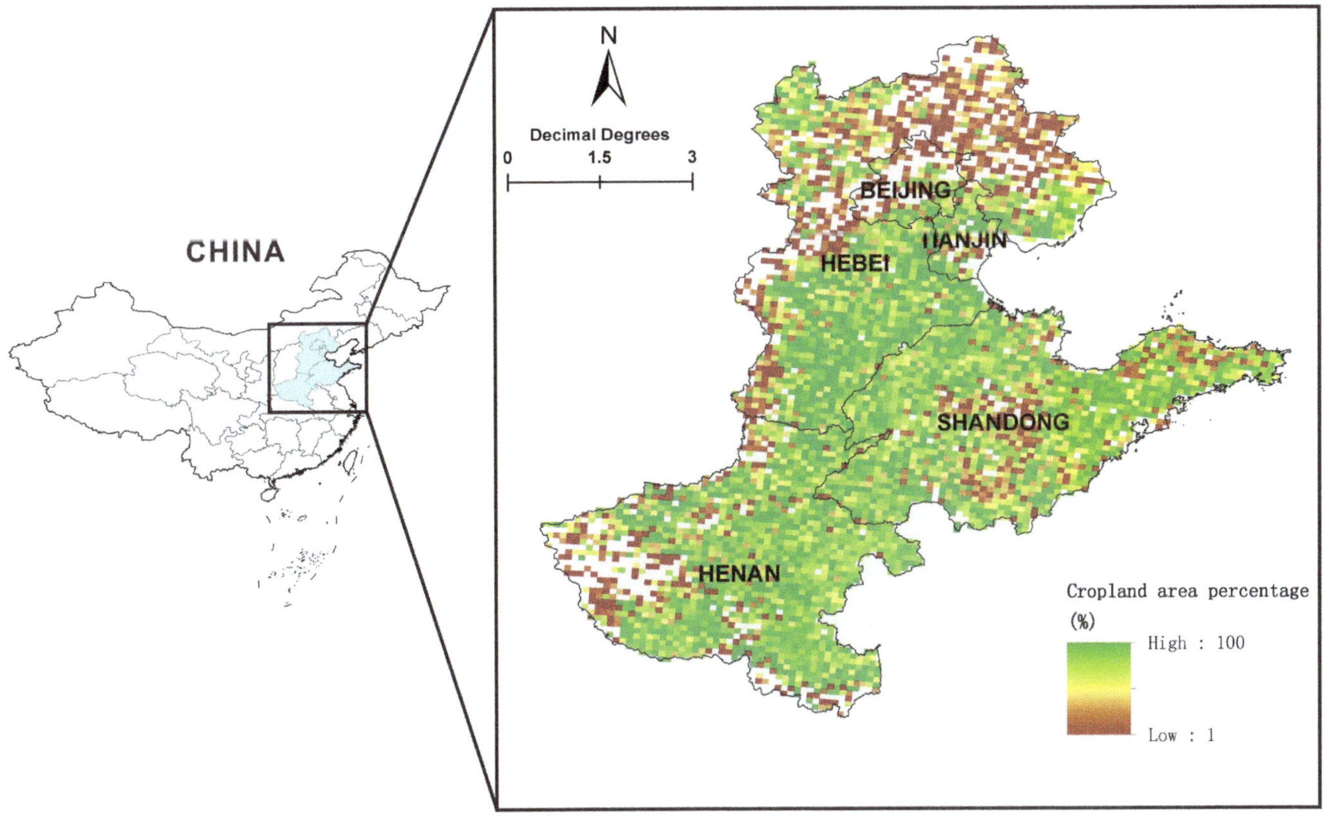

Figure 1. Spatial distribution of croplands in the North China Plain.

the heavy-C pool, and (iii) the decomposition of the heavy-C pool only produces CO_2. More detailed description of the Agro-C model was shown in Yu et al. [5] and Huang et al. [12].

In our previous studies [5,12], Agro-C has already been parameterized using observed data of several field measurements, and validated against independent datasets across a vast area representing different cropping systems, soil and climate conditions in China. The results indicate that the calibrated Agro-C model can reasonably capture observed aboveground biomass and changes in SOC under various agricultural management practices at different sites across china [5,12]. In the present study, we used the previously calibrated Agro-C model to simulate regional SOC dynamics of agro-ecosystems in North China Plain (NCP).

1.2 Spatial data

The changes in SOC under different climatic and management scenarios were simulated with a daily step from 2011 to 2100. Upscaling of the Agro-C model was accomplished by first rasterising the model inputs (e.g., climate data, initial soil properties and agricultural management information) with 10 km×10 km gridded datasets across the North China Plain and then running the model grid by grid across the study area.

Input soil properties in Agro-C include the concentrations of organic carbon and total nitrogen, bulk density, clay and sand fraction, and pH in the topsoil to 30 cm depth. These soil properties, with a 10 km×10 km spatial resolution, were firstly obtained from China Soil Scientific Database [19], representing more than 7000 soil profile measurements across China made in the late 1970s/early 1980s. Using Agro-C, we then simulated these soil properties grid by grid on a daily step from 1981 to 2010, based on the historical climate data and recorded agricultural management practices [5]. The model outputs of soil properties, including the amount of C in different pools, in the end of 2010 were further compiled as the initial model input of soil properties of the present study.

Gridded daily climate data such as maximum and minimum temperature, precipitation, and relative humidity from 2010 to 2100 were derived from the projections of the climate model FGOALS (Document S1). Due to the coarser spatial resolution (2.8° longitude×2.8° latitude) of outputs from the FGOALS model, the projections for the future GHG concentration scenarios [i.e., a set of four representative concentration pathways (RCPs) named RCP2.6, RCP4.5, RCP6.0 and RCP8.5 respectively; Figure S1 and Figure S2] of the FGOALS were therefore statistically downscaled [20,21] to a resolution of 10 km×10 km (Document S1). Additionally, as a driving variable in Agro-C, atmospheric carbon dioxide (CO_2) concentration affects the process of crop photosynthesis [12], which under each future climate scenario was obtained from Clarke et al. [22] and used as model inputs in this study.

The spatial distribution of croplands was computed based on the National Land Cover Data Sets (NLCD) of China that was developed from Landsat TM digital images around 2000 [23]. Due to a lack of yearly cropland data with sufficient spatial distribution, we assumed that the area of croplands in North China Plain did not significantly change over the study period.

1.3 Model initialization

In Agro-C model, the half-life residence time for labile-C, resistant-C, light-C and heavy-C were 0.1 y, 2.3 y, 7.6 y, and 105.4 y [5], respectively. To obtain initial fractions for the different C pools in 1981, a 'spin-up' procedure was performed by first setting the initial fractions of light-C and heavy-C as 0.25 and 0.75 [5], respectively, and then running the model until a steady state of SOC pool was achieved. In this, an equilibrium run was not stopped until the difference of SOC pools between two successive years was less than 0.1% of the pools themselves. The length of spin-up procedure differed across different places, and generally lasted about 100 years. The average time duration of the spin-up seemed relatively short but is reasonable. Most of Chinese croplands, particularly in North China Plain, have generally experienced a very long history of cultivation. When the spin-up procedure was performed, the average amount of crop stubble retention and manure amendment over the period between 1980 and 1989 was used as organic C input, due to the unavailability of organic C input before 1980. The average of 1960–1990 climate were used as driving data, because of the unavailability of climate data before 1960. The other driving datasets such as nitrogen fertilizer and irrigation were set to 1980–1989 average values. We then calculated the initial values of different C pools based on the total SOC information obtained from the above mentioned soil database and the spin-upped fractions of different sub-C pools. The dynamics of different C pools in each grid from 1981 to 2010 on a daily time step were then modeled based on the historical climate data and recorded agricultural management practices [5]. The simulated values of above-mentioned four C pools in the end of 2010 were further compiled as the model inputs of the present study.

1.4 Agricultural management scenarios

Nitrogen (N) fertilization, application of manure, and crop stubble retention through tillage are the major management practices that affect SOC dynamics in croplands of North China Plain (NCP). A database containing the county-level yearly cultivation acreage and total amount of synthetic N application (Chinese Academy of Agricultural Sciences [24]), shows that N fertilization rates in 2010 across croplands of NCP varied from 0 to 800 kg ha^{-1} (Figure S3A). The environmental pressures that are related to excessive mineral N fertilization in Chinese croplands decrease the possibility of increasing its use in the future [25]. Another province-level database [26] shows that farm manure application rates during the past 30 years varied from 1000 to 4000 kg ha^{-1} across the studied area (Figure S3B). In the croplands of NCP, the average stubble retention rate over the period of 2000–2010 varied from ~40% to ~60% across space [27]. In general, an increasing fraction of crop residues were being left in the fields instead of being taken away for household fuel and animal feed in China over the past several decades [25], which was mainly attributed to the financial motivations of the government policies. To quantify the effects of both climate change and agricultural management practices on SOC, we simulated SOC changes within the winter wheat-summer maize systems under 225 combinations of N fertilization, manure application and stubble retention rates, under each of the above-mentioned four climate scenarios.

In this study, management practices included 9 nitrogen application rates (0–800 kg N ha^{-1} yr^{-1} in 100 kg N ha^{-1} yr^{-1} increments, i.e., N:0, N:100, N:200, N:300, N:400, N:500, N:600, N:700, and N:800). Apart from N:0, we specified that 55% of the N fertilizers were applied during summer-maize growing seasons, with another 45% applied during winter-wheat growing seasons. For each growing season, 40% of the N fertilizers were applied at crop emergence, another 40% applied at tillering, and the rest applied at heading for wheat and silking for maize. We also specified five manure application rates (0–4000 kg ha^{-1} yr^{-1} in 1000 kg ha^{-1} yr^{-1} increments, i.e., M:0, M:1000, M:2000, M:3000, and M:4000), and five stubble retention rates (0–100% in 25% increments, i.e., R:0, R:25, R:50, R:75, and R:100). Apart from M:0, 72% of the manure was applied before sowing of summer maize, with the rest applied before sowing of winter wheat. Stubble retention rates denoted the percentage of aboveground straw and leaf biomass incorporated into the soil by tillage after harvesting, with the rest directly removed from the system. Combinations of management practices are abbreviated in this study as M:x, N:x, R:x. The M:0, N:0, R:0 denotes the management practice with no N fertilization, no manure application, and no stubble retention.

Because of a lack of detailed crop cultivar and phenology data on the regional scale, we simply assumed that the same wheat and maize cultivars were used throughout the studied area, and crop parameters were set as default values of the pre-calibrated Agro-C model. Sowing dates of maize and wheat were based on local crop calendars. From the south to north in NCP, the sowing date ranged from 6 June to 15 June for maize, and ranged from 5 October to 11 November for wheat [28,29]. Crop growth stops at physiological maturity as determined by the accumulated temperature. In addition, according to the local agronomic management, we assumed that the soil moisture would arrive at field capacity using irrigation module when the modeled soil moisture in the 30 cm depth drops to 70% of the field capacity during the growing seasons [5].

In total, we ran 4,086,000 (225 management scenarios ×4 climate scenarios ×4540 grids) Agro-C simulations. Each simulation quantified SOC content in the top 30 cm of soil from 2011 to 2100 on a daily step.

1.5 Identifying controls on SOC dynamics

In the present study, we assessed the impacts of management practices, climatic and soil variables on SOC change using Spearman's rank correlation coefficient (*rho*). Selected climatic variables included mean annual temperature (hereafter simply denoted as *temperature*) and mean annual precipitation (hereafter simply denoted as *precipitation*) since these two variables have been reported uncorrelated and significantly represent the spatial variation in a range of climate patterns [30]. For correlation analysis, the long-term daily climate variables were summarized to mean annual values under each climate scenario for each grid. Selected soil parameters included the main model edaphic inputs, e.g., pH, initial SOC, clay and total nitrogen content. Change in SOC of the top 30 cm (ΔSOC) is calculated as the difference in SOC between 2100 and 2011. Spearman's rank correlation coefficient was then calculated between ΔSOC and the management practices and environmental variables across the full set of Agro-C simulations. The sign of *rho*, positive or negative, indicates the direction of association between the independent and dependent variables. The absolute magnitude of *rho*, between 0 and 1, suggests the strength of correlation between the two variables.

We also investigated the effects of the initial SOC content and each of the three management practices on ΔSOC by producing boxplots characterizing the influence of each variable on ΔSOC including the variance calculated across agricultural management practices, grids, and the four climate scenarios. All analyses were performed using statistical and graphical software R 3.0.1 [31].

Results

2.1 Correlations between environmental variables, management practices and ΔSOC

Initial SOC was strongly but negatively correlated (median *rho* = −0.87) with ΔSOC (Figure 2). Initial soil clay content showed a weak negative correlation (median *rho* = −0.21) with ΔSOC (Figure 2). The climatic variables, e.g., temperature and precipitation, displayed a weak positive (median *rho* = 0.21) and negligible correlation with ΔSOC, respectively (Figure 2). Agricultural management such as stubble retention (median *rho* = 0.45) and manure application (median *rho* = 0.81) both showed a strong and positive correlation, whereas N fertilization showed a negligible correlation with ΔSOC (Figure 2).

Figure 3 presented the impacts of initial SOC content, manure application, N fertilization and stubble retention, respectively, on ΔSOC. On average, initial SOC content seemed to linearly and negatively correlate with ΔSOC (Figure 3A), whereas manure application and stubble retention had a relatively linear positive effect (Figure 3B and D) on ΔSOC. N fertilization increased ΔSOC most significantly ($P<0.05$) at rates between 0 and 300 kg ha^{-1} yr^{-1}, with a limited influence at higher fertilization rates. Across the sets involved with different climate scenarios, the median ΔSOC increased slightly, although not significantly ($P>0.05$), from RCP2.6 to RCP8.5 (Figure 4).

2.2 Spatiotemporal effects of agricultural practices on SOC

On average of the four climate scenarios, the response of ΔSOC to N fertilization, manure application and stubble retention varied over the study area, and cropland soils in Hebei and northern part of Shandong Province would accumulate more carbon than the rest of the study area if agricultural practices were optimized (Figure 5 and 6). Increasing the manure application rate led to soil across more areas turning into a net C sink, regardless of the other two management practices (Figure 5 and 6). N fertilization could not significantly halt the declining SOC trend unless more crop residues were incorporated into soils. For example, when stubble retention rate increased from 0 to 100% under M:2000, SOC increased in 11.4%, 27.7%, 40.7%, 53.9%, and 63.9% of the study area under N:0 (Figure 5), whereas increased in 13.5%, 44.5%, 62.5%, 71.3%, and 76.6% of the study area under N:300 (Figure 6). If the amount of manure application rate reached 4000 kg ha^{-1} yr^{-1} and all stubble incorporated into soils (i.e., M:4000 and R:100), SOC increased in

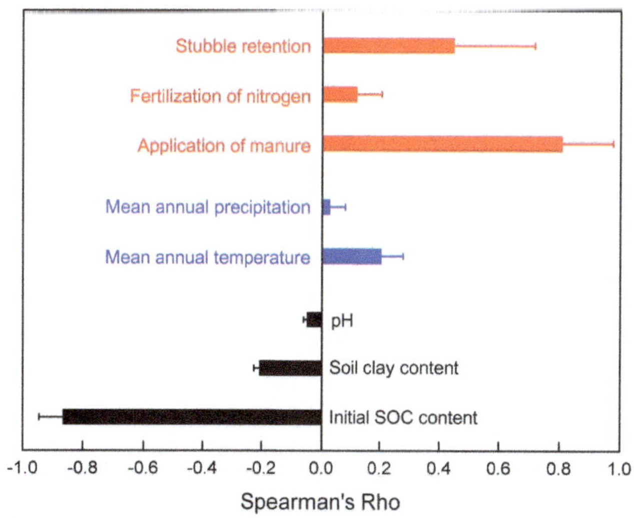

Figure 2. Spearman's rank correlation coefficients between ΔSOC (2011–2100, Mg C ha^{-1}) and climate (blue), soil (black), and management (red) variables. All the tests were significant ($P<0.001$).

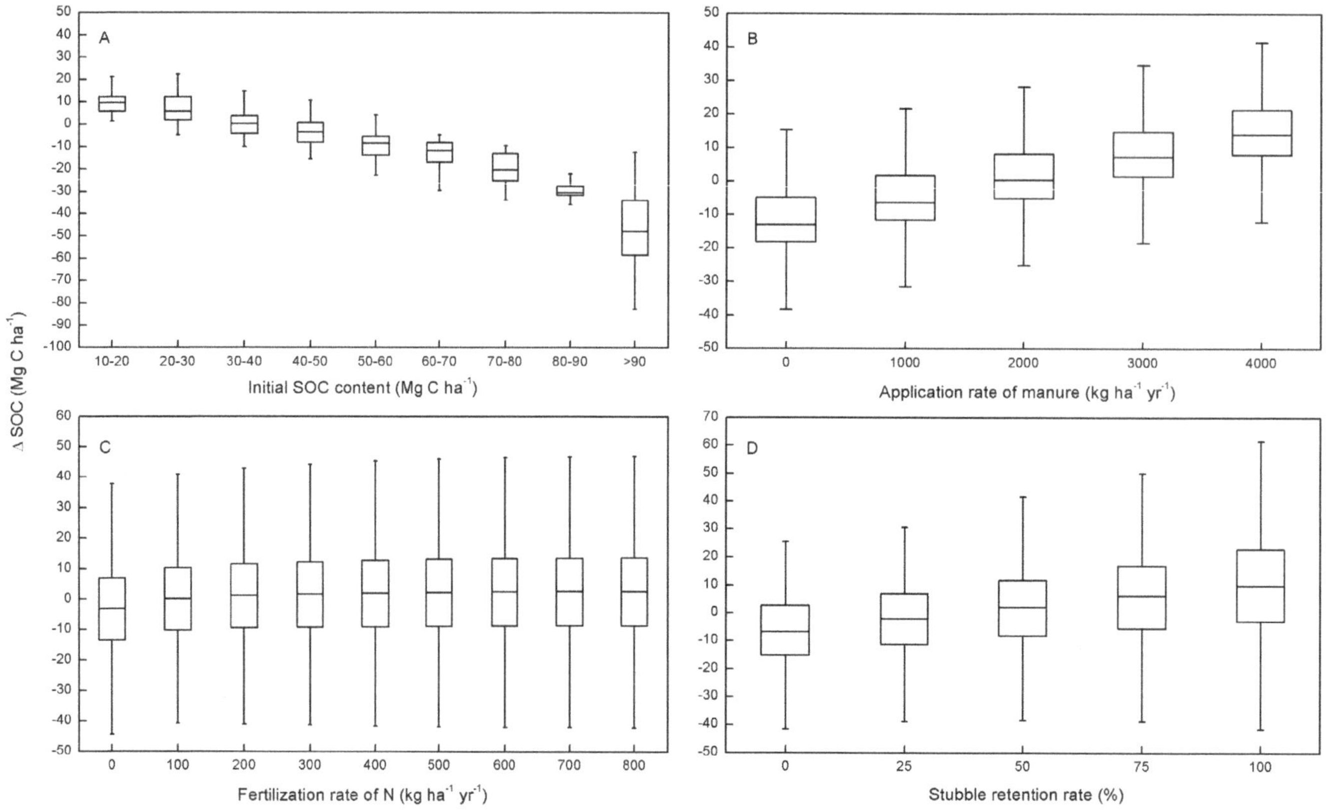

Figure 3. Response of ΔSOC (2011–2100, Mg C ha^{-1}) to initial SOC content (A), manure application (B), N fertilization (C), and stubble retention (D). Boxplots show the median and interquartile range, with whiskers extending to the most extreme data point within 1.5×(75–25%) data range.

more than 90% of the study area under both N:0 and N:300 (Figure 5 and 6).

Under M:0, N:0, R:0, the average SOC content across croplands of NCP kept decreasing from 35.7 Mg ha^{-1} during the 2010s to 16.1 Mg ha^{-1} during the 2090s, on average of the four climate scenarios (Figure 7). However, adopting a moderate management practice (e.g., M:2000, N:100, R:50) could maintain the SOC level during the next 90 years (Figure 7). Under M:4000, N:300, R:100, the average SOC density kept increasing from 39.5 Mg ha^{-1} during the 2010s to 54.6 Mg ha^{-1} during the 2060s, with a relatively lower increasing rate afterward (Figure 7).

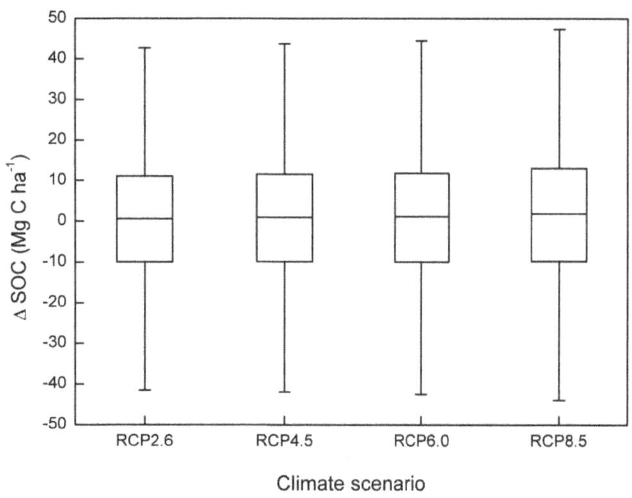

Figure 4. Response of ΔSOC (2011–2100, Mg C ha^{-1}) to different climate scenarios. Boxplots show the median and interquartile range, with whiskers extending to the most extreme data point within 1.5×(75–25%) data range.

Discussion

3.1 Interpretation and implication of the results

Initial SOC content was identified as strongly and negatively influencing SOC change (Figure 2). This is because, in model initialization, organic C input were taken to be the average of crop stubble and manure from 1980 to 1989. However, in the following model simulations with various kinds of management scenarios, great land use changes took place and the equilibrium of soil C pools got changed. In this, under otherwise similar environmental and managed conditions, soils with higher initial SOC content would experience greater SOC loss, and vice versa [32]. This negative correlation between ΔSOC and initial SOC content has also been documented in other studies [33–35]. Although higher soil clay content has been reported to benefit the stabilization of soil organic matter (SOM) in many studies [36–38], we identified a weak negative correlation between ΔSOC and soil clay content in the present study (Figure 2). This is because, other than the benefits of SOM stabilization, higher soil clay content could also reduce the amount of C inputs to soil [12], thereby showing a negative impact on total SOC variations in the present study.

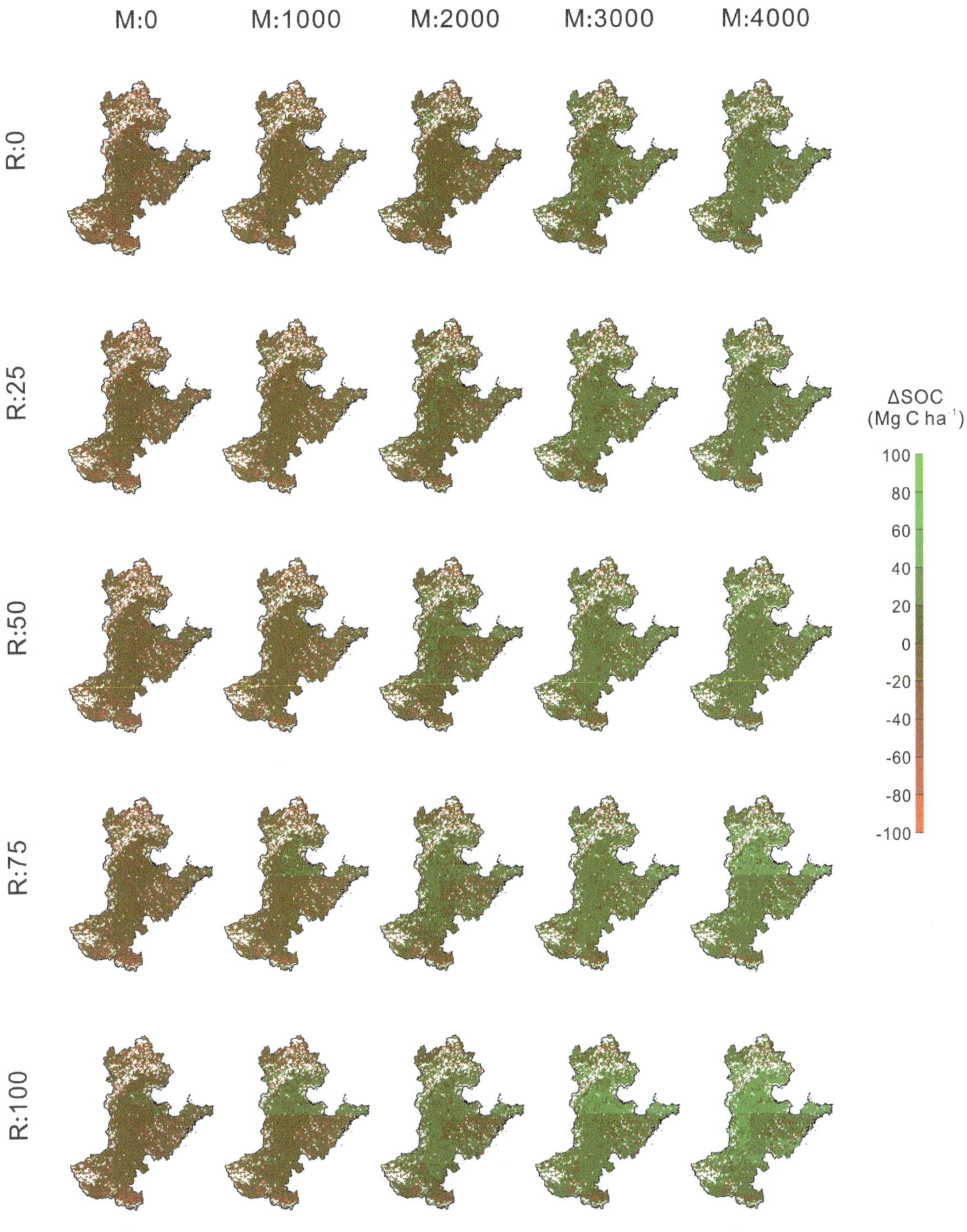

Figure 5. Spatial pattern of the effects of manure application and stubble retention on ΔSOC (2011–2100, Mg C ha^{-1}) under N:0. For a certain combination of management practices, ΔSOC in each grid represents the mean value of the results under four climate scenarios.

In Agro-C, high soil temperature and soil water content increase the SOC decomposition rate which underpins the negative relationship of SOC accumulation with both temperature and rainfall [12]. However, in NCP, we found that the mean annual temperature and precipitation were not strongly correlated with ΔSOC (Figure 2), and higher temperature seemed to benefit SOC accumulation (Figure 4 and Figure S1). This is because the influences of temperature and precipitation on SOC dynamics depend on other factors such as soil and management practices [39]. For example, artificial irrigation, which was adopted across all sets of scenarios in this study, would certainly overshade the possible impacts of precipitation on SOC change. And the elevated atmospheric CO_2 concentration accompanied with higher temperature increased the crop productivity, thereby compensating the elevated amount of decomposed SOC induced by climate warming [40]. Furthermore, Zhang and Huang [41] reported a positive correlation between air temperature and the production of cereal crops in China during the past 30 years. This

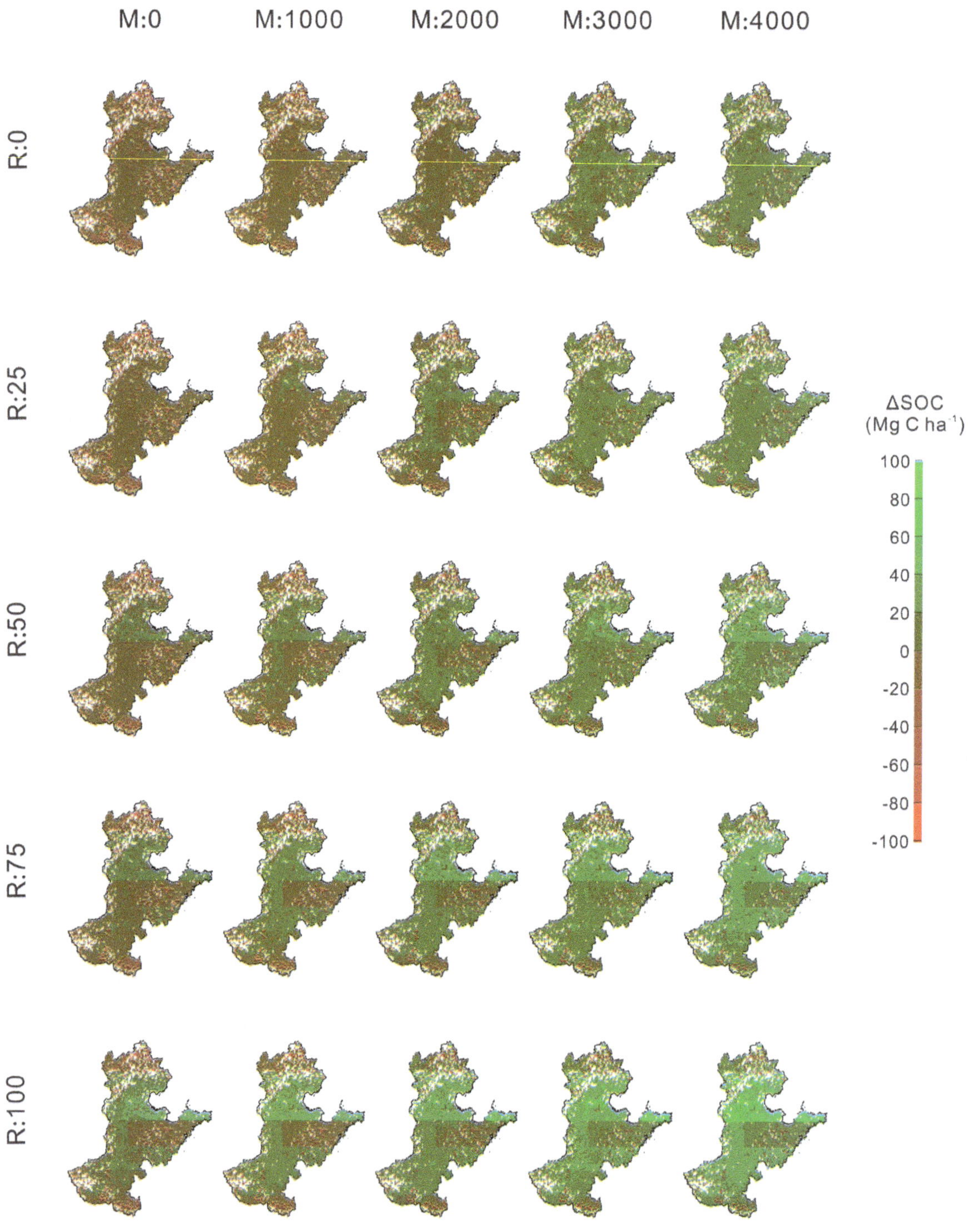

Figure 6. Spatial pattern of the effects of manure application and stubble retention on ΔSOC (2011–2100, Mg C ha^{-1}) under N:300. For a certain combination of management practices, ΔSOC in each grid represents the mean value of the results under four climate scenarios.

indicated that climate warming could potentially cause a higher amount of C input from crop residues to soils, with which our findings was consistent.

Manure application and stubble retention directly add C into soil, and were identified as the most two predominant agricultural management practices driving SOC (Figure 2 and 3). Although it has been reported that the manure application levels has been decreasing during the past 50 years, and the changing trends might continue in the future across Chinese croplands [25]. The increasing amount of incorporated crop residues, caused by both enhanced stubble retention rate [27] and elevated amount of crop NPP [42], could still potentially override the weakening manure application rates and increase SOC.

N fertilization enhances crop residue and root biomass production (Figure 8), thus increases the amount of C incorporated into soil. However, the impact of N fertilization depends on whether nutrient is limiting. In this study, we found that N fertilization showed a negligible correlation with changes in SOC

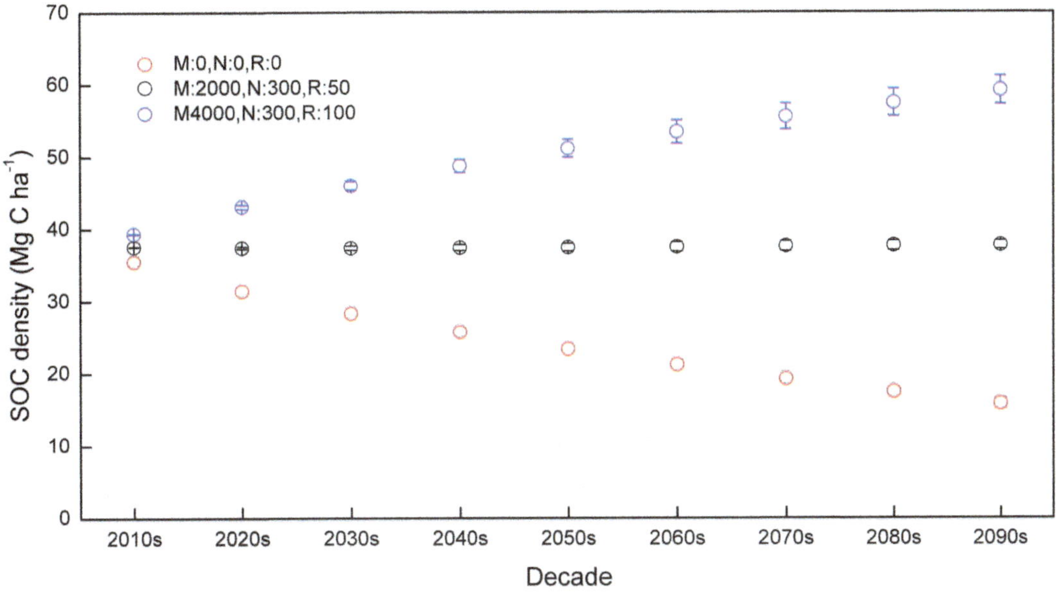

Figure 7. Temporal change of SOC (Mg C ha^{-1}) under three combinations of manure application, nitrogen fertilization and stubble retention. Open circles represent mean values of SOC under four climate scenarios, with standard deviations represented by error bars on the circles.

(Figure 2). This is mainly because changes in SOC is dominated by manure application and stubble retention in most of the scenarios. However, under scenarios with high stubble retention rates and low manure application rates, effects of N fertilization could be more obvious (Figure 5 and 6). Additionally, our results also showed that although N fertilization had a positive effect on SOC accumulation, the effect did not further increase if N fertilization rate exceeds 300 kg ha^{-1} yr^{-1} (Figure 3C). This is because with a N fertilization rate higher than 300 kg ha^{-1} yr^{-1}, nutrient was not a limiting factor of crop growth, and enhancing N inputs could not significantly elevate the amount of soil C inputs from crop residues and roots (Figure 8). Additionally, excessive use of N fertilizer is currently very common in NCP [43], which has aroused concerns about its negative impacts on ecosystems and environments [44]. We suggest that a N fertilization rate of about 300 kg ha^{-1} yr^{-1} should be adopted to both maintain SOC accumulation and reduce the environmental risks caused by excessive use of N fertilizers. As the benefit of N fertilization on SOC sequestration could be offset by N_2O emissions from the soil and greenhouse gas (GHG) emissions from fossil fuel combustion during the processes involving fertilizer production and transport [45], there remains a need to further investigate the proper rates and timing of N fertilization under different climate and soil conditions.

With a high (M:4000, N:300, R:100) C input management, the temporal change in SOC generally experienced a fast to low increase, with the most gain of soil C happened in the first 50 years of simulations (Figure 7). This is consistent with a number of other studies [1,46], which suggest that soil C would reach a new steady state after 20–50 years of cultivation if management practices were significantly changed. Moreover, our results revealed that under M:4000, N:300, R:100, the average SOC density of NCP croplands could achieve 55 Mg ha^{-1} during 2060s, which is similar to the current worldwide average agricultural SOC density [47]. However, the average SOC density would keep decreasing under M:0, N:0, R:0, and the "new" steady state of SOC was not attained even after 90 years of simulation (Figure 7). This could be attributed to the limited C input to balance the SOC loss by decomposition. With a much lower decomposition rate, the

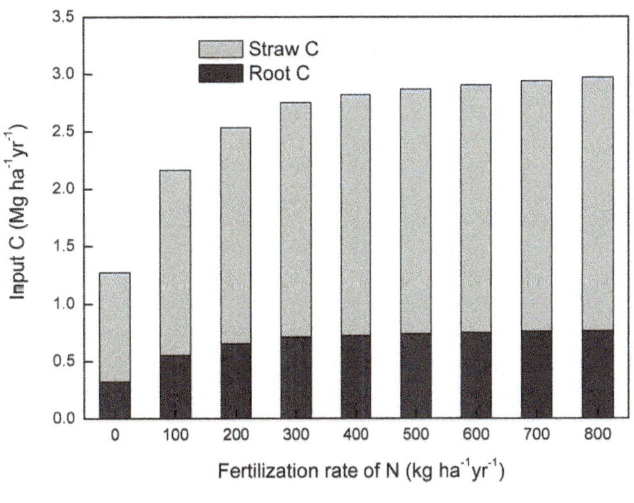

Figure 8. Response of mean annual C inputs from crop residues and roots (Mg C ha^{-1}) to different N fertilization rates.

proportion and absolute amount of heavy-C would have significant impact on long-term SOC balance.

3.2 Uncertainties and limitations

Several uncertainties and limitations should be noticed in interpreting the simulation results in this study. First, several management scenarios simulated in this study represent the most extreme conditions and may not happen in the real world. For example, zero inputs of N and C to soil, because it is improbable to extract all roots and residues from the fields, or prevent atmospheric N deposition. Second, the possible improvement in future crop varieties, which could potentially affect the quality and quantity of soil C inputs, was not considered in this study. For example, if the crop variety improvement increased the production of either crop residues or roots, this could lead to an underestimation of our results in the future SOC accumulation. Third, the effects of conservation tillage on SOC were not simulated in this study, due to a lack of the detailed information on annual conservation tillage practice. However, it could still be hypothesized that our results could represent most parts of the studied areas, because the average fraction of the area with conservation tillage practices in Chinese croplands is only about 9% in 2009 [5]. Moreover, a meta-analysis of global data on the effects of tillage on SOC change [48] indicates that conversion from conventional tillage to conservation tillage only redistributed more C into the top layer and less C into the deeper layer, and the total SOC stock in the upper 40 cm soil profile was not significantly changed. Fourth, the rainfed condition was not simulated because artificial irrigation has been traditionally adopted to support crop production in NCP. However, the current irrigation-supported agricultural production might be affected by future resource and environment related pressures [49]. In this, other environmental factors such as precipitation and temperature may play a more important role in regulating both crop productivity and SOC dynamics [50]. Last but not least, certain issues related to state-of-the-art model limitations should be mentioned. For instance, in a first-order decay model (e.g., Agro-C), there is a general linear relation between C input and SOC variation, which usually contradicts the fact that increase of C inputs by crop residues may have variable effects on SOC change in the real world [51]. Future efforts are needed from model developers to implement currently uncertain mechanisms, such as SOC saturation, priming, and different sensitivity of SOC pools to temperature, which will enable a better understanding the influence of environment on agricultural SOC dynamics.

Supporting Information

Figure S1 Climate change in North China Plain (NCP) from 2011 to 2100 under the RCP2.6 (A), RCP4.5 (B), RCP6.0 (C) and RCP8.5 (D) scenarios. The solid circles represent the annual temperature, and the open triangles represent annual precipitation.

Figure S2 Spatial distribution of mean annual temperature and precipitation increments in North China Plain (NCP) between 2011 and 2100 under four climate scenarios.

Figure S3 The range and density of N fertilization (A) and manure application (B) rates in North China Plain (NCP). The N fertilization data represent the year of 2010 on a county scale, and the manure application data represent that from 1981 to 2010 on a provincial scale. The datasets were derived from online database and published literatures. *n* shows the sample size.

Acknowledgments

Many thanks to Dr. Yongqiang Yu from the State Key Laboratory of Numerical Modeling for Atmospheric Sciences and Geophysical Fluid Dynamics (LASG), Institute of Atmospheric Physics, Chinese Academy of Sciences, for providing the access to the FGOALS modeling result datasets.

Author Contributions

Conceived and designed the experiments: GW TL. Performed the experiments: GW TL. Analyzed the data: GW TL. Contributed reagents/materials/analysis tools: TL WZ YY. Wrote the paper: GW.

References

1. Lal R (2004) Soil carbon sequestration impacts on global climate change and food security. Science 304: 1623–1627.
2. Wang G, Huang Y, Wang E, Yu Y, Zhang W (2013) Modeling Soil Organic Carbon Change across Australian Wheat Growing Areas, 1960–2010. PloS one 8: e63324.
3. Brady NC, Weil RR (2008) The nature and properties of soils (14th ed). Upper Saddle River: Pearson Education Publication.
4. Van Wesemael B, Paustian K, Meersmans J, Goidts E, Barancikova G, et al. (2010) Agricultural management explains historic changes in regional soil carbon stocks. Proceedings of the National Academy of Sciences 107: 14926–14930.
5. Yu Y, Huang Y, Zhang W (2012) Modelling soil organic carbon change in croplands of China, 1980–2009. Global and Planetary Change 82–83: 115–128.
6. Li C, Frolking S, Harriss R (1994) Modeling carbon biogeochemistry in agricultural soils. Global biogeochemical cycles 8: 237–254.
7. Jenkinson D, Andrew S, Lynch J, Goss M, Tinker P (1990) The Turnover of Organic Carbon and Nitrogen in Soil. Philosophical Transactions: Biological Sciences 329: 361–368.
8. Parton W, Scurlock J, Ojima D, Gilmanov T, Scholes R, et al. (1993) Observations and modeling of biomass and soil organic matter dynamics for the grassland biome worldwide. Global biogeochemical cycles 7: 785–809.
9. Li CS, Zhuang Y, Frolking S, Galloway J, Harriss R, et al. (2003) Modeling soil organic carbon change in croplands of China. Ecological Applications 13: 327–336.
10. Falloon P, Smith P, Bradley R, Milne R, Tomlinson R, et al. (2006) RothCUK–a dynamic modelling system for estimating changes in soil C from mineral soils at 1-km resolution in the UK. Soil Use and Management 22: 274–288.
11. Ogle SM, Breidt F, Easter M, Williams S, Killian K, et al. (2010) Scale and uncertainty in modeled soil organic carbon stock changes for US croplands using a process-based model. Global Change Biology 16: 810–822.
12. Huang Y, Yu Y, Zhang W, Sun W, Liu S, et al. (2009) Agro-C: A biogeophysical model for simulating the carbon budget of agroecosystems. Agricultural and Forest Meteorology 149: 106–129.
13. Ding Y, Ren G, Shi G, Gong P, Zheng X, et al. (2006) National assessment report of climate change (I): climate change in China and its future trend. Advances in Climate Change Research 2: 3–8 (in Chinese).
14. Huang Y, Sun W (2006) Changes in topsoil organic carbon of croplands in mainland China over the last two decades. Chinese Science Bulletin 51: 1785–1803.
15. Bond-Lamberty B, Thomson A (2010) Temperature-associated increases in the global soil respiration record. Nature 464: 579–582.
16. Yu Y, Huang Y, Zhang W (2013) Projected changes in soil organic carbon stocks of China's croplands under different agricultural managements, 2011–2050. Agriculture, Ecosystems & Environment 178: 109–120.
17. West TO, Post WM (2002) Soil organic carbon sequestration rates by tillage and crop rotation: A global data analysis. Soil Science Society of America Journal 66: 1930–1946.
18. Luo Z, Wang E, Sun OJ, Smith CJ, Probert ME (2011) Modeling long-term soil carbon dynamics and sequestration potential in semi-arid agro-ecosystems. Agricultural and Forest Meteorology 151: 1529–1544.
19. Yu D, Shi X, Wang H, Sun W, Chen J, et al. (2007) Regional patterns of soil organic carbon stocks in China. Journal of Environmental Management 85: 680–689.
20. Bertacchi Uvo C, Olsson J, Morita O, Jinno K, Kawamura A, et al. (2001) Statistical atmospheric downscaling for rainfall estimation in Kyushu Island, Japan. Hydrology and Earth System Sciences 5: 259–271.
21. Kidson JW, Thompson CS (1998) A comparison of statistical and model-based downscaling techniques for estimating local climate variations. Journal of Climate 11: 735–753.
22. Clarke L, Edmonds J, Krey V, Richels R, Rose S, et al. (2009) International climate policy architectures: Overview of the EMF 22 International Scenarios. Energy Economics 31: S64–S81.
23. Liu J, Liu M, Zhuang D, Zhang Z, Deng X (2003) Study on spatial pattern of land-use change in China during 1995–2000. Science in China Series D: Earth Sciences 46: 373–384.

24. Huang Y, Tang Y (2010) An estimate of greenhouse gas (N_2O and CO_2) mitigation potential under various scenarios of nitrogen use efficiency in Chinese croplands. Global Change Biology 16: 2958–2970.
25. Zhang W, Yu Y, Huang Y, Li T, Wang P (2011) Modeling methane emissions from irrigated rice cultivation in China from 1960 to 2050. Global Change Biology 17: 3511–3523.
26. National Development and Reform Commission (2007) The People's Republic of China, Initial National Communication on Climate Change (In Chinese). China Planning Press, Beijing, China.
27. Gao L, Ma L, Zhang W, Wang F, Ma W, et al. (2009) Analysis on the Quantities and Utilization of Crop Straw and its Nutrient in Huang-Huai-Hai Region. Chinese Agricultural Science Bulletin 25: 186–193 (in Chinese with English summery).
28. Wu D, Yu Q, Lu C, Hengsdijk H (2006) Quantifying production potentials of winter wheat in the North China Plain. European Journal of Agronomy 24: 226–235.
29. Wu D, Yu Q, Wang E, Hengsdijk H (2008) Impact of spatial-temporal variations of climatic variables on summer maize yield in North China Plain. International Journal of Plant Production 2: 71–88.
30. Bryan BA (2003) Physical environmental modeling, visualization and query for supporting landscape planning decisions. Landscape and urban planning 65: 237–259.
31. R Development Core Team (2013) R: A language and environment for statistical computing. Vienna, Austria: R Foundation for Statistical Computing.
32. Sanderman J, Baldock JA (2010) Accounting for soil carbon sequestration in national inventories: a soil scientist's perspective. Environmental Research Letters 5: 034003.
33. Saby N, Arrouays D, Antoni V, Lemercier B, Follain S, et al. (2008) Changes in soil organic carbon in a mountainous French region, 1990–2004. Soil Use and Management 24: 254–262.
34. Goidts E, Wesemael Bv, Van Oost K (2009) Driving forces of soil organic carbon evolution at the landscape and regional scale using data from a stratified soil monitoring. Global Change Biology 15: 2981–3000.
35. Zhao G, Bryan BA, King D, Luo Z, Wang E, et al. (2013) Impact of agricultural management practices on soil organic carbon: simulation of Australian wheat systems. Global change biology 19: 1585–1597.
36. Oades J (1988) The retention of organic matter in soils. Biogeochemistry 5: 35–70.
37. Amato M, Ladd J (1992) Decomposition of ^{14}C-labelled glucose and legume material in soils: Properties influencing the accumulation of organic residue C and microbial biomass C. Soil Biology and Biochemistry 24: 455–464.
38. Burke I, Yonker C, Parton W, Cole C, Schimel D, et al. (1989) Texture, climate, and cultivation effects on soil organic matter content in US grassland soils. Soil Science Society of America Journal 53: 800–805.
39. Giardina CP, Ryan MG (2000) Evidence that decomposition rates of organic carbon in mineral soil do not vary with temperature. Nature 404: 858–861.
40. Smith J, Smith P, Wattenbach M, Zaehle S, Hiederer R, et al. (2005) Projected changes in mineral soil carbon of European croplands and grasslands, 1990–2080. Global Change Biology 11: 2141–2152.
41. Zhang T, Huang Y (2012) Impacts of climate change and inter-annual variability on cereal crops in China from 1980 to 2008. Journal of the Science of Food and Agriculture 92: 1643–1652.
42. Huang Y, Zhang W, Sun W, Zheng X (2007) Net primary production of Chinese croplands from 1950 to 1999. Ecological Applications 17: 692–701.
43. Liu X, Ju X, Zhang F, Pan J, Christie P (2003) Nitrogen dynamics and budgets in a winter wheat–maize cropping system in the North China Plain. Field Crops Research 83: 111–124.
44. Richter A, Burrows JP, Nüβ H, Granier C, Niemeier U (2005) Increase in tropospheric nitrogen dioxide over China observed from space. Nature 437: 129–132.
45. Knorr W, Prentice I, House J, Holland E (2005) Long-term sensitivity of soil carbon turnover to warming. Nature 433: 298–301.
46. West TO, Six J (2007) Considering the influence of sequestration duration and carbon saturation on estimates of soil carbon capacity. Climatic Change 80: 25–41.
47. Batjes NH (1999) Management options for reducing CO_2-concentrations in the atmosphere by increasing carbon sequestration in the soil: National Institute of Public Health and the Environment.
48. Luo Z, Wang E, Sun OJ (2010) Can no-tillage stimulate carbon sequestration in agricultural soils? A meta-analysis of paired experiments. Agriculture, Ecosystems & Environment: 224–231.
49. Wang E, Yu Q, Wu D, Xia J (2008) Climate, agricultural production and hydrological balance in the North China Plain. International Journal of Climatology 28: 1959–1970.
50. Piao S, Ciais P, Huang Y, Shen Z, Peng S, et al. (2010) The impacts of climate change on water resources and agriculture in China. Nature 467: 43–51.
51. Powlson DS, Glendining MJ, Coleman K, Whitmore AP (2011) Implications for soil properties of removing cereal straw: results from long-term studies. Agronomy Journal 103: 279–287.

26

Soil Particle Heterogeneity Affects the Growth of a Rhizomatous Wetland Plant

Lin Huang, **Bi-Cheng Dong**, **Wei Xue, Yi-Ke Peng, Ming-Xiang Zhang, Fei-Hai Yu***

School of Nature Conservation, Beijing Forestry University, Beijing, China

Abstract

Soil is commonly composed of particles of different sizes, and soil particle size may greatly affect the growth of plants because it affects soil physical and chemical properties. However, no study has tested the effects of soil particle heterogeneity on the growth of clonal plants. We conducted a greenhouse experiment in which individual ramets of the wetland plant *Bolboschoenus planiculmis* were grown in three homogeneous soil treatments with uniformly sized quartz particles (small: 0.75 mm, medium: 1.5 mm, or large: 3 mm), one homogeneous treatment with an even mixture of large and medium particles, and two heterogeneous treatments consisting of 16 or 4 patches of large and medium particles. Biomass, ramet number, rhizome length and spacer length were significantly greater in the treatment with only medium particles than in the one with only large particles. Biomass, ramet number, rhizome length and tuber number in the patchy treatments were greater in patches of medium than of large particles; this difference was more pronounced when patches were small than when they were large. Soil particle size and soil particle heterogeneity can greatly affect the growth of clonal plants. Thus, studies to test the effects of soil heterogeneity on clonal plants should distinguish the effects of nutrient heterogeneity from those of particle heterogeneity.

Editor: Ben Bond-Lamberty, DOE Pacific Northwest National Laboratory, United States of America

Funding: This research was supported by the Forestry Commonwealth Project (201004078), the Fundamental Research Funds for the Central Universities (TD-JC-2013-1), the Program for New Century Excellent Talents in University (NECT-10-0234) and the National Science Foundation of China (31070371). The funders had no role in study design, data collection and analysis, decision to publish, or preparation of the manuscript.

Competing Interests: The authors have declared that no competing interests exist.

* E-mail: feihaiyu@bjfu.edu.cn

These authors contributed equally to this work.

Introduction

Spatial heterogeneity is an important feature of natural habitats [1–5], and plays major roles in shaping plant growth, species interactions, community structure and ecosystem functioning [6–15]. Heterogeneity occurs in various spatial scales [2,3,5], and the scale of heterogeneity is important because ecological processes that function at one spatial scale may not do so at other scales [16–21].

Clonal plants are very abundant in nature and often experience spatial heterogeneity in such a way that connected ramets of the same clone experience contrasting environmental conditions [22–24]. Many studies have addressed how spatial heterogeneity in essential resources such as light, soil water and mineral nutrients affects the growth of clonal plants, and these studies generally show that clonal plants can respond to heterogeneity in ways that increase their performance and likely fitness [7,19–20,25–29]. Moreover, clonal plants can respond differently to heterogeneity of soil and light at different spatial scales [18–21,30].

Particle size is an important feature of soil [31–34], and it can affect soil physical and chemical properties such as structure, oxygen concentration and water content [33,35–38]. Studies testing the effects of soil nutrient heterogeneity on the performance of clonal plants sometimes compare the growth in substrate with patches of fertile and infertile soil to the growth in a homogeneous mixture of the same amounts of fertile and infertile soil [6,20–21,39–44]. However, fertile and infertile soils generally differ in the size composition of particles, meaning that results of these studies might be partly due to effects of particle size rather than nutrient availability. To test the effects of soil nutrient heterogeneity, therefore, it would be good to create nutrient-rich and nutrient-poor patches with the same type of soil substrate but supplied with different amounts of nutrients [11,27,45–47]. On the other hand, it would be useful to test the effects of particle size separately. However, no published study appears to have tested the effects of soil particle heterogeneity on the growth of clonal plants.

To test the effects of soil particle heterogeneity on clonal growth, we conducted a greenhouse experiment in which we grew individual ramets of the rhizomatous, wetland plant *Bolboschoenus planiculmis* in three homogeneous soils each containing only one size class of quartz particles (large, medium or small), one homogeneous soil containing an even mixture of large and medium particles, and two patchy soils consisting of separate patches of large and medium particles and differing in patch scale (large and small patches). We asked the following questions. First, does soil particle size affect the growth (biomass, number of ramets, rhizome length and number of tubers) of *B. planiculmis*, as shown by comparing the treatments with only large, medium and small particles? Second, does patchiness or scale of patchiness of particle size affect the growth of *B. planiculmis* at the level of the whole clone, as shown by comparing the treatments with the homogeneously mixed and the two patchy large and medium particles? Based on effects of heterogeneity of other environmental factors on clonal growth, we expected that the growth in the homogeneously mixed treatment would be lowest and that the growth would differ between the two patchy treatments. Scaling theory predicts that organisms respond to spatial heterogeneity at scales related to their own size [16]. We therefore expected that *B. planiculmis* would show larger responses to soil particle heterogeneity at the scales closer to its spacer length (i.e. inter-ramet distance). Third, does patchiness or scale of patchiness of particle size affect the growth of *B. planiculmis* at the level of the patches? We expected that, in the patchy treatments, the growth of *B. planiculmis* would differ between patch types and that these differences would depend upon patch size.

Materials and Methods

Species and Sampling

Bolboschoenus planiculmis (F. Schmidt) T. V. Egorova (Cyperaceae), formerly called *Scirpus planiculmis* Fr. Schmidt, is a rhizomatous, perennial, herbaceous, wetland plant [48]. The species is native in China (e.g. in Beijing), and also occurs in India, Japan, Kazakhstan, Korea, Kyrgyzstan, Mongolia, Papua New Guinea, Russia, Tajikistan, southwestern Asia and Europe [48]. It attains heights of 0.6 to 1 m and forms solitary shoots from globose tubers. Shoots die back at the end of one growing season, but tubers can form new shoots and rhizomes that produce new tubers in the next growing season. Both tubers and rhizomes that connect them can survive for several years.

On 10 May 2011, more than 200 ramets, each consisting of a green shoot, a tuber at the base of the shoot and some roots, were collected from a natural population of *B. planiculmis* on the bank of the Beisha River in Beijing. Although the genetic background of these ramets is unknown, they were likely to originate from more than one genet because the longest distance between the sampled ramets was over 50 m. The sampling site did not belong to any farms or national parks and also did not involve any endangered or protected species, so we did not need any relevant permission for collecting plant samples. Each ramet was transplanted into a pot with sufficient nutrition to grow for 15 days. Then 48 ramets of similar size (10–15 cm high with 3–5 leaves) were chosen and used for the experiment.

Experimental Design

The experiment had three homogeneous treatments with a single type of particle size (termed "Small", "Medium" and "Large", respectively), two patchy treatments with different patch sizes ("Large patch" and "Small patch") and a mixture treatment ("Mixture"; Fig. 1). The substratum used in this experiment was pure quartz without any nutrients. In the Small, Medium and Large treatments, the particle sizes of the quartz were on average 0.75, 1.5 and 3.0 mm, respectively.

In a previous, eight-week, pilot experiment using the Small, Medium and Large treatments, it was found that *B. planiculmis* grew best in the Medium and worst in the Large treatment [49]. The medium and large particle sizes were, therefore, used in the Mixture and patchy treatments. In the Large patch treatment, each box (26 cm long×26 cm wide×13.0 cm deep) was divided equally into four parts (each 13 cm×13 cm), and two parts were filled with medium and two with large particles, in checkerboard fashion (Fig. 1). The Small patch treatment was similar except that each box was divided equally into 16 parts (each 6.5 cm×6.5 cm). In the Mixture treatment, each box was filled with a 1:1 (v: v) mixture of the medium and large particles (Fig. 1). There were no physical barriers between patches, and roots and rhizomes could grow freely between them. Each treatment had eight replicates (boxes) and there were a total of 48 boxes.

On 5 May 2011, the 48 ramets of *B. planiculmis* were randomly assigned to the six treatments and one ramet was planted in the center of each of the 48 boxes (Fig. 1). The boxes were closed at the bottom so that water and nutrients could not leak. We added 200 ml nutrient solution (1 g Peters Professional [Scotts Professional Limited, Nottingham, UK; N-P-K, 20-20-20] L^{-1} tap water) evenly to each box every week in the first eight weeks. The total amount of nutrients added corresponded to 14.201 g m^{-2} yr^{-1} of each of nitrogen, potassium and phosphorus. The concentration of the nutrient solution used was among the manufacturer's recommended application rate (Scotts Professional Limited, Nottingham, UK) and the nutrients are sufficient for the growth of the plants. Sufficient water was added to the boxes so that the soil in the boxes was flooded and there was always 1 to 3 cm deep water above the soil surface. These measures were to simulate typical wetland conditions experienced by *B. planiculmis* and also ensured that nutrients were evenly distributed in the boxes.

The experiment was conducted in a greenhouse at Forest Science Co., Ltd., of Beijing Forestry University, maintained at a mean temperature of 26.8°C and a mean relative humidity of 72% (measured by iButton DS 1923, Maxim Integrated Products, USA). The boxes were randomly positioned within an area of 18 m^2 and re-positioned systematically every two to three weeks to minimize effects of position.

Harvest and Measurements

The experiment lasted 16 weeks and was ended on 14 September 2011. One plant in each of the Small, Small patch and Large patch treatments died during the experiment, and these plants were excluded from harvest and subsequent analyses. In the homogeneous treatments (Small, Medium, Large and Mixture), the mother and offspring ramets were harvested separately. In the two heterogeneous treatments, the mother ramet and the offspring ramets in each type of patches were harvested separately. We counted number of ramets, number of non-sprouted tubers and measured rhizome length. These three variables are commonly used to measure the capacity of clonal growth, and tubers represent potential ramets. Roots, tubers, rhizomes, leaf blades and leaf sheaths were then separated, oven-dried at 70°C for at least 48 h, and weighed.

Data Analysis

We used one-way ANOVA to examine the effects of soil particle size on the growth (biomass, number of ramets, rhizome length and number of tubers), root-shoot ratio and spacer length (the distance between adjacent ramets) at the whole plant level. In these analyses, we included only the three homogeneous treatments with a single type of particle size (Small, Medium and Large). If a significant effect was detected, a Duncan test was used to compare individual means. We conducted one-way ANOVAs or non-parametric Kruskal-Wallis tests to examine the effects of spatial heterogeneity of soil particles on the growth, root-shoot ratio and spacer length at the whole plant level. In these analyses, three treatments (Small patch, Large patch and Mixture) were used.

We employed two-way ANOVA to examine the effects of particle size (large vs. medium) within patches and patchiness (Large patch vs. Small patch vs. the hypothesized "Null" treatment, see explanation below) on the growth of the plants at the patch level. Data from mother ramets were excluded because they were not located at any patches. The growth pattern

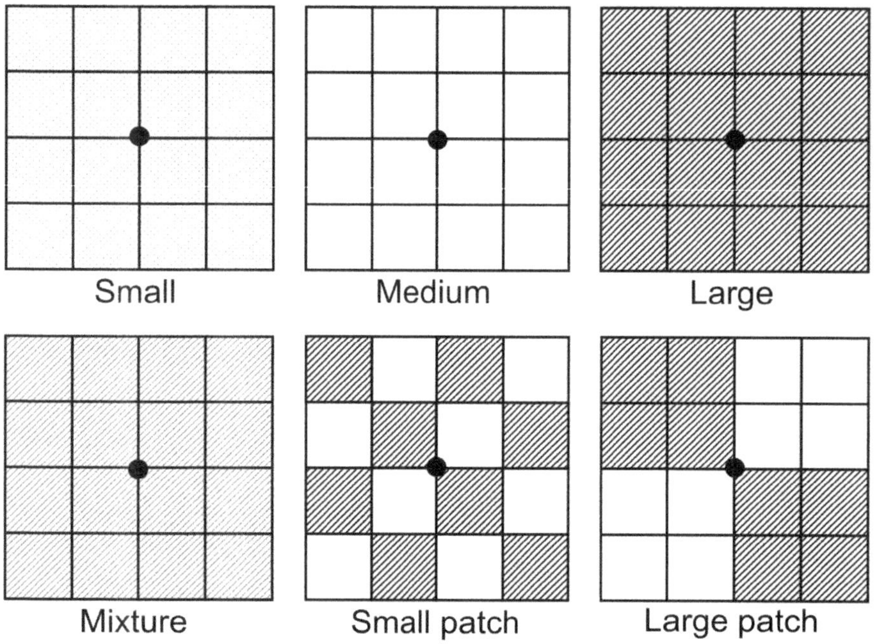

Figure 1. Schematic representation of the experimental design. The experiment had three homogeneous soil treatments with whole boxes filled with small (0.75 mm), medium (1.5 mm) or large (3 mm) quartz (coded as Small, Medium and Large, respectively), one homogeneous treatment with boxes filled with an even mixture of the same amount of large and medium quartz (coded as Mixture), and two heterogeneous treatments consisting of 16 or 4 patches of large and medium quartz (coded as Small patch and Large patch, respectively).

(distribution of biomass, ramets, rhizomes and tubers in the two types of soil particle patches) in the Null treatment represented the expected (null) growth pattern in soil without a real patch texture. For this treatment, half values of the corresponding growth variables in the treatment Large were used to represent the expected growth in the large particle patches, and half values of the corresponding growth variables in Medium were used to represent the expected growth in the medium particle patches. We used half values because in the Large patch and the Small patch treatment the plants in each type of patches occupied only half of the area of the whole box. The growth pattern in the Null treatment was used as a null model to show what the pattern was if the placement of biomass, ramets, rhizomes or tubers did not show selectivity (response). For the two heterogeneous treatments, if the growth pattern deviated significantly from the pattern of the Null treatment, as indicated by significant interaction effects of particle size and patchiness, then the plants showed significant selectivity in response to soil particle heterogeneity.

Before analysis, all data were checked for normality (by Shapiro-Wilk test) and homogeneity of variance (by Levene's test). When we tested the effects of particle heterogeneity on biomass and ramet number at the whole plant level, data could not meet the assumption of normality and/or homogeneity of variance even after transformation. Therefore, non-parametric tests were used for these analyses. Statistical analyses were conducted with SPSS 18.0 for Windows (SPSS, Chicago, IL, USA).

Results

Effects of Soil Particle Size at the Whole Plant Level

In the homogeneous treatments, biomass, number of ramets, rhizome length and spacer length of *B. planiculmis* were significantly greater in the Medium than in the Large treatment, but they did not differ between the Small and Medium treatments (Fig. 2A–C and F). Root-shoot ratio was higher in the Medium and Large treatments than in the Small treatments (Fig. 2E). Soil particle size did not affect number of tubers (Fig. 2D).

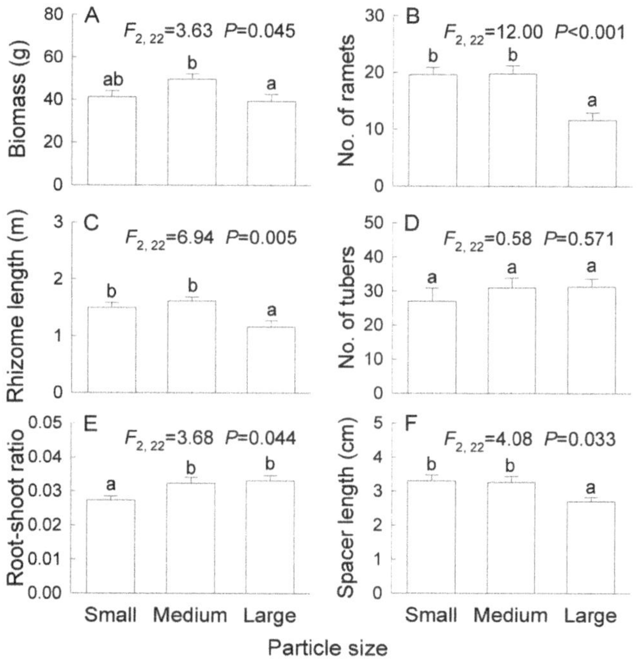

Figure 2. Effects of soil particle size on growth (A–D) and morphology (E, F) in the treatments with a single particle size. The treatment codes are described as in Fig. 1. Bars and vertical lines are means and SE. Bars sharing the same letters are not different at $P = 0.05$.

Effects of Soil Particle Heterogeneity at the Whole Plant Level

None of the growth or morphological measures of the whole plants differed significantly between the Mixture, Small patch and Large patch treatments ($\chi^2 = 2.47–3.16$, $P = 0.206–0.290$ for biomass and number of ramets; $F_{2,\ 21} = 0.501–2.186$, $P = 0.140–0.614$ for rhizome length, number of tubers, spacer length and root-shoot ratio). The result suggests that soil particle heterogeneity did not affect the growth or morphology at the whole plant level.

Effects of Soil Particle Heterogeneity at the Patch Level

At the patch level, particle size significantly affected biomass, number of ramets, rhizome length and number of tubers of *B. planiculmis* (Table 1). In general, biomass, number of ramets, rhizome length and number of tubers were greater in the medium than in the large particle patches (Fig. 3), and such effects were the largest in the Small patch and smallest in the Null treatment (Table 1: significant interaction effect of particle size and patchiness; Fig. 3).

Table 1. Effects of particle size within patches and patchiness on the growth at the patch level.

Effects	DF	Biomass	No. of ramets	Rhizome length	No. of tubers
Particle size	1,38	100.5***	70.0***	71.3***	35.5***
Patchiness	2,38	0.3^{ns}	0.3^{ns}	1.4^{ns}	1.4^{ns}
Interaction	2,38	23.3***	7.5**	14.0***	10.8***

F-values and degree of freedom (DF) of two-way ANOVAs are given. *Superscripts give P:* ***$P<0.001$, **$P<0.01$ and $^{ns}P\geq0.05$.

Discussion

Soil particle size significantly affected the growth of *B. planiculmis*: plants accumulated more ramets and biomass in soil composed of medium (diameter 1.5 mm) or small (0.75 mm) particles than of large (3.0 mm) particles. This result suggests that differences in soil particle size should be considered when we use different soil types to test the effects of soil nutrient availability. Soil particle size may greatly affect both chemical and physical properties of soil [33,36,50–51]. In the present study, the effects on soil chemical properties may be of little importance because sufficient nutrient solutions were supplied to each plant during the experiment. Previous studies have showed that changing soil physical properties such as soil mechanical resistance, infiltration capacity and adsorption capacity can significantly influence the growth and reproduction of plants [51–52]. For example, increasing soil mechanical resistance may slow down root growth and root expansion [52–56]. In the present study, each plant was supplied with the same, sufficient amount of nutrients and also grown in simulated wetland (flooded) conditions. Therefore, changing soil infiltration capacity or adsorption capacity caused by different soil particle sizes may also be of little importance. One likely explanation is that differences in soil mechanical resistance mediated by different soil particle sizes are responsible for the difference in the biomass production and clonal reproduction of *B. planiculmis*. Compared to soil composed of small or medium particles, soil composed of large particles is likely to have a higher

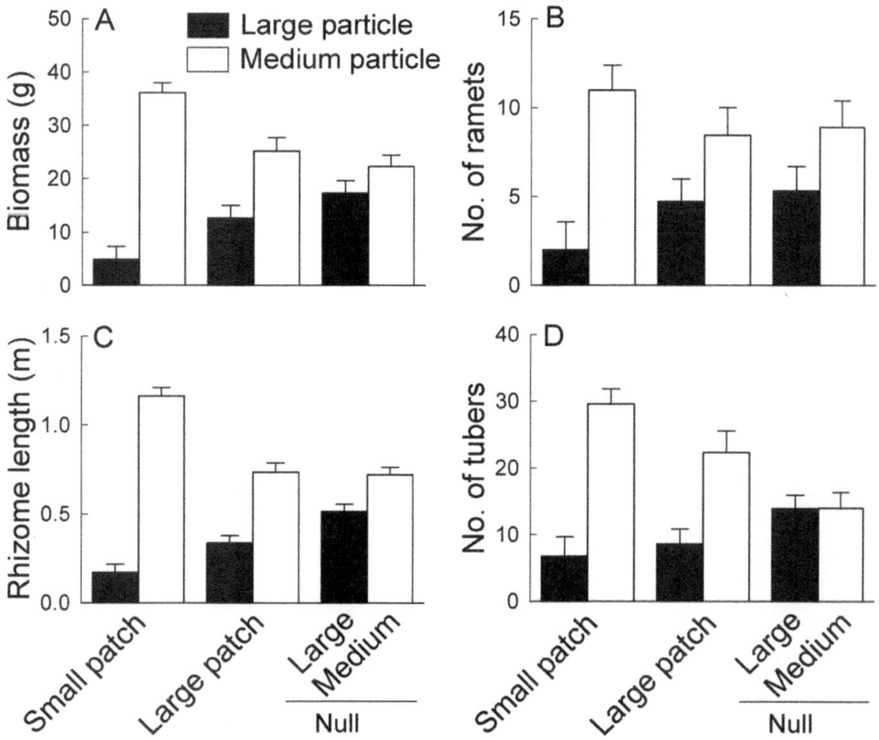

Figure 3. Effects of patchiness and particle size within patches on the growth within patch types. The treatment codes are described as in Fig. 1. In the "Null" treatment, half values of the corresponding variables in Large and Medium were used because in Large patch and Small patch the plants in each type of patches occupied only half of the area of the whole box. Bars and vertical lines are means and SE.

mechanical resistance, which significantly decreased the growth of *B. planiculmis*. Further studies are required to test the relationships between soil particle size and soil mechanical resistance.

Soil particle heterogeneity significantly affected the growth of *B. planiculmis* at the patch level. Compared to the homogeneous large particle treatment, *B. planiculmis* produced less biomass and tubers and shorter rhizomes in the comparable area in the large particle patches in the two heterogeneous treatments, and such effects were greater in the small patch than in the large patch treatment. Also, compared to the homogeneous medium particle treatment, *B. planiculmis* produced more biomass and ramets and longer rhizomes in the comparable area in the medium particle patches in the small patch treatment, but not in the large patch treatment. As a result, in the heterogeneous treatments, comparably more ramets, tubers, biomass and rhizomes were placed in the patches composed of medium particles, which are more favorable for the growth of *B. planiculmis* [49]. These responses are similar to the foraging responses of clonal plants when growing in resource-heterogeneous environments where clonal plants commonly position more ramets in favorable (resource-rich) patches and less in unfavorable (resource-poor) patches [21,24,41,57–58]. Such responses are thought to be adaptive since they potentially can increase resource uptake of the whole clone [57,59–62].

However, at the whole (clone) plant level, we found that soil particle heterogeneity did not affect the growth of *B. planiculmis*. In contrast, spatial heterogeneity consisting of patches of fertile and infertile soil could greatly increase the whole plant growth of some clonal plants such as *Glechoma hederacea* and *Alternanthera philoxeroides* [40–41,63]. Therefore, in these studies the effects of soil heterogeneity at the whole plant levels are very likely due to soil nutrient heterogeneity. When more ramets are located in nutrient-rich patches, both the efficiency and the total amount of nutrient uptake increase markedly [41]. This will result in a significant increase in nutrient uptake of the whole clone as well as its subsequent growth. In the present study, however, because differences in soil mechanical resistance might be the main mechanism underlying the effects of soil particle heterogeneity, the patch-level effects on the selectivity of where ramets are placed and their growth in the patches could not translate into the effects at the whole clone level.

The effects of soil particle heterogeneity were greater in the small patch than in the large patch treatment, suggesting that patch scale may have played an important role during the responses of *B. planiculmis* to soil particle heterogeneity. Previous studies have also shown that patch scale affected the responses of *Glechoma hederacea* [20–21] and *Colophospermum mopane* [16] to soil heterogeneity and *Duchesnea indica* to light heterogeneity [19]. In the present study, the smaller the patch scale was, the more the ramets were placed in the medium particle patches than in the large particle ones. The larger effects of soil particle heterogeneity at the smaller scale were likely because the mean inter-ramet distance (spacer length, 2.3–3.7 cm on average, Fig. 2F) of *B. planiculmis*, which partly determines the placement and density of ramets in a patch [57,60,64], was closer to the patch size (6.5 cm×6.5 cm) in the small patch treatment. This result agrees with scaling theory, which predicts that organisms respond to spatial heterogeneity at scales related to their own size [16,65]. Another possible interpretation is that in the small patch treatment plants had a higher probability of finding a favorable patch (i.e. consisting of medium particles) when such favorable patches were more spread out. However, this interpretation should be considered only when patch sizes are within a range that plants can sense them, such as the two patch sizes used in the present study. If patch size is set smaller than this range, plants will likely not be able to respond to spatial heterogeneity [65].

In conclusion, soil particle size and heterogeneity may greatly affect the growth of plants. Thus, when we address the effects of soil nutrient availability or the effects of soil nutrient heterogeneity, caution should be taken when different soil types are used to form low and high nutrient conditions, because different soil types may differ greatly in soil particle composition [31–32,66–67]. Further studies testing the effects of soil heterogeneity should distinguish the effects of soil nutrient heterogeneity from those of soil particle heterogeneity. The results also suggest that scale should always be considered when we address the effects of spatial heterogeneity [16,18–21].

Acknowledgments

We thank Prof. Peter Alpert and two anonymous reviewers for valuable comments on the manuscript, and Yu-Qiang Li, Qian Zhang, Xiong-Ying Wu and Ming-Lian Chen for assistance with the harvest of the experiment.

Author Contributions

Conceived and designed the experiments: BCD FHY. Performed the experiments: LH YKP WX. Analyzed the data: LH BCD FHY. Contributed reagents/materials/analysis tools: MXZ. Wrote the paper: LH BCD FHY.

References

1. Alpert P, Mooney HA (1996) Resource heterogeneity generated by shrubs and topography on coastal sand dunes. Vegetatio 122: 83–93.
2. Caldwell MM, Pearcy RW (1994) Exploitation of environmental heterogeneity by plants: ecophysiological processes above-and belowground. San Diego: Academic Press.
3. Jackson RB, Caldwell MM (1993) Geostatistical patterns of soil heterogeneity around individual perennial plants. Journal of Ecology 81: 683–692.
4. Li P-X, Wang N, He W-M, Krusi BO (2008) Fertile islands under *Artemisia ordosica* in inland dunes of northern China: Effects of habitats and plant developmental stages. Journal of Arid Environments 72: 953–963.
5. Chen YF, Yu F-H, Dong M (2002) Scale-dependent spatial heterogeneity of vegetation in Mu Us sandy land, a semi-arid area of China. Plant Ecology 162: 135–142.
6. Fransen B, de Kroon H, Berendse F (2001) Soil nutrient heterogeneity alters competition between two perennial grass species. Ecology 82: 2534–2546.
7. Day KJ, Hutchings MJ, John EA (2003) The effects of spatial pattern of nutrient supply on the early stages of growth in plant populations. Journal of Ecology 91: 305–315.
8. García-Palacios P, Maestre FT, Gallardo A (2011) Soil nutrient heterogeneity modulates ecosystem responses to changes in the identity and richness of plant functional groups. Journal of Ecology 99: 551–562.
9. Mommer L, Visser EJW, van Ruijven J, de Caluwe H, Pierik R, et al. (2011) Contrasting root behaviour in two grass species: a test of functionality in dynamic heterogeneous conditions. Plant and Soil 344: 347–360.
10. Maestre FT, Reynolds JF (2006) Spatial heterogeneity in soil nutrient supply modulates nutrient and biomass responses to multiple global change drivers in model grassland communities. Global Change Biology 12: 2431–2441.
11. Wijesinghe DK, John EA, Hutchings MJ (2005) Does pattern of soil resource heterogeneity determine plant community structure? An experimental investigation. Journal of Ecology 93: 99–112.
12. Maestre FT, Bradford MA, Reynolds JF (2005) Soil nutrient heterogeneity interacts with elevated CO_2 and nutrient availability to determine species and assemblage responses in a model grassland community. New Phytologist 168: 637–650.
13. Maestre FT, Reynolds JF (2007) Amount or pattern? Grassland responses to the heterogeneity and availability of two key resources. Ecology 88: 501–511.
14. Maestre FT, Reynolds JF (2006) Nutrient availability and atmospheric CO_2 partial pressure modulate the effects of nutrient heterogeneity on the size structure of populations in grassland species. Annals of Botany 98: 227–235.
15. García-Palacios P, Maestre FT, Bardgett RD, de Kroon H (2012) Plant responses to soil heterogeneity and global environmental change. Journal of Ecology 100: 1303–1314.

16. van der Waal C, de Kroon H, Heitkönig I, Skidmore AK, van Langevelde F, et al. (2011) Scale of nutrient patchiness mediates resource partitioning between trees and grasses in a semi-arid savanna. Journal of Ecology 99: 1124–1133.
17. Wiens JA (1989) Spatial scaling in ecology. Functional Ecology 3: 385–397.
18. Kume T, Sekiya N, Yano K (2006) Heterogeneity in spatial P-distribution and foraging capability by *Zea mays*: effects of patch size and barriers to restrict root proliferation within a patch. Annals of Botany 98: 1271–1277.
19. Wang P, Lei J-P, Li M-H, Yu F-H (2012) Spatial heterogeneity in light supply affects intraspecific competition of a stoloniferous clonal plant. PloS ONE 7: e39105.
20. Wijesinghe DK, Hutchings MJ (1997) The effects of spatial scale of environmental heterogeneity on the growth of a clonal plant: an experimental study with *Glechoma hederacea*. Journal of Ecology 85: 17–28.
21. Wijesinghe DK, Hutchings MJ (1999) The effects of environmental heterogeneity on the performance of *Glechoma hederacea*: the interactions between patch contrast and patch scale. Journal of Ecology 87: 860–872.
22. Kleijn D, van Groenendael JM (2001) The exploitation of heterogeneity by a clonal plant in habitats with contrasting productivity levels. Journal of Ecology 87: 873–884.
23. Klimes L, Klimesova J, Hendriks R, Van Groenendael J (1997) Clonal plant architecture: a comparative analysis of form and function. In: de Kroon H, van Groenendael J, editors. The ecology and evolution of clonal plants. Leiden: Backhuys Publishers. 1–29.
24. Roiloa SR, Retuerto R (2006) Small-scale heterogeneity in soil quality influences photosynthetic efficiency and habitat selection in a clonal plant. Annals of Botany 98: 1043–1052.
25. Alpert P (1999) Clonal integration in *Fragaria chiloensis* differs between populations: ramets from grassland are selfish. Oecologia 120: 69–76.
26. de Kroon H, Kreulen R, van Rheenen JWA, van Dijk A (1998) The interaction between water and nitrogen translocation in a rhizomatous sedge (*Carex flacca*). Oecologia 116: 38–49.
27. Mommer L, van Ruijven J, Jansen C, van de Steeg HM, de Kroon H (2012) Interactive effects of nutrient heterogeneity and competition: implications for root foraging theory? Functional Ecology 26: 66–73.
28. Roiloa SR, Retuerto R (2007) Responses of the clonal *Fragaria vesca* to microtopographic heterogeneity under different water and light conditions. Environmental and Experimental Botany 61: 1–9.
29. Zhang Y, Zhang Q, Luo P (2008) Photosynthetic response of *Fragaria orientalis* in different water contrast clonal integration. Ecological Research 24: 617–625.
30. Lamb EG, Haag JJ, Cahill JF (2004) Patch-background contrast and patch density have limited effects on root proliferation and plant performance in *Abutilon theophrasti*. Functional Ecology 18: 836–843.
31. Gee GW, Or D (2002) Methods of soil analysis: physical methods. Soil Science Society of America Journal 5: 255–278.
32. Shein EV (2009) The particle-size distribution in soils: problems of the methods of study, interpretation of the results, and classification. Eurasian soil science 42: 284–291.
33. Wang D, Fu B-J, Chen L-D, Zhao W-W, Wang Y-F (2007) Fractal analysis on soil particle size distributions under different land-use types: a case study in the loess hilly areas of the Loess Plateau, China. Acta Ecologica Sinica 27: 3081–3089.
34. Zhou X-F, Zhao R, Li Y-Y, Chen X-Y (2009) Effects of land use types on particle size distribution of reclaimed alluvial soil of the Yangtze Estuary. Acta Ecologica Sinica 29: 5544–5551.
35. Kettler T, Doran JW, Gilbert T (2001) Simplified method for soil particle-size determination to accompany soil-quality analyses. Soil Science Society of America Journal 65: 849–852.
36. Medinski TV, Mills AJ, Fey MV (2009) Relationships between soil particle size fractions and infiltrability. South African Journal of Plant and Soil 26: 147–156.
37. Montero E (2005) Rényi dimensions analysis of soil particle-size distributions. Ecological Modelling 182: 305–315.
38. Rawls WJ, Brakensiek DL, Saxton KE (1982) Estimation of soil water properties. Trans Asae 25: 1316–1320.
39. Poor A, Hershock C, Rosella K, Goldberg D (2005) Do physiological integration and soil heterogeneity influence the clonal growth and foraging of *Schoenoplectus pungens*? Plant Ecology 181: 45–56.
40. Zhou J, Dong B-C, Alpert P, Li H-L, Zhang M-X, et al. (2012) Effects of soil nutrient heterogeneity on intraspecific competition in the invasive, clonal plant *Alternanthera philoxeroides*. Annals of Botany 109: 813–818.
41. Birch CPD, Hutchings MJ (1994) Exploitation of patchily distributed soil resources by the clonal herb *Glechoma hederacea*. Journal of Ecology 82: 653–664.
42. Cahill JF, McNickle GG, Haag JJ, Lamb EG, Nyanumba SM, et al. (2010) Plants integrate information about nutrients and neighbors. Science 328: 1657.
43. McNickle GG, Cahill JF (2009) Plant root growth and the marginal value theorem. Proceedings of the National Academy of Sciences 106: 4747–4751.
44. Janecek S, Janeckova P, Leps J (2004) Influence of soil heterogeneity and competition on growth features of three meadow species. Flora 199: 3–11.
45. Gross KL, Peters A, Pregitzer KS (1993) Fine root growth and demographic responses to nutrient patches in four old-field plant species. Oecologia 95: 61–64.
46. Reynolds HL, Mittelbach GG, Darcy-Hall TL, Houseman GR, Gross KL (2007) No effect of varying soil resource heterogeneity on plant species richness in a low fertility grassland. Journal of Ecology 95: 723–733.
47. Day KJ, Hutchings MJ, John EA (2003) The effects of spatial pattern of nutrient supply on yield, structure and mortality in plant populations. Journal of Ecology 91: 541–553.
48. Wu Z-Y (2010) Flora of China, V 23. Beijing: Science Press. 180 p.
49. Huang L (2011) Effects of soil particle on the growth of the clonal plant *Scirpus planiculmis* [Dissertation]: Beijing Forestry University.
50. Bronick CJ, Lal R (2005) Soil structure and management: a review. Geoderma 124: 3–22.
51. Mando A, Ouattara B, Somado AE, Wopereis MCS, Stroosnijder L, et al. (2005) Long-term effects of fallow, tillage and manure application on soil organic matter and nitrogen fractions and on sorghum yield under Sudano-Sahelian conditions. Soil Use and Management 21: 25–31.
52. Tracy SR, Black CR, Roberts JA, Mooney SJ (2011) Soil compaction: a review of past and present techniques for investigating effects on root growth. Journal of the Science of Food and Agriculture 91: 1528–1537.
53. Bengough AG, McKenzie BM, Hallett PD, Valentine TA (2011) Root elongation, water stress, and mechanical impedance: a review of limiting stresses and beneficial root tip traits. Journal of Experimental Botany 62: 59–68.
54. Bengough AG, Mullins CE (1990) Mechanical impedance to root growth: a review of experimental techniques and root growth responses. Journal of Soil Science 41: 341–358.
55. Dexter AR (1987) Mechanics of root growth. Plant and Soil 98: 303–312.
56. Ehlers W, Köpke U, Hesse F, Böhm W (1983) Penetration resistance and root growth of oats in tilled and untilled loess soil. Soil and Tillage Research 3: 261–275.
57. de Kroon H, Hutchings MJ (1995) Morphological plasticity in clonal plants: the foraging concept reconsidered. Journal of Ecology 83: 143–152.
58. Evans JP, Cain ML (1995) A spatially explicit test of foraging behavior in a clonal plant. Ecology 76: 1147–1155.
59. Alpert P, Stuefer JF (1997) Division of labour in clonal plants. In: de Kroon H, van Groenendael J, editors. The ecology and evolution of clonal plants. Leiden: Backbuys Publishers. 137–154.
60. Hutchings MJ, de Kroon H (1994) Foraging in plants: the role of morphological plasticity in resource acquisition. Advances in Ecological Research 25: 159–238.
61. Hutchings MJ, Wijesinghe DK (1997) Patchy habitats, division of labour and growth dividends in clonal plants. Trends in Ecology & Evolution 12: 390–394.
62. de Kroon H, Visser EJ, Huber H, Mommer L, Hutchings MJ (2009) A modular concept of plant foraging behaviour: the interplay between local responses and systemic control. Plant, Cell & Environment 32: 704–712.
63. Hutchings M, Wijesinghe D (2008) Performance of a clonal species in patchy environments: effects of environmental context on yield at local and whole-plant scales. Evolutionary Ecology 22: 313–324.
64. Dong M (1995) Morphological responses to local light conditions in clonal herbs from contrasting habitats, and their modification due to physiological integration. Oecologia 101: 282–288.
65. Hutchings MJ, John EA, Wijesinghe DK (2003) Toward understanding the consequences of soil heterogeneity for plant populations and communities. Ecology 84: 2322–2334.
66. Keller T, Håkansson I (2010) Estimation of reference bulk density from soil particle size distribution and soil organic matter content. Geoderma 154: 398–406.
67. Miranda JGV, Montero E, Alves MC, Paz González A, Vidal Vázquez E (2006) Multifractal characterization of saprolite particle-size distributions after topsoil removal. Geoderma 134: 373–385.

PERMISSIONS

All chapters in this book were first published in PLOS ONE, by The Public Library of Science; hereby published with permission under the Creative Commons Attribution License or equivalent. Every chapter published in this book has been scrutinized by our experts. Their significance has been extensively debated. The topics covered herein carry significant findings which will fuel the growth of the discipline. They may even be implemented as practical applications or may be referred to as a beginning point for another development.

The contributors of this book come from diverse backgrounds, making this book a truly international effort. This book will bring forth new frontiers with its revolutionizing research information and detailed analysis of the nascent developments around the world.

We would like to thank all the contributing authors for lending their expertise to make the book truly unique. They have played a crucial role in the development of this book. Without their invaluable contributions this book wouldn't have been possible. They have made vital efforts to compile up to date information on the varied aspects of this subject to make this book a valuable addition to the collection of many professionals and students.

This book was conceptualized with the vision of imparting up-to-date information and advanced data in this field. To ensure the same, a matchless editorial board was set up. Every individual on the board went through rigorous rounds of assessment to prove their worth. After which they invested a large part of their time researching and compiling the most relevant data for our readers.

The editorial board has been involved in producing this book since its inception. They have spent rigorous hours researching and exploring the diverse topics which have resulted in the successful publishing of this book. They have passed on their knowledge of decades through this book. To expedite this challenging task, the publisher supported the team at every step. A small team of assistant editors was also appointed to further simplify the editing procedure and attain best results for the readers.

Apart from the editorial board, the designing team has also invested a significant amount of their time in understanding the subject and creating the most relevant covers. They scrutinized every image to scout for the most suitable representation of the subject and create an appropriate cover for the book.

The publishing team has been an ardent support to the editorial, designing and production team. Their endless efforts to recruit the best for this project, has resulted in the accomplishment of this book. They are a veteran in the field of academics and their pool of knowledge is as vast as their experience in printing. Their expertise and guidance has proved useful at every step. Their uncompromising quality standards have made this book an exceptional effort. Their encouragement from time to time has been an inspiration for everyone.

The publisher and the editorial board hope that this book will prove to be a valuable piece of knowledge for researchers, students, practitioners and scholars across the globe.

LIST OF CONTRIBUTORS

Yuying Wang, Chunsheng Hu, Hua Ming, Wenxu Dong, Yuming Zhang and Xiaoxin Li
Key Laboratory of Agricultural Water Resources, Center for Agricultural Resources Research, Institute of Genetics and Developmental Biology, Chinese Academy of Sciences, Shijiazhuang, Hebei, China

Oene Oenema
Department of Soil Quality, Wageningen University, Alterra, Wageningen, The Netherlands

Douglas A. Schaefer
Key Lab of Tropical Forest Ecology, Xishuangbanna Tropical Botanical Garden, Chinese Academy of Sciences, Menglun, Yunnan, China

Peter J. Edwards
Institute of Integrative Biology, Plant Ecology, ETH Zurich, Zurich, Switzerland

Gurbir S. Bhullar
Institute of Integrative Biology, Plant Ecology, ETH Zurich, Zurich, Switzerland Research Institute of Organic Agriculture (FiBL), Frick, Switzerland

Harry Olde Venterink
Institute of Integrative Biology, Plant Ecology, ETH Zurich, Zurich, Switzerland
Plant Biology and Nature Management, Vrije Universiteit Brussel, Brussels, Belgium

Baoming Du
School of Agriculture and Biology and Research Center for Low-Carbon Agriculture, Shanghai Jiao Tong University, Shanghai, China

Chunjiang Liu, Hongzhang Kang, Shan Yin and Guangrong Shen
School of Agriculture and Biology and Research Center for Low-Carbon Agriculture, Shanghai Jiao Tong University, Shanghai, China
Key Laboratory of Urban Agriculture (South), Ministry of Agriculture, People's Republic of China, Shanghai, China

Penghua Zhu
Key Laboratory of Urban Agriculture (South), Ministry of Agriculture, People's Republic of China, Shanghai, China

Jingli Hou
Instrumental Analysis Center of SJTU, Shanghai Jiao Tong University, Shanghai, China

Hannu Ilvesniemi
Finnish Forest Research Institute, Vantaa, Finland

Nianpeng He, Qiufeng Wang, Xinyu Zhang and Guirui Yu
Synthesis Research Center of Chinese Ecosystem Research Network, Key Laboratory of Ecosystem Network Observation and Modeling, Institute of Geographic Sciences and Natural Resources Research, Chinese Academy of Sciences, Beijing, China

Ning Zhao, Ruili Wang and Zhiwei Xu
University of Chinese Academy of Sciences, Beijing, China

Forough Aghili, Hannes A. Gamper and Emmanuel Frossard
Institute of Agricultural Sciences, Department of Environmental Systems Science, Swiss Federal Institute of Technology (ETH) Zürich, Switzerland

Jost Eikenberg
Paul Scherrer Institute (PSI), Radioanalytics Laboratory, Villigen, Switzerland

Amir H. Khoshgoftarmanesh and Majid Afyuni
College of Agriculture, Department of Soil Sciences, Isfahan University of Technology, Isfahan, Iran

Rainer Schulin
Institute of Terrestrial Ecosystems, Department of Environmental Systems Science, Swiss Federal Institute of Technology (ETH) Zürich, Switzerland

Jan Jansa
Institute of Microbiology, Academy of Sciences of the Czech Republic, Prague, Czech Republic

Tomonori Tsunoda, Naoki Kachi and Jun-Ichirou Suzuki
Department of Biological Sciences, Tokyo Metropolitan University, Hachioji, Tokyo, Japan

Simone Dealtry, Guo-Chun Ding, Viola Weichelt, Holger Heuer and Kornelia Smalla
Julius Kühn-Institut - Federal Research Centre for Cultivated Plants (JKI), Institute for Epidemiology and Pathogen Diagnostics, Braunschweig, Germany

Vincent Dunon and Dirk Springael
Division of Soil and Water Management, KU Leuven, Heverlee, Belgium

Andreas Schlüter
Center for Biotechnology (CeBiTec), Institute for Genome Research and Systems Biology, Bielefeld University, Bielefeld, Germany

María Carla Martini, María Florencia Del Papa and Antonio Lagares
IBBM (Instituto de Biotecnología y Biología Molecular), CCT-CONICET-La Plata
Departamento de Ciencias Biológicas, Facultad de Ciencias Exactas, Universidad Nacional de La Plata, La Plata, Argentina

Gregory Charles Auton Amos, Elizabeth Margaret Helen Wellington and William Hugo Gaze
School of Life Sciences, University of Warwick, Warwick, United Kingdom

Detmer Sipkema
Laboratory of Microbiology, Wageningen University, Wageningen, The Netherlands

Sara Sjöling
Sö dertörns högskola (Sodertorn University), Inst. för Naturvetenskap, Miljöoch medieteknik (School of Natural Sciences, Environmental Studies and media tech), Huddinge, Sweden

Jan Dirk van Elsas
University of Groningen, Groningen, The Netherlands

Christopher Thomas
School of Biosciences, University of Birmingham, Edgbaston, Birmingham, Warwick, United Kingdom

Huanping Lu
Key Laboratory of Vegetation Restoration and Management of Degraded Ecosystems, South China Botanical Garden, Chinese Academy of Sciences, Guangzhou, China
University of Chinese Academy of Sciences, Beijing, China

Jorge Paz-Ferreiro
Key Laboratory of Vegetation Restoration and Management of Degraded Ecosystems, South China Botanical Garden, Chinese Academy of Sciences, Guangzhou, China
Departamento de Edafología, Universidad Politécnica de Madrid, Madrid, Spain

Zhian Li and Shenglei Fu
Key Laboratory of Vegetation Restoration and Management of Degraded Ecosystems, South China Botanical Garden, Chinese Academy of Sciences, Guangzhou, China

Ana Méndez
Departamento de Ingeniería de Materiales, Universidad Politécnica de Madrid, Madrid, Spain

Gabriel Gascó
Departamento de Edafología, Universidad Politécnica de Madrid, Madrid, Spain

Biao Liu
Nanjing Institute of Environmental Sciences, Ministry of Environmental Protection of China, Nanjing, Jiangsu Province, China

Xiaogang Li
Nanjing Institute of Environmental Sciences, Ministry of Environmental Protection of China, Nanjing, Jiangsu Province, China
Key Laboratory of Soil Environment and Pollution Remediation, Institute of Soil Science, Chinese Academy of Sciences, Nanjing, Jiangsu Province, China

Xinshu Liang, Yinan Gao, Xiaoying Zhang, Yongqiang Tian, Zhenxian Zhang and Lihong Gao
Beijing Key Laboratory of Growth and Developmental Regulation for Protected Vegetable Crops, Department of Vegetable Science, China Agricultural University, Beijing, P.R. China

Christine Fischer and Anke Hildebrandt
Institute of Geosciences, Friedrich-Schiller-University Jena, Jena, Germany
Max Planck Institute for Biogeochemistry, Jena, Germany

Christiane Roscher
UFZ, Helmholtz Centre for Environmental Research, Department of Community Ecology, Halle, Germany

Britta Jensen and Stefan Scheu
J. F. Blumenbach Institute of Zoology and Anthropology, Georg August University of Göttingen Göttingen, Germany
Nico Eisenhauer
Institute of Ecology, Friedrich Schiller University Jena, Jena, Germany

Jussi Baade
Department of Geography, Friedrich Schiller University Jena, Jena, Germany

Sabine Attinger
Institute of Geosciences, Friedrich-Schiller-University Jena, Jena, Germany
UFZ, Helmholtz Centre of Environmental Research, Department of Computational Hydrosystems, Leipzig, Germany

Wolfgang W. Weisser
Technical University Munich, Terrestrial Ecology Research Group, Department of Ecology and Ecosystem Management, Center for Food and Life Sciences Weihenstephan, Freising, Germany

Jens Schumacher
Institute of Stochastics, Friedrich Schiller University Jena, Jena, Germany

Xiao-Tao Lü
State Key Laboratory of Forest and Soil Ecology, Institute of Applied Ecology, Chinese Academy of Sciences, Shenyang, China

Hai-Wei Wei
State Key Laboratory of Forest and Soil Ecology, Institute of Applied Ecology, Chinese Academy of Sciences, Shenyang, China
Graduate University of Chinese Academy of Sciences, Beijing, China

Fu-Mei Lü
State Key Laboratory of Vegetation and Environmental Change, Institute of Botany, Chinese Academy of Sciences, Beijing, China

Xing-Guo Han
State Key Laboratory of Forest and Soil Ecology, Institute of Applied Ecology, Chinese Academy of Sciences, Shenyang, China
State Key Laboratory of Vegetation and Environmental Change, Institute of Botany, Chinese Academy of Sciences, Beijing, China

Xiangyang Sun, Suyan Li, Tao Zhang, Wei Zhang and Penghui Zhai
College of Forestry, Beijing Forestry University, Beijing, China

Linlin Wang
College of Forestry, Beijing Forestry University, Beijing, China
Forestry Research Institute of Hulun Buir, Hulun Buir, Inner Mongolia, China

Desirée Rivera, Violeta Mejías, Ana Isabel López-Archilla and Begoña Peco
Departamento de Ecología, Universidad Autónoma de Madrid, Madrid, Spain

Berta M. Jáuregui
Dirección Técnica, OHL Construcción, Madrid, Spain

Marga Costa-Tenorio
Departamento de Biología Vegetal I, Universidad Complutense, Madrid, Spain

Xavier Le Roux, Franck Poly, Nadine Guillaumaud and Joana Falcao Salles
Universitéde Lyon, INRA, CNRS, Université Lyon 1, Microbial Ecology Centre (UMR 5557 CNRS, USC 1364 INRA), Villeurbanne, France

Bernhard Schmid
Institute of Evolutionary Biology and Environmental Studies, University of Zurich, Zurich, Switzerland

Romain L. Barnard and Pascal A. Niklaus
Institute of Plant Sciences, ETH Zurich, Zurich, Switzerland

Maike Habekost and Sibylle Steinbeiss
Max Planck Institute for Biogeochemistry, Jena, Germany

Yvonne Oelmann
Geoecology, University of Tübingen, Tübingen, Germany

Laurent Philippot
Laboratoire de Microbiologie des Sols, INRA, Dijon, France

Michael Schloter
Research Unit for Environmental Genomics, Helmholtz Zentrum München, Neuherberg, Germany

Alexandra Weigelt
Institute of Ecology, Friedrich Schiller University, Jena, Germany

Judith Prommer, Wolfgang Wanek, Florian Hofhansl and Rebecca Clare Hood-Nowotny
Department of Microbiology and Ecosystem Science, University of Vienna, Vienna, Austria

Daniela Trojan, Pierre Offre, Tim Urich and Christa Schleper
Department of Ecogenomics and Systems Biology, University of Vienna, Vienna, Austria

Stefan Sassmann
Core Facility of Cell Imaging and Ultrastructure Research, University of Vienna, Vienna, Austria

Barbara Kitzler
Institute of Forest Ecology and Soil, Federal Research and Training Centre for Forests, Natural Hazards and Landscape, Vienna, Austria

Gerhard Soja
Department of Health and Environment, Austrian Institute of Technology, Tulln, Austria

Benjamin D. Duval, Paul Dijkstra and Bruce A. Hungate
Department of Biological Sciences, Northern Arizona University, Flagstaff, Arizona, United States of America
Merriam-Powell Center for Environmental Research, Northern Arizona University, Flagstaff, Arizona, United States of America

J. Patrick Megonigal and Bert G. Drake
Smithsonian Environmental Research Center, Edgewater, Maryland, United States of America

Dale W. Johnson
Department of Natural Resources, University of Nevada-Reno, Reno, Nevada, United States of America

Michael E. Ketterer
Department of Chemistry and Biochemistry, Northern Arizona University, Flagstaff, Arizona, United States of America

Jiang Wang, Tong Chen, Yi Bai, Bao Ying Qian and Chong Bang Zhang
School of Life Science, Taizhou University, Linhai, China

Yuan Ge
Bren School of Environmental Science and Management/Earth Research Institute, University of California Santa Barbara, Santa Barbara, California, United States of America

Marleen de Blécourt, Rainer Brumme, Marife D. Corre and Edzo Veldkamp
Soil Science of Tropical and Subtropical Ecosystems, Büsgen Institute, Georg-August-Universität Göttingen, Göttingen, Germany

Jianchu Xu
World Agroforestry Centre, East Asia Node, Kunming, China

Ling Zhang and Evan Siemann
College of Resources and Environmental Sciences, Nanjing Agricultural University, Nanjing, China
Department of Ecology and Evolutionary Biology, Rice University, Houston, Texas, United States of America

Hong Wang
College of Resources and Environmental Sciences, Nanjing Agricultural University, Nanjing, China

Jianwen Zou
College of Resources and Environmental Sciences, Nanjing Agricultural University, Nanjing, China
Department of Horticulture, Cornell University, Ithaca, New York, United States of America

William E. Rogers
Department of Ecosystem Science and Management, Texas A & M University, College Station, Texas, United States of America

Yongsheng Wang, Huajun Fang, Guirui Yu, Xusheng Dang and Lei Wang
Key Laboratory of Ecosystem Network Observation and Modeling, Institute of Geographical Sciences and Natural Resources Research, Chinese Academy of Sciences
Beijing, China University of Chinese Academy of Sciences, Beijing, China

Linsen Li
Key Laboratory of Ecosystem Network Observation and Modeling, Institute of Geographical Sciences and Natural Resources Research, Chinese Academy of Sciences, Beijing, China

Shulan Cheng and Minjie Xu
University of Chinese Academy of Sciences, Beijing, China

W. H. Gera Hol and Freddy ten Hooven
Netherlands Institute of Ecology, Department of Terrestrial Ecology, Wageningen, The Netherlands

Wietse de Boer
Netherlands Institute of Ecology, Department of Microbial Ecology, Wageningen, The Netherlands
Wageningen University, Department of Soil Quality, Wageningen, The Netherlands

Wim H. van der Putten
Netherlands Institute of Ecology, Department of Terrestrial Ecology, Wageningen, The Netherlands
Wageningen University, Laboratory of Nematology, Wageningen, The Netherlands

Songze Wan, Zhanfeng Liu, Lixia Zhou and Shenglei Fu
Key Laboratory of Vegetation Restoration and Management of Degraded Ecosystems, South China Botanical Garden, Chinese Academy of Sciences, Guangzhou, China

Chenlu Zhang
State Key Laboratory of Cotton Biology, Key Laboratory of Plant Stress Biology, College of Life Sciences, Henan University, Kaifeng, China

Jie Zhao
Key Laboratory of Vegetation Restoration and Management of Degraded Ecosystems, South China Botanical Garden, Chinese Academy of Sciences, Guangzhou, China
Key Laboratory of Agro-Ecological Processes in Subtropical Region, Institute of Subtropical Agriculture, Chinese Academy of Sciences, Changsha, China

Peng Zhang, Ting Wei, Zhikuan Jia, Qingfang Han and Xiaolong Ren
The Chinese Institute of Water-Saving Agriculture, Northwest A&F University, Yangling, Shaanxi, China
Key Laboratory of Crop Physi-Ecology and Tillage Science in Northwestern Loess Plateau, Ministry of Agriculture, Northwest A&F University, Yangling, Shaanxi, China

Yongping Li
Guyuan Institute of Agricultural Sciences, Guyuan, Ningxia, China

Guocheng Wang, Tingting Li, Wen Zhang and Yongqiang Yu
State Key Laboratory of Atmospheric Boundary Layer Physics and Atmospheric Chemistry, Institute of Atmospheric Physics, Chinese Academy of Sciences, Beijing, China

Lin Huang, Bi-Cheng Dong, Wei Xue, Yi-Ke Peng, Ming-Xiang Zhang and Fei-Hai Yu
School of Nature Conservation, Beijing Forestry University, Beijing, China

Index

A
Aboveground Net Primary Production (anpp), 110
Agricultural Management, 4, 132-133, 148, 232-235, 237, 239-241
Altitudinal Patterns, 28-29, 31, 33-35
Altitudinal Transect, 20-21, 23, 25, 27-29
Anthropogenic Activities, 57, 65
Available Soil Nutrients, 122

B
Bacillus Thuringiensis, 72, 85-86
Bacterial Ammonia-oxidizers, 150
Belowground Herbivore, 50-55
Biochar Decelerates, 146-147, 149, 151, 153, 155, 157, 159, 161
Biodiversity Experiment, 98, 108, 136, 171, 173, 175, 177-179
Biotic Plant-soil, 207, 209, 211

C
Cadmium Remediation, 65, 67, 69, 71
Calculations Of Gas Fluxes, 2
Carbon Dioxide, 1, 3, 5, 7, 9, 11, 13, 35, 133, 155, 169, 189, 191, 193, 195, 234
Cation Exchange Capacity (cec), 117, 121
Centaurea Jacea, 15-17
Chemical Analysis, 21, 29, 35, 110, 119
Chemical Properties, 2, 117, 124, 126-127, 179, 190, 229, 231, 242, 245
Climatic Control, 20-21, 23, 25, 27
Coastal Saline Soil, 117-119, 121, 123-125
Cucumber (cucumis Sativus L.), 87
Cucumber Root Morphologic
Characters, 90
Cultivation-independent Screening, 57, 59, 61, 63

D
Daily Fertigation, 87-89, 91-95
Data Availability, 165
Different Environmental Habitats, 57
Different Plant Species, 15, 17, 19, 97, 115

E
Earthworm Density Manipulations, 98
Earthworms, 82, 84-85, 97-101, 103-108, 125
Ecological Restoration, 126-127, 129, 131, 133, 179
Eragrostis Pilosa, 189-190, 195
Eucalyptus Monocultures, 213

F
Field Experimental Design, 2
Fieldwork Permission, 183
Fundamental Ecosystem, 189

G
Green Manure Addition, 37-41, 43-47, 49

H
Harvest And Measurements, 243
Hevea Brasiliensis, 180-181, 183, 185, 187

I
Industrialisation, 65
Invaded Annual Grassland, 189, 191, 193, 195

J
Juncus Effuses, 16-17

L
Laboratory Analyses, 183
Litter Decomposition, 20, 23, 25-27, 35, 85, 113, 133, 189-195, 214
Long-term Planting, 72-73, 75, 77, 79, 81, 83, 85
Lushan Mountain, 20-27

M
Methane Emissions, 13, 15-19, 241
Microbial Activity, 11, 47, 86, 119, 124-133
Microbial Community Size, 150, 154

N
Natural Plant Communities, 172, 174
Nematode Community Analysis, 78
Nematode Diversities, 213, 215-217, 219
Nitrogen Addition, 12, 109, 111-115, 125, 197, 202, 205
Nitrous Oxide, 1, 3, 5, 7, 9, 11, 13-14, 136-137, 141, 143-144, 161, 197, 205-206
Non-native Plant Litter, 189, 191, 193, 195

O

Organic Amendments, 117-125, 132-133

P

Partial Soil Desalination, 38

Phytoextraction Efficiency, 69

Phytoextractors, 65-67, 69-71

Plant Composition, 97, 99, 101, 103, 105, 107, 144

Plant Diversity, 97-99, 101, 103, 105, 107, 127, 132, 135-145, 211, 213-214, 217, 219

Plant Growth, 2, 4, 6, 8, 10, 12, 14-16, 18, 22, 24, 26, 28-36, 38, 40, 42, 44, 46-48, 50-52, 54-56, 58, 60, 62, 64, 66, 68, 70, 74, 76, 78, 80, 82, 84, 86, 88, 90, 92, 94, 96, 98, 100, 102, 104, 106, 108-112, 114, 116-118, 120, 122-124, 126, 128, 130-132, 134-136, 138, 140, 142, 144, 148, 150, 152, 154-156, 158, 160, 162, 164, 166, 168, 170, 172, 174, 176-179, 182, 184, 186, 188, 190, 192, 194, 196, 198, 200, 202, 204, 206-208, 210-212, 214, 216, 218, 220, 222, 224, 226, 228, 230, 234, 236, 238, 240, 242, 244, 246

Plant Species Richness, 98-103, 105, 126, 136-141, 143-144, 172, 220, 247

Plasmid Occurrence, 57-61, 63

R

Restored Swiss Wetland, 15, 17, 19

S

Scrub-oak Ecosystem, 162-163, 165, 167, 169-170

Secondary Forests, 20, 180-187

Semiarid Regions, 221, 223, 225, 227, 229, 231

Simulated Nitrogen Deposition, 197, 199, 201, 203, 205

Soil Carbon Dioxide Emissions, 189, 191, 193, 195

Soil Carbon Dynamics, 20-21

Soil Carbon Stocks, 180-188

Soil Environmental Conditions, 135, 137, 139, 141, 143-145, 193

Soil Fertility, 34, 37, 124, 126-131, 133, 232

Soil Gas Sampling, 2

Soil Moisture Measurement, 52

Soil Nematodes, 72-73, 75, 78-80, 82, 85, 219

Soil Organic Carbon Dynamics, 232-233, 235, 237, 239, 241

Soil Organic Nitrogen, 146-147, 149, 151, 153, 155, 157, 159, 161

Soil Particle Heterogeneity, 242-247

Soil Profile, 1-3, 5, 7, 9-13, 52-53, 56, 98, 167, 169, 184, 188, 204, 234, 240

Soil Temperature, 2, 4, 10, 12, 95, 197-198, 200, 202-204, 206, 214-215, 218-219, 237

Solar-greenhouse, 87, 95

Statistical Analysis, 3, 22, 78, 90, 99, 110, 128, 173, 191, 212, 214, 224

Stoichiometry, 26-29, 31, 33-36, 110, 115, 169, 179

Straw Incorporation, 221-231

Subtropical Plantation, 197-199, 201-205

T

Temperate Steppe, 109-115, 212

Temporal Heterogeneity, 50-51, 53, 55-56

Trace Metal Removal, 171, 173, 175, 177-179

Transgenic Insect-resistant Cotton, 72-73, 75, 77, 79-83, 85-86

U

Understory Vegetations, 213-214, 219

V

Vegetation Cover, 98, 126, 186

Vertical Distribution, 27, 50-51, 53, 55, 136

W

Wheat (triticum Aestivum), 207-209, 211

Winter Wheat-summer, 1-3, 5-13, 232, 234

Z

Zinc Concentration, 37, 39, 41, 43, 45, 47-49